TABLE OF CONTENTS

W0227400

PLANETARY SYSTEMS:
FORMATION, EVOLUTION, AND DETECTION

First International Conference

December 8–10, 1992
Ramo Auditorium
California Institute of Technology
Pasadena, California

Sponsored by
National Aeronautics and Space Administration
Solar System Exploration Division
Planetary Astronomy Program

Hosted by
Jet Propulsion Laboratory
Office of Space Science and Instruments
California Institute of Technology

HONORARY CONFERENCE CHAIRMAN

Dr. Lew Allen, Jr., Pasadena, California

SCIENCE ORGANIZING COMMITTEE

Dr. Bernard F. Burke, Chairman, Massachusetts Institute of Technology

Dr. Michael J.S. Belton, Kitt Peak Observatory
Dr. David C. Black, Lunar and Planetary Institute
Dr. Alan P. Boss, Carnegie Institution of Washington
Dr. A.A. Boyarchuk, Russian Academy of Sciences
Dr. Robert A. Brown, Space Telescope Science Institute
Dr. Moustafa T. Chahine, Jet Propulsion Laboratory
Dr. Charles Elachi, Jet Propulsion Laboratory
Dr. Hans Elsaesser, Max-Planck-Institut für Astronomie
Dr. George H. Herbig, University of Hawaii
Dr. Wesley Huntress, National Aeronautics and Space Administration
Dr. Eugene Levy, University of Arizona
Dr. Peter G. Mezger, Max-Planck-Institut für Radio-astronomie

Dr. Minoru Oda, Riken Institute Physical Chemical Research
Dr. Tobias Owen, University of Hawaii
Dr. Franco Pacini, Instituto Astronomia Arcetri
Dr. Jürgen Rahe, National Aeronautics and Space Administration
Dr. Anneila I. Sargent, California Institute of Technology
Dr. Bradford A. Smith, University of Hawaii
Dr. Edward C. Stone, Jr., Jet Propulsion Laboratory
Dr. Daniel W. Weedman, National Aeronautics and Space Administration
Dr. Lodewijk Woltjer, Observatorie De Haute

LOCAL ORGANIZING COMMITTEE

Mr. Neil L. Nickle, Chairman, Jet Propulsion Laboratory
Ms. Patricia B. McLane, Logistics, Jet Propulsion Laboratory
Dr. David J. Diner, Jet Propulsion Laboratory
Dr. William D. Langer, Jet Propulsion Laboratory
Mr. Harvey Meyerson, President, International Space Year Association
Dr. Michael Shao, Jet Propulsion Laboratory
Dr. Richard J. Terrile, Jet Propulsion Laboratory

FOREWORD

The possibility of observing planetary systems about stars other than our own has intrigued many people for a long time. Recent observations and improved observational techniques have increased the enthusiasm for this challenging problem. Our greatly increased understanding of the solar system, due in large part to journeys by scientific spacecraft, has led to greater confidence in the theories of planetary formation and evolution. But we still cannot answer the fascinating questions: Are there other planetary systems like ours? Other planets like ours? Is there life elsewhere in the universe?

Some scientists are concerned that the very broad public interest in such questions leads to unscientific speculation and perhaps premature observational attempts. In spite of this, at the August 1991 IAU (International Astronomical Union) meeting in Buenos Aires, a group of scientists including Toby Owen, Brad Smith, Dave Morrison, and Jürgen Rahe discussed the need for a symposium to define a central paradigm of planetary formation and to place theories, models, observations, and plans into a broad intellectual context. NASA sponsored the conference, the Jet Propulsion Laboratory hosted it, and the 1992 International Space Year Association endorsed it. Interest was high: 129 individuals attended, 45 papers were accepted for publication, and 29 attendees were from outside the United States.

The sponsor and host judged the conference to be quite successful, and a second international conference is planned for December 13–15, 1993, in Hawaii. It seems to me that the broad intellectual context of these conferences does nothing to decrease the fascination of the topics and helps significantly to keep the fast-moving field on a sound scientific basis.

DR. LEW ALLEN, JR.
Honorary Conference Chairman
June 29, 1993

Astrophysics and Space Science **212**: ix, 1994.
© 1994 *Kluwer Academic Publishers.*

PLANETARY SYSTEMS:

FORMATION, EVOLUTION, AND DETECTION – INTRODUCTION

BERNARD F. BURKE

How our planetary system came to be and how it evolved to its present state are questions that have fascinated thinkers from antiquity to the present day. The literature on cosmogony forms a distinguished body of knowledge that traces the evolution of scientific thought, beginning with the intellectual framework created by the Greek philosophers as they struggled to understand the world around them. They set the tone that gave rise to the natural philosophy of the seventeenth century, leading to science as we know it today. The body of knowledge has grown through the work of planetary astronomers, geophysicists, and physicists and astronomers more generally, aided by important contributions from scientific missions in space.

We now have a fascinating but sketchy picture of how the solar system evolved, but we have very little knowledge of how often the phenomenon of planetary formation takes place elsewhere. The gaps in our knowledge and the shakiness of the intellectual structures are well-known to planetary scientists, and as we take satisfaction in what we do know, we also look forward to the adventure of exploring a territory that is still only half-familiar.

During the course of the last decade or so, the question of the existence of planetary systems beyond our own, bound to stars other than our sun, has passed from speculation to action. Observational techniques, both on the ground and in space, have reached the point where serious searches can be undertaken.

This work will not take place in isolation. Recent infrared and millimeter-wave observations, particularly, have sharpened our notions of how stars and planetary systems form. There is a scientific consensus that, rather than conducting a hunt for new planets, spectacular though the results might be, the planetary science community ought to contemplate a scientific program of far greater intellectual depth. The program is not a purely observational one because there has been a continual advance in theoretical sophistication and understanding, aided by the continually advancing scope of computational power.

In view of the widespread realization that a set of major new scientific opportunities is at hand, this seemed to be an opportune moment to convene a symposium that would address the issues from as many different perspectives as possible. The participants have been engaged in a process that will open a new era of planetary science. We can expect that there will be many gaps in our knowledge and as many wrong turns and blind alleys to be pursued as there are grand avenues and new vistas of knowledge. Every contributor to this symposium volume knows that we have been presenting works in progress, coupled with hopes for the future. Although there is a body of conjecture, there is also a core of factual knowledge.

Astrophysics and Space Science **212**: xi–xii. 1994.
© 1994 *Kluwer Academic Publishers.*

In the past, some have expressed skepticism that other planetary systems could even be detected, let alone studied, and there have been false alarms along the way. Science is a self-correcting process. However, and it is now clear that this is the time to begin the work on instruments and observing plans for serious planetary searches, coupled with broader scientific aims. This symposium is one more step along this new path.

The breadth of this symposium, therefore, is deliberate, and there is the expectation that the mutual interests of theorists and observers will stimulate one another. There are numerous instances of dramatic surprises in the course of studying our own solar system. Mercury was thought to keep the same hemisphere toward the sun, but its tidal lock turned out to be more complicated than that. Venus was supposed to be cold, but it turned out to be fiercely hot. Mars has great volcanoes, and a wealth of surface features (but no canals and no life). Jupiter has a strong dipolar magnetic field, with extensive radiation belts and a volcano-rich satellite that through its homopolar generator action stirs up violent radio bursts. And the surprises continue as we turn our attention to the planets beyond. Even greater surprises surely lie in store as the study of other planetary systems proceeds. In the spirit of combining present knowledge with future plans, a broad range of topics has to be covered in this symposium, and this holistic view must be maintained if effective progress and deep understanding are to follow.

One of the study activities that led to this symposium took the form of the TOPS (Toward Other Planetary Systems) report, released recently by NASA's Solar System Exploration Division. In this work, the result of four years of wide ranging study by a broad based team of scientists, the question of the proper range of inquiry had to be addressed. The group concluded that an effective program should not be limited to discovering other planetary systems, but should address the broader questions of how planetary systems form and evolve as well. In disciplinary terms, the study of how interstellar materials collapse to form stars is astrophysics, and the study of the planets that have formed is planetary science, but there are surely a large number of interesting phenomena that occur in between. Science knows no boundaries, and it is the duty of all of us to make sure that no artificial boundaries are constructed.

This symposium, therefore, combines the ancient topic of cosmogony with the new fields of planetary science and focuses on common aims. We do not understand fully how our own solar system was formed, and we do not know whether we are unique (or at least a rare anomaly). But we do know that our own solar system exists, and the null hypothesis has to be that planetary systems are not uncommon occurrences. We can reasonably expect a rich variety of other systems, and as the subject evolves, the successors to this symposium will create a fascinating record. Cosmogony and planetary science, once widely separated disciplines, will be welded into a new entity that we can, from this viewpoint in time, only perceive dimly, but with high expectations.

THE SEARCH FOR OTHER PLANETS:
CLUES FROM THE SOLAR SYSTEM *

T. OWEN

Institute for Astronomy, Honolulu, Hawaii, USA

Abstract. Studies of element abundances and values of D/H in the atmospheres of the outer planets and Titan support a two-step model for the formation of these bodies. This model suggests that the dimensions of Uranus provide a good index for the sensitivity required to detect planets around other stars. The high proportion of N_2 on the surfaces of Pluto and Triton indicates that this gas was the dominant reservoir of nitrogen in the early solar nebula. It should also be abundant on pristine comets. There is evidence that some of these comets may well have brought a large store of volatiles to the inner planets, while others were falling into the sun. In other systems, icy planetesimals falling into stars should reveal themselves through high values of D/H.

1. Introduction

The goal of this contribution is to see what evidence we can find from a study of volatile elements and compounds in our own solar system that could guide us in the search for other, similar planetary systems. We begin with a consideration of outer planet atmospheres, move on to the condensed volatiles we find on the surfaces of Pluto and Triton (at the inner edge of the Kuiper Comet Belt), and then consider the possible delivery of this condensed matter to the inner planets.

The focus of this essay is thus on the icy planetesimals that formed in the solar nebula prior to the formation of the planets. What was the composition of these bodies? How important were they in the formation of planets? What can they tell us about conditions in other pre-planetary nebulae? How could we recognize the existence of such bodies in forming systems around other stars?

2. Outer Planet Atmospheres

2.1. MODELS FOR ATMOSPHERIC COMPOSITION

There have been two basic paradigms for the formation of planetary atmospheres:

1. Gravitational "condensation" of gases from the solar nebula — the formation of giant protoplanets. For many years, this was the accepted mode for forming the atmospheres of the outer planets.

2. Outgassing from the solid material in the planet itself. In this case, one must be concerned about the origin of that solid material — did it come only from the region of the nebula presently occupied by the planet's orbit or was there a significant contribution from more distant sources? The details of the accretion process also become important. Planetesimals colliding with the planet can

---

* Paper presented at the Conference on *Planetary Systems: Formation, Evolution, and Detection* held 7–10 December, 1992 at CalTech, Pasadena, California, U.S.A.

add volatiles but they can also remove them. This paradigm has traditionally been used to explain the atmospheres of the inner planets, but it is now seen to have applications to Pluto, Triton, and Titan as well.

Following the work of Mizuno (1980) and others on the formation of the giant planets a third paradigm has emerged that is essentially the sum of the first two:

3. The idea is that the giant planets originated in a two stage process which might be regarded as nucleation and capture: first a solid core was accreted, and as this grew massive enough (to a size on the order of $10M_E$), it captured gases from the nebula through its gravitational attraction. This is the presently accepted view for the formation of giant planet atmospheres. A variation of this approach calls for significant contributions to the atmospheres through the dissolution of subsequently impacting planetesimals (Pollack *et al.*, 1986).

It is possible to test these three alternatives by studying the present composition of planetary atmospheres. Formation of giant protoplanets should produce atmospheres in which the elemental abundances and isotopic ratios are identical to those in the sun. In contrast, outgassing from icy planetesimals (a special case of paradigm #2 that corresponds to giant planet cores, Titan, Triton and Pluto) will release gases that were trapped in the ices during their formation. We can get some insight into what those gases were by studying comets and through laboratory experiments in which ice is deposited at low temperatures in the presence of various gas mixtures (Bar-Nun *et al.*, 1988, 1989). We expect CO, N_2, Ar, plus CH_4 and NH_3 if they were present, but **not** Ne, H_2 or He, since these three gases are not trapped in ice unless the temperature is below 25 K (Ne) or even lower (H_2, He).

Hence paradigm #3 should lead to atmospheres in which the heavier volatile elements — C, N, O, etc. — are enriched relative to H, He, and Ne, since these atmospheres will be a blend of outgassed volatiles from the cores plus captured gas from the nebula. He/H should be equal to the solar value in these atmospheres (Ne/H should also be solar but this ratio is nearly impossible to measure by remote investigations).

2.2. OBSERVATIONS OF ATMOSPHERIC COMPOSITION AND THE RATIO OF D/H

The observations of the outer planets made by the IRIS and Radio Occultation experiments on the Voyager spacecraft, supplemented by ground-based studies of Uranus and Neptune, demonstrate the validity of paradigm #3 (Gautier and Owen, 1989). The atmospheres of Jupiter and (especially) Saturn are actually depleted in helium as a result of the solution of this element in metallic hydrogen in the planets' interiors (Stevenson and Salpeter, 1977). Uranus exhibits a proto- solar value of He/H, as does Neptune, once the observed value is corrected for the probable presence of significant N_2 in Neptune's atmosphere (Conrath *et al.*, 1993). These two outer giants have insufficient mass to produce metallic hydrogen in their interiors, so they have retained the composition of the nebular gas they captured.

We can use carbon as an index of the heavy element concentration in these atmospheres because we can measure the carbon abundance by observing methane, which does not condense except on Uranus and Neptune, where only tenuous methane clouds are formed. Ratioing CH_4 to H_2 gives us C/H, with the following results:

Planet	Core Mass/Total Mass	CH (Sun = 1.0)
Jupiter	0.03–0.06	2.3 ± 0.3
Saturn	0.11–0.22	4.3 ± 2.0
Uranus	~0.75	25 ± 10
Neptune	~0.75	25 ± 10

We can see that the trend is exactly as predicted by paradigm #3: as the relative mass of the planetary core increases, the enrichment of carbon also increases.

We can further test the models by using the isotope deuterium as a probe of the physical conditions that played a role in planetary formation, and as a trace of interstellar material incorporated in the planets (Geiss and Reeves, 1972, 1981). It is well known that deuterium is highly enriched in some interstellar molecules compared with its value in hydrogen gas (e.g., Irvine and Knacke, 1989). My colleagues and I have demonstrated that deuterium in the outer solar system exhibits a similar division into two reservoirs, viz. condensed matter, where the average D/H $\sim 2 \times 10^{-4}$ and hydrogen gas, where D/H $\sim 2 \times 10^{-5}$ (de Bergh et al., 1986, 1988, 1990; Owen et al., 1986; Owen, 1992). Both reservoirs were evidently established in the interstellar medium prior to solar system formation. The smaller one in condensed matter must actually consist of a variety of different values for different molecules, dominated by the value in the most abundant molecule, H_2O.

Our results for D/H are summarized in Figure 1, which also includes values for Halley's Comet (Eberhardt et al., 1987) the Interplanetary Dust Particles (IDP's) (Bradley et al., 1988), the meteorites (Robert and Epstein, 1982; Epstein et al., 1987), the local interstellar hydrogen (Linsky et al., 1993) and the original solar nebula hydrogen value (protosolar D/H) deduced from ^3He in the solar wind and meteorites (Geiss, 1993). Here we see that Uranus and Neptune are again clearly separated from Jupiter and Saturn on the basis of the values of D/H in their atmospheres. This is another example of the influence of the core material on atmospheric composition, since the higher values of D/H we find in the atmospheres of these two planets reflect the enrichment of D/H in the condensed matter that formed their cores (Hubbard and MacFarlane, 1980). Jupiter and Saturn, in contrast, appear to exhibit the values of D/H that were set in the hydrogen gas that constituted the bulk of the mass of the original solar nebula, and which now dominates the masses of these two giant planets.

The mass spectrometer on the Galileo probe will be able to test this conclusion further by analyzing both D/H and $^3He/^4He$ as well as He/H, Ne/H and a host of

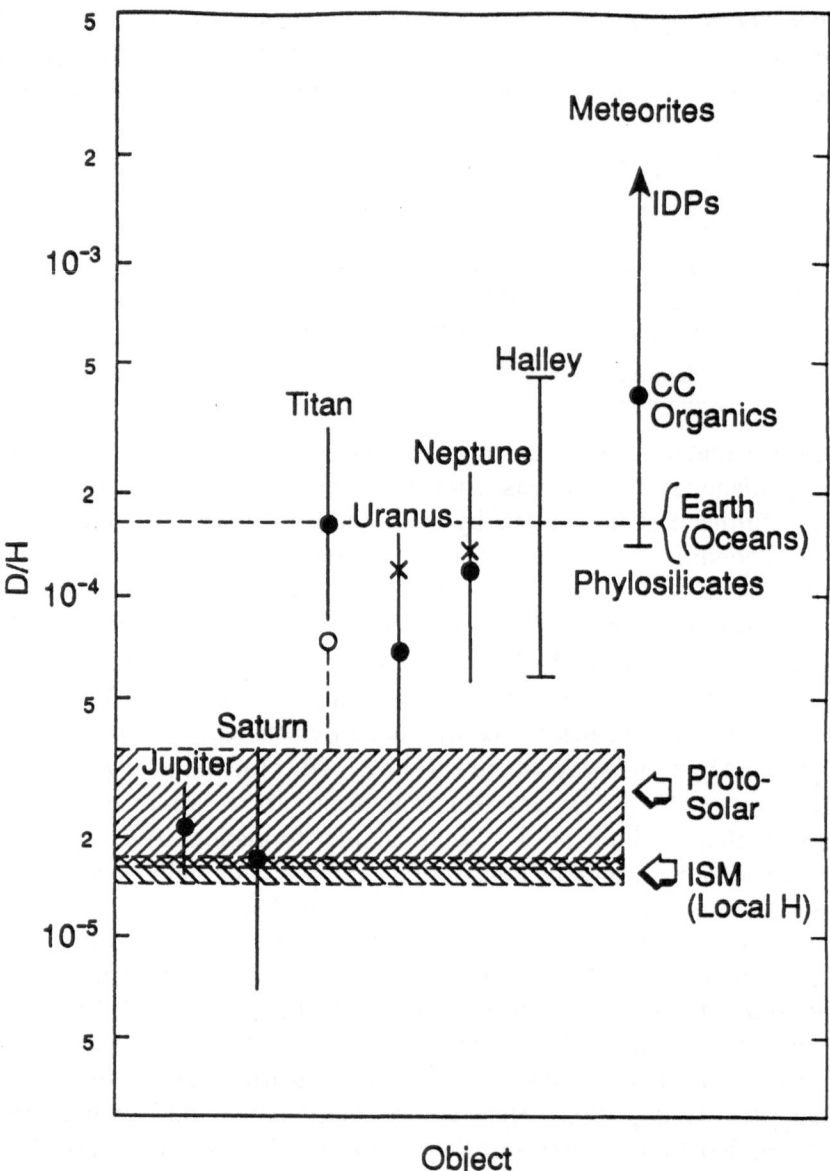

Fig. 1. This plot shows measurements of D/H in the atmospheres of the outer planets, Saturn's satellite Titan (the empty circle shows the correction for deuterium enrichment through atmospheric escape), Halley's comet and various measurements in meteorites and Interplanetary Dust Particles (IDPs). The values of D/H measured in atomic hydrogen in the local Interstellar Medium (ISM), the Earth's oceans, and the Protosolar value deduced from studies of helium in meteorites and the solar wind are provided for reference.

other ratios in Jupiter's atmosphere during its entry in December, 1995 (Niemann *et al.*, 1992).

2.3. CONCLUSIONS — PART 2

The data that we have obtained so far on the composition of outer planet atmospheres clearly favor paradigm #3. This means that we can set a lower limit for the size of outer planet we should be able to detect if we want to find planets around other stars. That limit will be about the size of Uranus ($M = 14.5 \, M_E$, $R = 4R_E$), since this will be about the minimum size for a planet forming in this two-stage process. Obviously there will be many smaller bodies in such a system, corresponding to our inner planets, but these are very difficult targets. If we have the sensitivity to find something the size of Uranus orbiting a nearby star, we will have some assurance that we are making a serious test for the existence of another planetary system.

The icy planetesimals required to make the four giant planets are equivalent to about 50 Earth masses of material. This plus the 50 to 100 Earth masses of icy planetesimals lost to the Oort Cloud (Weissman, 1991) should be dispersed throughout the pre-planetary nebula surrounding a forming star. The study of deuterium in our system indicates that D/H in this condensed matter will be higher by about a factor 10 than the background value in the hydrogen gas. Elevated D/H is therefore something to look for in instances where material from a disk is falling into a star, as is the case for β Pic (Lagrange-Henri *et al.*, 1988).

The fact that enriched values of D/H from interstellar gas are preserved through the formation of the solar nebula suggests that other interstellar chemistry may also be invariant to this process, (Geiss and Reeves, 1981). In particular, we might expect to find CO, N_2 and various organic compounds mixed in with the water ice, sublimating according to the local temperature. This expectation is supported by the new observations of Pluto and Triton described in the next section.

3. Icy Planetesimals

3.1. PLUTO AND TRITON

Another approach to the elucidation of conditions in the primitive solar nebula is to try to examine the most primitive solid objects that are presently available to see what volatiles they contain. The traditional targets of this investigation have been the comets, but our understanding of their composition has been severely limited by the fact that we have only been able to study the sublimation and dissociation products of the gases that are actually frozen in cometary nuclei. The nature of the so-called parent molecules therefore remained obscure until the relatively recent application of radio astronomical methods and the missions to Comet Halley in 1986.

Pluto and Triton also represent primitive solid bodies that formed and remained at temperatures that are low enough to preserve most of the original volatiles. Both objects were known from ground-based observations to have frozen methane on their surfaces, with frozen N_2 also identified on Triton (Cruikshank and Brown, 1986). The Voyager 2 flyby of Triton revealed an icy surface at a temperature of 37 K, decorated with splotches of wind-blown dark material beneath a predominantly nitrogen atmosphere with a surface pressure of only 16 microbars (Smith *et al.*, 1989, Broadfoot *et al.*, 1989, Tyler *et al.*, 1989). Plumes of the dark material rose 8 km into the atmosphere at some places, producing trails of dark aerosols that stretched some 150 km downwind.

Pluto has not yet been visited by spacecraft, but extensive ground-based observations including a stellar occultation and a series of mutual eclipses and occultations between Pluto and its satellite Charon have taught us much about this distant world (Stern, 1993).

With the development of cooled array spectrometers that can be used in the near-IR, it has become possible to study the spectra of Pluto and Triton with a resolution some 10 times better than previous work. Using the CGS4 spectrometer with the United Kingdom Infrared Telescope on Mauna Kea, we have recorded spectra of Triton and Pluto from 1.4–2.5 μm (Cruikshank *et al.*, 1993, Owen *et al.*, 1993). These data revealed the presence of solid CO and CO_2 on Triton, and solid N_2 and CO on Pluto. Analyses of the data with the help of laboratory spectra provided by B. Schmitt and R.H. Brown and models developed by T. Roush demonstrated that the surfaces of both objects consist of ~99% N_2 ice, with just traces of the other constituents: CH_4 and CO on Pluto and these ices plus solid CO_2 on Triton. No evidence for other hydrocarbons or nitriles was found in these spectra.

3.2. HALLEY'S COMET

This is very different from the situation for Halley's comet, where nitrogen appears to be deficient relative to cosmic abundances, whereas carbon exhibits the cosmic ratio (Geiss, 1987, 1988; Encrenaz *et al.*, 1992). The solution to this apparent paradox may be found in the interstellar medium where 70% of the expected nitrogen is undetected. The solution for the missing nitrogen in the ISM has been to suggest that it is present in the form of N_2, and thus unobservable. The observations of Pluto and Triton indicate that most of the nitrogen in the outer solar system was probably also in the form of N_2. Conversion of N_2 to NH_3 without a catalyst is very difficult (Schlesinger, 1950). As a result, a comet can only be expected to exhibit a cosmic abundance of nitrogen if it formed at a temperature below the freezing point of N_2 (63 K) and was then preserved at temperatures well below that until the time that we observe it.

Halley's comet in its present, "captured" orbit, never gets farther from the sun than Pluto, and thus N_2 will be subliming from the nucleus all the way around the orbit. Although a small amount of N_2^+ has been detected in Halley's spectrum, most of the nitrogen that was found in this comet at its last apparition was in the form of NH_2 or the CHON particles (Krankowsky, 1991), consistent with this interpretation. It is therefore to be expected that the total nitrogen budget in the comet is deficient relative to cosmic standards: the abundant, highly volatile N_2 has mostly disappeared.

The studies of Pluto and Triton also raise some interesting questions. If N_2 is the dominant form of nitrogen on these bodies, we would expect CO to be the dominant volatile carbon compound, again by analogy with the interstellar medium. In the ISM, $CO/CH_4 > 10^2$ (Irvine and Knacke, 1989). Yet CH_4 is present on both of these bodies, and is also abundant in the atmosphere of Titan. what is its origin?

At the moment, the best solutions appear to be the formation of CH_4 from CO and H_2O and/or reactions between amorphous carbon and water and/or the release of CH_4 from trapped organic matter, all energized by the heat generated during accretional impacts.

Why is there CO_2 on Triton but not (so far!) on Pluto? We know that Pluto exhibits a marked light curve, indicating a variable surface albedo as it rotates, and we have only seen one hemisphere. Thus CO_2 may be there. We expect it from the reaction

$$CO + H_2O \rightarrow CO_2 + H_2$$

which should also occur during impacts. If CO_2 is really absent, we could postulate that it was lost from Pluto during the formation of Charon. Further observations are scheduled.

3.3. CONCLUSIONS — PART 3

At present, we can conclude that both Pluto and Triton have N_2 as their major surface ice and their major atmospheric constituent. It seems logical that this gas dominated the nitrogen budget of the outer solar nebula and represents an unmodified contribution from the interstellar cloud from which the solar system formed, just like the unmodified D/H that we found in the outer planet atmospheres. And like the deuterium, it appears that nitrogen was supplied to the solar system in two distinctly different reservoirs: the dominant one as nitrogen gas and the smaller one as condensed compounds. It will be interesting to see if the nitrogen in these two reservoirs exhibits different isotopic ratios. This investigation might shed some light on the problem posed by the secular increase in $^{15}N/^{14}N$ observed in lunar soils (Kerridge, 1989; Geiss and Bochsler, 1991).

4. Radial Transport in the Nebula: Is There a Cometary Contribution to Inner Planet Atmospheres?

4.1. THE IMPORTANCE OF NOBLE GASES

The formation of the giant planets is generally assumed to have led to the expulsion of icy planetesimals both inward and outward from their original orbits (e.g., Weissman, 1991). Many, perhaps most of the outward bound planetesimals were on hyperbolic orbits that caused them to leave the solar system, but a large fraction ended up in the Oort Cloud. Many of the inward-bound objects would have collided with the inner planets, thereby serving as a source of volatiles for these bodies (Ip and Fernandez, 1988). We would expect the same situation to exist in other forming planetary systems, so it is useful to see if we can prove that this is really what happened here.

To do this, we need to find evidence for a specific signature of cometary volatiles in the inventories of gases that are now found in planetary atmospheres. The best place to look for such a signature is in the noble gases, since they will not participate in the chemical reactions that will drastically change the proportions of other gases as they interact with planetary surface material over the 4.5 AE since the solar system formed. (We exclude He because of its low mass.)

Even the abundances and isotopic ratios of noble gases can be changed however, if there was sufficient hydrogen present on these planets to produce hydrodynamic escape of the early atmospheres. Pepin (1991) has constructed a detailed model that accounts for the various isotopic ratios and abundances using this approach. Even in his model, however, it is necessary to invoke cometary impact to explain the current volatile inventory on Venus.

4.2. A MIXING MODEL FOR THE ORIGIN OF INNER PLANET ATMOSPHERES

We have suggested that cometary delivery of the heavy noble gases may account directly for their abundances on all the inner planets, if one assumes that the atmospheres of these objects represent a mixture of volatiles released from the rock composing the planets and volatiles delivered by icy planetesimals (Owen *et al.*, 1992; Owen and Bar-Nun, 1993). This mixing line is shown in Figure 2, which is a log-log plot of $^{36}Ar/^{132}Xe$ versus $^{84}Kr/^{132}Xe$. The key to the icy planetesimal interpretation is provided by a series of laboratory experiments in which mixtures of gases were trapped in ice formed at low temperatures (e.g., Bar-Nun *et al.*, 1988), since noble gases have not yet been detected in comets. Abundances of these gases as measured in the SNC meteorites, now generally assumed to have come from Mars, and in terrestrial submarine basalts also fit this mixing line (see Figure 2). Ozima and Wada (1993) have pointed out a potential ambiguity in this interpretation, noting that a line representing equilibrium partitioning between gas and melt will also fit the noble gas data in Figure 2. Owen and Bar-Nun (1993) argue that the variation in xenon isotopic abundances in the SNCs precludes gas-melt partitioning on Mars, although it may have played a role on Earth. We suggest that the noble gases on Venus must have come from one or more icy planetesimals formed at temperatures ≤ 30 K, meaning objects originally from the Kuiper Belt, whereas the icy planetesimals that brought these gases to the Earth and Mars would have formed in the Uranus-Neptune region, near 50 K.

4.3. CONCLUSIONS — PART 4

This model obviously requires a number of additional tests before it can be accepted, starting with a search for noble gases in comets. We expect that the pristine comets that carry abundant N_2 will also have argon, krypton and xenon, but no neon,

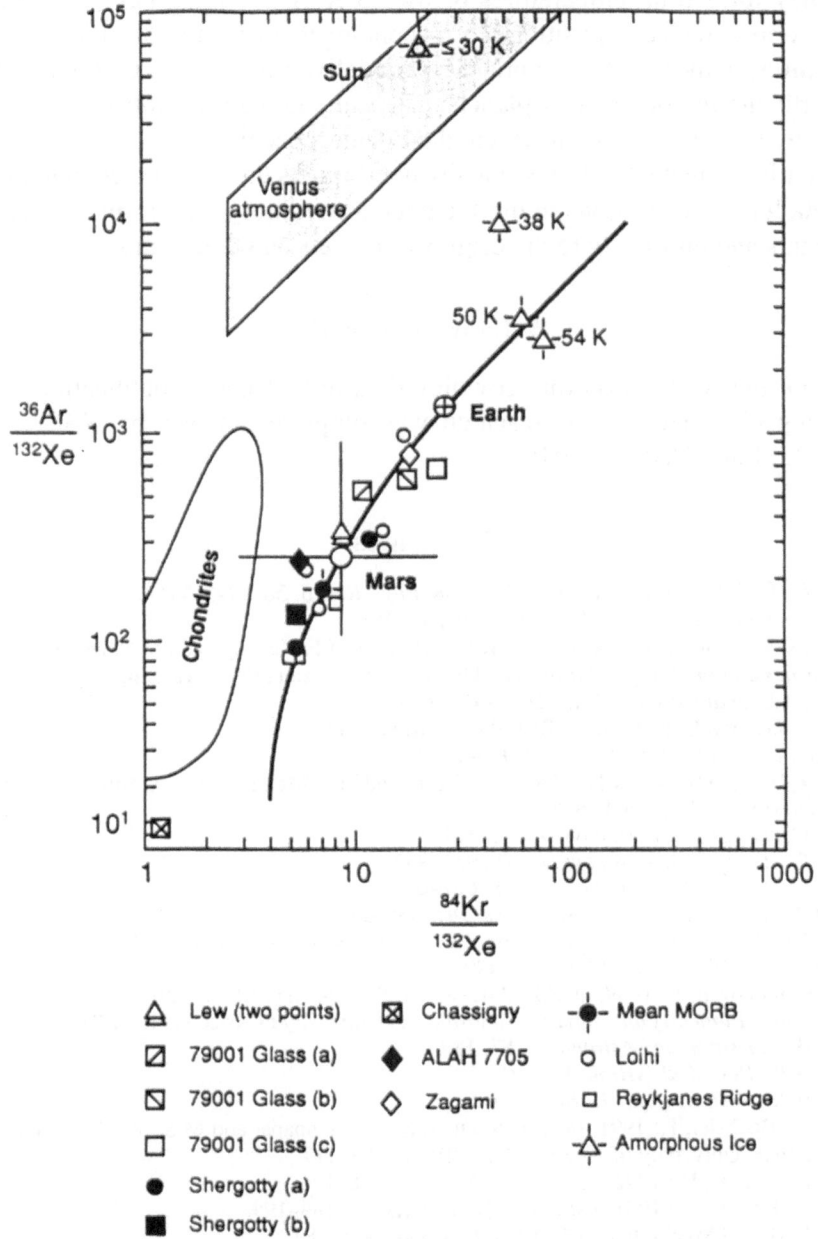

Fig. 2. The heavy noble gases measured in the atmospheres of the Earth and Mars (Owen *et al.*, 1977) define a mixing line that passes through values determined in the SNC (Shergotty–Nakhla–Chassigny) meteorites and submarine basalts (Loihi Seamount, Reykjanes Ridge, Mid-Ocean Ridge Basalts [MORB]), and laboratory samples of amorphous ice formed at low temperatures in a solar mixture of gases (Owen and Bar-Nun, 1993). The SNC meteorites are identified in the left and central columns of the figure legend.

unless they formed at temperatures below 25 K. There must have been several types of comets, with compositions corresponding to the formation of ice at various temperatures, so there is the potential here for delivering a variety of volatiles to the early Earth and the other inner planets. The same radial transport process should exist in any forming system in which giant planets occur.

Further investigations of dynamically new comets thus seem certain to provide basic data linking conditions in the Interstellar Medium to the formation of disks around stars and ultimately to the origins of planets and their atmospheres.

Acknowledgements

The author thanks J. Geiss for providing data in advance of publication and for stimulating discussions. This research was supported in part by NASA grants NAGW 2631 and NAGW 2650.

References

Bar-Nun, A., Kleinfeld, J. and Kochavi, E.: 1988, *Phys. Rev. B.* **38**, 7749–7754.
Bar-Nun, A. and Kleinfeld, I.: 1989, *Icarus* **80**, 243–253.
Bradley, J.P., Sandford, S.A. and Walker, R.M.: 1988, in: J.F. Kerridge and M.S. Matthews, ed(s)., *Meteorites and the Early Solar System*, University of Arizona Press, Tucson, 861–898.
Broadfoot, A.L. *et al.*: 1989, *Science* **246**, 1459–1465.
Conrath, B., Gautier, D. and Owen, T.: 1993, *Icarus* submitted.
Cruikshank, D.P. *et al.*: 1993, *Science* **261**, 742–745.
Cruikshank, D.P. and Brown, R.H.: 1986, in: J. Burns and M.S. Matthews, ed(s)., *Satellites*, University of Arizona Press, Tucson, 836–874.
de Bergh, C. *et al.*: 1986, *Astrophys. J.* **311**, 501–510.
de Bergh, C. *et al.*: 1988, *Astrophys. J.* **329**, 951–955.
de Bergh, C. *et al.*: 1990, *Astrophys. J.* **355**, 661–666.
Eberhardt, P. *et al.*: 1987, *Astron. Astrophys.* **187**, 435–437.
Encrenaz, Th., Puget, J.L. and D'Hendecourt, L.: 1991, *Space Sci. Rev.* **56**, 83–92.
Epstein, S., *et al.*: 1987, *Nature* **326**, 477–479.
Gautier, D. and Owen, T.: 1989, in: S.K. Atreya, J.B. Pollack and M.S. Matthews, ed(s)., *Origin and Evolution of Planetary and Satellite Atmospheres*, University of Arizona Press, Tucson, 487–512.
Geiss, J.: 1987, *Astron. and Astrophys.* **187**, 189.
Geiss, J.: 1988, *Rev. Mod. Astron.* **1**, 1–27.
Geiss, J.: 1993, *Festschrift for H. Reeves* in press.
Geiss, J. and Bochsler, P.: 1991, in: C.P. Sonnett, M.S. Grampapa and M.S. Matthews, ed(s)., *The Sun in Time*, University of Arizona Press, Tucson, 98–117.
Geiss, J. and Reeves, H.: 1972, *Astron. and Astrophys.* **18**, 126.
Geiss, J. and Reeves, H.: 1981, *Astron. and Astrophys.* **93**, 189–199.
Hubbard, W.B. and MacFarlane, J.J.: 1980, *Icarus* **44**, 676–682.
Ip, W.-H. and Fernandez, J.A.: 1988, *Icarus* **74**, 47–61.
Irvine, W.M. and Knacke, R.F.: 1989, in: S.K. Atreya, J.B. Pollack and M.S. Matthews, ed(s)., *Origin and Evolution of Planetary and Satellite Atmospheres*, University of Arizona Press, Tucson, 3–34.
Kerridge, J.F.: 1989, *Science* **245**, 480–486.
Krankowsky, D.: 1991, in: R.L. Newburn, Jr., M. Neugebauer and J. Rahe, ed(s)., *Comets in the Post-Halley Era*, Kluwer, Dordrecht, 855–878.
Lagrange-Henri, A.M., Vidal-Madjar, A. and Ferlet, R.: 1988, *Astron. and Astrophys.* **190**, 275.
Linsky, J. *et al.*: 1993, *Astrophys. J.* in press.
Mizuno, H.: 1980, *Prog. Theor. Phys.* **60**, 544–557.

Niemann, H.B., Harpold, D.N., Atreya, S.K., Carignan, G.R., Hunten, D.M. and Owen, T.C.: 1992, *Space Sci. Rev.* **60**, 111–142.

Owen, T.: 1992, in: P.D. Singh, (ed.), *Astrochemistry of Cosmic Phenomena*, Kluwer, Dordrecht 97–101.

Owen, T. and Bar-Nun, A.: 1993, *Nature* **361**, 693–694.

Owen, T. and Gautier, D.: 1989, in: S.K. Atreya, J.B. Pollack and M.S. Matthews, ed(s)., *Planetary Atmospheres*, University of Arizona Press, Tucson, 487–512.

Owen, T., Bar-Nun, A. and Kleinfeld, I.: 1992, *Nature* **358**, 6381.

Owen, T., Lutz, B.L. and de Bergh, C.: 1986, *Nature* **320**, 244–246.

Owen, T. *et al.*: 1977, *J. Geophys. Res.* **82**, 4635–4639.

Owen, T. *et al.*: 1993, *Science* **261**, 745–748.

Pepin, R.O.: 1991, *Icarus* **92**, 2–79.

Pollack, J.B. *et al.*: 1986, *Icarus* **67**, 409–443.

Robert, F. and Epstein, S.: 1982, *Geochim. et Cosmochim. Acta* **46**, 81–95.

Schlesinger, H.I.: 1950, *General Chemistry (4th ed.)*, Longmans, Green and Co., New York, 492.

Smith, B.A., *et al.*: 1989, *Science* **246**, 1422.

Stern, A.: 1992, *Ann. Rev. Astron. Astrophys.* **30**, 185–233.

Stevenson, D.J. and Salpeter, E.E.: 1977, *Astrophys. J. Suppl.* **35**, 221–261.

Tyler, G.L., *et al.*: 1989, *Science* **246**, 1466–1472.

Weissmann, P.R.: 1991, in: R.L. Newburn, Jr., M. Neugebauer and J. Rahe, ed(s)., *Comets in the Post-Halley Era*, Kluwer, Dordrecht, 463–486.

FORMATION AND EVOLUTION OF PLANETS *

V.S. SAFRONOV and E.L. RUSKOL

O. Yu. Schmidt Institute of Physics of the Earth, Russian Academy of Sciences, Moscow, Russia

Abstract. The basic stages of the formation of the planets are considered. Decay of turbulence in the solar nebula allows the precipitation of dust particles to the solar nebula's equatorial plane, their growth occurring by sticking at collisions and transformation into kilometer-sized bodies (planetesimals) partly by direct growth and partly via gravitational instability in a dense dust layer formed near the central plane of the nebula. Analytical study of the dynamical evolution of a rotating swarm of bodies with quasi-equilibrium values of velocity and mass distributions gave a reasonable description of the final stage of accumulation. This is a robust result in that the time scale of formation of the terrestrial planets is 10^8 yr. Numerical simulation of the earlier stage revealed an accelerated (runaway) growth of the largest bodies—planet embryos. More careful study of the intermediate stage is needed. The two-step formation of giant planets is considered: accumulation of icy-rock cores of about 10–15 Earth masses, similar to that in the terrestrial zone, and accretion of gas onto cores of Jupiter and Saturn require time scales not much longer than 10^7 yr (mostly for the growth of their cores). Problems connected with a slow time of growth of Uranus and Neptune are discussed. Small bodies of the Solar System are byproducts of the formation of the planets. The fall of large bodies onto the planets at the last stage of accumulation brought much of stochastics into the process (planetary obliquities, etc.). It led to a much higher initial temperature of the Earth's mantle, early beginning of partial melting and differentiation, which determined its pre-geological evolution.

1. Introduction

It occurred that our study of the origin and evolution of planets from a disk-like protoplanetary cloud began long before circumstellar disks were discovered and even much earlier than surfaces of other planets in the solar system became visible thanks to the space missions. The problem of the origin of the Earth and the planets was formulated in the 1940's as a complex astronomical-geophysical problem by the founder of our Institute, Academician O. Yu. Schmidt (1891–1956). He was the first distinguished scientist who began regular systematic work on a theory of the accumulation of planets from solid particles and bodies (1944, 1957). Schmidt emphasized that the problem of formation of the planets in the already existing solar nebula is relatively independent of the problem of formation of the nebula and should be studied without waiting for the solution of the latter. Thus many of our results were published in the 50's–60's whereas western scientists began to study the problem closely only in the 70's. Main stages of the process were studied: settling of particles (as well as their growth) to the central plane of the nonturbulent solar nebula (SN), formation of a dense dust layer (subdisk) in this plane, its gravitational instability and formation of dust condensations, coagulation of condensations and their transformation into planetesimals, dynamical evolution of a rotating swarm of preplanetary bodies, distribution of their masses and random velocities established during their growth, a moderate runaway growth of planet embryos, formation

---

* Paper presented at the Conference on *Planetary Systems: Formation, Evolution, and Detection* held 7–10 December, 1992 at CalTech, Pasadena, California, U.S.A.

Astrophysics and Space Science **212**: 13–22, 1994.
© 1994 *Kluwer Academic Publishers.*

of the planets (Safronov, 1969). As a first approximation the results remain valid until now, though the bulk of new investigations by many scientists discovered important new features of the process as well as new difficulties connected with its complexity.

2. Formation of the Solar Nebula

Laplace's idea of the common origin of the sun and planets is expressed currently in the assumption that the solar system was formed as a result of a collapse of a protosolar nebula (PSN)—the dense core of a molecular cloud. Numerical calculations have shown that consequences of the collapse are highly dependent on the initial value of the angular momentum of the PSN. Contraction of a rapidly rotating PSN leads to the formation of a binary system whereas a slowly rotating PSN collapses into a single star. Formation of a planetary system can be expected only if the angular momentum of the star mass PSN is on the order of 10^{52}–10^{53} $g\,cm^2\,s^{-1}$. The theory of this process has not yet reached its maturity. In order to concentrate almost total mass inside the star and at the same time almost all angular momentum inside the planets (i.e., in the solar nebula) large scale redistribution of angular momentum is needed. Turbulent viscosity is usually supposed to be a main mechanism. However, it is not yet clear how favorable convective turbulence for physical conditions were in the nebula and how efficient this turbulence could be. To the end of collapse (10^5–10^6 yr) young sun (like T Tau stars) and a circumsolar disk with the Keplerian rotation have been formed. But we do not know if this disk was already the size of the present planetary system or was considerably smaller and continued to grow due to turbulent viscosity. The discovery of pre-main sequence T Tau stars and especially the recent discovery of gas-dust disks around some of them (Beckwith et al., 1990; Strom et al., 1989) begin to give very important observational data for investigation of the most obscure stage of the formation of the solar system. Not less important will be detection of planet-size objects in such disks.

3. Formation of Planetesimals

The largest planets of the solar system, Jupiter and Saturn, consist mainly of hydrogen and helium, and some scientists have suggested that these planets (and possibly all planets) have been formed via gravitational instability in a massive gaseous solar nebula (Kuiper, 1951; Fesenkov, 1951; Cameron, 1979). For differentially rotating systems of a finite thickness (disk-like systems) the classic Jeans' criterion is not applicable. Safronov (1960, 1969) has found a dispersion equation for such systems in the form:

$$n^2 = 4\pi G\rho(1 + 2/kH)^{-1} - k^2c^2 - \kappa^2 \tag{1}$$

where $\kappa = \mathrm{d}(\omega R^2)/R^3\,\mathrm{d}R$ is epicyclic frequency (for the Keplerian rotation $\omega \sim R^{-3/2}$; $\kappa^2 = \omega^2$), $k = 2\pi/\lambda$-wave number, c is the sound speed; $H \approx (2/\pi G\rho_c)^{1/2}c$ at $\rho_c \sim \rho_{cr}$ is the thickness of the disk, ρ_c is the density in the equatorial plane of the disk. Estimates have shown that minimum critical density necessary for gravitational instability in the disk, with respect to axisymmetric (radial) perturbations $\xi \sim \exp(nt + \mathrm{i}kx)$, equals

$$\rho_{cr} \approx 2\rho^* = 3M_\odot/2\pi R^3 \tag{2}$$

when the perturbation length is

$$\lambda_{cr} \approx (8 \div 10)H \tag{3}$$

It was also found that in moderate mass disks ($\lesssim 10^{-1}M_\odot$), like our protoplanetary cloud, the maximum density ρ_c reached only 10^{-1} of the critical value ρ_{cr}, so that gravitational instability in such gaseous disks was implausible. This result was strongly in favor of the formation of the planets by accumulation of solid matter.

Interstellar clouds contain only about 1% of their mass in the form of solid particles. Hence, the first important step in the accumulation process was separation of dust from gas. It began when initial turbulence decayed in the SN, probably soon after the collapse (Safronov and Ruskol, 1957; Safronov, 1982). Different scenarios can be expected from different assumptions about efficiency of turbulence, and of the ability of particles to stick. In the absence of turbulence, particles precipitated towards the central plane and grew sticking at collisions. During 10^3–10^4 periods of revolution around the sun, a dense dust layer (subdisk) formed near $z = 0$, which became gravitationally unstable when its density ρ_p reached criticality. If particles were equal or there was a prevailing size containing the major part of the mass, differential gas drag did not prevent development of axisymmetric ring-like perturbations, their instability and disintegration. Numerous condensations then appeared which after a number of collisions transformed into kilometer-sized planetesimals. The qualitative picture of evolution of the gas-dust disk, with the account of our early results, was presented in the popular book by B. Yu. Levin (1964) as a schematic sequence of eight stages (Figure 1).

However, due to gas lag from the Keplerian rotation a large vertical gradient of the dust density $\mathrm{d}\rho_p/\mathrm{d}z$ near the upper and lower boundaries of the layer caused a considerable vertical gradient of rotation velocity in this region. This induced a shear turbulence which could stop further settling of particles and prevent gravitational instability in the subdisk (Weidenschilling, 1984; Weidenschilling and Cuzzi, 1993). Being rather weak, the turbulence did not prevent growth of particles. Moreover, it accelerated their growth, making collisions more frequent. When particles reached a meter or a few meters in size they decoupled from the gas, settled down and made the layer unstable.

If a global (throughout the disk) turbulence remained after the collapse of the PSN but was so moderate that particles could stick at collisions the situation was

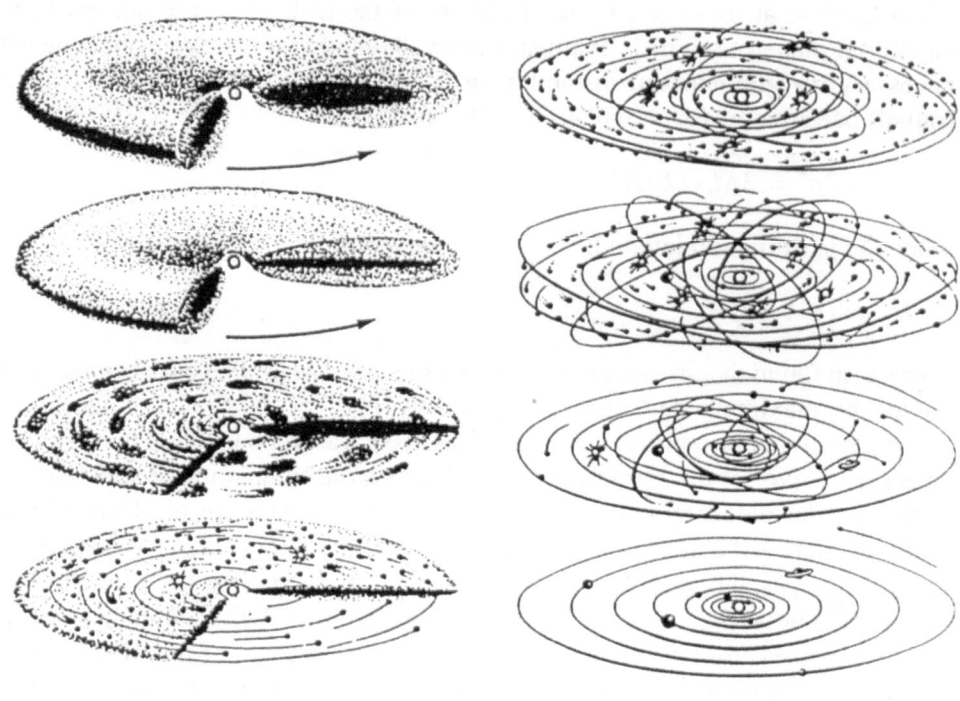

Fig. 1. See text.

like that described above, only conditions for gravitational instability in the subdisk were more severe. Dispersion equation (1) can be applied to the dust subdisk if we take into account the interaction of particles with the gas. There were four types of such interactions: (a) due to turbulent stirring of the subdisk, particles acquired random velocities v_p, (therefore term $k^2 c^2$ should be changed to $k^2 v_{pz}^2$), (b) the term taking into account the resistance of gas to a perturbation motion of particles n/t_e should be added where $t_e = v/(\mathrm{d}v/\mathrm{d}t)$ is the response time to the gas drag force, which depends on sizes of bodies and on their velocities relative to gas, (c) initially gas perturbs together with particles, but its perturbation does not develop and it returns soon to equilibrium state. It could be carried by the dust only at an implausibly high dust density $\rho_p/\rho_g > 10^6$. Therefore, without appreciable error perturbations in the gas can be neglected, (d) gas lag from the Keplerian rotation caused a radial motion of particles toward the sun, which depended on their sizes. When particles are equal this motion does not influence the development of axisymmetric perturbations. Then we have the following dispersion relation

$$n^2 = 4\pi G\rho_p(1 + 2/kH_p)^{-1} - k^2 v_{pz}^2 - \omega - n/t_e \tag{4}$$

If turbulence is weak or absent, term $k^2 v_{pz}^2$ is small (or zero) and the critical density required for gravitational instability does not differ much from that given by Equation (2).

When particles are of different sizes, gravitational instability of the subdisk takes place at the same critical density. However, the differential radial drift of particles leads to spreading of axisymmetric ring-like perturbations. Spreading of the ring can be prevented if a certain prevailing size of particles containing a major fraction of the ring mass appears. One can expect that at some conditions such a prevailing size could really arise (Safronov, 1991). Otherwise, sticking of particles allowed the growth of planetesimals – bodies large enough to decouple from the turbulence and the gas drag. In the region of giant planets, conditions necessary for gravitational instability were more favorable than in the region of the terrestrial planets.

4. Accumulation of the Planets

Investigation of the dynamical evolution of a rotating swarm of planetesimals and larger bodies (Safronov, 1969) has shown that the rate of growth of bodies depended first of all on their random velocities. In turn, velocities depended on mass distribution. These key parameters were estimated for quasi-stationary conditions. In the present planetary system, mass distributions for different objects (asteroids, impact craters, etc.) can be well approximated by an inverse power law

$$n(m) = c \cdot m^{-q} \tag{5}$$

where $q \approx 1.8 \div 2$. It was reasonable to assume a similar law as a first approximation for our study. The power laws with $q \lesssim 2$ were supported by a strict solution of the coagulation equation (Safronov, 1962) with the coagulation efficiency proportional to the sum of masses of colliding bodies

$$A(m, m') = A_1(m + m')) \tag{6}$$

which represents intermediate value between the geometrical cross-section of collisions ($\sim m^{2/3}$) and the gravitational one ($\sim m^{4/3}$) for larger bodies. For the initial function $n(m, 0) = (N_0/\bar{m}_0)e^{-m/\bar{m}_0}$ the solution has the form

$$n(m, \tau) = \frac{N_0(1 - \tau)}{m\sqrt{\tau}} e^{-(1+\tau)m/\bar{m}_0} I_1\left(\frac{2m\sqrt{\tau}}{\bar{m}_0}\right), \tag{7}$$

where $I_1(x)$ is a modified Bessel function, and $\tau = 1 - e^{-A_1\rho t}$. For not too small τ

$$n(m, \tau) = \frac{N_0(1 - \tau)}{2\sqrt{\pi}\tau^{3/4}} m_0^{-1/2} m^{-3/2} e^{-(1-\sqrt{\tau})^2 m/\bar{m}_0}. \tag{8}$$

The exponential term is important only in the range of the largest bodies. With the increase of mass, the power law $m^{-3/2}$ extends to a larger m. A qualitative study of the coagulation equation also allowed some power law asymptotic solutions (Safronov, 1969, Vityazev et al., 1990) for $A(m, m') = A_0(m^\alpha + m'^\alpha)$, $(m^\beta + m'^\beta)$: at $\beta = 0$, $\alpha < 2$: $q = 1 + \alpha/2$, at $\alpha = \beta = 1/3$: $q = 5/3$ (geometrical cross-section) at $\alpha < 1$, $\alpha \to 1$, $\beta = 1/3$: $q \to 2$ (gravitational cross-section). It was concluded that for equal random velocities of bodies their mass distribution tends to an inverse power law (5) with $3/2 < q < 2$, when the main mass is concentrated in larger bodies.

When a viscous medium moves with a gradient of velocity, the energy of its regular motion transforms into "heat", i.e., into random motions of particles. In the differentially rotating disk, velocities of planetesimals increased at the expense of potential energy of the disk. They were found from a balance of the "thermal" energy gained due to gravitational perturbations of bodies ("viscous stirring") and the energy lost at inelastic collisions (Safronov, 1969). On average, velocities tend to the "first cosmic" velocity at the surface of the most effective perturbing bodies. They can be expressed by the mass m and the radius r of the largest body

$$v = Gm/\theta z \tag{9}$$

where θ is a dimensionless parameter showing how much less $(m/r)_{\text{eff}}$ is than the ratio m/r for the largest body. When bodies are equal, $\theta \approx 1$. When they have a power law mass distribution (7) with $q \lesssim 2$, parameter θ is of the order of a few. Dependence of velocities of bodies on their masses was found to be very weak.

A possibility of accelerated ("runaway") growth of largest bodies was revealed (Safronov, 1969). As the gravitational cross-section of collisions πl^2 is proportional to m the first largest body m_1 grows relatively faster than the second largest body m_2 in the same feeding zone, i.e., $dm_1/m_1 > dm_2/m_2$, and the ratio m_1/m_2 increases with time. In this way the largest bodies (planet embryos) separate from the distribution of masses of other bodies. The growth of these embryos was studied:

$$\frac{dm}{dt} = \pi l^2 \rho v = 4\pi r^2 \delta \, dr/dt, \tag{10}$$

$$\rho = \sigma/H, \ H = Pv/4 = \pi v/2\omega, \ \rho v = 4\sigma/P, \ \sigma_z = \sigma_0(1 - r^2/r_p^2), \tag{11}$$

where m, r and δ are the mass, radius and density of the growing planet, σ, H and ρ are surface density, thickness and volume density of the swarm of bodies,

$P = 2\pi/\omega$ is the period of revolution around the sun, r_p is the present radius of the planet. Hence $dr/dt = (1 + 2\theta)\sigma_0(1 - r^2/r_p^2)/P\delta$ and the time scale of formation of the planet is

$$T = \int_0^{0.99r_p} \frac{P\delta\, dr}{(1 + 2\theta)\sigma_0(1 - r^2/r_p^2)} \sim \frac{\delta r_p}{\theta\sigma_0} P \tag{12}$$

Assuming $\theta \approx 3 - 5$, such a quasi-stationary ("orderly") scenario leads to the time-scale of the formation of the Earth (and other terrestrial planets) of $\sim 10^8$ yr.

Later on, non-stationary coupled evolution of mass and velocity distributions was studied by numerical modeling of the process (Greenberg et al., 1978; Wetherill, 1990; Lissauer and Stewart, 1993). At an early stage of accumulation, "runaway" growth of larger bodies was found. Mass distribution of bodies did not tend to a power law with $q < 2$. The largest bodies (planet embryos) contained a small part of the total mass and could not effectively perturb other bodies. Their velocities remained small, and the gravitational cross-section of larger bodies rapidly increased ($\sim r^4$). In the region of terrestrial planets, runaway growth changed to orderly growth when bodies reached lunar size. The final stage was much more lengthy and the total time scale of growth remained equal to $\sim 10^8$ yr.

Formal application of Equation (11) to the giant planets gives a disappointing result—the time scale of growth of Uranus and Neptune should be longer than the age of the solar system. However, the formation of the giant planets was very complicated and Equation (11) may not be applied directly in this case. One can only see that the product $\theta\sigma$ should be at least 10^2 times larger than in the case of terrestrial planets. To construct a reliable theory of the formation of giant planets a number of important additional processes should be studied: accretion of gas onto Jupiter and Saturn (may be partly onto Uranus and Neptune), ejection of bodies into the asteroid belt and cometary cloud as well as out of the solar system, and dissipation of gas from the disk. Due to low temperatures in the region of giant planets, the abundant volatiles H_2O, CH_4, NH_3 were condensed and the ratio of dust to gas was about three times higher than in the terrestrial region. Formation of these planets proceeded with two major steps. Initially, solid cores accumulated in the same way as planets formed in the terrestrial region. When the cores of Jupiter and Saturn reached a critical mass ($\sim 10 \div 15$ masses of the Earth) they began to accrete gas (mainly hydrogen and helium). Therefore, they should have grown to this size before the dissipation of gas from their region ($\sim 10^7$ yr?). At such large masses these planets began to effectively eject the bodies from their zones. Hence, the initial mass of solids in the zones could have been several times larger than they are now in these planets (Safronov, and Ruskol, 1982). The character of accretion changed with the increase of mass: after a slow beginning it accelerated, gravitational radius r_g increasing $\propto m$. When r_g reached the thickness of the gaseous disk as well as the radius of the Hill sphere, the accretion slowed down considerably. More full data on the lifetime of gas in the circumstellar disks is very desirable. Some models permitting to avoid difficulties with the long time scale of the formation of Uranus and Neptune were recently suggested by Pechernikova (1991) and by Lissauer et al. (1993).

5. Evolution of Planets

Our study of the origin of planets began first of all from geophysical problems. It is generally agreed that the inner structure of the Earth cannot be understood if we do not know its formation process. O. Yu. Schmidt, in the 1940's, looked for a mechanism to explain the formation of the Earth in a cold, non-molten state, and his hypothesis led just to such an initial state.

However, with time, the study of pregeologic history of the Earth gradually changed the sense of a "cold initial condition" of the Earth. In a model of its accumulation from bodies many orders of magnitude smaller than the planet (Safronov, 1959), the greatest part of the energy release was radiated from the surface and the initial temperature of the Earth's interior was found to be $\sim$ 1000 K, much lower than the melting point of rock material inside the Earth. However, at the beginning of the 1960's, it was found that the mass distribution of preplanetary bodies was characterized by a presence of large planetesimals capable of contributing considerably more heat into the Earth due to deeper penetration (Safronov, 1969, 1982; Kaula, 1980). The result was also supported by the study of inclinations of the axes of rotation of planets resulting from large impacts (Safronov, 1969; Lissauer and Safronov, 1991). The initial temperature of the Earth is now evaluated as approaching the melting temperatures, and favorable for the early differentiation in a vast region of the Earth's mantle (Vityazev et al., 1990). See Figure 2.

The results for the Earth may be applied (with the corresponding numerical change) to the four terrestrial planets.

During the accretion, photospheric temperatures of Jupiter could have reached a few thousand degrees (Safronov and Ruskol, 1982). At the final stage of accretion ($m_J \sim 300 m_\oplus$, $m_S \sim 90 m_\oplus$), which was more prolonged, the photospheric temperatures dropped considerably, and T_J reached about 1500–1000 K, and T_S – several hundred degrees.

This result agrees with the chemical composition of Jupiter's and Saturn's satellites which formed during the final stages of growth of the giant planets.

Acknowledgements

We are indebted to the Organizing Committee of the Conference on the formation, evolution and detection of planetary systems, in Pasadena, for the invitation to present this talk. Our great gratitude is to the Institute for Theoretical Physics at UCSB that organized the research program, "Planet Formation," the members of which we have been during autumn 1992. This research was supported in part by the NSF under Grant No. PHY89-04035.

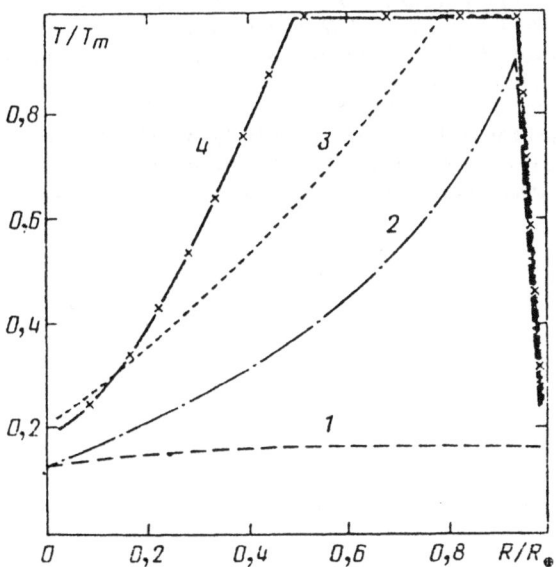

Fig. 2. Initial temperature of the Earth: (1) Heating by accumulation of small bodies (Safronov, 1959). (2) Heating by accumulation of large bodies (Safronov, 1969). (3) The same as (2) with the larger upper limit of mass (Safronov, 1982). (4) Heating by large bodies (Kaula, 1980).

References

Beckwith, S.V.W., Sargent, A.I., Chini, R.S. and Güsten, R.: 1990, *Astronomical Journal* **99** (N3), 924–945.

Cameron, A.G.W.: 1979, *The Moon and the Planets* **21**, 178– 183.

Fesenkov, V.G.: 1951, *Astronomicheskij Zhurnal* **28** (N3), 492–517 (Russ.).

Greenberg, R., Wacker, J., Hartman, W. and Chapman, C.: 1978, *Icarus* **35**, 1–26.

Kaula, W.M.: 1980, in : *The Continental Crust and Its Mineral Deposits*, New York, 25–31.

Kuiper, G.P.: 1951, in: J.A. Hynek, ed(s)., *Astrophysics*, New York, Chap. 8, 357.

Levin, B.Yu.: 1964, *Origin of the Earth and Planets*, 4th edition, Nauka, Moscow, Russia.

Lissauer, J.J. and Safronov, V.S.: 1991, *Icarus* **93**, 288–297.

Lissauer, J.J., Pollack, J.B., Wetherill, G.W. and Stevenson, D.J.: 1993, in *Neptune and Triton*, University of Arizona Press, Tucson, Arizona, in press.

Lissauer, J.J. and Stewart, G.R.: 1993, in: E.H. Levy and J. Lunine, ed(s)., *Protostars and Planets III*, University of Arizona Press, Tucson, Arizona, in press.

Pechernikova, G.V.: 1991, *Fizika Zemli* **N8**, 59–72 (Russ.).

Safronov, V.S.: 1959, *Fizika Zemli* **N1**, 139–143 (Russ.).

Safronov, V.S.: 1960, *Annals of Astrophysics* **23**, 979–982.

Safronov, V.S.: 1962, *Doklady Akademii Nauk SSSR*, **147(N1)**, 64–67 (Russ.).

Safronov, V.S.: 1969, *Evolution of Protoplanetary Cloud and the Origin of the Earth and Planets*, Moscow, Nauka, Russia, English translation, NASA TTF- 667, 1972.

Safronov, V.S.: 1982, *Fizika Zemli* **N6**, 5–24 (Russ.).

Safronov, V.S.: 1991, *Icarus* **94**, 260–271.

Safronov, V.S. and Ruskol, E.L.: 1957, *Voprosy Kosmogonii* **5**, 22–46 (Russ.).

Safronov, V.S. and Ruskol, E.L.: 1982, *Icarus* **49**, 284–296.

Schmidt, O.Yu.: 1944, *Doklady Akademii Nauk SSSR* **45** (N6), 245–249 (Russ.).

Schmidt, O.Yu.: 1957, *Four Lectures on the Origin of the Earth*, 3rd Edition, Moscow, Nauka, Russia, English translation 1959, Lawrence and Wishart, London.
Vityazev, A.V., Pechernikova, G.V. and Safronov, V.S.: 1990, *Planets of Terrestrial Group. The Origin and Early Evolution*, Nauka, Moscow.
Weidenschilling, S.J.: 1984, *Icarus* **60**, 553–567.
Weidenschilling, S.J. and Cuzzi, J.N.: 1993 , in: E.H. Levy and J. Lunine, ed(s)., *Protostars and Planets III*, University of Arizona Press, Tucson, Arizona, in press.
Wetherill, G.W.: 1990, *Annual Review Earth Planetary Sciences* **18**, 205–256.

POSSIBLE CONSEQUENCES OF ABSENCE OF "JUPITERS" IN PLANETARY SYSTEMS *

GEORGE W. WETHERILL

*Dept. of Terrestrial Magnetism, Carnegie Institution of Washington,
Washington, DC, USA*

Abstract. The formation of the gas giant planets Jupiter and Saturn probably required the growth of massive ~ 15 Earth-mass cores on a time scale shorter than the $\sim 10^7$ time scale for removal of nebular gas. Relatively minor variations in nebular parameters could preclude the growth of full-size gas giants even in systems in which the terrestrial planet region is similar to our own. Systems containing "failed Jupiters," resembling Uranus and Neptune in their failure to capture much nebular gas, would be expected to contain more densely populated cometary source regions. They will also eject a smaller number of comets into interstellar space. If systems of this kind were the norm, observation of hyperbolic comets would be unexpected. Monte Carlo calculations of the orbital evolution of region of such systems (the Kuiper belt) indicate that throughout Earth history the cometary impact flux in their terrestrial planet regions would be ~ 1000 times greater than in our Solar System. It may be speculated that this could frustrate the evolution of organisms that observe and seek to understand their planetary system. For this reason our observation of these planets in our Solar System may tell us nothing about the probability of similar gas giants occurring in other planetary systems. This situation can be corrected by observation of an unbiased sample of planetary systems.

1. Introduction

During the past decade there has been major progress in establishing an observational and theoretical basis for the "planetesimal hypothesis" in which planet formation is thought to proceed by accumulation of smaller bodies. In the past, principal attention has been directed to the development of qualitative scenarios and more quantitative models for the formation of our own particular Solar System. The time is now ripe to begin addressing broader questions of the formation of general planetary systems, and the possible range in the variety of these systems. This will require understanding how such general models are to be observationally tested. At present, by far the greatest source of relevant observations are those obtained from our own Solar System. It has often been stated that the usefulness of these observations is severely limited by providing only "statistics of one," which makes it difficult to distinguish features of our Solar System that a valid theory of planetary formation should be able to explain from those which are idiosyncratic to our Solar System.

This matter of statistics of one is a real problem. However a more serious problem will be addressed here: that of distinguishing between moderately useful statistics of one and completely useless "statistics of zero." Consider the observation that our Solar System contains a warm wet planet that supports an abundant variety of life, including sentient creatures that observe the external universe and

---

* Paper presented at the Conference on *Planetary Systems: Formation, Evolution, and Detection* held 7–10 December, 1992 at CalTech, Pasadena, California, U.S.A.

Astrophysics and Space Science **212**: 23–32, 1994.
© 1994 *Kluwer Academic Publishers.*

develop theories to explain its origin. Should it be required that a good general theory of planet formation predict the formation of such systems to be reasonably probable, or would it be just as satisfactory if a theory predicted this probability to be very small, e.g,. 10^{-10}? Actually there is no observational basis for rejecting the latter theory. If there are $\sim 10^{20}$ planetary systems in the universe, there would still be $\sim 10^{10}$ that contain warm wet planets. Scientists dwelling on all of them would observe that they live in a planetary system containing a warm wet planet, without any regard to the probability of this fortunate situation arising.

On theoretical grounds it does seem that there is a reasonable expectation that Earth-like planets may indeed be fairly common (Wetherill, 1991). Laskar *et al.* (1992) have suggested that this may not be sufficient, and that Earth's moon may be necessary to provide the climatic stability responsible for life. Are there other basic observational data from our Solar System which also effectively represent "statistics of zero" and which are quite possibly much more rare elsewhere? The possibility that the existence of the gas giant planets Jupiter and Saturn in our Solar System may be an example of this will be considered here. At times, the presence of such planets has been considered the *sine qua non* of a real planetary system, but maybe they aren't very common, and the reason we see gas giants in our Solar System may simply be that, otherwise, we just wouldn't be here.

The concept discussed above is sometimes dismissed as simply being a variation of the "anthropic principle" (Barrow, 1988), especially by those who do not hold that concept in high regard. This is not the case. The anthropic principle raises similar questions regarding the entire universe. That is a very different matter. There is a straightforward remedy to the problem of our Solar System representing statistics of one or zero. This is simply to observe a random sample of planetary systems. The problem is an ordinary example of biased sampling. In contrast, it is hard to understand how one could begin to go about obtaining a random sample of universes.

2. Formation of Terrestrial and Giant Planets

Currently fashionable models of planet formation are conceptual descendants of the "planetesimal hypothesis" (Chamberlin, 1905; Safronov, 1969; Hayashi *et al.*, 1977), a variant of the older Kant–Laplace monistic theory of formation of the Solar System from a primordial circumsolar nebula.

According to these models, planets form in a residual centrifugally supported circumstellar gas–dust disk containing $\sim 10\%$ of the stellar mass in which turbulence has decreased sufficiently to permit settling and coalescence of dust grains in a central $\sim 10^3$ km thick dust layer.

Within the framework of this approach, formation of the terrestrial planets can be divided into three more or less distinct stages (reviewed by Wetherill, 1990a):

1. Growth of ~ 1 μm dust grains into very many ~ 1 to 10 km diameter planetesimals.

2. Accumulation of these planetesimals into a much smaller number of $\sim 10^{26}$g planetary "embryos" by runaway growth on an $\sim 10^5$ year time scale.

3. Accumulation of embryos into final planets on an $\sim 10^8$ year time scale.

Simple extension of this chain of events to the problem of formation of the outer planets has not been very successful (Lissauer *et al.*, 1993). The giant planets, Jupiter and Saturn, must accumulate not only a very large planetary core, but also gravitationally capture a much more massive H_2-He envelope. For these planets a somewhat different series of events is usually considered necessary:

1. As for the terrestrial planets, growth of dust grains into ~ 10 km planetesimals.

2. Formation of ~ 15 Earth mass cores either as single runaway embryos, or by merger of embryos.

3. For Jupiter and Saturn, massive capture of nebular gas to form gas giant planets. This stage must be completed while there is still sufficient nebular gas remaining, i.e., probably within $\sim 10^7$ years.

The formation of Jupiter and Saturn is therefore more complex than that of the terrestrial planets, requiring synchronism of time scales for the formation of very massive embryos and the removal of nebular gas. As a consequence, there are several ways in which Jupiter (and Saturn) may fail to form in systems for which terrestrial planets may occur.

These are:

1. Systems for which the surface density at > 5 AU falls off as, for example, $a^{-3/2}$ (Weidenschilling, 1977). As a consequence the mass of the embryos at 5 AU is limited to $\sim 10^{27}$g, according to the usual relationship between embryo mass and surface density (Lissauer, 1987). This will result in a distribution of final planets in the terrestrial planet region and asteroid belt not very different from that found in our Solar System, with the exception that "terrestrial planets," probably rich in volatile elements, will also be formed not only in the terrestrial planet region but also in the asteroid belt. Such systems will resemble the "no Jupiter" case illustrated by Figure 4 in Wetherill (1991), except that small planets will also form in the outer planet region.

2. Formation of runaway 10 to 15 Earth mass embryos, as a result of an enhanced surface density, but formed too late to capture the nebular gas required to form a gas giant.

3. Accumulation of intermediate mass (~ 1 to 5 Earth mass) embryos to form bodies as large as 10 to 15 Earth masses, but too late to capture much nebular gas. These embryos may be expected to scatter one another into highly eccentric orbits as illustrated by Figure 6 in Lissauer *et al.*, (1993).

3. Development of Cometary Source Regions in Our Solar System

At least in a general way, the locations of cometary sources in our Solar System are now understood (Duncan *et al.*, 1987). Short-period comets from the Kuiper belt and long-period comets from the Oort cloud constitute the observed population of bodies of typical cometary appearance. Estimates of the relative importance to terrestrial cratering of comets from these sources vary widely (Weissman, 1982, 1990; Shoemaker and Shoemaker, 1993). If only active comets are considered, it seems likely that the contribution from active short-period comets is only $\sim 10\%$ that of long-period comets, which in turn are probably somewhat less significant than impacts by bodies originating in the asteroid belt. Because the dynamic lifetime of short-period comets probably greatly exceeds their active lifetime, both for Jupiter family comets and comets in the guise of "near-Earth asteroids," almost all short-period comets may be expected to lack an overt cometary appearance (Wetherill, 1990b). When these "asteroidal bodies" are included, the respective contributions of short-period comets, long-period comets, and asteroids to the terrestrial cratering flux may be expected to be similar to one another to within about a factor of three (Wetherill, 1989; Weissman, 1990). The contribution of shower comets from the inner Oort cloud is more hypothetical, but theoretical considerations (Duncan *et al.*, 1987) suggest that on the average its significance to cratering may also be of similar magnitude, but more episodic.

Duncan *et al.* (1987) have also provided satisfactory quantitative mechanisms for the formation of the inner and outer Oort clouds from residual planetesimals in the region of the outer planets. In brief, they show that an initial population of small bodies with perihelia in the vicinity of Uranus–Neptune's orbits and semi-major axes $\lesssim 100$ AU will evolve to form a swarm of comets extending from ~ 3000 AU to 2×10^5 AU. The outer portion of this swarm is vulnerable to perturbation into Earth-crossing inner Solar System orbits as a result of galactic tidal perturbations, as well as close encounters of the Solar System with stars and giant molecular clouds. This evolution is found to occur in two fairly discrete stages:

1. Random walk of the semimajor axis to ~ 2000 AU by planetary perturbations. During this stage the perihelia remain near 25 AU.

2. When their semimajor axis reaches ~ 2000 AU, galactic tides are sufficiently strong to usually increase their perihelia beyond the range of strong planetary perturbations, permitting transfer of the bodies into the more distant inner and outer Oort clouds.

There is a "leak" in this evolutionary path whereby a large fraction of the initial population is ejected into Solar System escape orbits by close encounters with the outer planets. This is especially serious for bodies having Jupiter- or Saturn-crossing initial orbits. In addition, objects initially in purely Uranus- or Neptune-crossing orbits will be perturbed into Jupiter and Saturn crossing, again usually leading to their ejection from the Solar System (Wetherill, 1975; Fernandez and Ip, 1983).

The "Kuiper belt" with semimajor axes beyond Neptune but less than several hundred AU, has been shown to be the only dynamically satisfactory source region for the short-period comets (Duncan *et al.*, 1988). In the context of the above scheme of events, this region may be expected to contain objects from two original sources.

1. Residual planetesimals that were initially in trans-Neptunian orbits. Calculations by Levison and Duncan (1992) indicate that many bodies in low eccentricity orbits with $a < 40$ AU will be marginally stable on a 1.0 b.y. time scale. As a consequence, a significant fraction of these bodies may be expected to not only have survived for 4.5 b.y., but to be also available for transfer back into planet-crossing orbits throughout Solar System history.

2. Objects that were initially planet-crossing and that participated in the same evolution that transferred a much larger number of similar bodies to the inner and outer Oort clouds. A small fraction of this population may be expected to evolve into orbits for which planetary perturbations are sufficiently weak or infrequent to permit their survival for 4.5 b.y., but with semimajor axes too small to lead to transfer to the Oort clouds by galactic tidal perturbations. It may be expected that objects of this kind will be in more highly eccentric orbits than the primordial trans-Neptunian residual population.

4. A Conjectural Discussion of the Evolution of the Cometary Source Regions in Planetary Systems with "Failed" Jupiters

Within the context of the discussion of the previous section, it is possible to consider, at least semi-quantitatively, the effect of substitution of 10 to 15 Earth-mass "failed" Jupiters and Saturns for these gas giant planets.

Several differences may be expected:

1. The "leak" whereby Jupiter and Saturn encounters ejected most of the original planetesimals into interstellar space will be much less effective. For this reason, the total residual cometary population remaining gravitationally bound to the Solar System will be much greater.

2. The mechanism by which Uranus and Neptune perturbations and galactic tides combine to transfer bodies to the Oort clouds and Kuiper belts will remain intact. Jupiter and Saturn can now primarily contribute constructively to this transfer mechanism, rather than primarily short-circuiting it by ejecting comets into hyperbolic escape orbits.

3. The closing of the Jupiter–Saturn leak would increase the population of the evolved component of the Kuiper belt by a factor similar to the factor by which the population of the Oort clouds was increased. In addition, if in our Solar System long-range Jupiter and Saturn perturbations made a significant contribution to the planetary perturbations that drove aphelia to ~ 2000 AU, then the relative increase in the Kuiper belt population might be greater, because the weaker Jupiter and Saturn perturbations would be less effective in driving the evolution to the degree at which galactic tidal forces dominate.

4. The effective removal of the "Jupiter barrier" that must be penetrated by both Oort cloud and Kuiper belt comets if they are to achieve Earth-crossing orbits when they are perturbed back into the inner Solar System.

5. In the present Solar System, Jupiter family comets, mostly extinct or dominant, are added to the population of near-Earth "asteroids" by gravitational decoupling of their aphelia from close approaches to Jupiter for 0.1 to 1 percent of the comets that achieve Earth-crossing perihelia (Wetherill, 1990). The decoupling mechanism is a "tug of war" in which Jupiter and Earth perturbations compete for dominance of the orbital evolution. Because of the much greater mass of Jupiter, Earth rarely wins this contest. Decrease of Jupiter's mass by a factor of 20 will increase the efficiency of this decoupling, and thereby permit more Jupiter family comets to extend their lifetime in Earth-crossing orbits to $\sim 10^7$ - 10^8 years, rather than the $\sim 10^5$ years found for bodies that remain Jupiter–crossing.

5. Some Relevant Calculations

A somewhat quantitative indication of the effect of this weakening of Jupiter's role in cometary evolution has been obtained by carrying out a series of Öpik–Arnold Monte Carlo calculations of the orbital evolution of 8000 "cometary" test bodies of negligible mass initially in eccentric ($e = 0.6$), low inclination ($i = 0.06$) outer Solar System orbits. Initial semimajor axes were distributed uniformly between 5 and 75 AU. The numerical techniques used were essentially those used previously for a large number of simulations of planet formation in the terrestrial planet region and asteroid belt, as discussed in Wetherill (1992) and previous references given therein. The fundamental basis of this approach consists of following the orbital evolution of close-encountering bodies by Monte Carlo iteration of Öpik's (1951) expressions for gravitational scattering and collision. This technique was first used by Arnold (1965) in connection with an investigation of meteoritic orbital evolution, and has been used by the author in numerous investigations of orbital evolution and planet formation in the terrestrial planet region.

This approach has also been used by Fernandez and Ip (1984), as well as by the author (Wetherill, 1991; Lissauer et al., 1993) for the outer Solar System. As previously discussed (Wetherill, 1990b), existing techniques of this kind are of limited validity for calculating encounters with giant planets. Nevertheless, it was found that these techniques did reproduce quite well the results of Duncan et al. (1988) for the expected distribution of the inclinations of Oort cloud and Kuiper belt comets, and the calculations of Fernandez and Ip have proven to be of value in discussions of outer Solar dynamics. It is thought that they may be of use in approximately quantifying the foregoing qualitative discussion.

Calculations of expected cometary perihelion passages within 1 AU, as a consequence of the orbital evolution of the initial outer Solar System populations described above have been carried out. Attention was confined to the Kuiper belt

TABLE I

Perihelion passages of short-period comets at 1 AU/10^9 year (normalized to 1000 test bodies) lost from Kuiper belt when aphelion $> 10^3$ AU.

Initial configuration	Calculation no.	Test bodies	Time before present (m.y.)		
			0–3500	3500–4000	4000–4500
A. Present	1	8000	1.4×10^3	2.2×10^4	1.6×10^6
	2	2000	3.6×10^3	3.8×10^4	1.9×10^6
	3	8000	2.4×10^3	4.2×10^4	1.5×10^6
B. "Failed Jupiters" in low eccentricity orbits	1	2000	9.1×10^5	4.0×10^7	7.2×10^8
	2	8000	2.4×10^6	4.9×10^7	7.5×10^8
	3	8000	4.3×10^6	5.2×10^7	7.2×10^8
C. 1 $M_\oplus$ Neptune and Uranus	1	2000	6.1×10^4	1.3×10^5	8.4×10^5
D. "Failed Jupiter" eccentric embryos	1	200	1.8×10^8	6.7×10^9	5.8×10^9
E. Neptune, Uranus have Saturn mass	1	2000	0	0	2.0×10^6

source because the effect of galactic tides was not explicitly introduced. Instead it was simply assumed that when the aphelion of a body exceeded 10^3 AU, it was securely in the nearly "leakproof pipeline" connecting the planetary region to the comet cloud region as described by Duncan *et al.* (1987), and the calculation was therefore terminated at that point. The sensitivity of the results to the assumption was tested by using a maximum aphelion of 10^4 AU as well. Comparison was made of the results of calculations of this kind for our Solar System and an otherwise similar planetary system in which the positions of Uranus and Neptune were occupied by failed Jupiters and Saturns of 15 Earth masses. The results of these calculations are given in parts A and B of Table I.

The calculated number of cometary perihelion passages within 1 AU is an approximate indicator of the relative expected number of cratering impacts on Earth, assuming the distribution of short-period comet orbital elements is the same for the two cases compared.

Primarily as a consequence of closure of the "Jupiter leak" into escape orbits, the number of perihelion passages is always much higher for the "failed Jupiter" case. The magnitude of this increase varies from a factor of ~ 100 for the period of "heavy bombardment" ranging from 4500 to 4000 m.y. ago, to ~ 1000 times during the epoch between 4000 and 3500 m.y. B.P. in which life probably arose on Earth, and the subsequent 3500 million years of biological evolution on Earth. Occurrence of impact events similar to the one that caused the mass extinction at the Cretaceous–Tertiary boundary would then be expected to occur on a $\sim 10^5$ year

TABLE II

Perihelion passages of short-period comets at 1 AU/10^9 year (normalized to 1000 test bodies) lost from Kuiper belt when aphelion > 10^4 AU.

Initial configuration	Calculation no.	Test bodies	Time before present (m.y.)		
			0–3500	3500–4000	4000–4500
A. Present	1	2000	7.0×10^3	4.4×10^4	1.8×10^6
	2	8000	1.1×10^4	8.7×10^4	1.8×10^6
	3	8000	7.8×10^3	5.3×10^4	1.7×10^6
B. "Failed Jupiters" in low eccentricity orbits	1	2000	2.6×10^6	1.7×10^7	7.2×10^8
	2	8000	2.9×10^6	6.7×10^7	6.9×10^8
	3	8000	2.5×10^6	2.6×10^7	5.0×10^8
C. 1 $M_\oplus$ Neptune and Uranus	1	2000	5.0×10^4	1.6×10^5	6.4×10^5
D. "Failed Jupiter" eccentric embryos	1	200	2.1×10^7	5.0×10^8	9.4×10^9
E. Neptune, Uranus have Saturn mass	1	2000	0	0	2.2×10^6

time scale, rather than on a $\sim 10^8$ year scale. In addition, if cometary impacts are responsible for a major fraction of Earth's volatile inventory, its volatile content would be much larger if gas giant planets were absent.

Similar calculations are given in Table II, parts A and B for the assumption that loss of comets from the Kuiper belt to the Oort clouds does not occur until their aphelia exceed 10^4 AU. According to the results of Duncan *et al.* (1987), this value is probably too high. Nevertheless, the terrestrial planet flux is still ~ 300 times greater than for the present Solar System.

In addition, it should be expected that this increased flux should lead to a larger value for the largest impactor. Because of the size distribution of these large bodies, this statement is difficult to quantify. If it is assumed that the largest body to impact the Earth since 3500 m.y. ago had had a diameter of 25 km, about twice that of the Cretaceous–Tertiary boundary event, and the cumulative radius distribution was a power law with an exponent of -2, the largest impactor would have a diameter of about 750 km, comparable to that of Ceres.

The Part C portions of Tables I and II show the results of similar calculations in which Jupiter and Saturn had their present mass, but the mass of Uranus and Neptune was set equal to that of Earth. In this case Jupiter and Saturn ejected bodies in a similar way to that found for the present outer planetary configuration, but Uranus and Neptune were less effective in scattering them into the Oort clouds. The resulting post-3500 m.y. Kuiper belt fluxes are about 20 times greater than that for the present configuration. This is less than the "failed Jupiter" fluxes, but still high enough to provide a greater environment stress to terrestrial evolution.

An alternative type of "failed Jupiter" is considered in part D of Tables I and II. In this case it is assumed that the initial configuration of embryos in the outer Solar System was an assemblage of twenty-four 0.2 to 5 Earth-mass embryos, extending from 3.5 AU to 10 AU, similar to the case considered in Figure 6 of Lissauer *et al.* (1993). In addition, "Earth" and "Venus" are assumed to exist in their present orbits, as well as 200 test bodies between 2.0 and 9.0 AU.

Comparisons of the results of these calculations with the present Solar System are less straightforward than for the previous cases, because the overall evolution of the system is unlike the present system in other ways than simply an increased cometary flux. The large outer-planet embryos do not grow much but scatter one another into highly eccentric orbits, often with aphelia as large as that permitted by the 10^3 AU (Table I) or 10^4 AU (Table II) boundary for escape from the Kuiper belt. In addition, the original "Earth" and Venus grew to form objects about 8 times their initial masses, and with eccentricities of 0.2 to 0.35. These circumstances alone would be sufficient to cause the geological and biological evolution of Earth to be quite different from our Solar System. In addition, the short-period cometary flux shown in Tables I and II, part D, is about a factor of 10 higher than the more simple failed Jupiters shown in part B of these tables.

The case of a system in which four gas giant planets formed is considered in part E of Tables I and II. It is assumed that the mass of Uranus and Neptune were equal to that of Saturn. As might be expected, this system was much more effective than even the present Solar System in ejecting bodies into hyperbolic orbits, leading to a much smaller cometary flux at 1 AU.

It may also be mentioned that a planetary system in which gas giants do not occur would be much less effective in populating interstellar space with cometary bodies. An abundance of such systems could explain the absence of observations of bodies of this kind (McGlynn and Chapman, 1989; Weissman, 1990; Stern, 1990).

6. Conclusions

Although difficult to quantify, differences between planetary systems may be expected to have a profound effect on terrestrial ecology and the evolution of organisms with a relatively lengthy reproductive cycle. Failure to observe interstellar comets in markedly hyperbolic orbits could be a consequence of gas giant planets being unusual elsewhere in the Galaxy. The observation in our Solar System of the gas giant planets, Jupiter and Saturn, may tell us nothing about their frequency in other planetary systems. It may be that systems lacking such gas giants may simply not be self-observable. As discussed in the Introduction, there may be no obligation for an adequate general theory of planet formation to predict the existence of gas giant planets until a less grossly biased sample of planetary systems can be studied. Until then, emphasis should instead be concentrated on the ability of such theories to "link" to one another predictions for one observed Solar System. For example, it seems quite likely that further development of current theory will permit prediction of giant impacts and high initial temperatures to accompany terrestrial planet formation, or a depleted asteroid belt to accompany systems containing an equivalent to Jupiter, however rare such gas giants may be.

Acknowledgements

I wish to thank Martin Duncan and Paul Weissman for valuable discussions, and Jan Dunlap for invaluable assistance in preparing the manuscript for publication. This work was partially supported by NASA Grant NAGW-1969.

References

Arnold, J.R.: 1965, *Astrophys. J.*, **141**, 1536–1556.
Barrow, J.D.: 1988, *The Anthropic Cosmological Principle*, Oxford University Press.
Chamberlin, T.C.: 1905, *Fundamental problems of geology*, Carnegie Institution of Washington, Year Book #3 for 1904, 195–234.
Duncan, M., Quinn, T. and Tremaine, S: 1987, *Astron. J.*, **94**, 1330–1338.
Duncan, M., Quinn, T. and Tremaine, S.: 1988, *Astrophys. J.*, **328**, L69–L73.
Fernandez, J.A. and Ip, W.-H.: 1983, *Icarus*, **54**, 377–387.
Fernandez, J.A. and Ip, W.-H.: 1984, *Icarus*, **58**, 109–120.
Hayashi, C., Nakazawa, K. and Adachi, I.: 1977, *Publ. Astron. Soc. Japan*, **29**, 163–196.
Laskar, J., Joutel, F. and Robutel, P.: 1992, The moon: a climate regulator for the Earth, submitted to *Nature*.
Levison, H.F. and Duncan, M.J.: 1992, The gravitational sculpting of the Kuiper belt, in press, *Astrophys. J. Letters*.
Lissauer, J.J.: 1987, *Icarus*, **69**, 249–265.
Lissauer, J.J., Pollack, J.B., Wetherill, G.W. and Stevenson, D.J.: 1993, submitted to *Neptune*, D. Cruikshank, ed., University of Arizona Press, Tucson.
McGlynn, T.A. and Chapman, R.D.: 1989, *Astrophys. J.*, **346**, L105.
Öpik, E.J.: 1951, *Proc. Roy. Irish Acad.*, **54A**, 165–199.
Safronov, V.S.: 1969, *Evolution of protoplanetary cloud and formation of the Earth and planets*, Nauka, Moscow (Translated 1992, NASA TT F-677).
Shoemaker, E.M. and Shoemaker, C.S.: 1993, *Program of Conference on Hazards due to Comets and Asteroids*, Univ. of Arizona Press, Tucson.
Stern, S.A.: 1990, *Proc. Astron. Soc. Pap.*, **102**, 793–795.
Weidenschilling, S.J.: 1977, *Astrophys. Space Sci.*, **51**, 153–158.
Wetherill, G.W.: 1975, *Proc. Lunar Sci. Conf. 6th*, , 1539–1559.
Wetherill, G.W.: 1989, *Meteorites*, **24**, 15–22.
Wetherill, G.W.: 1990a, *Ann. Rev. Earth Planet Sci.*, **18**, 205–256.
Wetherill, G.W.: 1990b, in: R.L. Newburn, J. Rahe and M. Neugebauer (eds.), *Comets in the Post-Halley Era*, I.A.U. Coll. 116, Kluwer Acad. Publ., Dordrecht, 537–556.
Wetherill, G.W.: 1991, *Science*, **253**, 535–538.
Wetherill, G.W.: 1992, *Icarus*, **100**, 307–325.
Weissman, P.R.: 1982, *Terrestrial impact rates for long- and short-period comets*, Geological Implications of Impacts of Large Asteroids and Comets on the Earth, L.T. Silver and P.H. Schultz, eds., 15–24, *Geol. Soc. Am. Special Paper 190*.
Weissman, P.R: 1990, V.L. Sharpton and P.D. Ward (eds.), *The cometary impactor flux at the Earth: Global Catastrophes in Earth History*, Geol. Soc. Am. Special Paper **247**, pp. 171–180.
Weissman, P.R.: 1990, *Nature*, **344**, 825–830.

DUST BLOBS IN THE SOLAR NEBULA — PRIMARY DISTENDED ATMOSPHERE *

SHO SASAKI

Planetary Science Group, Geological Institute,
University of Tokyo, Bunkyo-ku, Tokyo, Japan

Abstract. When planetary accretion proceeds in the gas disk–solar nebula, a protoplanet attracts surrounding gas to form a distended H_2-He atmosphere. The blanketing effect of the atmosphere, hampering the escape of accretional energy, enhances the surface temperature of planets. Furthermore, evaporation of ice or reduction of surface silicate and metallic oxide can supply a huge amount of water vapor into the atmosphere, which would raise the temperature and promote evaporation. Evaporated materials can be efficiently conveyed outward by vigorous convection, and condensed dust particles should keep the atmosphere opaque during accretion. The size of this opaque atmosphere "dust blob" is defined by the gravitational radius, which exceeds 3×10^8 m when the planetary mass is the Earth's mass (5.97×10^{24} kg). This is larger than the radii of present Jovian planets and so-called brown dwarfs. The expected lifetime of "dust blobs" is 10^6–10^7 yr, which is longer than that of the later gas accreting and cooling stages of Jovian planets. The number of "dust blobs" could exceed that of Jovian planets. If the gas disk is rather transparent, the possibility of observing such objects with a distended atmosphere may be higher than that of detecting Jovian planets. Contamination of the gas disk by the dust from primary atmospheres is negligible.

1. Introduction: Planetary Growth in a Gas Disk

The current theoretical scenario of planetary formation starts from the formation of a gas disk surrounding the main star. When turbulent motion in the disk becomes weak enough, settling of dust grains should form a dust-rich zone at the disk mid-plane, and then, gravitational instability of the dust layer should produce numerous planetesimals 10–100 km (Safronov, 1969; Goldreich and Ward, 1973; Hayashi *et al.*, 1985). Planets are formed through mutual collisions between planetesimals. The presence of massive H_2-He gas in Jupiter, Saturn, Uranus, and Neptune implies that further planetary growth should have proceeded in a gas at least in the outer part of the solar system. In the gas disk–solar nebula, a growing protoplanet attracts nebular gas to form a distended atmosphere. Finally some of the gaseous objects should become present Jovian planets through the collapse of steady-state structure (so-called "core instability") and following rapid gas accretion (Mizuno, 1980; Bodenheimer and Pollack, 1986; Sasaki, 1989; Wuchterl, 1991).

Prior to the core instability, however, the size of the gravitated atmosphere has become quite large. In the present study, we propose the possibility that the distended atmosphere should be opaque by evaporation and transport of ice or silicate/metal from the planetary surface. If the gas disk which involves planets is rather transparent, this gaseous object (we call "a dust blob") is observable.

---

* Paper presented at the Conference on *Planetary Systems: Formation, Evolution, and Detection* held 7–10 December, 1992 at CalTech, Pasadena, California, U.S.A.

2. Size of the Primary Atmosphere*

In early studies, the outer boundary of the atmosphere was implicitly assumed
to be Hill radius where planetary gravity and solar gravity are balanced in a
rotating coordinate (e.g., Mizuno, 1980). But in the outer zone of the Hill sphere,
gas density is nearly equal to that of the gas disk, and the gas there should be
influenced by the flow in the disk. Here the atmospheric size r_q is defined by a
sphere where gas density is $e \sim 2.7$ times as large as the nebular density, and it
is determined by the balance between gravitational energy and gas thermal energy
($r_q \sim GM\mu m_H/k_B T$, where G is the gravitational constant, M is planetary mass,
μ is gas molecular weight, m_H is mass of hydrogen atom, k_B is the Boltzmann
constant, and T is temperature) (Bodenheimer and Pollack, 1986; Sasaki, 1989).
In other words, the sphere inside r_q has little effect from gas flow in the nebula.
Numerical simulations of gravitating fluid by Takeda *et al.* (1985) showed that
the gas drag coefficient should be proportional to the cross section of the attracted
atmosphere πr_q^2. Figure 1 shows the increase of the atmospheric size in comparison
with the planetary radius and the Hill radius. When the planetary mass is 1 M_E (M_E
being the present Earth's mass 5.97×10^{24} kg), r_q becomes larger than 3×10^8 m,
which is comparable with the present orbital radius of the Moon.

3. High Surface Temperature and Evaporation

Owing to the blanketing effect of the atmosphere, accretional energy cannot freely
escape from the planetary surface. As a result, a protoplanet should have high
surface temperature (Hayashi *et al.*, 1979; Mizuno and Wetherill, 1984; Sasaki,
1990). In the outer domain of the solar nebula (heliocentric distance: $a > 4$ AU),
accreting materials consist of ice-rock-metal mixture. During the accretional process, evaporation of ice from the planetary surface should supply a huge amount
of water vapor into the H_2-He atmosphere. Because temperature gradient in the
atmosphere is proportional to gas molecular weight

$$\nabla T = -(1 - \frac{1}{\gamma} + \delta)\frac{GM\mu m_H}{k_B r^2} \tag{1}$$

where $\gamma \equiv (d \ln P/d \ln \rho)_s$ and $\delta \equiv (d \ln \mu/d \ln P)_s$ (Sasaki, 1990), heavy vapor-
rich gas should raise the surface temperature and, in turn, promote evaporation.
Water vapor can be conveyed outward and mixed by convection, and in the outer
cooler zone, recondensed ice particles should render the atmosphere opaque.

   In the inner region of the nebula, ice is no longer the major component of
accreting materials. Once the surface temperature exceeds the melting temperature
of silicate, however, reduction of surface silicate and/or iron oxide by atmospheric

---

* We use the term "primary atmosphere" in order to discriminate this from the secondary
(degassed) atmosphere that would consist of the present atmospheres of Earth, Venus and Mars.

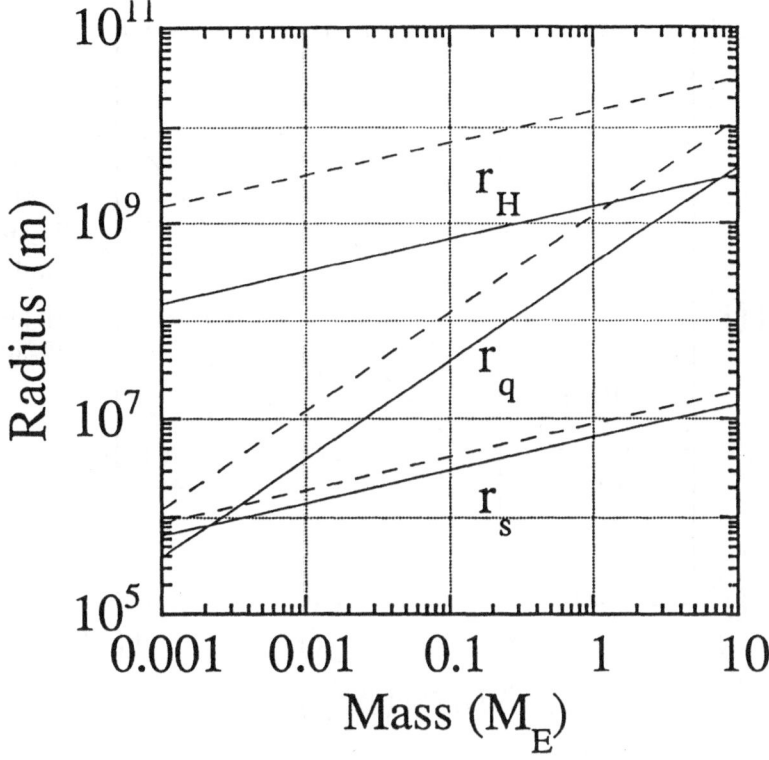

Fig. 1. The size of the atmosphere r_q in comparison with the planetary radius r_s and the Hill radius r_H. The horizontal axis is planetary mass. Solid lines denote the inner region of the solar system (at 1 AU, without ice, 280 K). Dashed lines denote the outer region (at 10 AU, with ice, 100 K).

hydrogen can supply a large amount of water vapor into the atmosphere and enhance gas molecular weight. Thus the surface temperature is raised greatly to be much higher than the evaporation temperature of silicate and the atmosphere becomes convective (Sasaki, 1990). A part of evaporated silicate should be transported by convection and condensed dust should increase the opacity of the atmosphere.

Figure 2(a) shows an increase in the surface temperature T_s of a protoplanet due to the blanketing effect of the primary atmosphere. Because of the ice evaporation, T_s of an icy protoplanet should start increasing earlier than that of a silicate-metal protoplanet. Even in the latter case, since the reduction of silicate and metallic oxide produces water vapor, the temperature at the bottom atmosphere is enhanced greatly to be higher than the evaporation temperature of silicates. Figure 2(b) shows dependence of the surface temperature on mass accretion rate. Whereas the equilibrium temperature due to blackbody radiation is proportional to $(dM/dt)^{1/4}$, the enhanced temperatures by the blanketing effect of the primary atmosphere do

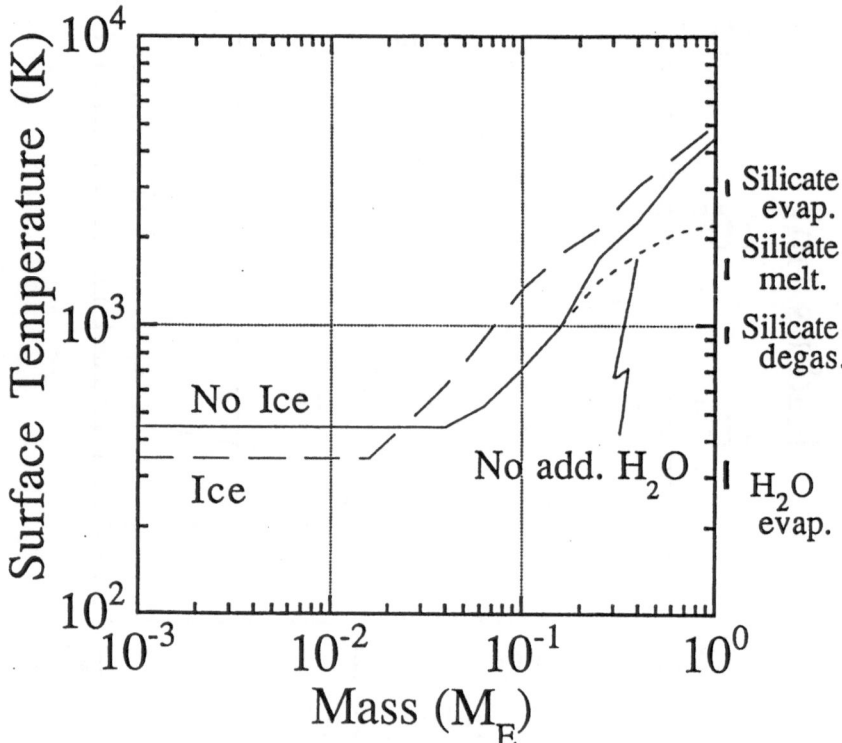

Fig. 2a.   Variation of the surface temperature of a protoplanet under the blanketing effect of the primary atmosphere. The horizontal axis denotes planetary mass. Water abundance in the atmosphere is enhanced to be about 100 times as much as its solar abundance (Sasaki, 1990). We assume that the dust abundance parameter f is unity and the mean dust size is 10 μm.
Dependence on ice existence. Typical accretion time $t_{accr} = 1\ M_E/(dM/dt)$ is 10^7 yr. The dashed curve corresponds to a protoplanet in the outer region of the solar nebula (at 10 AU) where accreting materials contain ice. T_s starts increasing at $M \sim 0.02\ M_E$ and exceeds 1000 K when $M \sim 0.08\ M_E$. The solid curve corresponds to the inner region (at 1 AU) where ice no longer exists in the accreting materials. When the temperature surpasses 1000 K, reduction of surface silicate by atmospheric hydrogen can supply additional H_2O which enhances the temperature to be higher than the silicate evaporation temperature. The case without additional H_2O is shown by the dotted curve for comparison.

not differ greatly. Even if gravitational energy release by mass accretion is small, T_s should finally exceed the evaporation temperature of surface silicates.

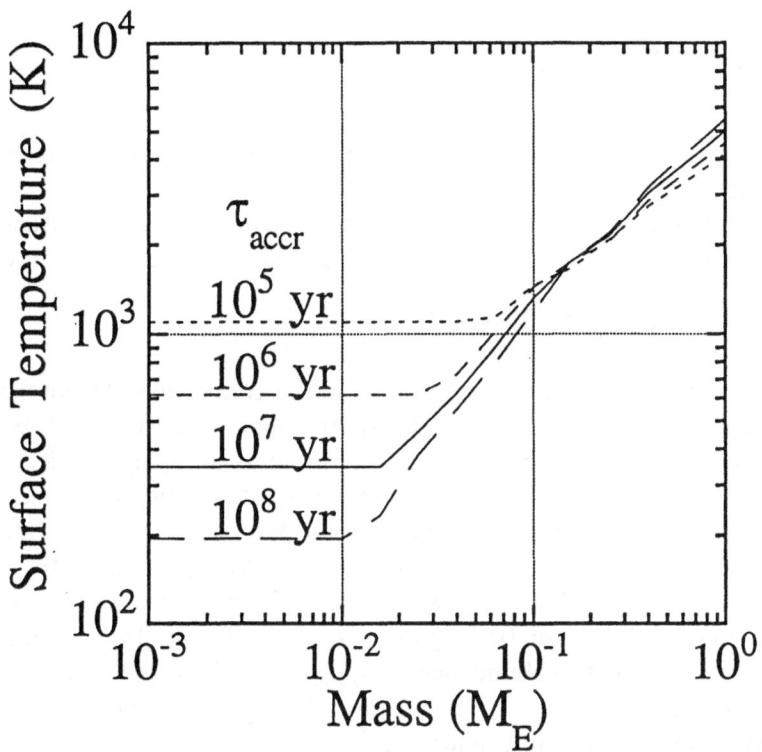

Fig. 2b. As Figure 2a. Dependence on mass accretion rate. Typical accretion time $t_{accr} = 1\ M_E/(dM/dt)$ is $10^5, 10^6$, or 10^7 yr. Temperatures T_s at small planetary masses are determined by the balance between the release of gravitational energy $GM/r_s (dM/dt)$ and black-body radiation $4\pi r_s^2 \sigma_{SB} T_s^4$. In the later stage of accretion, however, dependence of T_s on mass accretion rate is small.

4. Dust Transport in the Atmosphere

From the dust condensation zone $(r \sim 3r_s \sim 2 \times 10^7$ m, when $M = 1\ M_E)$, condensed dust can be transported outward by convection. Using a simple mixing length theory, convective velocity in the atmosphere is then estimated:

$$v_{conv} \sim \left(\frac{\ell F}{H_T \rho} \right)^{\frac{1}{3}} \sim \left(\frac{F}{\rho} \right)^{\frac{1}{3}} \sim 10\text{--}100 \ \text{m s}^{-1} \qquad (2)$$

where ℓ is mixing length, H_T is temperature scale height, ρ is gas density, F is the heat flux from accretional energy. The variation of velocity value comes from the density structure of the atmosphere, $\rho(r, M)$. The timescale of dust transport is estimated by

$$t_{\text{trans}} \sim \frac{r_q^2}{D} \sim \frac{r_q^2}{\nu_{\text{conv}}\ell} \sim 10\text{--}100 \text{ yr} \qquad (3)$$

where D is the diffusion coefficient due to convective motion. This is much shorter than the accretional timescale, 10^6–10^7 yr.

The outward mass flux of dust transport by the convection can be also estimated. Using the above diffusion coefficient controlled by the convection, we have

$$I_{\text{dust}} = 4\pi r^2 \rho D \nabla C \sim 10^{14} \text{ kg yr}^{-1} \qquad (4)$$

where ∇C is the dust concentration gradient in the atmosphere. Using the accretional time 10^6–10^7 yr, we find the net transported mass of dust is 10^{20}–10^{21} kg.

So long as the primary atmosphere can be seen as an opaque object, optical length ($\ell_{\text{opt}} \sim 1/\kappa\rho$, κ being the opacity) at $r = r_q$ should be much smaller than r_q itself. Using relative dust mass abundance normalized by the solar value, f (f being unity when $\rho_{\text{dust}}/\rho_{\text{gas}} = 3 \times 10^{-3}$ kg/kg), we have $\ell_{\text{opt}} = 2 \times 10^6 \, f^{-1}$ m at $a = 1$ AU. So long as $f \geq 0.1$, the whole atmosphere can be opaque. Since the atmospheric mass is less than 10^{23} kg at $M = M_E$ (Sasaki, 1990), atmospheric convection can enhance dust abundance $\rho_{\text{dust}}/\rho_{\text{gas}}$ larger than 10^{-3}, corresponding to $f \geq 0.3$. Therefore the convective transport of evaporated materials can supply enough dust for the opaque atmosphere.

Here we have neglected dust growth and sedimentation, which would decrease dust opacity in the atmosphere. The above I_{dust} might be the upper limit. Since convective velocity is much higher than the sedimentation velocity of small dust particles, however, the convective atmosphere should sustain a large amount of dust. Note that the ablation or disruption of incoming planetesimals can also supply dust grains into the atmosphere. Even if dust growth and inward sedimentation are taken into account, planetesimal ablation can keep the atmosphere opaque ($f > 0.1$) (Sasaki and Nakazawa, 1990).

5. Dust Blobs in a Gas Disk

If a protoplanetary gas disk is rather transparent, these "dust blobs" are perceptible in the disk. From the standard model of the solar nebula by Hayashi *et al.* (1985), the vertical optical depth of the gas disk is expressed as

$$\tau_{\text{opt}} \sim 0.8 \left(\frac{f}{10^{-2}}\right)\left(\frac{s_{\text{dust}}}{100 \ \mu m}\right)^{-1}\left(\frac{a}{5 \text{ AU}}\right)^{-\frac{3}{2}} \qquad (5)$$

where s_{dust} is the dust size and f is the normalized dust abundance. Dependence on s_{dust} is valid only if $s_{\text{dust}} \geq 1 \ \mu$m. We can expect the low optical depth since dust growth should increase s_{dust} and planetesimal-planet formation should lower f. Note that tubulence in the gas disk should be quiet so that planetesimals are formed through the gravitational instability of a dust layer at the disk equatorial plane. Using the turbulent viscosity parameter α (the ratio of effective turbulent viscosity to the product of gas thermal speed and the vertical scale height of the disk), the above condition is expressed as $\alpha < 10^{-5}$ (S. Watanabe, in preparation). If the turbulence was operated on by thermal convection, turbulent motion should cease when the disk's optical depth has decreased.

The number of "dust blobs" depends on the scenario of planetary accretion. If runaway growth dominates the whole stage of accretion, the number of dust blobs should correspond to the number of final planets (S. Ida, personal communication). But if multiple growth is predominant in the later stage of planetary accretion, a few 10–100 protoplanets with a distended atmosphere should subsist in the gas disk. In our solar system, the present tilted rotation axes of giant planets suggest that collisions among large ($> 1 \ M_E$) protoplanets should have taken place in the later stage of accretion, which would support the multiple growth scenario.

Since atmospheric gas is supplied from a surrounding gas disk, the life time of dust blobs should be basically compatible with that of the gas disk, probably 10^6–10^7 yr according to recent observations of T Tauri stars. If gas disks still exist around so-called weak-line T Tauri stars, the upper limit of the life time should be longer ($\sim 10^8$ yr).

Because of the large size of a dust blob, the outward energy flux per unit area is decreased at the outer boundary and the brightness temperature at $r = r_q$ is low. We find

$$
T_p = \left\{ \begin{matrix} 40 \\ 60 \end{matrix} \right\} \left(\frac{M}{1 \ M_E} \right)^{-\frac{1}{3}} \left(\frac{\dot{M}}{1 \ M_E/10^7 \ \text{yr}} \right)^{\frac{1}{4}} K \left\{ \begin{matrix} 5 \ \text{AU...70} \ \mu\text{m} \\ 1 \ \text{AU...50} \ \mu\text{m} \end{matrix} \right\} \tag{6}
$$

where corresponding radiation wavelengths are also shown. The above temperature is much lower than the insolation temperature from the central sun (~ 100 K at 5 AU and 270 K at 1 AU). Therefore, unless core protoplanets were small ($< 0.1 \ M_E$) and existed at large heliocentric distance (> 10 AU), irradiation by the main star should be the principal source of the emission from the dust blobs.

6. Contamination of the Gas Disk

Let us examine the possibility that dust transport from primary atmospheres should enhance opacity of the circumstellar gas disk. If the dust supply into the surrounding disk is efficient, the gas disk should become opaque again and we cannot expect to observe "dust blobs" embedded in the disk.

As discussed above, the dust transport timescale in the convective atmosphere is shorter than 100 yr and the outward dust mass flux is at most 10^{14} kg/yr. In 1×10^6 yr, then, 50 protoplanets ($\sim 1 \ M_E$) with distended atmospheres can supply dust mass 5×10^{21} kg $\sim$ 10^{-3} M_E. This is only 10^{-5} of the original solid mass in the solar nebula. Assuming $f \sim 10^{-5}$ in Eq. (5), we find that the contribution of the atmospheric dust on the disk opacity is small even if $s_{dust} \sim 1$ μm.

Dust transport time from the atmosphere to the surrounding nebula is estimated from Keplerian shear velocity $\Delta v_K \sim (r_q/2a)v_K$, where v_K is Keplerian velocity. Then the timescale ($t \sim r_q^2/D \sim r_q/\Delta v_K \sim 2a/v_K$) is comparable to the Keplerian orbital period $t_K = 1(a/1 \text{ AU})^{1.5}$ yr. Dust transport in the gas disk is, however, a fairly long process. Because of the relative velocity between a protoplanet and the disk ($\Delta v \sim (e^2 + i^2 + \eta^2)^{1/2}$ $v_K \sim 10^{-3}$ v_K, where e and i are eccentricity and inclination of the protoplanet, η is the ratio of radial pressure gradient to solar gravity), emitted dust particles can be scattered along the protoplanet's orbit in 10^3 $t_K \sim 10^3 (a/1 \text{ AU})^{1.5}$ yr. Further dust transport depends on the turbulence state in the nebula. For the initial dust settling leading to planetesimal formation, the turbulent viscosity parameter α in the gas disk should have decreased to be very small ($< 10^{-5}$). Then the dust diffusion time through 1 AU distance is estimated to be longer than $10^8(a/1 \text{ AU})^{-1}$ yr.

Therefore, because of the small total supplied mass and the long diffusion time in the disk, contamination of the gas disk by dust blobs would be rather negligible.

7. Discussion: Comparison Among Unseen Planetary Objects

In Table I, we compare various probable objects (which would be observed in future) in protoplanetary disks. The size of dust blobs of the primary atmosphere is as large as that of proposed gas accretion disks of Jovian planets and much larger than that of Jovian planets or brown dwarfs. Moreover, if multiple growth is a dominant mechanism in the late stage of planetary formation, we would expect a larger number of dust blobs ($\gg 10$).

Although the gas accretion disk of a Jovian planet would produce a large amount of luminosity from gravitational energy, its lifetime is fairly small (Sekiya et al., 1987). There is a possibility that the initial cooling stage of Jovian planets (or brown dwarfs) should still have a large size (10^6–10^7 km), but numerical simulations suggest that the size becomes smaller than a few 10^5 km in 10^5 yr because of rapid cooling (Bodenheimer and Pollack, 1986). In the multiple growth scenario, the impact between protoplanets would produce a tenuous disk which would form the moon (Thompson and Stevenson, 1983) and Uranian satellites. However, the cooling time of such disks would be very short. Therefore among objects in Table I, dust blobs would have relatively high probability of future detection.

Acknowledgements

This work started when the author stayed at the Lunar and Planetary Laboratory of University of Arizona. The author also thanks Dr. D.M. Hunten for his support. The author also thanks Dr. Y. Nakagawa and Dr. S. Ida for discussions. This work

TABLE I

Comparison of prospective large objects in the protoplanetary disk.

	Size	Distance	Lifetime	Number
Dust blobs (primary atmospheres)	$\leq 10^6$ km	< a few 100 AU	$\leq 10^7$ yr	10 ~ 100
Disks around Jovian planets	~ 10^6–10^7 km	0 ~ 100 AU?	< 10^4–10^5 yr	< 10
Jovian planets, brown dwarfs	~ 10^5 km	0 ~ 100 AU?	> 10^9 yr	< 10
Disks at giant impacts	~ 10^5 km	< a few 100 AU	10^2–10^3 yr?	a few 10

is partly supported by the Grants-in-Aid for Scientific Research of the Ministry of Education, Science and Culture (No. 04835012, 05231103 and 05740307).

References

Bodenheimer, P. and Pollack, J.B.: 1986, *Icarus* **67**, 391–408.
Goldreich, P. and Ward, W.R.: 1973, *Astrophys. J.* **183**, 1051–1061.
Hayashi, C. *et al.*: 1979, *Earth Planet. Sci. Lett.* **43**, 22–28.
Hayashi, C. *et al.*: 1985, in: D.C. Black and M.S. Matthews, ed(s)., *Protostars and Planets II*, University of Arizona Press, Tucson, 1100–1153.
Mizuno, H.: 1980, *Prog. Theor. Phys.* **64**, 544–557.
Mizuno, H. and Wetherill, G. W.: 1984, *Icarus* **59**, 74–86.
Safronov, V. S.: 1969, *Evolution of the Protoplanetary Cloud and Formation of the Earth and the Planets*, NASA TTF-677 1972, Trans.
Sasaki, S.: 1989, *Astron. Astrophys.* **215**, 177–180.
Sasaki, S.: 1990, in: H.E. Newsom and J.H. Jones, ed(s)., *Origin of the Earth*, Oxford Univ. Press, New York, 195–209.
Sasaki, S. and Nakazawa, K.: 1990, *Icarus* **84**, 21–42.
Sekiya, M. *et al.*: 1987, *Earth Moon Planets* **39**, 1–15.
Takeda, H. *et al.*: 1985, *Prog. Theor. Phys.* **74**, 272–287.
Thompson, T.W. and Stevenson, D. J.: 1983, *Lunar Planet. Sci.* **XIV**, 787–788.
Wuchterl, G.: 1991, *Icarus* **91**, 39–52.

TABLE

References

...

FORMATION AND PROPERTIES OF FLUFFY PLANETESIMALS *

BERTRAM DONN

*Department of Materials Science and Engineering, University of Virginia,
Charlottesville, Virginia, USA*

and

J. MARK DUVA

*Department of Applied Mathematics, University of Virginia,
Charlottesville, Virginia, USA*

Abstract. A mechanism of accumulation of grains in the primordial solar nebula is described. This process produces porous, low density compressible aggregates. Compaction of the aggregates in a collision between them dissipates the kinetic energy of the collision and can result in efficient growth. A simple analysis of such collisions is developed and applied over a range of aggregate sizes and relative velocities. The results indicate that large planetesimals could grow through collisions rather than fragment if the conditions are favorable. Our modelling suggests that primordial asteroids and comets on the order of a kilometer in size will have low densities and irregular shapes.

1. Accretion of Grains

Although there do not appear to have been any controlled experiments on the accretion of submicron grains, agglomeration in laboratory experiments is difficult to prevent. There is extensive information on the structure of small clusters (Arnold, 1977; Forrest and Witten, 1979; Stephens and Russell, 1979; Weitz and Oliveria, 1984; Samson *et al.*, 1987). One conclusion that can be drawn is that grains in the 10-nm to 1-μm size range stick efficiently. Brownian velocities leading to collisions are of the order of cm/s (Fuchs, 1964). Relative velocities for grains in the primordial solar nebula are similar (Weidenschilling and Cuzzi, 1993; Markiewicz and Volk, 1988), thus, the collision characteristics of submicron grains observed in the laboratory apply to the early solar nebula.

A striking and significant result reported in the references cited above is that grain clusters have low densities and are very porous and often filamentary. Several analyses have demonstrated the fractal structure of these clusters (Forrest and Witten, 1979; Weitz and Oliveria, 1984; Samson *et al.*, 1987). It was pointed out by Meakin and Donn (1988) that accretion in the solar nebula would have been expected to produce fractal aggregates as well. A relevant property of fractals is that $N(r) = cr^d$ where $N(r)$ is the number of grains within a volume of radius r, c is a constant and d is the fractal dimension. The fractal dimension is always less than the geometric dimension; for aggregates grown by cluster–cluster accretion, $d \approx 2$. Hence, the density varies approximately as $1/r$. For $d \leq 2$ the cluster is transparent, that is, essentially all of the grains are exposed. The cross-section per grain for such a cluster $\sigma(N)/N$, was estimated by numerical simulation

---

* Paper presented at the Conference on *Planetary Systems: Formation, Evolution, and Detection* held 7–10 December, 1992 at CalTech, Pasadena, California, U.S.A.

Astrophysics and Space Science **212**: 43–47, 1994.

by Meakin and Donn (1988) for d= 1.95 and found to asymptotically approach 0.3σ (1) as $N \rightarrow \infty$. The cross-section per grain for a very large aggregate with $d = 1.95$ is only slightly smaller than that of a single grain, whereas for a compact body the cross-section per grain approaches zero. This result is useful in predicting drag dependent relative velocities among the grains and aggregates in the planet-forming stage of the nebula.

Fractal aggregates cannot grow larger than 10^4 grains (Kantor and Witten, 1984) because of instabilities inherent in their filamentary structure. By simulating a microgravity environment for gas evaporation, Slobodrain, *et al.* (1991) obtained metallic fractal particles about 1 mm in size. The somewhat larger bodies considered here formed by the accumulation of fractal particles will not survive the necessary collisions as fractal bodies. Nevertheless, we will assume that larger bodies initially have the low densities common to the fractal particles from which they are formed.

2. Properties of Fluffy Planetesimals

Two properties of low density, uniform "non-fractal" aggregates relevant in modelling their deformation in a collision are the relative density (the density of the fluffy body divided by the density of the bulk solid) and the resistance to compaction. Measurements of these properties exist for fluffy terrestrial materials, in particular for Cab-O-Sil (pyrogenic silica), as discussed below.

Evidence that the relative density of small, uniform aggregates should not exceed 0.1 comes from two sources. A ballistic, Monte Carlo accretion simulation for small spheres was performed by Daniels and Hughes (1981) in which they calculated the relative density (or packing fraction) of an aggregate as a function of aggregate size. The mean value of the relative density of an aggregate 130 times the size of the mean particle radius was about 0.1. An estimate of the relative density of clusters of grains that have undergone many collisions is given by the density of a column of fine powder compacted under its own weight. Oudemans (1965) measured a relative density of 0.11 for iron oxide powder in this way.

The resistance to penetration by low density rigid spheres has been measured for three powders, Cab-O-Sil, talc and titanium oxide, with relative densities of 0.02, 0.20 and 0.21 respectively, by Peak *et al.* (1991). By dropping the spheres into large containers of powder in vacuum, they found that the crater depth S is proportional to the square root of the change in potential energy per unit area U of the dropped sphere. As the powder is stationary, U can be interpreted as the energy per unit area of the collision. The craters in these experiments have steep, smooth walls with little or no ejecta, and material to the side of the craters is relatively undisturbed. This suggests that the compaction of fluffy agregates in collision can be treated realistically as a one-dimensional process.

3. Collisions Between Planetesimals

The measurements of Peak *et al.* (1991) can be put to use in specifying the conditions under which collisions will lead to aggregation instead of fragmentation, if some simple idealizations are made. We suppose that there is a plug or core of compacted material at the bottom of a crater caused by the penetrator, and that this plug has the same diameter as the penetrator, a length L_C, and a uniform relative density D_C. Thus, at the end of the plug of compacted material, the relative density changes suddenly from D_C to D_0. As the process is one dimensional, conservation of mass requires that

$$D_0 L_0 = D_C L_C \tag{1}$$

where L_0 is the original length of the compacted material, that is, the distance from the top of the crater to the bottom of the compacted plug. The penetration depth S is $L_0 - L_C$, so , from the measurements of Peak *et al.* (1991) on Cab-O-Sil powder,

$$S^2 = (L_0 - L_C)^2 = 1.66 \times 10^{-4} U \tag{2}$$

with distances in meters and U in joules per meter squared. The measurements of Peak *et al.* (1991) suggest the the coefficient on the right-hand side of Equation (2) is dependent on the relative density, but we have made no attempt to include that dependence in our model. Solving (1) and (2) for L_0 and L_C gives

$$L_C = 1.29 \times 10^{-2} \sqrt{U} \, D_0/(D_C - D_0), \quad L_0 = L_C D_C/D_0 \tag{3}$$

We idealize a collision between planetesimals as a head-on, face-to-face, collision of two identical fluffy cubes with edge length L approaching each other with a relative velocity of V. The kinetic energy per unit area T of the collision is

$$T = \mu V^2/2L^2 \tag{4}$$

where μ is the reduced mass of the system. Half of T is to be identified with U in the equations above, because both bodies in the collision deform and consume energy. If D_S is the *actual* density of the solid material (in $kg \, m^{-3}$)

$$T = \frac{1}{4} D_0 D_S L V^2 \tag{5}$$

T can be computed from Equation (5) if the collision parameters and the initial relative density are known, and this last quantity was estimated to be 0.1 for small bodies in the previous section. Equation (3) can then be used to compute L_0 and L_C if D_C is known. The maximum relative density for random packing of hard spheres is observed to be about 0.65; our calculations will be based on values limited by this upper bound.

Aggregation through collisions is possible if the colliding bodies are sufficiently large to absorb the energy of collision. The quantity L_0 is an estimate of the size cube required to do the job. Table I shows the calculated values of L_0 and L_C for collisions of the type described above for a variety of collision parameters. For Cab-O-Sil, D_S is approximately 2500 kg m^{-3}. The dependence of relative velocities on size used in the table is estimated from a figure presented by Weidenshilling and Cuzzi (1993). Note that relative velocity V is not a monotonically increasing function of size L.

TABLE I

L (m)	D_0	D_C	V (m s^{-1})	$\frac{1}{2}T$ (J m^{-2})	L_0 (m)	L_C (m)
0.01	0.05	0.15	0.01	1.56×10^{-5}	7.64×10^{-5}	2.55×10^{-5}
	0.10	0.30		3.13×10^{-5}	1.08×10^{-4}	3.61×10^{-5}
0.10	0.05	0.15	0.05	7.81×10^{-3}	1.71×10^{-3}	5.70×10^{-4}
	0.10	0.30		1.17×10^{-2}	2.09×10^{-3}	6.98×10^{-4}
1.0	0.10	0.30	1.0	3.13×10^{1}	1.08×10^{-1}	3.61×10^{-2}
	0.15	0.45		4.69×10^{1}	1.33×10^{-1}	4.42×10^{-2}
10.0	0.15	0.45	5.0	1.17×10^{4}	2.10×10^{0}	6.98×10^{-1}
	0.20	0.60		1.56×10^{4}	2.42×10^{0}	8.06×10^{-1}
100.	0.30	0.60	2.0	3.75×10^{4}	3.75×10^{0}	1.25×10^{0}
	0.40	0.60		5.00×10^{4}	4.33×10^{0}	1.44×10^{0}

For the combinations of the collision parameters chosen, L_0 is always substantially less than L, so the energy of the collision *can* be absorbed. It is easy to check that if $D_0 = 0.4$ and $D_C = 0.6$ for a 10.0-m body, then a velocity of about 8.0 m s^{-1} will produce a collision sufficiently violent that not all of the energy can be absorbed. If accretion and not fragmentation is to take place as the bodies increase in size, the (average) relative density of the growing body must increase slowly so that the energy per unit area of subsequent collisions remains modest and the energy-dissipating capacity of the aggregate is retained. The relatively small values computed for L_0 suggest that this is the case; the relative densities after collision are largest for the 10-m bodies (with the largest relative velocities), but still the increase is less than 25%. The reasonableness of our estimates for "pre-collision" relative densities of the larger bodies can be judged, in the context of the model, by following such a body as it grows to larger sizes and tracking the average relative density of the body as a function of size. This is now under investigation. Finally, we note that an aggregate formed by the mechanism described here will be irregularly shaped and will have a position-dependent relative density with a low mean value. It appears these properties can persist into the kilometer size range.

Acknowledgements

The authors would like to thank graduate research assistant H.G. Wood for help in creating Table I and R.E. Johnson and D. Peak for their thoughtful criticisms of the manuscript.

References

Arnold, J.R.: 1977, in: A. Delsemme, (ed.), *Comets, Asteroids and Meteorites*, Toledo University Press, Toledo, Ohio.
Daniels, P.A. and D.W. Hughes: 1981, *Mon. Nat. Roy. Ast. Soc.* **195**, 1001.
Donn, B.: 1991, *Astron. Astrophys.* **235**, 441.
Forrest, S.R. and T.A. Witten: 1979, *J. Phys.* **A12**, L109.
Fuchs, N.A.: 1964, *The Mechanics of Aerosols*, McMillan, New York, 408.
Kantor, Y. and T.A. Witten: 1984, *J. Physique Lett.* **45**, L675.
Markiewicz, W.J. and H.J. Völk: 1988, in: A.K. Dupree and M.T.V.T. Lago, (eds.), *Formation and Evolution of Low Mass Stars*Kluwer Academic Publ., Dordrecht.
Meakin, P. and B. Donn: 1988, *Astrophysic J. Lett.* **329**, L39.
Oudemans, G.J.: 1965, *Science of Ceramics Vol. 2*, G.H. Stewart, ed., Academic Press, New York p. 139
Peak, D., S.J. Kusiak and B. Donn: 1991, *Lun. Planet. Sci. Conf.* **XXIII**, Abs.
Samson, R.J., G.W. Mullholland and J.W. Gentry: 1987, *Langmuir* **3**, 372.
Slobodrain, R.J., M. Cossette, B. Larouche, L. Potvin and C. Rioux: 1991, *Chaos, Solitons and Fractals* **1**, 529.
Stephens, J.R. and R.W. Russell: 1979, *Astrophys. J.* **228**, 780.
Weidenshilling, S.J. and J.N. Cuzzi: 1993, in: E. Levy, J. Lunine and M. Mathews, (eds.), Protostars and Planets, III, University of Arizona Press, Tucson, Arizona.
Weitz, D.A. and M. Oliveria: 1984, *Phys. Rev. Lett.* **52**, 1433.

RADIAL MIGRATION OF PLANETESIMALS [*]

R. NEUHÄUSER

*Max-Planck-Institut Extraterrestrische Physik,
Garching-bei-München, Germany*

and

J.V. FEITZINGER

*Ruhr-Universität Bochum, Astronomisches Institut,
Bochum, Germany*

Abstract. Planetesimals orbiting a protostar in a circumstellar disk are affected by gravitational interaction among themselves and by gas drag force due to disk gas. Within the Kyoto model of planetesimal accretion, the migration rate is interpreted as the inverse of the planetary formation time scale. Here, we study time scales of gravitational interaction and gas drag force and their influence on planetesimal migration in detail. Evaluating observations of 86 T Tauri stars (Beckwith *et al.*, 1990), we find the mean radial temperature profile of circumstellar disks. The disk mass is taken to be 0.01 $M_\odot$ in accordance with minimum mass models and observed T Tauri disks. The time scale of gravitational interaction between planetesimals is studied analogously to Chandrasekhar's stellar dynamics. Hence, Chandrasekhar's coefficient Λ, defined as the fraction between the mean separation of planetesimals and the impact parameter, plays an important role in determining the migration rate. We find ln Λ to lie between 5 and 10 within the protosolar disk. Our result is that, at the stage of disk evolution considered here, gas drag force affects the radial migration of planetesimals by a few orders of magnitude more than gravitational interaction.

1. Introduction

Planetary systems like the one we live in are believed to form in circumstellar disks around low-mass pre-main-sequence stars. According to the inside-out collapse model for star formation (Shu *et al.*, 1987), a circumstellar gas and dust disk should be a frequent by-product of low-mass star formation. In a solid subdisk, planetesimals of a few kilometers in diameter accrete (Safranov, 1969). By often colliding, merging, and breaking apart, they get fewer in number and larger in size and mass. After some time, planets and other objects like comets, asteroids, and meteoroids are left, i.e., a planetary system is formed.

The detailed processes by which planetesimals accrete and grow to planets and the time scales involved are unknown to a significant degree. In particular, the theoretically given planetary formation time scale is up to a few orders of magnitude larger than observational limits allow. Those observational limits are given by the age of the solar system and by the lifetime of gas and dust disks observed around T Tauri stars (Strom *et al.*, 1989; Beckwith *et al.*, 1990). The lifetime of disks around young stars is limited to approximately 10^6 to 10^7 years, because their major component, disk gas, is dispersed by T Tauri star wind (bipolar outflow) and/or evaporated due to stellar UV radiation (disk outflow).

---

[*] Paper presented at the Conference on *Planetary Systems: Formation, Evolution, and Detection* held 7–10 December, 1992 at CalTech, Pasadena, California, U.S.A.

Planetesimal motion within the primordial disk is affected by the gravity of the protosun, self-gravity of the disk, gravitational interaction with other planetesimals, and friction due to disk gas.

The motion of planetesimals in a gaseous disk has been studied in detail by Hayashi, Nakagawa, Nakano, and co-workers (Hayashi *et al.*, 1977 and 1985; Takeda *et al.*, 1985; Nakano, 1987); their model is sometimes called the Kyoto model. The aim of our work is to study in greater detail dynamics of planetesimals, their radial migration, and time scales of processes involved. We restrict ourselves to the period of disk evolution, during which planetesimal sizes range between a few kilometers, i.e., just after planetesimal formation (Safranov, 1969), and 10^3 km in diameter, i.e., until the first appearance of runaway seeds (Wetherill, 1989).

2. The Kyoto Model of Planetesimal Dynamics

As with Chandrasekhar's stellar dynamics, the diffusion time scale of gravitational interaction among planetesimals is

$$t_d = t_r \left(\frac{v_K}{v}\right)^2 = \frac{v v_K^2}{6.5\pi G^2 m^2 N \ln \Lambda} \tag{1}$$

with relaxation time t_r, Chandrasekhar's coefficient Λ, mean mass m, velocity v, Keplerian velocity v_K, and number density N of planetesimals (Chandrasekhar, 1942; Hayashi *et al.*, 1977; Nakano, 1987).

The flow time scale due to gas friction is given by

$$t_f = \frac{t_g}{2\eta + \frac{5}{8}e^2 + i^2} = \frac{2m}{\pi C_D a^2 \varrho v \left(2\eta + \frac{5}{8}e^2 + i^2\right)} \tag{2}$$

with dissipation time $t_g = mv/F$, friction $F = -\frac{1}{2}C_D\pi a^2 \varrho v$, gas drag coefficient C_D, volume mass density ϱ of the disk, eccentricity e, inclination i, and radius a of the planetesimals. The deviation of real planetesimal motion from Keplerian motion is given by η (Nakano, 1987).

Planetesimals accreting at wave number κ have a mean mass

$$m = 4\pi\sigma_s\kappa^{-2} = \frac{4\pi^4\sigma_s^3 r^6}{M_z^2} \tag{3}$$

where M_z is the mass of the central object (the protosun), ζ is the solid mass fraction of the gas and dust disk, and the surface mass density of the solid subdisk $\sigma_s = \zeta\sigma$ (Nakano, 1987; Hayashi *et al.*, 1985).

3. The Coefficient of Migration

Due to gravitational interaction and gas friction, a planetesimal with a semi-major axis r migrates a distance Δr in the radial direction. Let r_n and r_{n+1} be semimajor axes of planetesimals (with $r_n < r < r_{n+1}$) which eventually become neighbouring planets. Then the mean migration distance in the radial direction is

$$\Delta r = \frac{r_{n+1} - r}{2} \tag{4}$$

because planetesimals either coalesce or cease to grow. Almost every planetesimal's material will eventually become part of a planet. Assuming $\Delta r \propto r$, a dimensionless coefficient of migration c_m can be introduced (Nakano, 1987):

$$\Delta r = c_m r \tag{5}$$

$$\Rightarrow \quad \text{rate of migration} = \frac{1}{c_m^2 t_d} + \frac{1}{c_m t_f}$$

As a crucial assumption within the Kyoto model, the planetary formation time scale t is assumed to be given by the inverse of the migration rate (Nakano, 1987):

$$t = \left(\frac{1}{c_m^2 t_d} + \frac{1}{c_m t_f} \right)^{-1} \tag{6}$$

The coefficient c_m of planetesimal migration will be evaluated in detail below. A resulting formula gives c_m as a function of constants and properties of the disk and planetesimals, almost all of which are known by theory. From the statement above, we easily get a quadratic equation for c_m:

$$c_m^2 - c_m S_f - S_d = 0 \tag{7}$$

with solutions

$$c_m = \frac{1}{2} S_f \pm \sqrt{\frac{1}{4} S_f^2 + S_d} \tag{8}$$

where $S_f(t) = t/t_f$ and $S_d(t) = t/t_d$ are due to gas friction and gravitational interaction of planetesimals, respectively:

$$S_f = \frac{t}{t_f} = \frac{\pi C_D a^2 v \left(2\eta + \frac{5}{8} e^2 + i^2 \right) \varrho t}{2m} \tag{9}$$

$$S_d = \frac{t}{t_d} = \frac{6.5 \pi G m^2 N r \left(\ln \Lambda \right) t}{v M_z} \tag{10}$$

We will now evaluate S_f and S_d in detail in order to find a solution for c_m. Radial migration of planetesimals accelerates planet formation, because empty feeding zones are refilled by migrating planetesimals.

4. The Temperature Profile of Disks

Equilibrium of radiative losses and energy production yields for an axisymmetric, thin, self-luminous, circumstellar accretion disk (e.g., Adams *et al.*, 1987):

$$T \propto r^{-q} \quad \text{with} \quad q = 3/4 \tag{11}$$

But, to the contrary, sources with q between $1/2$ and $3/4$ have been observed (e.g., Beckwith *et al.*, 1990). Though this problem has been addressed by many authors, it has not been resolved so far.

We did some additional statistics and correlation analyses on the sample of 86 low-mass pre-main-sequence stars observed in the Taurus–Auriga dark cloud complex (Beckwith *et al.*, 1990), of which approximately one-half are embedded in gas and dust disks believed to be evolving planetary systems.

Out of the disk properties being observed, the property with the least variation is q, the average being $q = 0.580 \pm 0.062$. The disk temperature T_1 at the distance 1 AU is significantly linearly correlated with the star mass $M_\star$. If $M_\star = 1 M_\odot$, linear regression gives T_1 for the protosolar disk as (206 ± 43) K.

If the temperature profile of the protosolar disk was a mean of Beckwith's sample, then

$$T = (206 \pm 43) \text{ K} \left(\frac{r}{[\text{AU}]} \right)^{-(0.580 \pm 0.062)} \tag{12}$$

5. Evaluating the Coefficient of Migration

For the coefficient of planetesimal migration c_m, we found in equations 7 and 8:

$$c_m^2 - c_m S_f - S_d = 0 \quad \Leftrightarrow \quad c_m = \frac{1}{2} S_f \pm \sqrt{\frac{1}{4} S_f^2 + S_d}$$

with $S_f \propto t$ and $S_d \propto t$ due to gas friction and gravitational interaction, respectively.

In order to find the two mathematically possible solutions to the quadratic equation above, we evaluate the quotient S_f/S_d, for which we get, by inserting the time scales t_d and t_f from the Kyoto model given in equations 9 and 10

$$\frac{S_f}{S_d} = \frac{C_D a^2 v^2 M_z \varrho \left(2\eta + \frac{5}{8} e^2 + i^2 \right)}{13 m^3 G N r \ln \Lambda} \tag{13}$$

We insert the following properties of a mean planetesimal and the protosolar disk:
- Radius a with a mean mass m and volume mass density ϱ_s of planetesimals:

$$a = \left(\frac{3m}{4\pi \varrho_s} \right)^{1/3} \tag{14}$$

- The planetesimal velocity v with the Safranov number θ defined as $\theta = (Gm)/(av^2)$:

$$v = \left(\frac{4\pi}{3\varrho_s}\right)^{1/6} m^{1/3} \left(\frac{G}{\theta}\right)^{1/2} \tag{15}$$

- Number density N of planetesimals with a solid mass fraction ζ and surface mass density σ of the disk:

$$N \simeq \frac{\zeta\sigma}{2mri} \tag{16}$$

- For the mean mass m of a planetesimal accreted in a solid subdisk, we insert equation 3.
- Small orbital inclination i with the half-width H of the disk:

$$\sin i \simeq i \simeq \frac{H}{r} \tag{17}$$

- Solving the disk structure equation analytically yields (by neglecting self-gravity) the disk's half-width:

$$H = \sqrt{\frac{kT}{\mu u}\frac{2r^3}{GM_z}} \tag{18}$$

with the mean molecular weight μ of the disk, atomic mass unit u, and Boltzmann constant k.

- Volume mass density ϱ of the gas and dust disk integrated in the z direction (perpendicular to the disk plane):

$$\sigma = \int_{-\infty}^{\infty} \varrho\,dz \simeq \sqrt{\pi}\varrho H \quad \Rightarrow \quad \varrho \simeq \frac{\sigma}{\sqrt{\pi}H} \tag{19}$$

- Surface mass density σ of the disk from integrating the disk mass M_d from the inner ($r \simeq 0$) to the outer ($r = R$) edge of the disk:

$$\sigma = \frac{M_d}{4\pi\sqrt{R}}r^{-3/2} \tag{20}$$

- For the temperature profile $T(r)$, we insert the mean temperature profile of Beckwith's sample given in equation 12.

All unknown properties in equation 13 can be eliminated by inserting the equations given above. Still, we have to insert numbers for values the quotient S_f/S_d depends on. Some of those values, e.g., ζ, the solid mass fraction, may vary depending on further assumptions and/or with radial distance from the sun. In order to find the lower limit for this quotient, we insert upper or lower limits for those values depending on whether they appear in the numerator or the denominator of the quotient, respectively. Hence, we insert the following values:

(1) For reasons given above, we have to insert a lower limit for the Safranov number. According to a recent review of planet formation time scales and the role of the Safranov number in this context (Lissauer, 1987), we insert $\theta = 1$. This value may increase in later disk evolution stages, when the difference in planetesimal sizes increases. But we restrict our study to a limited time

interval, starting at the time when planetesimals form and ending as soon as runaway growth seeds appear.

(2) For similar reasons, we choose for the solid mass ratio $\zeta = 0.01$, because the gas to dust ratio observed in the interstellar medium is approximately 100 : 1. Within the protosolar disk, ζ varies with temperature, i.e., with distance r. A transition between water ice and water vapour occurred at 170 K, probably at a distance of $r \simeq 5$ AU, where Jupiter formed as a runaway planetesimal. From the composition and masses of the planets and the Sun, it can be concluded that at $T > 170$ K, the solid mass fraction was 0.0042 (water in gaseous phase), whereas at $T < 170$ K the fraction was 0.018. The value we adopted (0.01) is approximately the average of these two figures.

(3) For the coefficient C_D of the gas drag force, it was found by numerical simulations (Takeda et al., 1985) that C_D varies slowly with θ, v, and the local sound speed, if $C_D = 3$ is chosen. We could confirm this approximation to be sufficiently precise in this context, as long as motion in the disk is subsonic and $\theta \simeq 1$ (Neuhäuser, 1992).

(4) In accordance with both minimum mass models of the protosolar accretion disk and observed T Tauri disks, we assume the approximative value for the disk mass to be $M_d = 0.01\ M_\odot$.

(5) For Chandrasekhar's coefficient Λ, we insert $\ln \Lambda = 10$ (see the next chapter).

(6) For the planetesimal mass density, we insert $\varrho_s = (2 \pm 1)\mathrm{g\,cm^{-3}}$, for the mean molecular weight of the disk $\mu = \mu_\odot$, and for the outer disk edge $R = 50$ AU. All of these values are more or less well accepted.

Inserting all these values and equations 14 to 20 into equation 13, we find:

$$\frac{S_f}{S_d} > 75 \pm 20 \quad \Rightarrow \quad S_f \gg S_d \tag{21}$$

According to equation 8, we get two solutions:

$$c_m \simeq 0 \quad \text{or} \quad c_m \simeq S_f \tag{22}$$

Since the first solution, $c_m = 0$, means no migration at all, there is a physically unique solution for the coefficient of migration: $c_m \simeq S_f$.

We conclude that radial migration of planetesimals is much more due to gas friction than to gravitational interaction. Evaluating S_f in greater detail, c_m can be found as a function of theoretically known or well observed disk and planetesimal properties. Since this study is still in progress, we refer to a later paper (Neuhäuser and Feitzinger, in preparation).

6. Chandrasekhar's Coefficient

The time scale of gravitational interaction of planetesimals depends on Λ (the mean planetesimal separation divided by the impact parameter). By deriving $S_f \gg S_d$ we have to insert a value for Λ. We chose $\ln \Lambda = 10$, because from the definition of Chandrasekhar's coefficient Λ given by

$$\Lambda = \frac{v^2}{2GmN^{1/3}} \tag{23}$$

we derive (by inserting for planetesimal and disk properties as above) $\ln \Lambda$ as a function of r and get $\ln \Lambda(R) \simeq 10.33 \pm 0.83$ at $r = R = 50$ AU (Neuhäuser, 1992).

Since $\ln \Lambda < \ln \Lambda(R)$ for all $r < R = 50$ AU, $\ln \Lambda \simeq 10$ is the upper limit for $\ln \Lambda$ for all regions where planetesimals form. At a Roche limit $r = 0.01$ AU for planetesimals forming close to the protosun, we get $\ln \Lambda \simeq 5$. Therefore, the condition $\Lambda > 1$ is fulfilled wherever planetesimals form.

7. Comparison With Observation

From the definition of c_m given above, it can be deduced (Nakano, 1987) that

$$c_m = \frac{1}{2} \ln \left(\frac{r_{n+1}}{r_n} \right) \tag{24}$$

with semi-major axes r_n, r_{n+1} of neighbouring planetesimals. In the present solar system, only a few big planetesimals are left, the known planets. Given the semi-major axes of the planets (and Ceres), we can insert the average of the quotients r_{n+1}/r_n into the equation for c_m given above and get as an observational value:

$$c_m = 0.262 \pm 0.068 \quad \text{by observation} \tag{25}$$

This gives the migration coefficient for later disk evolution stages, when only a few planetesimals are left. Equations 5 and 25 give a migration distance of a few AUs within the inner solar system. Since c_m may be varying with time, this may not hold during earlier stages of disk evolution. The time variation of c_m has still to be determined.

8. Summary

From time scales of processes that determine the radial migration of planetesimals, we deduce an equation giving c_m as a function of planetesimal and disk properties. For the quadratic equation having two mathematically possible solutions, we find one physically unique solution.

Since $S_f \gg S_d$, we conclude that radial migration of planetesimals is much more a result of gas friction than of gravitational interaction. This is correct at least during the evolutionary stage considered here, i.e., from the time of formation of planetesimals until the appearance of a runaway seed. Gas friction may vary with the height above the midplane, if a distinct solid subdisk has formed at the midplane surrounded by a gaseous envelope with a different gas to dust density ratio. This effect was neglected here, but should be included in the future.

The migration of planetesimals is an important factor in accelerating planet formation, because of the refilling of empty feeding zones by migrating planetesimals. Therefore, it is imperative to consider gaseous disks (e.g., the so-called Kyoto model) instead of gasless disks (e.g., Safranov, 1969). It should be noted that, due to meteoritic evidence, planetesimal (= meteoritic parent body) migration should be limited to a few AUs, maybe even less.

The question of whether or not migration of planetesimals can solve the problem of too large planetary formation time scales remains to be solved. Also, the variation of S_f and c_m with disk aging due to loss of gas has to be determined. After some time, when a few large planetesimals dominate the disk, gravitational interaction will affect planetesimal motion more than gas drag.

Acknowledgements

R. Neuhäuser wishes to thank the Hanns-Seidel-Foundation, Munich, in conjunction with the federal ministry for education and science, Germany, for financial support.

References

Adams, F.C., Lada, C.J. and Shu, F.H.: 1987, *Astrophysical Journal* **312**, 788–806.
Beckwith, S.V.W., Sargent, A.I., Chini, R.S. and Güsten, R.: 1990, *Astronomical Journal* **99**, 924–945.
Chandrasekhar, S.: 1942, *Principles of Stellar Dynamics*, University of Chicago Press, Chicago.
Hayashi, C., Nakazawa, K. and Adachi, I.: 1977, *Publications of the Astronomical Society of Japan* **29**, 163–196.
Hayashi, C., Nakazawa, K. and Nakagawa, Y.: 1985, in: D.C. Black and M.S. Matthews, (eds.), *Protostars and Planets II*, University of Arizona Press, Tucson, 1100–1153.
Lissauer, J.J.: 1987, *Icarus* **69**, 249–265.
Nakano, T.: 1987, *Monthly Notices of the Royal Astronomical Society* **224**, 107–130.
Neuhäuser, R.: 1992, *Die radiale Wanderung von Planetesimalen in der präsolaren Planetenentstehungsscheibe*, Diplom thesis, Ruhr-Universität Bochum.
Safranov, V.S.: 1969, *Evolution of the Protoplanetary Cloud and Formation of the Earth and the Planets*, Nauka, Moscow (translated by Israel program for scientific translations, Keter press, Jerusalem. (Also by NASA TTF-677).
Shu, F.H., Adams, F.C. and Lizano, S.: 1987, *Annual Review Astronomy and Astrophysics* **25**, 23–81.
Strom, K.M., Strom, S.E., Edwards, S., Cabrit, S. and Skrutskie, M.F.: 1989, *Astronomical Journal* **97**, 1451–1470.
Takeda, H., Matsuda, T., Sawada, K. and Hayashi, C.: 1985, *Progress in Theoretical Physics* **74**, 272–287.
Wetherill, G.W.: 1989, in: H.A. Weaver and L. Danly (eds.), *The Formation and Evolution of Planetary Systems*, Cambridge University Press, Cambridge, 1–24.

VERY LOW MASS STARS, BLACK DWARFS AND PLANETS *

SHIV S. KUMAR

Department of Astronomy, University of Virginia,
Charlottesville, Virginia, USA

Abstract. Hydrogen-rich stars of very low mass ($M \lesssim 0.08\ M_\odot$) never go through hydrogen-burning thermonuclear reactions and, in a time scale much shorter than the age of the Galaxy, become completely degenerate objects or black dwarfs. The number of the very-low-mass (VLM) black dwarfs is expected to be very large and they are likely to make a significant contribution to the total mass of the Galaxy. Processes of star and planet formation are discussed and it is concluded that the luminous and dark objects of mass $0.001\ M_\odot$–$0.08\ M_\odot$ beyond the solar system are not likely to be planets. Formation of Jupiter is discussed and it is suggested that the mass of Jupiter at the time of formation was smaller than its present mass.

1. Introduction

The existence of the hydrogen-rich stars that do not go through hydrogen-burning thermonuclear reactions was postulated by the author in 1963 (Kumar 1963a, b, c). It was pointed out in these papers that there exists a minimum hydrogen-burning mass limit of approximately $0.08\ M_\odot$ which gives the lower end-point of the hydrogen-burning main sequence in the H-R diagram. The structure and evolution of the stars with mass below the minimum hydrogen-burning mass (called here the MHB limit) was studied in these papers and the main result of this theoretical investigation was that a star with mass below the MHB limit becomes a completely degenerate (electron degeneracy) object or a black dwarf in a time scale much shorter than the age of the Galaxy.

Since 1963, there has been a great deal of activity in the areas of theoretical and observational studies of the stars below the MHB limit. Observational evidence is emerging which shows that the stars with mass below the MHB limit (called here the VLM stars) do exist in the universe, but there has been some confusion in the scientific community concerning the basic nature of the VLM stars. In Section 2, this confusion will be cleared up and comments will be made on the relevance of the VLM stars for the dark matter problem. In Section 3, processes of planet formation will be discussed. Special attention will be paid to the formation and evolution of Jupiter.

2. VLM Stars and Black Dwarfs

The MHB limit of approximately $0.08\ M_\odot$ is a consequence of stellar evolutionary processes. When a star with mass below the MHB limit is formed from interstellar or primordial clouds, it collapses to the stage of complete electron degeneracy or

---

* Paper presented at the Conference on *Planetary Systems: Formation, Evolution, and Detection* held 7–10 December, 1992 at CalTech, Pasadena, California, U.S.A.

Astrophysics and Space Science **212**: 57–60, 1994.
© 1994 *Kluwer Academic Publishers.*

the black-dwarf stage in a time scale less than 1 billion years. The VLM stars that become completely degenerate objects (called here the VLM black dwarfs) are not non-stellar objects. It has been reported in the literature (see, for example, McCarthy *et al.*, 1985) that the VLM black dwarfs are planets. This is incorrect, for the minimum stellar mass is not equal to 0.08 $M_\odot$ (Kumar, 1990). Stars are formed by instabilities in interstellar or primordial clouds and the minimum stellar mass, in general, is smaller than the MHB limit. The exact value of the minimum stellar mass is unknown at present, but several workers (see Kumar, 1990 for references) have studied the fragmentation processes and have arrived at numerical values in the range 0.001–0.01 $M_\odot$. Consequently, the VLM black dwarfs are stars and not non-stellar objects.

The physical conditions in the dense cores of molecular clouds ($T \lesssim 10$ K, $\rho \gtrsim 1 \times 10^{-15}$ gm cm^{-3}) favor the formation of gaseous stellar objects with masses much lower than the MHB limit (Kumar, 1990). Planet formation processes are different from the star formation processes and these differences will be discussed in the next section.

It should be noted that if the minimum stellar mass is less than 0.01 $M_\odot$, and if the stellar mass function has the form $F(M) \propto M^{-\alpha}$, where $\alpha \simeq 2$, then most of the mass in the Milky Way galaxy (or any other galaxy) will be in the form of VLM stars. Since the formation of a VLM black dwarf requires a time scale much shorter than the age of the Galaxy, the dark matter in the Galaxy may be explained by the presence of a very large number ($\sim 1 \times 10^{12}$) of these objects.

3. Planet Formation Processes and the Origin and Evolution of Jupiter

For many years (1950–1980), the idea that planets (in the solar system) are formed by gravitational instability in a gaseous cloud ("solar nebula") around the Sun was quite popular. According to this picture, a planet is formed as an extended, gaseous object ("protoplanet") which then collapses under its own gravity. This model was particularly popular for the giant planets and detailed numerical calculations were carried out for a "protoplanet" of mass 0.001 $M_\odot$ or 1 Jupiter mass (M_J). In 1972, the author (Kumar, 1972) pointed out that an extended gaseous object of mass 1 M_J is tidally unstable in the presence of the Sun, and the author (Kumar, 1974) proposed that Jupiter (and other planets in the solar system) have been formed by collisional-accretion processes in the material existing around the Sun 4.5×10^9 years ago.

The collision mechanism for planet formation has been studied in great detail by many workers and a great deal of progress has been made in this area (Wetherill, 1989; Kolvoord and Greenberg, 1992). However, there exist a few unresolved problems with the accretion models. One of the problems concerns the formation of the giant planets, especially Jupiter. Wetherill reports that his calculations do not produce a massive planet at the location of Jupiter. He is able to form an object with a mass close to that of the rocky core present in Jupiter ($M_{core} \simeq 20$ Earth masses), but the origin of the rest of the mass in the present Jupiter ($M_J = 318$ Earth masses) remains a mystery.

The author would like to propose a possible solution to this mystery. Most of the mass inside the present Jupiter is in the form of a hydrogen/helium envelope surrounding the rocky core and it is possible that some or all of this mass was acquired by Jupiter during the last 4.4×10^9 years by accretion from the interstellar medium. Our solar system is constantly moving through interstellar space and, over a period of 4.4 billion years, objects in the solar system in general and Jupiter in particular, must have accreted some material from the interstellar clouds. Calculations are under way to study the feasibility of this proposal, but it is entirely possible that the Jupiter we observe today has changed a great deal since its formation. A large fraction of the mass that is found today in Jupiter may have been acquired by the planet after its formation early in the history of the planetary system. If this turns out to be the case, then we will have to rethink our ideas concerning the origin of the solar system.

4. Concluding Remarks

As we embark on our searches for planets around other stars, it should be kept in mind that the luminous and dark objects in the mass range 1–80 M_J beyond the solar system are not likely to be planets. The detection of additional VLM stars in this mass range will have important implications for the theories of star formation and may also be of great significance for the resolution of the dark matter problem. As far as the planetary masses are concerned, they are likely to be smaller than 1 M_J and, because of the general paucity of dust—the raw material needed for planet building—in the Galaxy, small planets like the Earth are likely to be more common than large planets like Jupiter.

Note

The detection of a few dark objects by the microlensing technique has been recently announced by Alcock et al. (1993, Nature, **365**, 621) and Aubourg et al. (1993, Nature **365**, 623). It appears likely that these microlensing events were caused by dark objects of mass $\sim 0.1\ M_\odot$. If the masses of the lensing objects are indeed smaller than the MHB limit ($\simeq 0.08\ M_\odot$), then the discovery by Alcock et al. and Aubourg et al. provides additional evidence for the existence of very low mass stars as postulated by the author in 1963. It may also be said that this discovery supports the author's claim that the VLM black dwarfs are responsible for some of the dark matter in the Galaxy. It is expected that many more of such objects will be detected by the microlensing technique in the near future.

SHIV S. KUMAR

References

Kumar, S.S.: 1963a, "The Structure of Stars of Very Low Mass", *Astrophysical Journal*, **137**, 1121–1125.

Kumar, S.S.: 1963b, The Helmholtz–Kelvin Time Scale for Stars of Very Low Mass", *Astrophysical Journal*, **137**, 1126–1128.

Kumar, S.S.: 1963c, *Models for Stars of Very Low Mass*, NASA Technical Note D-1907.

Kumar, S.S.: 1972, "On the Formation of Jupiter", *Astrophysics and Space Science*, **16**, 52–54.

Kumar, S.S.: 1974, "The Formation of Jupiter and Saturn", *Astrophysics and Space Science*, **28**, 173–177.

Kumar, S.S.: 1990, "The Nature of the Luminous and Dark Objects of Very Low Mass", *Comments on Astrophysics*, **15**, 55–62.

Kolvoord, R.A. and Greenberg, R.: 1992, "A Critical Reanalysis of Planetary Accretion Models", *Icarus*, **98**, 2–19.

McCarthy, D.W., Probst, R.A. and Low, F.J.: 1985, "Infrared Detection of a Close Cool Companion to Van Biesbroeck 8", *Astrophysical Journal*, **290**, L9–L13.

Wetherill, G.W.: 1989, *The Formation and Evolution of Planetary Systems*, Cambridge University Press, Cambridge.

INTERACTION OF PLANETARY NEBULAE WITH PRENEBULAE DEBRIS *

J. FIERRO

Instituto de Astronomía, UNAM, México, D.F.

Abstract. In the present poster we suggest that some of the structures observed in the envelopes of planetary nebulae are caused by the interaction of central star wind and radiation with preplanetary nebula debris: planets, moons, minor objects and ring and ring arcs.

Recently considerable amount of planetary material has been reported to exist around solar type stars, this debris could be evaporated during the envelope ejection and alter the chemical abundance and produce some of the envelope inhomogeneities.

If there are massive enough rings of material surrounding the progenitor and planets in their vicinity, arc rings could be formed. If the rings are viewed pole on when the envelope is detached from the central star, it will interact with the arc ring material and produce "ansae" and pedal and garden-hose-shape structures observed in some planetaries.

1. Introduction

Planetary Nebulae are solar type stars that have released their extended atmospheres after the red giant phase during one or several episodes. In consequence they are composed of a central star, the previous nucleus, and a surrounding envelope. The envelope has great amount of structure, it can be ionized or neutral and has been found to possess dust. Many planetary nebulae envelopes (PN) present a double shell structure, others present bipolarity, and others ansae, and "garden hose arms", of the two later kinds NGC 6720 and NGC 6309 are typical examples. Others show extremely clumpy structures like NGC 7008. (See Chu *et al.*, 1991, for further examples).

In this paper we suggest that solid particles existing prior to the detachment of the stellar envelope may affect its appearance. The solid particles could be planets, moons, asteroids, comets, rings and arc rings (rock and ice). Each of these solid bodies will affect in a different way, or will be affected in a different way by, the envelope.

We think three possibilities could exist:

1. Even if protoplanetary nebulae debris exists it would be so insignificant that it would not affect the structure of the nebulae.
2. The planetary nebula would destroy the debris completely.
3. The debris survives enough time and has the correct mass and distribution to be able to influence the appearance of the nebula.

In this paper we discuss the possibility of having such solid particles in planetary nebulae, the observed morphologies that would be affected by the presence of such debris and finally the way in which the debris would be affected by the nebula.

---

* Paper presented at the Conference on *Planetary Systems: Formation, Evolution, and Detection* held 7–10 December, 1992 at CalTech, Pasadena, California, U.S.A.

Astrophysics and Space Science **212**: 61–65, 1994.

We will use the following notation: NP (nebulae plus central star), PNe (envelope), and PNn (nucleus).

It should be noted that many of the structures suggested to be produced by debris could be ejecta of the red giant, prior to the PN stage, for instance a dust ring. Many authors have described how such matter would influence the appearance of the PNe. (See for instance the review paper by Roche, 1989, he describes how dust can be in the shape of a torus or else mixed with the envelope).

1.1. STRUCTURES

As reported by Chu *et al.* (1987) in their compilation of PN at least half show multiple-shells and many of the resolved ones complex structures.

1.2. EXISTENCE OF DEBRIS

Several observations have led us to believe that large enough quantities of preplanetary nebula phase debris could exist prior to its formation.

1. IRAS observations have shown that all PN observed have IR emission probably due to dust.

2. Beta pictoris-like stars have been recently reported in the literature to be quite common. If this is so, when the planetary nebula is formed the envelope and the ionizing radiation will interact with the debris.

3. Molecules. Rodríguez (1989), has reported on molecular emission in PNe. We propose that their origin could be due to evaporation of material surrounding the progenitor before the ejecta of the PNe. Terzian (1989) has also reported on vast amounts of neutral material surrounding PN.

1.3. ASSUMPTIONS

In this paper the supposition is made that in some PN vast amounts of debris from the original star can exist surrounding the PNn, and as mentioned earlier it could be in the form of planets, moons, rings and ring arcs. If the amount of material would be similar to that of the solar system, 0.02 M, it would be difficult to detect, but if it would be ten times larger it would account for about 1/3 of the nebular mass.

Maybe some of this debris has been built up continuously during the lifetime of the progenitor the same way rings in the solar system are being continuously replenished.

Another supposition is that the debris are at varied distances from the central stars and that some of it has survived the nebula ejection at least long enough to perturb its spherical structure.

Since many PN have multiple shells it could be possible that the first ejection did not interact with the ejecta, whereas the second one did (Chu *et al.*, 1991).

If there is this kind of debris around the PN it could be: Planets and/or moons, comets and rocks, rings and ring arcs.

1.4. EVIDENCE FOR INFLUENCE OF BINUCLEARITY AFFECTING NEBULAR STRUCTURE

Lutz and Lame (1989) have found that three PN in the southern hemisphere with binary nuclei show peculiar morphologies like enhancements in the [NII] images and "jet-like" structures. This suggests that PNn that have invisible small mass companions could also have an envelope whose structure could be affected by the companion.

Of course higher progenitor stars could have more massive companions (in the form of planets or minor objects) and in consequence they would have a greater influence on the structure of the PNe. Peimbert (1978) has designated PN of type I as those objects having massive progenitors and a filamentary structure. So if debris are substantially present in PN I they might be responsible for at least part of the structure observed.

1.5. PLANETS

If planets exist around central stars they could either survive or not the nebula ejection. Calculations for the future of earth-like planets have been made by Goldstein (1987) and by Choi and Vila (1981), their destruction time would be of the order of 3000 years and they would leave no observational evidence.

If a rock rich planet would be farther from the sun the intense heating could fracture it and could produce comet like structures, see following section. The planet could also disintegrate completely and one would expect to find chemical enhancements in the nebula in accordance with the planet's original chemical composition and its mass.

1.6. COMETS

Peimbert (1990) has suggested that the O enhancement found in solar vicinity stars compared to H II regions could be due to contamination of the stars' atmospheres due to the fall of cometary material on its surface. Several authors have determined that comet-like structures observed in some PNe could actually be the Oort cloud of the central star lit up by the central star and the tails would be produced by its wind.

1.6.1. Rings
If massive rings really exist around a PN progenitor, when the central star starts sublimating the surrounding material eventually an ansae would form like the one observed in PNe.

1.6.2. Ring Arcs
We claim that some of the features observed in PNe could be produced by ring arcs. It seems that ring arcs, like those of Neptune, are produced when the ring material enters in resonance with several satellites. If there exists ring material surrounding

the PNn it could produce arc structures as long as there would be planets in the right places to produce the right resonances.

As mentioned previously the ring material could be an original ring, a disintegrated planet or else a torus of material ejected by the PNn. This last possibility seems to us realistic since there exist so many bipolar PN.

As discussed by Goldreich *et al.* (1986), virtually any kind of ring structure is possible, as long as one has massive enough satellites orbiting in eccentric orbits with adequate inclinations.

1.7. ORIENTATION OF THE RING ARCS

If our ideas are correct and there is enough protoplanetary material surrounding the PNn in order to influence the PNe appearance the orientation would have to be practically face on, otherwise we would observe a bipolar nebula which is very common.

1.8. OBSERVATIONAL "EVIDENCE"

Once one believes ring arcs around PNn could exist several features observed in the envelopes could be explained.

1.9. IC 1295

This planetary nebula is a double shell. In the picture presented by Chu *et al.* (1991) one observes an arc protruding from the second shell. If IC 1295 had a broken ring around it, spinning around the central star the first ejection would go through the opening, its rotation would produce the arc. After that the second ejection seems to have filled in the space and the ring arc could have been destroyed. See Figure 1.

This kind of arc or arm is also present in nebulae like NGC 6309. Two symmetric openings in the arc system would be necessary to explain its appearance, which is like a garden hose.

Planetaries like NGC 6720 and 6369 have structures that resemble volcanic plumes suggesting material coming from the central star has been pushed through a small opening.

The case of planetaries like NGC 2440 that present pedal structures or butterflies (Balick, 1989) would be produced by PNe matter being forced through the openings of the rings. If this would happen while they were spinning, then the appearance would resemble more NGC 6543. (We have been able to simulate such an appearance with water in a tank; now we are purchasing a Scheising machine to try it out with smoke.)

If our ideas are correct the ring material could dissipate completely after it has affected the morphology of the envelope in order for the stellar radiation to be able to reach it and ionize it.

1.9.1. Dynamical Models

Many authors have constructed models that try to explain the appearance of the PNe. In particular Balick *et al.* (1987a, b) have proposed models to explain the morphologies of three characteristic PN, their models use several episodes of winds to explain the observed shapes. So as Chu has proposed (1989) the structures of the envelopes can be explained by several alternative models and one needs accurate observations including high spectral resolution observations that permit one to measure velocities in order to discern amongst possible models.

2. Discussion

In conclusion we propose that many of the features observed in planetary nebulae envelopes could be explained by preplanetary nebulae material. For instance in Peimbert's type I the parlamentary structures could be due to evaporation of smaller bodies like comets and ring material. If the companions would be ice rich and massive or abundant enough then after evaporation they could mimic a C and O enhancement in the envelope (Dopita and Liebert, 1989).

References

Balick, B.: 1989, in: S. Torres-Peimbert (ed.), *Planetary Nebulae*, IAU Symposium **131**, p. 83.
Balick, B., Bignel, C. R., Hjellming, R. M. and Owen, R.: 1987, *Astron. J.* **94**, 948.
Balick, B. and Preston, L.: 1987, *Astron. J.* **94**, 958.
Choi, K. H. and Vila, S. C.: 1981, *Astroph. Space Sci.* **77**, 325.
Chu, Y.-H.: 1989, in: S. Torres-Peimbert (ed.), *Planetary Nebulae*, IAU Symposium **131**, p. 105.
Chu, Y.-H., Manchado, A., Jacoby, G. H. and Kwitter, K. B.: 1991, *Astroph. J.* **376**, 150.
Dopita, M. A. and Liebert, J.: 1989, *Astroph. J.* **347**, 1989.
Goldreich, P., Tremaine, S. and Borderies, N.: 1986, *Astron. J.* **92**, 490.
Goldstein, J.: 1987, *Astron. Astrophys.* **178**, 283.
Kahn, F. D.: 1989, in: S. Torres-Peimbert (ed.), *Planetary Nebulae*, IAU Symposium **131**, p. 411.
Lutz, J. and Lame, W. J.: 1989, in: S. Torres-Peimbert (ed.), *Planetary Nebulae*, IAU Symposium **131**, p. 462.
Peimbert, M.: 1978 in: Y.Terzian (ed.), *Planetary Nebulae*, IAU Symposium **76**, p. 215.
Peimbert, M.: 1990, *Reports on Progress in Physics* **53**, 1559.
Roche, P.F.: 1989, in: S. Torres-Peimbert (ed.), *Planetary Nebulae*, IAU Symposium **131**, p. 117.
Rodríguez, L.F.: 1989, in: S. Torres-Peimbert (ed.), *Planetary Nebulae*, IAU Symposium **131**, p. 129.
Shu, Y., Jacoby, G.H. and Arendt, R.: 1987, *Astroph. J. Suppl.* **64**, 529.
Terzian, Y.: 1989, in: S. Torres-Peimbert (ed.), *Planetary Nebulae*, IAU Symposium **131**, p. 29.

TOWARD PLANETS AROUND NEUTRON STARS *

A. WOLSZCZAN

*Department of Astronomy and Astrophysics, Pennsylvania State University,
University Park, Pennsylvania, USA*

Abstract. The two Earth-like mass objects orbiting a 6.2-ms pulsar, PSR1257+25, have survived more than one year of close scrutiny aimed at verifying their existence and remain the most serious candidates to become the first planets detected beyond the Solar System. The analysis of systematic timing measurements of the pulsar made over a 2.5-year period continues to require the presence of two planets with the minimum masses of 3.4 $M_\oplus$ and 2.8 $M_\oplus$ and the corresponding distances from PSR1257+12 of 0.36 AU and 0.47 AU to correctly predict the pulse arrival times. The presently available 3 μs rms accuracy of this procedure leaves little room for significant contributions to the pulsar's timing from any mechanisms other than the Keplerian motion. A detection of the effect of planetary perturbations on pulse arrival times which is commonly accepted as the most convincing way to furnish a "100% proof" of the reality of pulsar planets is already possible at a $\sim 2\sigma$ level. Intensive searches for millisecond pulsars now under way at various observatories are expected to address a very intriguing question of the frequency of occurrence of neutron star planetary systems.

1. Introduction

Searches around Sun-like stars aided by a variety of indirect evidence in favor of planetary formation obtained, for example, from the observations of star forming regions (Backman and Gillett, 1987) and from the IRAS studies of main sequence stars (Beckwith *et al.*, 1990), have yet to produce a solid detection of a planetary system other than our own. Although the concept of planetary systems around "normal" stars has been understandably more popular than a possibility of planets around neutron stars, the history of pulsar timing observations provides several interesting, if not dramatic instances of the reported detections of planet-like objects, all of which have either remained unconfirmed (Demiański and Prószyński, 1979) or have been retracted (Bailes *et al.*, 1991). In this climate, an announcement of the detection of two or more Earth-like objects in orbit around the millisecond pulsar, PSR1257+12, by Wolszczan and Frail (1992) has generated a wide spectrum of reactions in the astronomical community and beyond, with the skeptical ones being fairly common. The following list of the most typical attitudes toward the news of pulsar planets appears to be instructive:

- Pulsar planets? **NO**, thanks!!!
- They must be some Solar System-related effects!
- But a precessing neutron star will produce exactly the same timing residuals!
- Don't call these things planets!

---

* Paper presented at the Conference on *Planetary Systems: Formation, Evolution, and Detection* held 7–10 December, 1992 at CalTech, Pasadena, California, U.S.A.

Astrophysics and Space Science **212**: 67–75, 1994.

- Only one pulsar planetary system is statistically just as good as none at all!
- Confirming planetary perturbations would make me feel a lot better!
- Let's find some more!
- **EXCITING**, this confirms that the X-ray resistant ETI's do exist!!!

A careful analysis of the continuing timing measurements of the 1257+12 system with the Arecibo telescope (Wolszczan, 1993) has already removed all the major technical concerns apparent in the above list. In addition, it has been shown by Backer *et al.* (1992) that the independent timing observations of PSR1257+12 made at Green Bank are also best explained by postulating a two-planet system in orbit around the pulsar. A very small, 3 μs residual from the fit of a two-planet timing model to the data spanning the entire 2.5-year observing period (Figure 1) and an apparent lack of any significant changes in the pulse profile of PSR1257+12 (Kaspi and Wolszczan, 1993) make it doubtful that the neutron star precession-based models (Gil and Jessner, 1993; Dolginov and Stepinski, 1993) are plausible alternatives to the Keplerian motion hypothesis.

Clearly, much more analysis will have to be done, before the pulsar planets become a well established class of astrophysical objects. This obviously includes searches for further examples of pulsar planetary systems and theoretical attempts to understand their creation and evolution. Furthermore, the question of whether the term "planets" is astrophysically and philosophically correct in the case of objects orbiting neutron stars appears to be a non-trivial matter deserving a thorough consideration! However, the most urgent task at hand appears to be a detection of planetary perturbations predicted by Rasio *et al.* (1992) and Malhotra *et al.* (1992) to be measurable, provided that the pulsar planets are indeed planets, rather than some hitherto unanticipated effect. The value of such measurement would be enormous: most importantly, it would neutralize all but the most stubborn elements of the remaining opposition and it would open up a way to perform a detailed dynamical study of the 1257+12 system by means of a precision pulse timing technique.

So far, the published analyses of PSR1257+12 (Wolszczan and Frail, 1992; Wolszczan, 1993) include the timing data acquired before June 1992. In this paper, a further update of the timing model of the 1257+12 planetary system based on pulse arrival times measured between September 1990 and March 1993 is presented. In addition to these considerations, the discussion includes a description of the first attempts to detect planetary perturbations and a possibility that PSR1257+12 has a Moon-like inner planet in addition to the two already well-established companions.

2. The Updated Timing Model

The measurements of pulse arrival times from PSR1257+12 have been made since September 1990, typically at 2–3 week intervals, with the 305-m Arecibo telescope and the 430 MHz and 1400 MHz receiving systems using the Princeton pulsar timing backend. The details of data acquisition and analysis have been given in

TABLE I
Parameters of the PSR1257+12 System

Pulsar parameters

Rotational period, P	0.006218531931908(2) s
Period derivative, $\dot{P}$	1.1427(5) $\times 10^{-19}$ s s^{-1}
α (B1950.0)	$12^h\ 57^m\ 33^s.1212(4)$
δ (B1950.0)	$12°\ 57'\ 06''.565(9)$
μ_α (B1950.0)	46(3) mas yr^{-1}
μ_δ (B1950.0)	$-82(5)$ mas yr^{-1}
Epoch	JD 2448088.9
Dispersion measure	10.188(2) pc cm^{-3}
Flux density (430 MHz)	20(5) mJy
Flux density (1400 MHz)	1.0(2) mJy
Surface magnetic field, B	8.8×10^8 G
Characteristic age, τ_c	0.8×10^9 yr

Keplerian orbital parameters

Projected semi-major axis, $a_1 \sin i$	1.311(3) light ms	1.413(3) light ms
Eccentricity, e	0.018(4)	0.024(4)
Epoch of periastron, T_o	JD 2448703(1)	JD 2448686(1)
Orbital period, P_b	5749137(500) s	8486223(1000) s
Longitude of periastron, ω	246°(10)	108°(10)

Parameters of the planetary system

Planet mass, $m_{2,3}$ ($M_\oplus$)	3.4/$\sin i$	2.8/$\sin i$
Distance from the pulsar, d (AU)	0.36	0.47
Orbital period, P_b (days)	66.54	98.22

Wolszczan and Frail (1992) and Wolszczan (1993) and will not be repeated here. A timing model to predict the pulse arrival times from PSR1257+12 consists of the rotational and astrometric parameters of the pulsar (including proper motion) and two Keplerian, non-interacting orbits (Figures 1b,c) which account for large residuals resulting from a fit of the standard model to the timing data (Figure 1a). The two-planet model continues to produce an excellent fit to data, characterized by parameters listed in Table I and a 3 μs post-fit residual (Figure 1d).

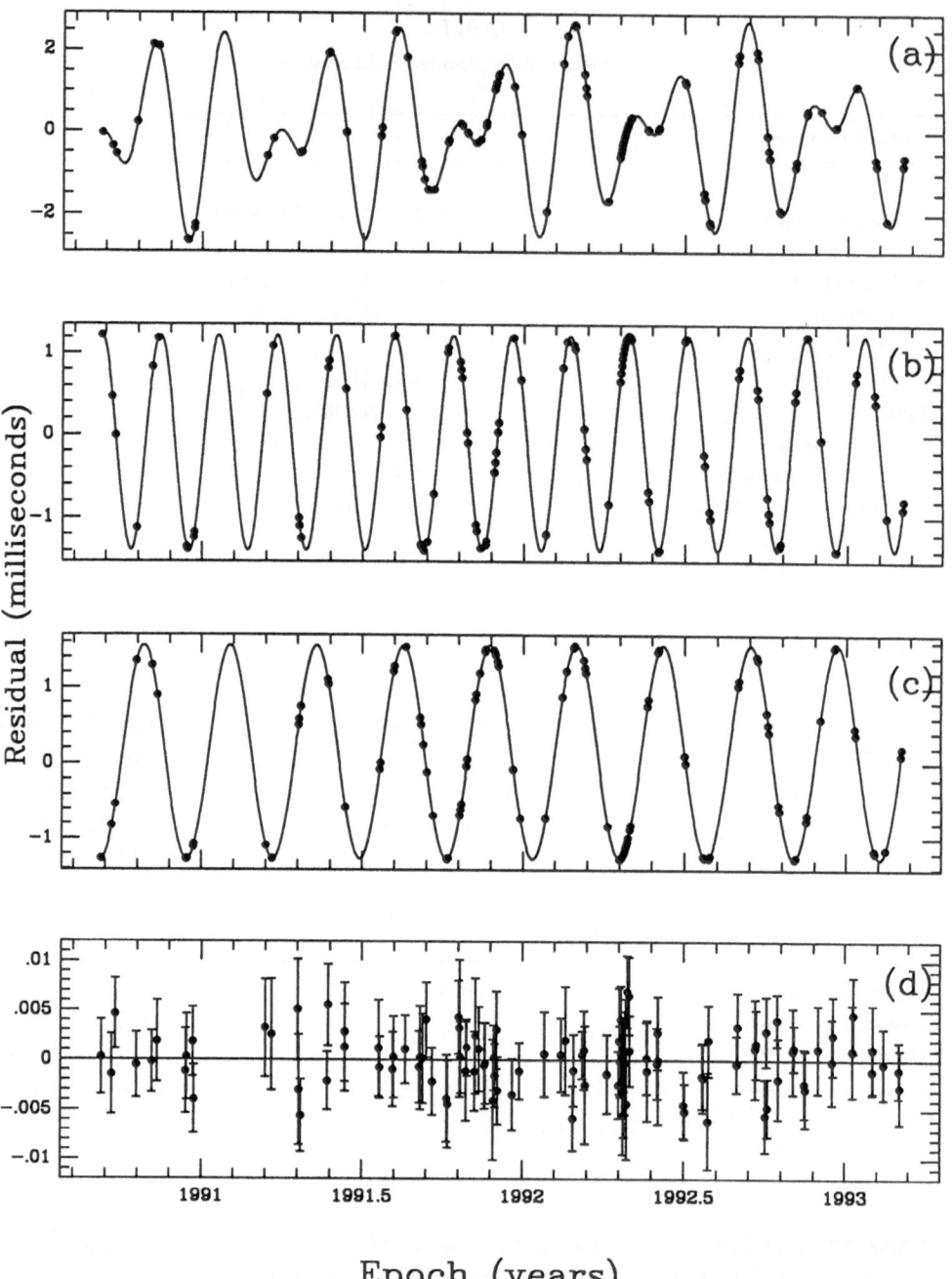

Fig. 1. The post-fit residuals of pulse arrival times from PSR1257+12. (a) Fit for rotational parameters only, (b) fit for a 98.2-day Keplerian orbit (leaves a 66.6-day periodicity as residual), (c) fit for a 66.6-day Keplerian orbit (leaves a 98.2-day periodicity as residual), (d) fit for all parameters of (a–c).

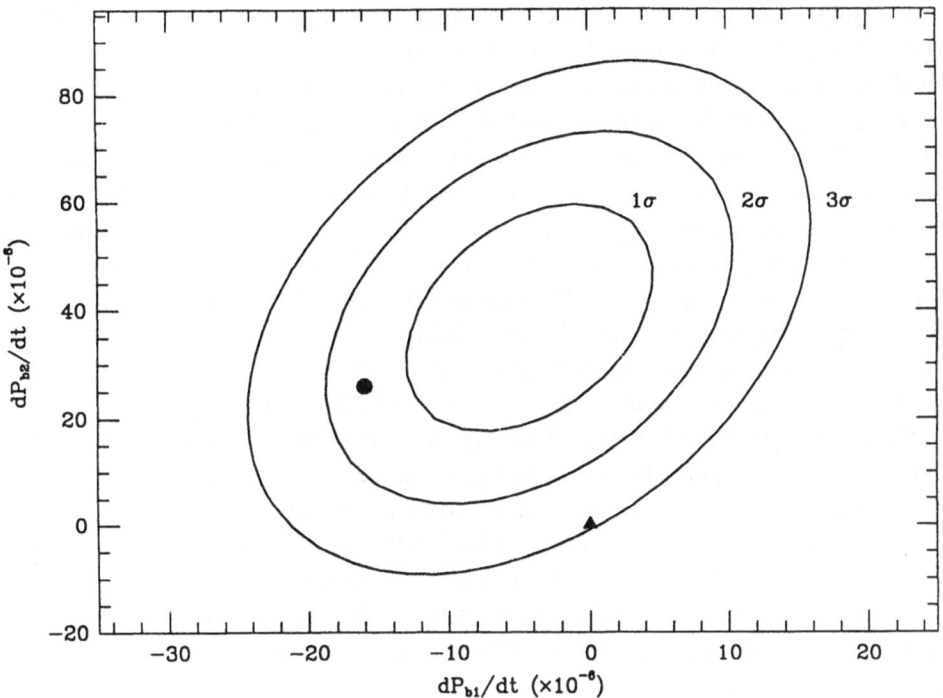

Fig. 2. Contours of χ^2 in the vicinity of its global minimum, displayed as a function of $\dot{P}_{b1}$ and $\dot{P}_{b2}$. The contours delimit areas of 68.3%, 95.4% and 99.73% confidence for these parameters. The circle denotes predicted values of $\dot{P}_{b1}$ and $\dot{P}_{b2}$ for $1/\sin i = 1$ and a 2.5-yr data span. The triangle corresponds to a timing model without perturbations.

Any other phenomena which may influence the measured pulse arrival times on timescales that are much shorter than the present data span cannot significantly exceed this limit.

3. Searching for Planetary Perturbations

Soon after the announcement of the 1257+12 planets, Rasio *et al.* (1992) and Malhotra *et al.* (1992) have pointed out that an approximately 3:2 resonance condition of the system leads to accurately predictable periodic perturbations of the two orbits and that the effect of these perturbations on pulse arrival times should be measurable. In fact, a detection of planetary perturbations has been commonly accepted as the most unambiguous and perhaps the only way to produce a final and irrefutable proof that the timing behavior of PSR1257+12 is indeed caused by Keplerian motion of two planet-sized bodies.

Detailed analyses of the dynamics of the 1257+12 planetary system and discussions of its effects on pulsar timing can be found in Malhotra (1993), Rasio *et al.* (1993), and Peale (1993), and these proceedings. For the purpose of this paper, it is sufficient to note that the most important consequence of a gravitational interaction between the two planets is a long-term, periodic variation of the elements of their orbits. If other short-term, small-amplitude effects are averaged out and planetary masses are assumed to be close to their minimum values (i.e., $1/\sin i \approx 1$), this variation can be adequately represented by smooth, periodic oscillations of the orbital elements about their mean Keplerian values. The relevant period is simply given in terms of the mean angular velocities of the planets, n_1 and n_2, as $2\pi/(2n_1 - 3n_2) = 5.56$ years and the predicted maximum amplitude of the corresponding timing residuals (after fitting out the two non-interacting orbits) is a function of planetary masses (a proximity of the system to the 3:2 resonance) and the total time span of observations (e.g., Peale, these proceedings).

The presently available timing data (Figure 1) do not contain any direct evidence for large amplitude, periodic residuals that could be due to perturbations caused by the two planets with large enough masses to be locked in a 3:2 resonance. Instead, the observed small residuals indicate that Earth-like mass objects and the near-resonance conditions are much more likely characteristics of the 1257+12 system. Consequently, the 5.5-year period, small-amplitude perturbations of the planetary orbits are most probably the largest effect that can be detected by means of the pulse timing analysis.

Although it is still too early for a meaningful confrontation of the timing model including a full set of necessary parameters with the arrival time data, a useful estimate of a contribution of the perturbations to the 1257+12 timing residuals can be made. The test relies on the assumption that changes in the orbital periods should be the easiest to detect, because they result in the largest orbital phase deviations accumulated over time. Furthermore, one can exploit the fact that, over a 2.5-year interval, these changes can be described by means of constant time derivatives of orbital periods to a sufficient degree of accuracy. Approximate numerical values of these derivatives are $\dot{P}_{b1} = -16 \times 10^{-6}$ s s^{-1} and $\dot{P}_{b2} = 26 \times 10^{-6}$ s s^{-1} for the inner and the outer planet, respectively.

Since the effect of non-zero orbital period derivatives on timing residuals is still too small to be reliably isolated by means of a direct χ^2 minimization technique, it is more enlightening to map the χ^2 surface around its global minimum by holding $\dot{P}_{b1}$ and $\dot{P}_{b2}$ at fixed values and fitting for all other model parameters. This procedure has been successfully employed in modelling the relativistic effect of Shapiro delay in binary pulsar systems (e.g., Taylor and Weisberg, 1989). The results for PSR1257+12 are shown in Figure 2 in the form of a contour map of confidence levels for both period derivatives. Clearly, a timing model with constant orbital periods of the two planets is already becoming untenable and the best-fit values of $\dot{P}_{b1} = -5 \times 10^{-6}$ s s^{-1} and $\dot{P}_{b2} = 38 \times 10^{-6}$ s s^{-1} represent a tentative, 2σ detection of planetary perturbations in the 1257+12 system. Further work on this exciting problem, which includes incorporating a more realistic perturbation model in the fitting process, is in progress.

4. Are There More Planets in the 1257+12 System?

Since there are no particular reasons why PSR1257+12 should not have more than two planets, a purposeful search for possible signatures of their presence in the timing data is of considerable interest. For example, a power spectrum analysis of the timing residuals provides some indication that the pulsar may have a very low-mass inner companion with a 25.3-day orbital period. A formal fit of an appropriate Keplerian orbit to data along with the two original planets leads to a $0.015M_{\oplus}/\sin i$ object about 0.2 AU away from the pulsar. Further analysis of this interesting possibility is in progress.

There is no indication in the data for a presence of more planets with orbital periods comparable to the total data span (2.5 years). This includes a possibility of a $\sim$ 1-year period planet briefly considered by Wolszczan and Frail (1992). Most likely, this planet-like signature in the timing data was caused by the initially unrecognized large proper motion of the pulsar (Table I).

A possible existence of any long-period planets in the 1257+12 system would affect the measured values of pulse period derivatives. PSR1257+12, along with the original millisecond pulsar, PSR1937+21 (Backer *et al.*, 1982), and the newly discovered PSR0437-47 (Johnston *et al.*, 1993), has one of the largest spindown rates ($\dot{P}$, Table I) among the galactic millisecond pulsars for which this parameter has been measured. Although the values of $\dot{P}$ of the remaining seven objects are typically an order of magnitude smaller, this is probably not a sufficient basis to claim that $\dot{P}$ of PSR1257+12 is affected by orbital dynamics. Since the measured value of $\ddot{P}$ is not significantly different from zero, no evidence for "outer planets" in the 1257+12 system is presently available.

5. Are There More Neutron Star Planetary Systems?

Aside from the important task of timing the 1257+12 system, a question of the frequency of occurrence of planetary systems around neutron stars has to be addressed. Undoubtedly, a detection of another such system would be enormously beneficial from the point of view of our understanding of pulsar planets, which, in addition to some lack of confidence, remains frustratingly (but understandably) ambiguous (Podsiadlowski, 1993; Phinney and Hansen 1993). So far, a systematic reexamination of large pulsar timing databases created by the long-term monitoring programs (Thorsett and Phillips, 1992; Bailes *et al.*, 1993; Thorsett *et al.*, 1993) has not led to any new detections. Similarly, a high precision timing of the recently discovered millisecond pulsars (Nice *et al.*, 1993; Foster *et al.*, 1993; Camilo *et al.*, 1993) has not revealed any periodicities that could be caused by orbiting objects with substellar masses, in spite of a clear superiority of the timing method over standard techniques of stellar astronomy (e.g., Black, 1980).

An intriguing possibility is that planets may also be detected around millisecond pulsars in globular clusters. As suggested by Sigurdsson (1993), in addition to forming around pulsars, they could also be "scavenged" from main sequence stars in the process of exchange encounters. One possible candidate that has already been proposed (Backer, 1993) and extensively discussed (Backer, 1993; Phinney, 1993; Sigurdsson, 1993) is PSR1620-26 in M4.

Of course, it is still much too early to attempt a meaningful statistical evaluation of these initial results. In the past, most of the pulsar timing data have been acquired with goals other than searches for planetary systems in mind. This obviously impairs the sensitivity of any analysis involving the archival data. On the other hand, the statistics of millisecond pulsars, which appear to provide more feasible "breeding environment" for planetary systems than the so-called "normal pulsars" (e.g., Podsiadlowski, 1993), indicate that among the ten well-timed galactic millisecond pulsars seven are in binary systems with stellar companions, two appear to be solitary objects, and one is accompanied by planet-sized bodies. If the solitary millisecond pulsars have disposed of their stellar companions (e.g., Bhattacharya and van den Heuvel, 1991) and if this process plays an important role in the formation of protoplanetary disks around pulsars (Phinney and Hansen, 1993), the above comparison provides grounds for an optimism concerning the results of the ongoing millisecond pulsar surveys and the pulse timing observations that will inevitably follow.

Acknowledgements

I am grateful to D. Frail, J. Taylor and A. Vázquez for their contributions to the research described in this paper.

References

Backer, D.C., Kulkarni, S.R., Heiles, C., Davis, M.M. and Goss, W.M.: 1982, *Nature* **300**, 615.

Backer, D.C., Sallmen, S. and Foster, R.S.: 1993, *Nature* **358**, 24.

Backman, D.E. and Gillett, F.C.: 1987, in: J. Linsky and R.E. Stencel (eds.), *Cool Stars, Stellar Systems and the Sun*, Berlin: Springer-Verlag, 340.

Bailes, M., Lyne, A.G. and Shemar, S.L.: 1991, *Nature* **352**, 311.

Bailes, M., Lyne, A.G. and Shemar, S.L.: 1993, in: J.A. Phillips, S.E. Thorsett and S.R. Kulkarni (eds.), *Planets Around Pulsars*, ASP Conf. Ser. **36**, 19.

Beckwith, S.V.W., Sargent, A.I., Chini, R. and Gusten, R.: 1990, *Astron. J.* **99**, 924.

Bhattacharya, D. and van den Heuvel, E.P.J.: 1991, *Phys. Rep.* **203**, 1.

Black, D.C.: 1980, *Space Sci. Rev.* **25**, 35–81.

Camilo, F., Nice, D.J. and Taylor, J.H.: 1993, *Astrophys. J. (Letters)* **412**, L37.

Demiański, M. and Prószyński, M.: 1979, *Nature* **282**, 383.

Dolginov, A.Z. and Stepinski, T.F.: 1993, in: J.A. Phillips, S.E. Thorsett and S.R. Kulkarni (eds.), *Planets Around Pulsars*, ASP Conf. Ser. **36**, 61.

Foster, R.S., Wolszczan, A. and Camilo, F.: 1993, *Astrophys. J. (Letters)* **410**, L91.

Gil, J.A. and Jessner, A.: 1993, in: J.A. Phillips, S.E. Thorsett and S.R. Kulkarni (eds.), *Planets Around Pulsars*, ASP Conf. Ser. **36**, 71.

Johnston, S. et al.: 1993, *Nature* **361**, 613.

Kaspi, V.M. and Wolszczan, A.: 1993, in: J.A. Phillips, S.E. Thorsett and S.R. Kulkarni (eds.), *Planets Around Pulsars*, ASP Conf. Ser. **36**, 81.

Malhotra, R., Black, D., Eck, A. and Jackson, A.: 1992, *Nature* **356**, 583.

Malhotra, R.: 1993, in: J.A. Phillips, S.E. Thorsett and S.R. Kulkarni (eds.), *Planets Around Pulsars*, ASP Conf. Ser. **36**, 89.

Nice, D.J., Taylor, J.H. and Fruchter, A.S.: 1993, *Astrophys. J. (Letters)* **402**, L49.

Peale, S.J.: 1993, *Astron. J.* **105**, 1562.

Phinney, E.S. and Hansen, B.M.S.: 1993, in: J.A. Phillips, S.E. Thorsett and S.R. Kulkarni (eds.), *Planets Around Pulsars*, ASP Conf. Ser. **36**, 371.

Phinney, E.S.: 1993, in: S. Djorgovski and G. Meylan (eds.), *Dynamics of Globular Clusters*, ASP Conf. Ser., in press.

Podsiadlowski, P.: 1993, in: J.A. Phillips, S.E. Thorsett and S.R. Kulkarni (eds.), *Planets Around Pulsars*, ASP Conf. Ser. **36**, 149.

Rasio, F.A., Nicholson, P.D., Shapiro, S.L. and Teukolsky, S.A.: 1992, *Nature* **355**, 325.

Rasio, F.A., Nicholson, P.D., Shapiro, S.L. and Teukolsky, S.A.: 1993, in: J.A. Phillips, S.E. Thorsett and S.R. Kulkarni (eds.), *Planets Around Pulsars*, ASP Conf. Ser. **36**, 107.

Sigurdsson, S.: 1993, in: J.A. Phillips, S.E. Thorsett and S.R. Kulkarni (eds.), *Planets Around Pulsars*, ASP Conf. Ser. **36**, 173.

Taylor, J.H. and Weisberg, J.M.: 1989, *Astrophys. J.* **345**, 434.

Thorsett, S.E. and Phillips, J.A.: 1992, *Astrophys. J. (Letters)* **387**, L69.

Thorsett, S.E., Phillips, J.A. and Cordes, J.M.: 1993, in: J.A. Phillips, S.E. Thorsett and S.R. Kulkarni (eds.), *Planets Around Pulsars*, ASP Conf. Ser. **36**, 31.

Wolszczan, A. and Frail, D.A.: 1992, *Nature* **355**, 145.

Wolszczan, A.: 1993, in: J.A. Phillips, S.E. Thorsett and S.R. Kulkarni (eds.), *Planets Around Pulsars*, ASP Conf. Ser. **36**, 3.

ON THE DETECTION OF MUTUAL PERTURBATIONS AS PROOF OF
PLANETS AROUND PSR1257+12 *

S.J. PEALE
Department of Physics, University of California,
Santa Barbara, USA

Abstract. Unambiguous detection of the consequences of mutual perturbations of the hypothesized planets about the pulsar PSR1257+12 would be unassailable proof of their existence. Nearly all of the residuals in the times of arrival (TOA) of the pulses after subtraction of the TOA predicted from the best fit constant period model are accounted for by including the effects of two orbiting planets with constant orbital parameters. The nature and magnitude of additional residuals in the TOA due to the gravitational interactions between the planets are determined by numerically calculating the TOA residuals for the orbital motion including the perturbations and subtracting the TOA residuals from analytic expressions of the orbital motion with orbital parameters fixed at averaged values. The TOA residual differences so obtained oscillate with periods comparable to the orbital periods with the oscillations varying in amplitude as a function of epoch within any given observational period. The signature of the perturbations is thus a quasiperiodic modulation of the residual differences obtained after removal of the effects of the orbital motion with best fit, constant orbital parameters. The amplitudes of this modulation reach about 10 μsec for observational periods exceeding 1000 days for the minimum planetary masses with $\sin i = 1$, and they increase as $1/\sin i$ for $1/\sin i < 5$, where i is the inclination of the orbit plane to that of the sky. Greater accumulated phase differences between the effects of perturbed and unperturbed orbital motions are available in the times of zero values in the observed and predicted TOA residuals and these comprise a second signature of the perturbations. The perturbation signatures should become detectable as the observation interval approaches 1000 days.

1. Introduction

Wolszczan and Frail (1992) have shown that the periodic residuals in the times of arrival (TOA) of radio pulses from the pulsar PSR1257+12 could be accounted for by the pulsar's reaction to two orbiting planets. Rasio *et al.* (1992) and Malhotra *et al.* (1992) pointed out that the mutual perturbations of the planetary orbits would be revealed by additional variations in the pulsar's radial motion whose detection would prove the hypothesis. Malhotra (1992) has determined differences between the TOA residuals resulting from perturbed orbits and those resulting from unperturbed motion of the planets using constant initial values of the orbital parameters for the unperturbed case. She finds the oscillatory residual differences to grow monotonically to about 40 μsec after 10 years of observation for $\sin i = 1$. However, initial conditions are inherently unknown, and only averaged values of the orbital parameters are approximated by the least squares fit to the TOA data. Peale (1993) chooses the averaged values of the orbit parameters over particular observational intervals to represent the least squares best fit values obtained from the data. The following outlines the results of this latter analysis, where the additional residuals from the perturbations do not continue to grow as the observational

---

 * Paper presented at the Conference on *Planetary Systems: Formation, Evolution, and Detection* held 7–10 December, 1992 at CalTech, Pasadena, California, U.S.A.

interval increases, and amplitudes as high as 40 μsec are reached only for values of $1/\sin i \gtrsim \sim 3$.

No alternative to orbiting planets can so easily reproduce the observations. In particular, attempts to account for the periodicities through a wobbling of an asymmetric rigid body in non principal axis rotation are frustrated by the fact that all such wobble frequencies must have the rotation frequency of 160 Hz as a factor. The spin axis would have to make very unlikely close approaches to the axis of intermediate moment of inertia during its precession in the body frame of reference in order to produce the 66.6 or 98.2 day periods of variation observed. In addition, there is no variation in pulse shape that should be evident if wobble were the source of the variation in the TOA (Wolszczan and Frail, 1992). Still, we desire the proof of the planetary existence that would result from the detection of the effects of the mutual perturbations.

The motivation for this analysis is to estimate the magnitude and nature of signature expected in the TOA as a function of the observational time span so that an observing program can perhaps be planned to optimize the detection of a perturbation signature. Section 2 describes the numerical and analytic procedures for evaluating the residuals followed by a description of the results in Section 3. A recognizable signature is shown to be a modulation of the amplitude of the TOA residual differences (after removal of the effects of the best fit orbital motions) as a function of both epoch and observational interval. A supplementary approach involving intensive, closely spaced observations over selected time intervals to find the time of a particular zero crossing of the TOA residuals is also mentioned. A summary follows in Section 4.

2. Numerical and Analytical Procedures

Table I shows the orbital parameters of the planets deduced from nearly 2 years of observations (Wolszczan, 1992) (See Wolszczan, this volume, for more precise values.)

TABLE I

Initial parameter values for numerical solution.

$P_1 = 5749068.0 \pm 500.0$ s		$P_2 = 8484936.0 \pm 1000.0$ s
$e_1 = 0.020 \pm 0.002$		$e_2 = 0.024 \pm 0.002$
$\omega_1 = 247° \pm 10°$		$\omega_2 = 105° \pm 10°$
$f_1 = 0°.0$		$f_2 = 25°.53$
$m_1 = 7.25 \times 10^{-6}/\sin i$		$m_2 = 5.97 \times 10^{-6}/\sin i$
	$t = 0$ at JD 2448105.3	

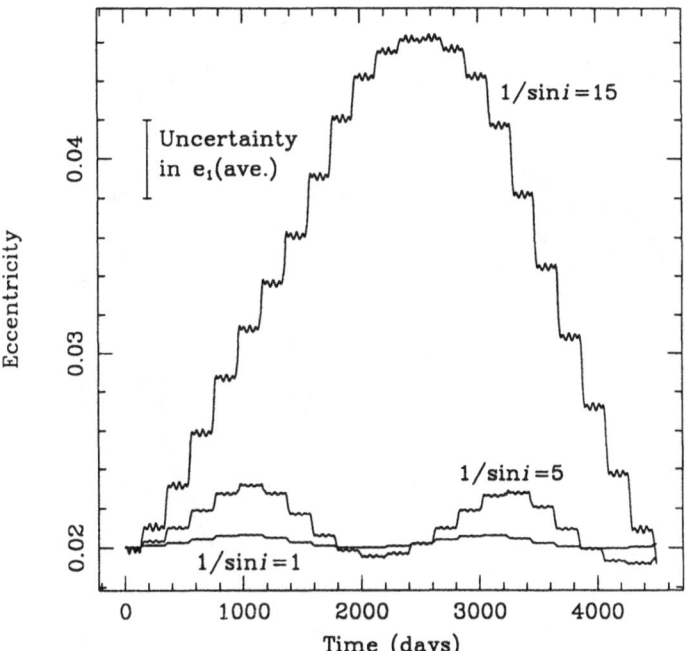

Fig. 1. Variation of the orbital eccentricity of the outer planet for non-resonance orbital motions ($1/\sin i = 1$, 5) and resonant orbital motions ($1/\sin i = 15$).

As the orbital parameters are solved for in the least squares data analysis, one might first seek to monitor the variation in the orbital parameters as the signature of the perturbation. However, the variation of e_1, for example, shown in Figure 1 is larger than the uncertainty in the averaged value of e_1 given in Table I only for large values of $1/\sin i$. The increase in the magnitude of the perturbations with $1/\sin i$ in Figure 1 follows from the larger inferred planetary masses for smaller inclination of the orbits relative to the plane of the sky. The largest variation in e_1 in Figure 1 is due to the enormous amplification caused by libration of both planets in 3:2 mean motion orbital resonances (Malhotra et al., 1992).

Figure 2 shows the resonance angles that librate (ϕ_i are true orbital longitudes of the planets defined in Figure 3.) and the nature of that libration with its enormous variation in the e_i, The change in e_1 (and also other parameters) is easily seen only if the resonance angles are librating within the resonance. For such libration to occur, $1/\sin i > 10$, which has a probability of only 0.005. For likely values of $1/\sin i$, we must therefore look for the signature of the perturbations in relatively small changes in the TOA.

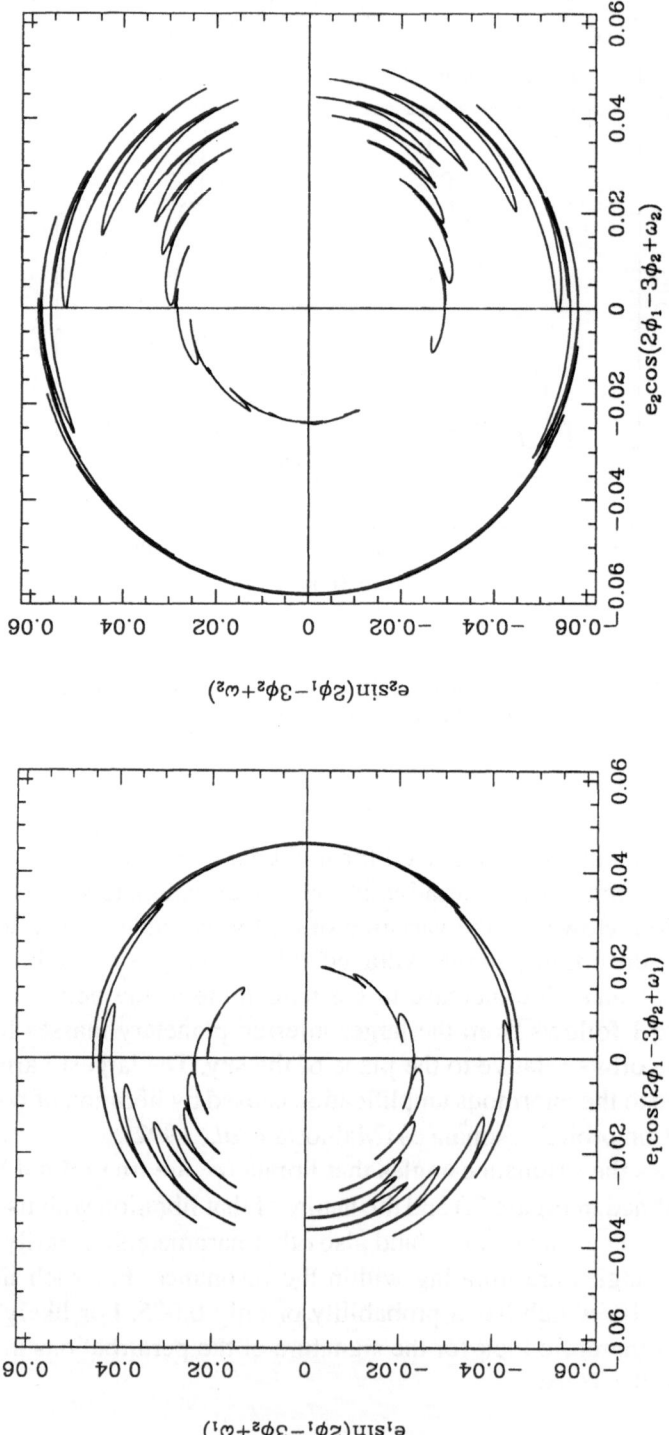

Fig. 2. Libration of resonance variables showing large variations in the orbital eccentricities.

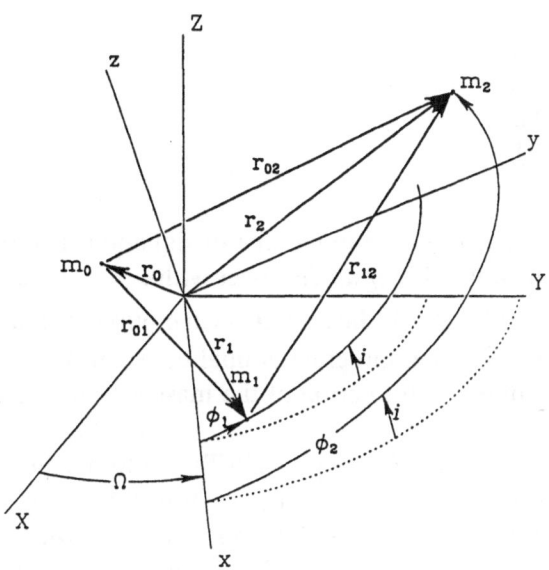

Fig. 3. Coordinate systems, variable and vector definitions. (From Peale, 1993)

In Figure 3, the XYZ system is inertial with the XY plane being the plane of the sky and with the Z axis pointing away from the observer. The X axis orientation is arbitrary, but it is customarily directed to the north. Both planets are assumed to be orbiting in the xy plane as the most likely configuration, where the z axis is parallel to the angular momentum of the system and the x axis is along the ascending node of the orbit plane on the XY plane. The orbital inclination is i and Ω is the longitude of the ascending node. The positions of all three bodies are referred to the center of mass of the system by the vectors $\mathbf{r}_0$, $\mathbf{r}_1$, $\mathbf{r}_2$. These vectors are located in the orbit plane by true longitudes ϕ_i, where $\phi_i = \omega_i + f_i$ are measured from the ascending node (x axis) with ω_i and f_i being the argument of periapse and the true anomaly respectively. All the vectors in Figure 3 can be expressed in terms of $\mathbf{r}_1$ and $\mathbf{r}_2$.

$$
\begin{aligned}
\mathbf{r}_0 &= -m_1\mathbf{r}_1 - m_2\mathbf{r}_2 \\
\mathbf{r}_{01} &= (1 + m_1)\mathbf{r}_1 + m_2\mathbf{r}_2 \\
\mathbf{r}_{02} &= m_1\mathbf{r}_1 + (1 + m_2)\mathbf{r}_2 \\
\mathbf{r}_{12} &= \mathbf{r}_2 - \mathbf{r}_1
\end{aligned}
\tag{1}
$$

where m_1, m_2 are the masses of the inner and outer planets respectively normalized by the neutron star mass m_0.

The equations of motion in dimensionless form can be written

$$
\begin{aligned}
\ddot{\mathbf{r}}_1 &= -\frac{\mathbf{r}_{01}}{r_{01}^3} + m_2\frac{\mathbf{r}_{12}}{r_{12}^3}, \\
\ddot{\mathbf{r}}_2 &= -\frac{\mathbf{r}_{02}}{r_{02}^3} - m_1\frac{\mathbf{r}_{12}}{r_{12}^3}
\end{aligned}
\tag{2}
$$

where the motion of m_0 is found from the first of Equations (1). Equations (2) are made dimensionless by normalizing all distances with $a = 1$ AU and replacing time t with nt, where $n = \sqrt{Gm_0/a^3}$ is the orbital mean motion of an infinitesimal mass at 1 AU from m_0 with G being the gravitational constant. Hence, dimensionless time increases by 2π in one orbit period of the mass at 1 AU or in 307.908 days for $m_0 = 1.4\ M_\odot$.

We use the values of the parameters given in Table I for the initial conditions for the numerical solution of Equations (2). The periods in Table I are used to determine the semimajor axes of the orbits near 0.360 and 0.467 AU, where the pulsar mass is assumed to be the nominal 1.4 $M_\odot$. The values of $m_j/\sin i$ correspond to 3.4 $M_\oplus/\sin i$ and 2.8 $M_\oplus/\sin i$ respectively. The zero of time is chosen to be the epoch of the periapse passage of the inner planet such that the initial value of $f_1 = 0°0$. The outer planet passed periapse 6.7 days earlier leading to the initial value of $f_2 = 25°53$. The initial Cartesian coordinates and velocities of the planets are determined from Table I after converting the semimajor axes derived from the periods to values relative to the center of mass of the system. The Burlisch and Stoer (1966) algorithm was used in the numerical integration of Equations (2). Energy and angular momentum are conserved to at least 12 significant figures over 4500 days of integration indicating the precision of the algorithm.

In order to detect changes in the variations of the TOA residuals, a reference curve of periodic variations must be established. The orbital parameters are deduced from a least squares fit to the observational data (Wolszczan and Frail, 1992), and the best fit orbits will change as the observations are extended. Essentially the only reference curve available for the line-of-sight position of the pulsar relative to the system center of mass will be that determined from the orbits found from the least squares analysis, and these orbits will generally depend on the particular observational interval used. With the orbits thus fixed, the observable consequences of the mutual planetary perturbations will be contained in the residuals in the TOA data after the effects of the motion with fixed orbital parameters are removed. We shall represent these least square determinations of the orbital parameters by numerically determined averaged values for each observational interval considered. Even if there is a cause other than orbiting planets for the periodic residuals in the observed radial velocity, the hypothesized planets produce variations in excellent agreement with the observations, and we can therefore use the planets to establish the reference curves due to some cause not involving planets.

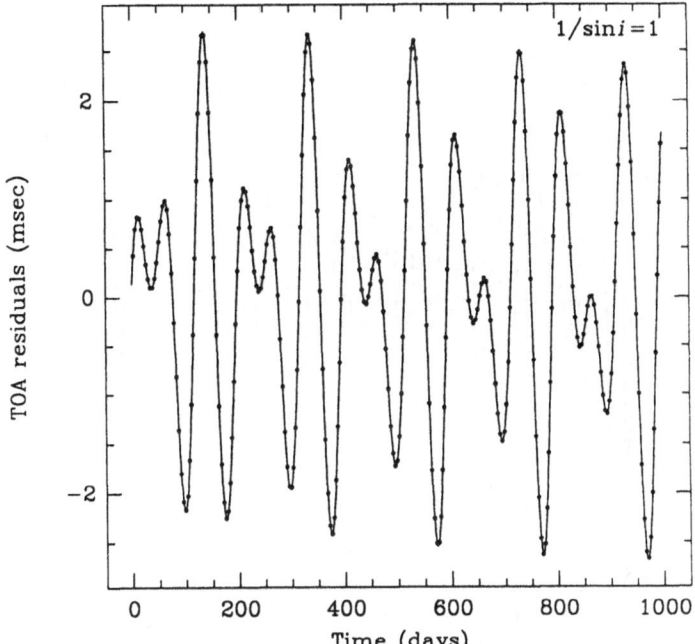

Fig. 4. Residuals in pulse TOA determined from the numerical determination of the line-of-sight positions of the pulsar relative to the system center of mass (solid line) and from analytic expressions for the line-of-sight position with constant planetary orbital parameters. The consequences of the perturbations are below the resolution of the graph.

Equations (2) are numerically integrated over some maximum observing interval to determine averaged values of the orbital parameters a_i, e_i, ω_i, t_{0i}, as a function of observing intervals less than the maximum according to

$$\langle \xi_j \rangle = \sum_{k=0}^{N-1} \frac{(\xi_j^k + \xi_j^{k+1})d}{2T}. \qquad (3)$$

Here t_{0i} are the epochs of periapse passage, ξ_j is any of the parameters, $N = T/d$ is the number of evaluations of the orbital elements at equal time intervals of length d over the observational time span T, and ξ_j^k is the kth evaluation of the parameter. The averaged values of the orbital parameters are then used in the analytic expression for the line of sight displacement of the neutron star from the center of mass.

$$Z_0^{\text{ref}} = - \sum_{j=1}^{2} \frac{m_j}{1 + m_j} \frac{a_j(1 - e_j^2)}{1 + e_j \cos f_j} \sin i \sin (\omega_j + f_j), \qquad (4)$$

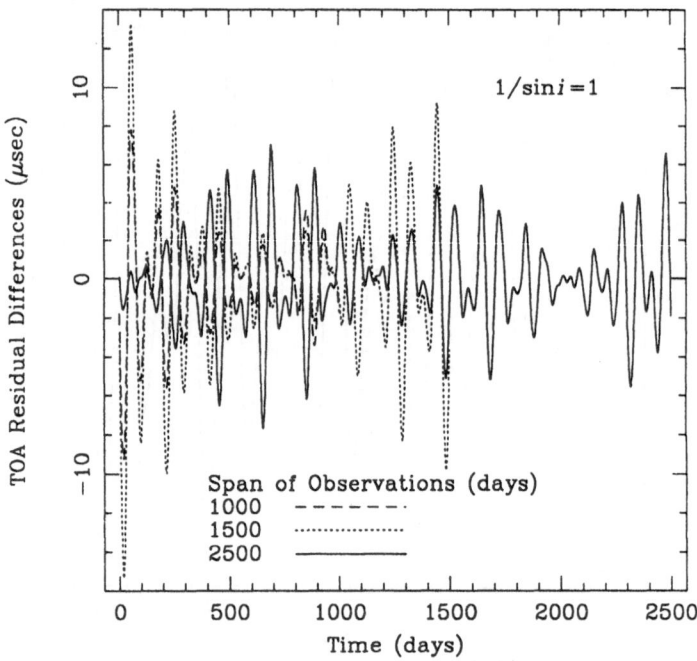

Fig. 5. Differences in the residuals in pulse TOA between numerically determined perturbed orbits and orbits with constant orbital parameters obtained by averaging the numerically determined parameters over the indicated observational time intervals. (From Peale, 1993)

where Equation (4) represents the reference pulsar position for no mutual planetary perturbations deduced for the particular observational interval. Equations (2) are integrated a second time but now over a particular observational interval to deduce the line of sight position of the neutron star relative to the center of mass through the first of Equations (1).

$$Z_0 = -m_1 Z_1 - m_2 Z_2 = -(m_1 y_1 + m_2 y_2)\sin i, \tag{5}$$

where the last form follows from Figure 3. The TOA residual is then given by Z_0/c, where c is the velocity of light. The expression $(Z_0 - Z_0^{ref})/c$ gives the TOA residual differences between the effects of the perturbed and unperturbed planetary orbits at each time step, where Z_0^{ref} is evaluated with the averaged values of the orbital parameters appropriate for the particular observational interval considered. More details of this procedure are given in Peale (1993).

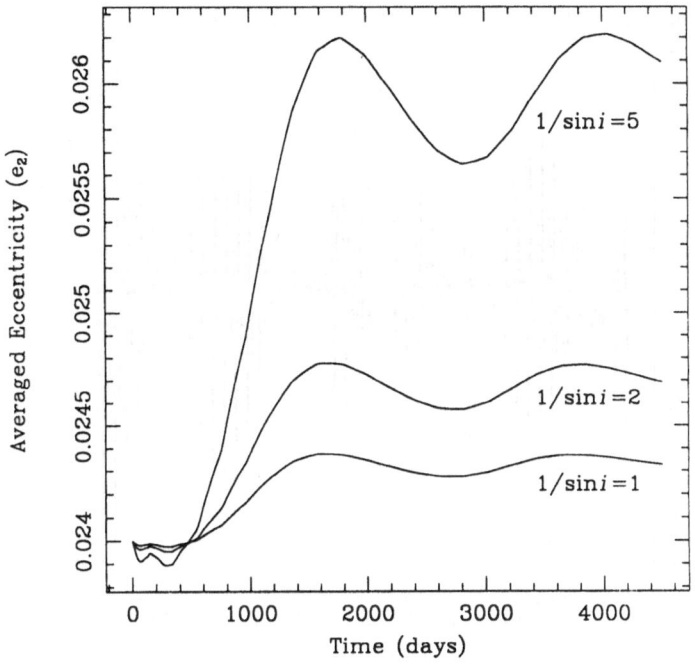

Fig. 6. Averaged orbital eccentricity for the outer planet.

3. Results

The numerical solution of Equations (2), with the initial conditions of Table I and $1/\sin i = 1$, when substituted into Equation (5) and the result multiplied by $1000 \times (1\ AU)/c$ yields the residuals in the time of arrival shown in Figure 4. The points on the curve are equally spaced evaluations of the analytic expressions (Equation (4)) with averaged (over 1500 days) orbital parameters. The points verify the analytic expression and show that the consequences of the perturbations are not distinguishable at the resolution of the graph for $1/\sin i = 1$. Figure 5 shows the perturbation induced TOA residual differences in microseconds for $1/\sin i = 1$ for several time spans of observation. The residual differences are oscillatory and grow in amplitude up to 1500 days, but at 2500 days the residuals have decreased in amplitude and the modulation of the residual differences has changed in those regions where the 2500 day curve overlaps the curves for shorter time spans. This behavior is understood from Figure 6, which shows that the averaged value of e_2 (and also the other parameters) reaches a peak deviation from its initial value at 1500 days. The reference curves deviate most from the real motion in this interval. At 2500 days, averaged parameters reach a local minimum leading to a reduction

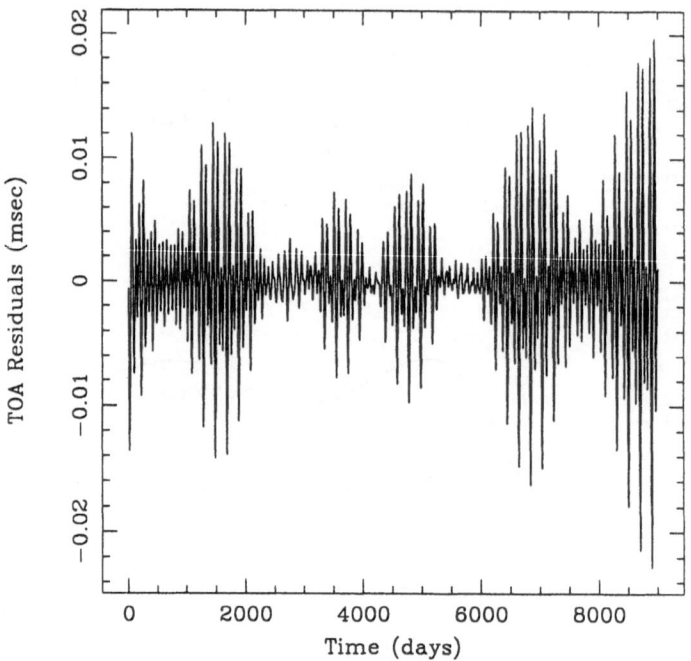

Fig. 7. Same as Figure 5 except for the longer observational period of 9000 days. This displays the persistence of the amplitude of the TOA residual differences as the observational interval is increased and the change in the positions of the maximum and minimum amplitudes when compared to the shorter observational intervals in Figure 5.

in the residual differences from the 1500 day peak. The quasiperiodic modulation of the residual differences is typical as is illustrated in Figure 7 for a 9000 day period of integration. The mutual perturbations are quasiperiodic, so one expects the residual differences to grow and diminish as major perturbing terms cycle through 2π. The terms with arguments $2\phi_1 - 3\phi_2 + \omega_{1,2}$ are the most important since the system is near the 3:2 mean motion orbital resonance, and these terms have a period near 2000 days as indicated by the variation of e_2 in Figure 1. Some of the modulation cycles approach 2000 days, but others are about half of this period. Finally, the change in the modulation envelope with averaging interval is due to the fluctuations in the averaged parameters with averaging interval. The amplitudes of the residual difference fluctuations are proportional to $1/\sin i$ for $1/\sin i < 5$, and they grow with observational interval up to 1500 days. Figure 8 shows the span of fluctuation amplitudes as a function of the observational interval. After almost 700 days of observation, the residual differences in the TOA after removal of the effects of the best fit constant parameter orbits have peak amplitudes less

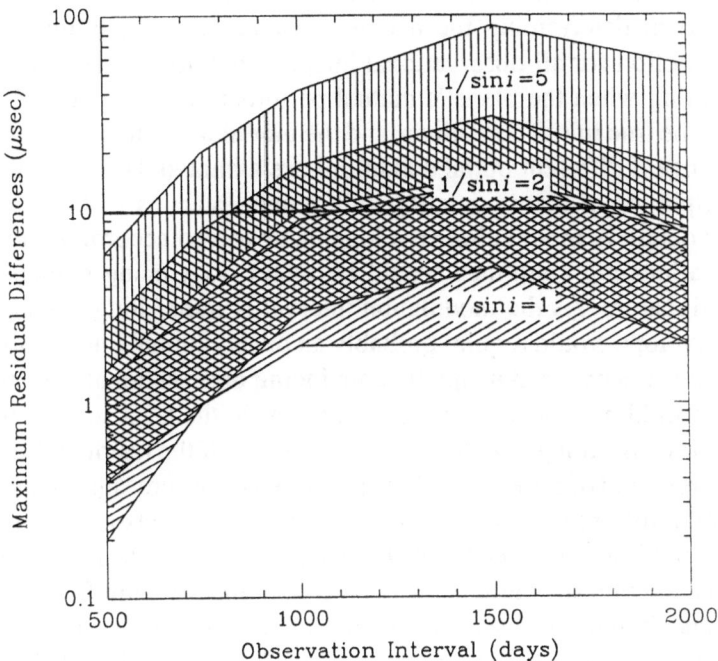

Fig. 8. Ranges of amplitudes in the quasiperiodic TOA residual differences like those displayed in Figure 5 as a function of observational interval and $1/\sin i$. (From Peale, 1993)

than 10 μsec (rms = 3 μsec) (Wolszczan, 1992) which, from Figure 8, already rule out $1/\sin i \gtrsim 4$.

Larger phase accumulation between the effects of perturbed and unperturbed orbits than is available in the TOA residuals can be obtained between the predicted and observed times when the TOA residuals are zero. The difficulty of determining the precise zero crossing of the TOA residuals with many closely spaced observations near the crossing decreases the advantages of sometimes large differences between the real zero crossing times and those predicted with the constant parameter orbits. Still, the reader is referred to Peale (1993) for a more complete discussion of this method and for an indication of those situations where the time differences should be a measurable proof of the perturbations.

4. Summary

The signature of the mutual perturbations of the planets orbiting PSR1257+12 is a characteristic modulation of the TOA residual differences (after removal of the effects of the best fit constant parameter orbits) with amplitude $\geq$ 10 μsec, where the envelope of modulation changes with the observational interval. We used the same initial conditions (the least squares averaged values after 700 days of observation) in determining the averaged values of the parameters used in the reference motion for all observational intervals and for determining the TOA residuals resulting when the planet's mutual perturbations are included. These of course are not the true initial conditions, and the modulation envelopes in Figures 5 and 7 will change if different initial conditions are assumed. However, the general nature and amplitude of the modulation will remain intact as will its evolution in form if not in detail. A modulation envelope that evolved according to prediction as the observations are continued would be rather convincing evidence for the perturbation signature. As the modulation interval is rather long, a change in the modulation envelope would require probably an observational interval longer than most will have patience for. An equally convincing observational detection of the perturbations would be to include the perturbations in the least squares model and perhaps solve for $\sin i$ along with the other parameters. If the residual differences in the TOA after removal of the effects of the perturbed orbits are significantly smaller than the residual differences after removal of only the effects of the unperturbed orbits, most would be convinced that the perturbations would have indeed been detected, and $\sin i$ would simultaneously be constrained. Figure 8 shows that any of these effects should be observable as the interval of observation approaches 1000 days for $i \gtrsim \sim 30°$. Values of $1/\sin i \gtrsim \sim 4$ are already ruled out by the magnitude of the peak residual differences less than 10 μsec obtained after slightly less than two years of observation (Wolszczan, 1992).

It would also be interesting to determine one or two zero crossings of the TOA residuals with many closely spaced observations spanning the time of the zero TOA. These could be compared with the predicted zero crossings with or without the perturbations included. A behavior of the differences in the zero crossings times according to predictions would also serve as a verification of the perturbations.

Note Added in Proof

The Julian dates of the planets' periapse passages given in Table I (JD 2448105.3 for m_1 and 6.7 days earlier for m_2) were inadvertently taken from Wolszczan and Frail (1992). The more accurate values based on a longer time span of observation (Wolszczan, 1992) are JD 2448104 for m_1 and 7.8 days earlier for m_2. The initial true anomaly f_2 would then be increased from $25°53$ to $29°94$. This change in the initial value of f_2 does not change the figures noticeably.

Acknowledgements

It is a pleasure to thank Robert Antonucci, Carl Gwinn and Alex Wolszczan for pointing out some errors and otherwise greatly improving an earlier version of this manuscript. Phil Nicholson's comments in a review of this manuscript clarified the presentation at several points. I specially thank Alex Wolszczan for furnishing the results of his latest data reductions before their publication. This work is supported in part by the Planetary Geology and Geophysics Program of NASA under grant NAGW 2061.

References

Burlisch, R. and Stoer, J.: 1966, *Numerical Mathematics* **8**, 1–13.

Malhotra, R.: 1992, paper presented at the 'Planets Around Pulsars' Workshop, California Institute of Technology, Pasadena, CA, April 30–May 1, 1992, to appear in J.A. Phillips, S.E. Thorsett and S.R. Kulkarni, ed(s)., *Planets Around Pulsars*.

Malhotra, R., Black, D., Eck, A. and Jackson, A.: 1992, Constraints on the Putative Companions to PSR1257+12, *Nature* **355**, 583–585.

Peale, S.J.: 1993, On the Verification of the Planetary System Around PSR1257+12, *Astronomical Journal* in press.

Rasio, F.A., Nicholson, P.D., Shapiro, S.L. and Teukolsky, S.A.: 1992, Planetary System in PSR1257+12: A Crucial Test, *Nature* **355**, 325–326.

Wolszczan, A.: 1992, PSR1257+12 and Its Planetary Companions, paper presented at the 'Planets Around Pulsars' Workshop, California Institute of Technology, Pasadena, CA, April 30–May 1, 1992, to appear in J.A. Phillips, S.E. Thorsett and S.R. Kulkarni, ed(s)., *Planets Around Pulsars*.

Wolszczan, A. and Frail, D.A.: 1992, A Planetary System Around the Millisecond Pulsar PSR1257+12, *Nature* **355**, 145–147.

PLANETS AROUND PULSARS: A REVIEW *

J.A. PHILLIPS and S.E. THORSETT

*Owens Valley Radio Observatory, California Institute of Technology,
Pasadena, California, USA*

Abstract. The discovery last year of a planetary system orbiting a millisecond pulsar raises important questions in pulsar evolution, planet formation, and planetary dynamics. We review the literature concerning pulsar-planetary systems, emphasizing particularly the contributions to the meeting *Planets around Pulsars* held at Caltech in 1992.

1. Introduction

This review covers a relatively new part of planetary studies: the formation, detection, and characterization of planets orbiting pulsars. Interest in pulsar planets has grown considerably since 1992, when Alex Wolszczan and Dale Frail announced the discovery of a planetary system around a neutron star. This was not the first report of planets orbiting pulsars, but it was by far the most convincing. Motivated by their discovery, we organized an international meeting last April at Caltech on *Planets around Pulsars*, where experts could debate the evidence in favor of planets, propose critical tests, and speculate on formation scenarios. The meeting was well attended by planetary scientists as well as pulsar astronomers, and the proceedings have just been published by the Astronomical Society of the Pacific (Phillips *et al.*, 1993). In this contribution we summarize the most important results to emerge from that meeting and discuss the more recent literature.

We first review a few basic facts about pulsars. (Excellent texts on pulsar astronomy include Manchester and Taylor (1977) and Lyne and Smith (1990).) A pulsar is a rapidly-spinning, highly-magnetized neutron star formed in the supernova explosion of a massive star (or possibly the collapse of a massive white dwarf). Radio emission from pulsars is directed in narrow beams parallel to the magnetic dipole axis of the star. We see pulses because the magnetic and spin axes are misaligned, so the emission beam sweeps past the Earth once per revolution. The radio emission mechanism is still unknown, but the emission process and the pulsar magnetosphere are believed to be extremely stable on time scales of interest here.

There are two general classes of pulsars:

1. "Normal" pulsars, of which about five hundred are known, have spin periods near one second, are nine hundred to about ten million years old, and have surface magnetic fields of $\sim 10^{12}$ G, and

2. Millisecond pulsars, of which only about thirty are known, have periods between 1.5 and a few tens of milliseconds, are $\sim 10^9$ years old, and have magnetic fields of $\sim 10^9$ G.

---

* Paper presented at the Conference on *Planetary Systems: Formation, Evolution, and Detection* held 7–10 December, 1992 at Caltech, Pasadena, California, U.S.A.

Astrophysics and Space Science **212**: 91–106, 1994.
© 1994 *Kluwer Academic Publishers.*

Of greatest interest here are the millisecond pulsars. They are extremely accurate clocks (Section 2): typically their periods change only through a small spin-down at a rate $\sim 10^{-19}$ s s^{-1}. It is believed that millisecond pulsars rotate so quickly because they underwent a period of mass and angular momentum transfer from a binary companion. Indeed, most known millisecond pulsars still have (non-accreting) binary companions, either white dwarfs or other neutron stars.

We know the companions exist because the orbital motion of the pulsar around the center of mass of the binary system Doppler-shifts the radio pulse train. In this way a "binary pulsar" is analogous to a single-line spectroscopic stellar binary. At present one can measure the radial velocity of optical lines of normal stars with a precision between 5 and 10 m s^{-1} (e.g., Cochran and Hatzes, 1993), which is sufficient to detect Jupiter-like planets. Pulsar observations are much more sensitive. The doppler shift of a millisecond pulsar can be measured with a precision better than 0.05 m s^{-1} in a single day, and Earth-mass planets should be easily detectable at the $\sim 100\sigma$ level in long-term timing residuals (Section 2).

In 1990 Alex Wolszczan discovered PSR B1257+12, a very unusual millisecond pulsar. Millisecond pulsars are widely known for their frequency stability—indeed, no intrinsic "timing noise" has been observed in any of them. However, the period of 1257+12 varied on a timescale of months in a highly ordered but indecipherable pattern. Eventually Wolszczan, in collaboration with Dale Frail, realized that the doppler shifts could be explained if the pulsar were orbited by not just one, but two companions. The companions have very low mass (2.8 and 3.5 $M_\oplus$) and are in nearly circular orbits within an AU of the neutron star. Wolszczan and Frail had apparently discovered the first known extra-solar planetary system.

The planets around 1257+12 were identified by pulse timing techniques, so in Section 2 we review pulsar timing before discussing PSR B1257+12 in detail. In Section 3, we present the observational tests that have been proposed and conducted to test the hypothesis that the period variations of 1257+12 are caused by the orbital motion of planets. In Section 4, we summarize the surprisingly large number of pulsar-planet formation scenarios proposed since the announcement of planets orbiting PSR B1257+12. None of the models answers all its critics, but the variety of plausible models suggests that even more planets may be found around other pulsars. The prospects for finding more such systems are considered in Section 5.

2. Pulsar Timing and Planets Orbiting PSR B1257+12

The huge rotational inertia of a spinning neutron star makes radio pulsars remarkable clocks. Some young pulsars show wander in their spin periods, probably caused by fluctuations in the coupling between the core and crust of the neutron star (Cordes, 1993). But if these are the quartz watches of pulsar timing, then millisecond pulsars are the atomic clocks. Indeed, timing of the millisecond pulsar PSR B1937+21 has shown that it is *more* stable than the best terrestrial long-term time standards, with phase jitter of about a microsecond in the last decade (Stinebring *et al.* , 1990). Other millisecond pulsars appear similarly free of "timing noise."

Although pulsar binary systems are analogous to single line spectroscopic binaries, there is one important difference. While spectral line observations measure doppler shifts directly, pulsar timing can measure the integral of the doppler-shifted pulsar frequency: the pulse phase. This ability to count the number and fraction of pulse cycles between two observations which may be days or years apart allows shifts in the the pulsar period to be measured to very high precision even though the fundamental frequency (hundred of hertz for millisecond pulsars) is much lower than typical spectral line frequencies.

The first step in timing a pulsar is to go to a radio observatory and—with the aid of an atomic clock—measure the arrival time of a pulse (actually the "average arrival time" of hundreds to millions of pulses). This yields a topocentric time of arrival, t_{obs}. The observatory is not in an inertial reference frame, but to a good approximation the center of mass of the solar system defines a good inertial frame, so the next step is to convert the topocentric time of arrival to an arrival time at the solar system barycenter (t_{ssb}). From, for example, Manchester and Taylor (1977):

$$t_{ssb} = t_{obs} + \frac{\mathbf{r} \cdot \hat{\mathbf{n}}}{c} - \frac{\alpha DM}{\nu^2} + \Delta t_r \tag{1}$$

There are three correction terms in Equation (1). The first gives the geometric delay between the arrival of a pulse front at the observatory and at the barycenter. In that term, $\mathbf{r}$ is the vector from the observatory to the barycenter and $\hat{\mathbf{n}}$ is a unit vector pointing from the barycenter in the direction of the pulsar. To compute the barycentric arrival time, we must have a good solar system ephemeris ($\mathbf{r}$) as well as a good estimate of the coordinates of the pulsar ($\hat{\mathbf{n}}$). In practice, the pulsar's location is solved for as part of the timing process. The next term in Equation (1) is a frequency-dependent correction for interstellar dispersion included because it is customary to refer all arrival times to infinite frequency. The final term, Δt_r, is a relativistic clock correction to account for the earth's motion in an elliptical orbit around the sun.

Because the pulsar signal is periodic, or nearly so, we can express the arrival times in terms of phase, ϕ. The measured phases can then be compared with a polynomial model for the pulsar spin down:

$$\phi(t) = \phi_0 + \Omega(t - t_o) + \frac{1}{2}\dot{\Omega}(t - t_o)^2 + \ldots \tag{2}$$

where ϕ_0 is the phase at some initial epoch t_o and $\Omega = 2\pi/P$. If we know Ω and $\dot{\Omega}$ with sufficient accuracy we can predict the pulse phase at any time t. The differences between predicted and observed phases are referred to as "residuals":

$$\Re(t) = \phi_{predicted}(t) - \phi_{observed}(t) \tag{3}$$

The goal of pulsar timing is to minimize these residuals, in a least-squares sense, by finding the best-fit model for the pulsar period (P), period derivative ($\dot{P}$),

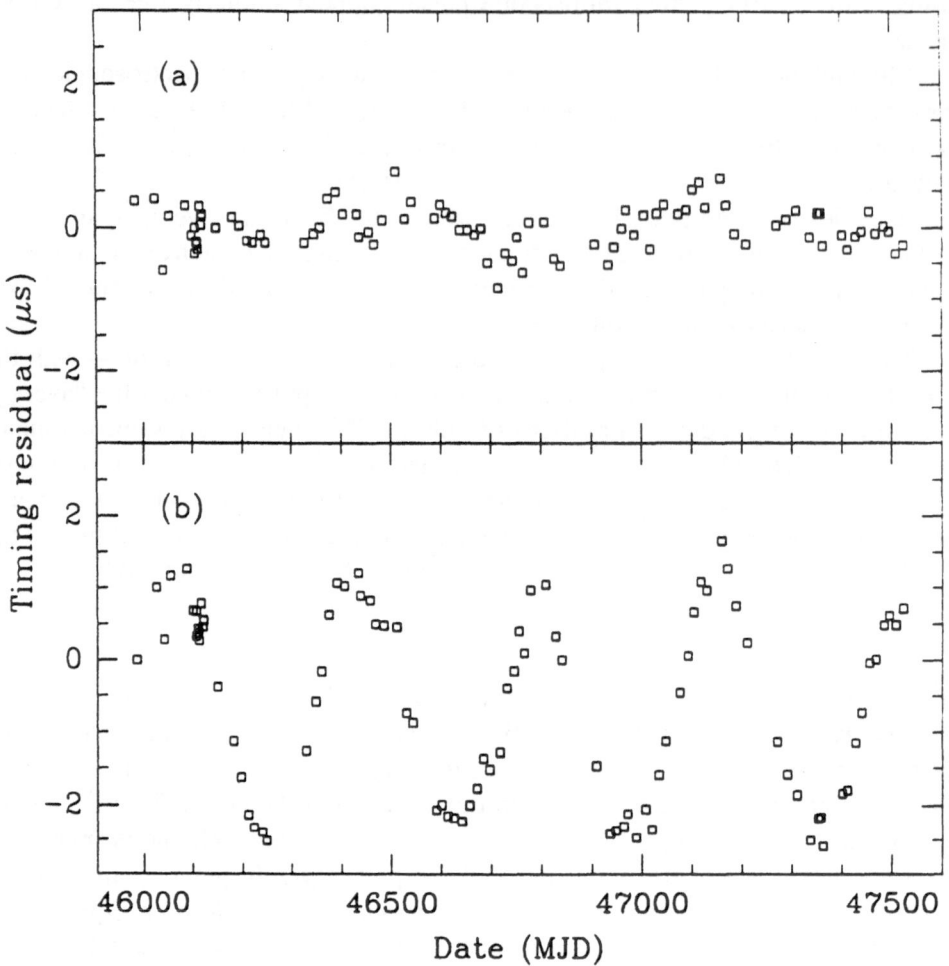

Fig. 1.   (a) Timing residuals of PSR B1937+21, after removal of an astrometric and spin-down model. (b) Timing "residuals" with a model including a 1 mas position error for the pulsar. Data are from J. Taylor (private communication).

right ascension, declination, and, if necessary, higher-order period derivatives and proper motion terms. If the pulsar has a binary companion, the model must include the projected semi-major axis of the orbit ($a \sin i$), the eccentricity (e), the orbital period (P_b), the longitude of periastron (ω), and the epoch of periastron (T_0).

Figure 1a shows the timing residuals of PSR B1937+21, a millisecond pulsar. A spin-down model was fit for P, $\dot{P}$, α, δ and proper motion, and the resulting rms scatter in the residuals is less than 1μs. For illustration we show the same data in Figure 1b reduced with a one milliarcsecond error in the assumed position of the pulsar. The position error introduces a sinusoidal oscillation in the residuals with a one-year period due to the changing geometric delay ($\mathbf{r} \cdot \hat{\mathbf{n}}/c$) as Earth moves

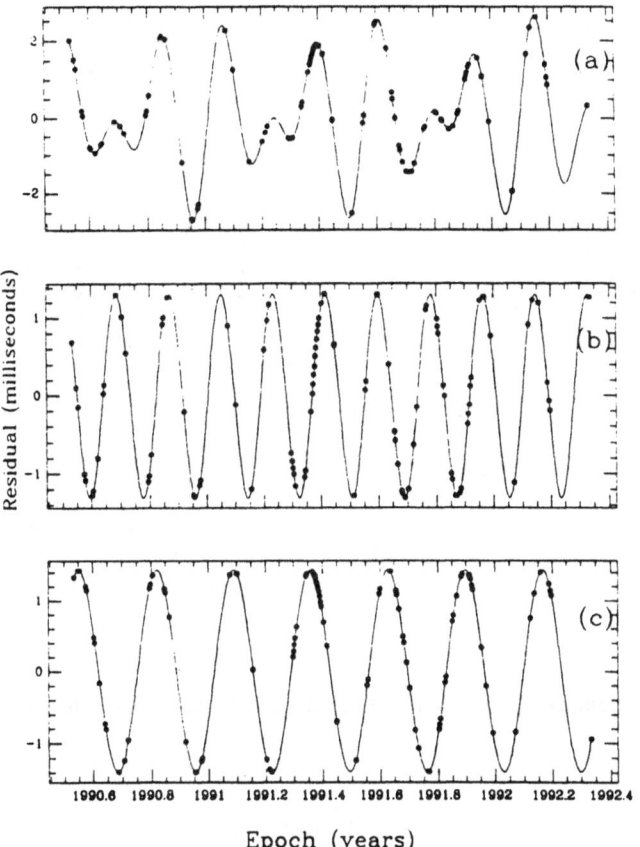

Fig. 2. (a) Timing residuals of PSR B1257+12, after removal of an astrometric and spin-down model (from Wolszczan, 1993). The data were obtained at the Arecibo Observatory. (b and c) Oscillations in part (a) can be decomposed into the sum of two near sinusoids corresponding to the Newtonian orbits of two planetary mass bodies.

around the sun. Fitting for the position of the pulsar removes the 1-year ripple but it also absorbs the signature of any object orbiting the pulsar with the same period. Timing measurements are therefore not very sensitive to orbital companions with $P_b \approx 1$ year.

It is interesting to compare the timing residuals of PSR 1937+21, an isolated pulsar, with those of PSR 1257+12 (Figure 2a). The residuals of 1257+12 show a large quasi-periodic oscillation which Wolszczan and Frail showed could be decomposed into two nearly-perfect sinusoids (Figures 2b and 2c). They proposed that these were the gravitational signatures of two planet-size objects orbiting the pulsar. A fit to the data including two non-interacting Keplerian orbits yielded the residuals shown in Figure 3 and orbital parameters for the planets in Table I.

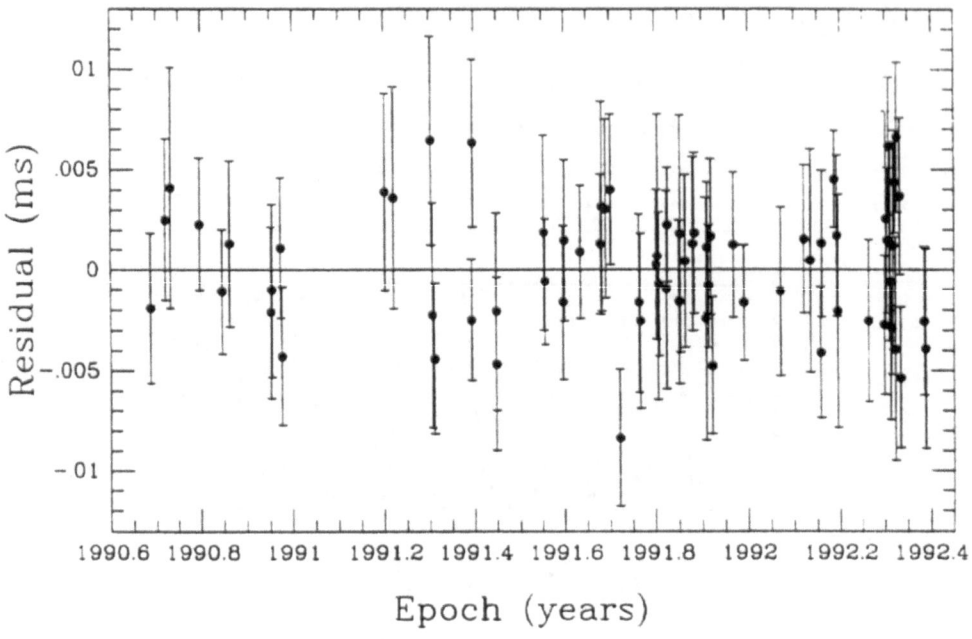

Fig. 3.   Timing residuals for PSR B1257+12 after a two planet model has been subtracted. From Wolszczan (1993).

TABLE I

Parameters of the planets orbiting PSR B1257+12 (from Wolszczan, 1993). The inclinations of the orbital planes to the line of sight, $i_{1,2}$, are unknown.

	Planet 1	Planet 2
Planet mass ($M_\oplus$)	$3.4/\sin i_1$	$2.8/\sin i_2$
Pulsar-planet distance (AU)	0.36	0.47
Orbital period (days)	66.54	98.20

3. Proving the Planet Hypothesis

To prove that Wolszczan and Frail have actually found a planetary system, it is necessary to first confirm that the double-sinusoidal oscillation is a real property of the timing residuals (that is, not an observer error or data reduction problem), and then show that the interpretation of the signal as a planetary system is correct.

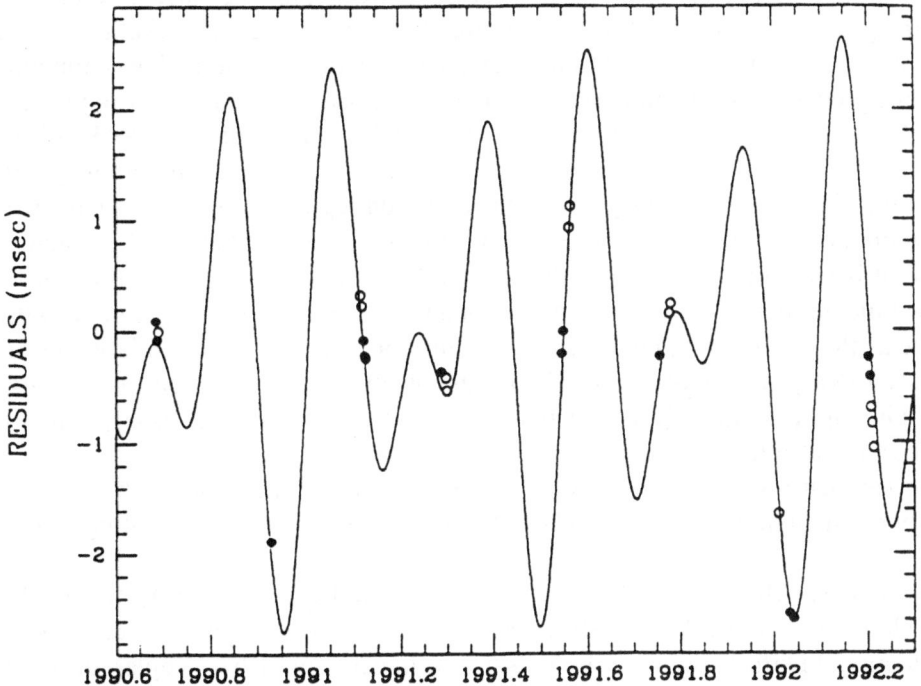

Fig. 4. Timing residuals for PSR B1257+12 from NRAO Green Bank observations (Backer, 1993). Solid points were obtained at 1330 MHz and open points at 800 MHz. The solid curve represents the predicted residuals for a two-planet model. The agreement of the planetary model with data from independent telescopes (Arecibo and Green Bank) suggests that the timing fluctuations reported by Wolszczan and Frail (1992) are real.

The first test is straightforward, and has been done by Backer (1993). Using a different telescope and observing hardware, Backer and collaborators have made timing measurements of PSR B1257+12 that confirm the double-sinusoidal signal (see Figure 4). To avoid the chance that software errors could mimic a planetary system, as in the case of PSR 1829−10 (Lyne and Bailes, 1992), the data were reduced with two independent analysis packages and with solar system ephemerides from both the Center for Astrophysics and the Jet Propulsion Laboratory (JPL). As a further test, other millisecond pulsars observed at the same epochs were shown to have no unusual timing residuals. There now seems little doubt that the signal itself is real.

The interpretation is more problematic. Several alternative explanations have been suggested. The first, and simplest, is the signal is simply random rotational fluctuations similar to those seen in many young pulsars (Cordes, 1993), which in this case just happen to be well fit by a double sinusoid. This suggestion has several problems, however: no other well-timed millisecond pulsar shows evidence for phase wander of more than about a microsecond in a decade, while PSR B1257+12's phase varies by several milliseconds over a few weeks; timing noise is stochastic, but the planetary model of Wolszczan correctly predicts future observations; and a two planet model improves the timing residuals by a factor of more than a hundred.

Free precession of the neutron star has also been suggested as a way to produce the observed timing residuals (Cordes, 1993; Dolginov and Stepinski, 1993; Gil and Jessner, 1993). While it appears possible to arrange the pulsar spin and magnetic axes, the observer's line of sight, and the emission region geometry in such a way as to produce a double sinusoidal signal when precession and nutation or a relative rotation between the core and crust are invoked, again several difficulties arise: some unidentified driving force must be found (Cordes, 1993); changes in the pulse shape as the line of sight cuts across the emission region at different latitudes are not seen (Kaspi and Wolszczan, 1993); and at times the oscillations are of large enough amplitude that the pulsar is actually *speeding up* (Rees, 1993), which is difficult to arrange with precession.

In summary, we agree with Rees (1993) that "at the moment it is wise to bet firmly on the hypothesis that 1257+12 has planetary-mass bodies orbiting around it."

Still, it is difficult to rule out all other possibilities, so it is fortunate that this claim is testable. To a fairly good approximation, it is possible to treat the gravitational perturbations on the pulsar as a linear sum of two Keplerian orbits. However, there will be mutual interactions between the two planets which cause departures from this simple model. The orbital periods of the two planets are nearly commensurate: $P_2/P_1 = 1.4758 \pm 0.004 \approx 3 : 2$. As a result, the orbital semi-major axes, eccentricities, and longitudes of periastron undergo nearly sinusoidal variations with period $(2/P_1 - 3/P_2)^{-1} \approx 5.6$ years. Also, because the semi-major axis of the orbits are similar $((a_2 - a_1)/a_2 \ll 1)$, the two planets can perturb one another significantly during individual close encounters. If the planets orbit the neutron star in the same direction the time interval between encounters will be $(1/P_1 - 1/P_2)^{-1} \approx 200$ days. If they orbit in opposite directions the interval would be ≈ 40 days.

Renu Malhotra, Fred Rasio, and their collaborators have analyzed three-body effects and concluded that they should be detectable by pulsar timing over the next few years. (Malhotra *et al.* , 1992; Rasio *et al.* , 1992; Malhotra, 1993a; Malhotra 1993b; Rasio *et al.* , 1993, also Stan Peale's contribution at this meeting). Figure 5 shows the pulse arrival time residuals due to three body interactions over a five-year period. The departures from a simple, non-interacting Keplerian model grow rapidly with time and—given the precision of arrival time measurements for 1257+12—should be detectable at a statistically significant level before the end of 1994. The detection of these three-body effects will make the planetary interpretation of Wolszczan and Frail's data nearly inescapable. Further, the accurate measurement of the perturbations, combined with analytic or numeric orbital modeling, may allow the direct determination of the planet masses and the inclinations of the orbital planes. As Wolszczan reported at this meeting, there is already some

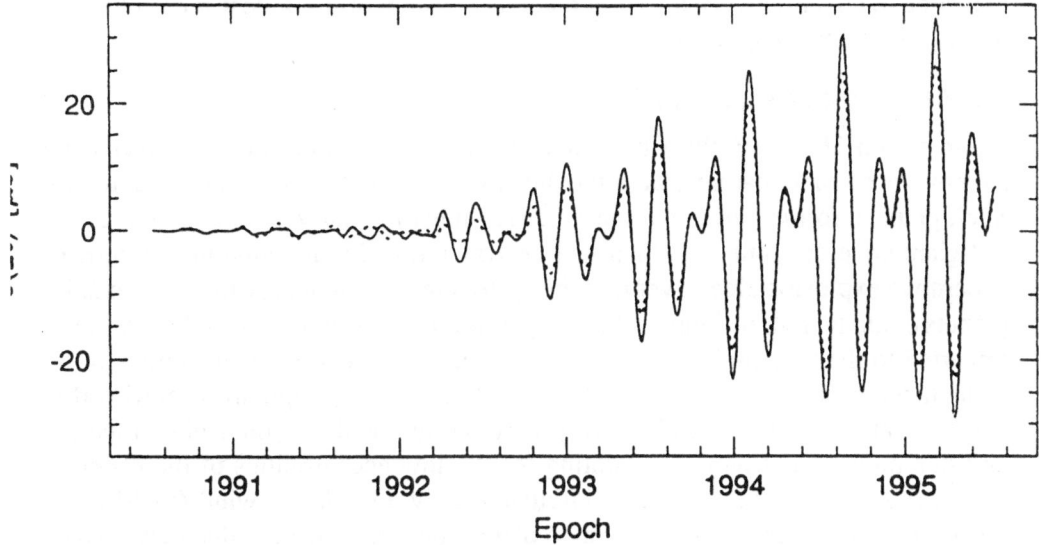

Fig. 5. Pulse arrival time residuals due to mutual interactions of the planets, from Malhotra (1993b). The solid line shows the residuals predicted by Malhotra's analytical solution for the case of co-planar, edge-on orbits. The dotted line shows the same predicted by a full numerical integration of the three-body equations. The initial conditions are those deduced from the best fit orbital parameters of Wolszczan and Frail (1992).

evidence for the synodic timing fluctuations. There is every reason to hope that observations within the next few months or years will settle the question of the reality of the planets orbiting PSR B1257+12 once and for all.

4. Creating Planets Around Pulsars

Assuming for the present that the planets orbiting 1257+12 are real, it is natural to ask how they were formed. In fact, this question of origin is crucial to the ultimate importance of the discovery, since planetary scientists are primarily interested in the origin of planetary systems like our own solar system. Were the planets formed in a way peculiar to their pulsar primary, or did their formation parallel the formation of a more "typical" planetary system enough that useful clues to generic planet formation can be gleaned from their masses and orbits?

Planet production processes have been reviewed in detail by Phinney and Hansen (1993) and Podsiadlowski (1993). They divide into two broad classes. In a few, the planets are formed around a "normal" star, either the pulsar progenitor or other, and then either survive the supernova explosion or are captured by the pulsar. In most scenarios, the planets actually form after the neutron star. Some models were introduced to explain the since retracted "discovery" of a planet orbiting the normal pulsar PSR B1829−10. Although many of these cannot easily be extended to cover the very different planetary system orbiting PSR B1257+12, we include them for completeness.

4.1. SALAMANDER SCENARIOS

Phinney and Hansen (1993) have named models in which planets survive the supernova of their parent star after the mythical salamander that lived in fire. Such planets, formed around a massive star, must first survive destruction by the red-giant envelope, but even then will be gravitationally unbound in a symmetric supernova explosion since over half the system mass is suddenly lost. Pre-existing planets could be retained either if the explosion is non-symmetric and the velocity imparted to the resulting neutron star is matched in direction or magnitude to the instantaneous planetary velocities (Hills, 1983), or if they begin in eccentric orbits (Postnov and Prokhorov, 1992). Only very rarely could two planets be retained, and this model has trouble accounting for the low eccentricities of the observed orbits, though the orbits could be circularized by an ablative wind (Field *et al.*, 1992). The most important objection to planet survival models is that they provide no mechanism for spinning the resulting pulsar up to millisecond periods.

This latter objection is met if the planets are captured in an exchange interaction between a millisecond pulsar and a normal star, but such interactions will be rare and will leave the planets in eccentric orbits which must be circularized.

Another possibility is that the planets were formed originally around a binary system, which survived the first supernova explosion because less than half the mass was lost. Wijers *et al.* (1992) have introduced a scenario in which the resulting neutron star then spirals into its companion, forming a Thorne-Zytkow object. Mass accretion during this phase spins the pulsar to millisecond periods, while the companion's wind circularizes the planetary orbits.

A variant of the salamander scenarios is one in which a companion star survives the supernova of the primary, but is then reduced to a "planet" by the evaporative pulsar wind (Bailes *et al.*, 1991; Krolik, 1991; Rasio *et al.*, 1992). There are at least two millisecond pulsars that appear to be ablating low-mass companions (Fruchter *et al.*, 1990; Lyne *et al.*, 1990), but it is difficult to stop the ablation in time to save a planetary mass remnant, and this model cannot create systems like PSR B1257+12 with more than one companion.

4.2. MEMNONIDES SCENARIOS

Most models that have been suggested for the creation of planets around a pulsar after the supernova explosion first create a disk around the pulsar, and then use standard theory (Levy, 1993; Ruden, 1993; Lissauer and Stewart, 1993; Strom and Edwards, 1993) to produce the planets themselves. This led Phinney and Hansen (1993) to name these scenarios after the Memnonides, birds which annually rose from the ashes of the warrior Memnon to circle his funeral pyre. Unlike the disk that arose from the solar nebula, the disks in the Memnonides scenarios tend to have fairly low mass and low angular momentum. A typical problem that these theories must overcome is the efficient outward transfer of mass through the disk (Phinney and Hansen, 1993). Because of the similarities of these models, and the fact that they have been reviewed in detail elsewhere (Phinney and Hansen, 1993; Podsiadlowski, 1993), we merely summarize them here.

There are two promising sources for material to form a protoplanetary disk around a neutron star: the outer regions of the progenitor star, and a binary companion. Although most of the original massive star is lost in the supernova explosion, it is possible that a small fraction of the material fell back into a close (~ 1000 km) disk orbiting the neutron star, expanding by viscous forces eventually into an AU-sized disk from which planets formed (Lin et al., 1991). This model does not produce a millisecond pulsar, however, so it is not a likely explanation for PSR B1257+12.

Because the spin-up of the pulsar is thought to require long-duration mass transfer from a companion star, it is more promising to look at models where the companion star is disrupted after spin-up to form a protoplanetary disk. Tavani and Brookshaw (1992), and Banit et al. (1993) have suggested models based on the complete evaporation of the companion by the pulsar wind. The ablated material forms an excretion disk around the binary system, from which planets form. It is interesting to note, however, that neither of the known ablating binaries have formed planets from their putative excretion disks with mass greater than $0.01 - 0.1 M_{\oplus}$, despite the fact that they both have orbital evolutionary timescales greater than ~ 30 Myr and have therefore probably been ablating systems for at least that long (Nice and Thorsett, 1993). If in some systems mass transfer continues to late stages, it may become unstable when the companion mass has been reduced to a few hundredths of a solar mass, and the final core of the companion may be disrupted into a disk (Ruderman and Shaham, 1983; Tutukov, 1991).

A more violent variation of this scenario occurs if the thermal timescale for the companion exceeds the timescale for radiation-driven mass loss (Stevens et al., 1992). In certain circumstances, a fast millisecond pulsar can induce rapid and complete destruction of a low-mass ($\sim 0.5 M_{\odot}$) companion into a circumstellar disk.

Van den Heuvel (1992) has suggested another variation. If the pulsar's companion is a white dwarf rather than main sequence star, then orbital decay due to gravitational radiation will eventually bring the white dwarf into contact with its Roche surface, and it will start losing mass. Because of the inverse mass-radius relation for degenerate dwarfs, this mass transfer is unstable and the companion is rapidly reduced into a massive circumstellar disk.

Other scenarios include the merger of two white dwarfs to form a neutron star and disk (Podsiadlowski et al., 1991); the creation of an accretion disk around the pulsar from a massive companion that later explodes (Fabian and Podsiadlowski, 1991); and the tidal destruction of the companion when an asymmetric supernova kick drives the nascent neutron star near or through it (C. Thompson, as quoted by Phinney and Hansen).

5. The Search Continues

The goal of most workers studying planet formation is to understand how planets are created around main sequence stars like our sun. Planets around exotic objects like pulsars are definitely outside the mainstream. Nevertheless, there are several good reasons for the community at large to take an interest in pulsar planets. A few are enumerated below.

1. *First known extrasolar planetary system.* Most astronomers believe that planets are common around main sequence stars but, despite considerable effort to detect such planets, PSR B1257+12 is the first known planetary system other than our own.

2. *Universal patterns in planet formation?* There are a number of regularities in our solar system that hold important clues to the planet forming process: planetary orbits are nearly coplanar; planets orbit the sun in the same direction; most rotate in the same counter-clockwise sense; massive planets are farther from the sun than the less-massive terrestrial planets, etc. By identifying similar regularities in pulsar planetary systems we will better understand which are fundamental signatures of the planet formation mechanism.

3. *Planetary orbital dynamics.* Because of the high precision radial velocity measurements possible through pulsar timing, it will be straightforward to conduct studies of orbital dynamics that would be impossible with ordinary single-line spectroscopic binaries. The mutual interactions described in section 3 are a good example.

4. *Pulsar evolution.* The unexpected detection of planets around a millisecond pulsar has important implications for evolutionary scenarios of recycled pulsars.

However interesting PSR B1257+12 and its planetary system may be, some may argue that it remains little more than a curiosity until other, similar systems are found. It is important for pulsar astronomers to carefully study the detectability of planets orbiting pulsars, if only to allow statistical arguments about the probability for creating them. This has not yet been carefully done.

There have been, however, several unsuccessful searches for planets orbiting other pulsars (Lamb and Lamb, 1976; Bailes *et al.*, 1993; Thorsett *et al.*, 1993). Because most of the targets are "normal" young pulsars, not millisecond pulsars like PSR B1257+12, the resulting null results are difficult to interpret; it is, however, clear that in this class of pulsars, planetary systems are unusual. There is some hint of an 1100-day periodicity in the timing data for PSR B0329+54 (Demiański and Prószyński 1979; Bailes *et al.*, 1993, though note Cordes and Downs, 1985), a pulsar which also shows significant random phase wander; many further observations are needed to confirm this result.

Far fewer millisecond pulsars are known than their slower cousins. The majority have binary companions, and most are found (largely because of observational selection effects) in globular clusters. Lacking a full theory of pulsar-planet formation, it is hard to predict whether the binary pulsars will have planets in wide orbits around the binary system or in tight orbits around the pulsar or its companion. (Note that planets orbiting a pulsar's companions are in principle detectable through three body interactions or resonance effects.) As mentioned above, there are no Earth-mass planets orbiting the two windy binary pulsars PSRs B1957+20 and B1744−24 (Nice and Thorsett, 1993), but while many other binary systems could be checked for planets, no systematic survey has been done.

Globular cluster environments introduce new problems and possibilities for pulsar-planetary systems (Sigurdsson, 1992, 1993). The high stellar density makes star–neutron-star collisions more likely, and the subsequent destruction of the star by the neutron star might provide material for a proto-planetary disk. Also, pulsars may "scavenge" planets from other systems. However, planets in all but the tightest orbits may be easily ionized from the pulsar by close encounters with other stars. One interesting observation has been of a significant acceleration-derivative (or jerk) on the binary pulsar PSR B1620−26, in the globular cluster M4 (Backer, 1993). This can be attributed to a massive ($\gtrsim 50 M_\oplus$) planet in a wide ($\gtrsim 10$ yr) orbit or to a close encounter by another cluster star. Continued observations over several more years will distinguish between these possibilities.

The only known millisecond pulsars in the galactic disk without stellar mass companions are PSRs B1257+12, B1937+21, and J2322+2057. The first has two planetary companions, the second has no planets with mass greater than $0.01 M_\oplus$ and orbital period less than several years (Thorsett and Phillips, 1992) (see Figure 6), and the third has no planets with mass greater than $0.1 M_\oplus$ and periods less than a few hundred days (Nice, 1992). The jury is therefore still out on the (im)probability of forming planets around millisecond pulsars. Several all-sky surveys for fast pulsars are currently underway, together with deeper searches at the Arecibo Observatory. Beginning in 1995, the Green Bank Telescope will be a superb instrument for very sensitive all-sky surveys. The discovery of more objects like PSR B1257+12 will turn pulsar-planetary systems into exciting laboratories for the study of planet formation and planetary dynamics.

Acknowledgements

We thank all the participants in the *Planets around Pulsars* meeting at the California Institute of Technology, especially those whose review and research papers made this report possible.

References

Backer, D. C.: 1993, in *Planets around Pulsars*, eds. J. A. Phillips, S. E. Thorsett and S. R. Kulkarni, Astron. Soc. Pac. Conf. Ser. Vol. 36, 11–18.
Bailes, M., Lyne, A.G. and Shemar, S.L.: 1991, *Nature*, **352**, 311–313.
Bailes, M., Lyne, A.G. and Shemar, S.L.: 1993, in: J.A. Phillips, S.E. Thorsett and S.R. Kulkarni (eds.), *Planets around Pulsars*, Astron. Soc. Pac. Conf. Ser. Vol. 36, 19–30.

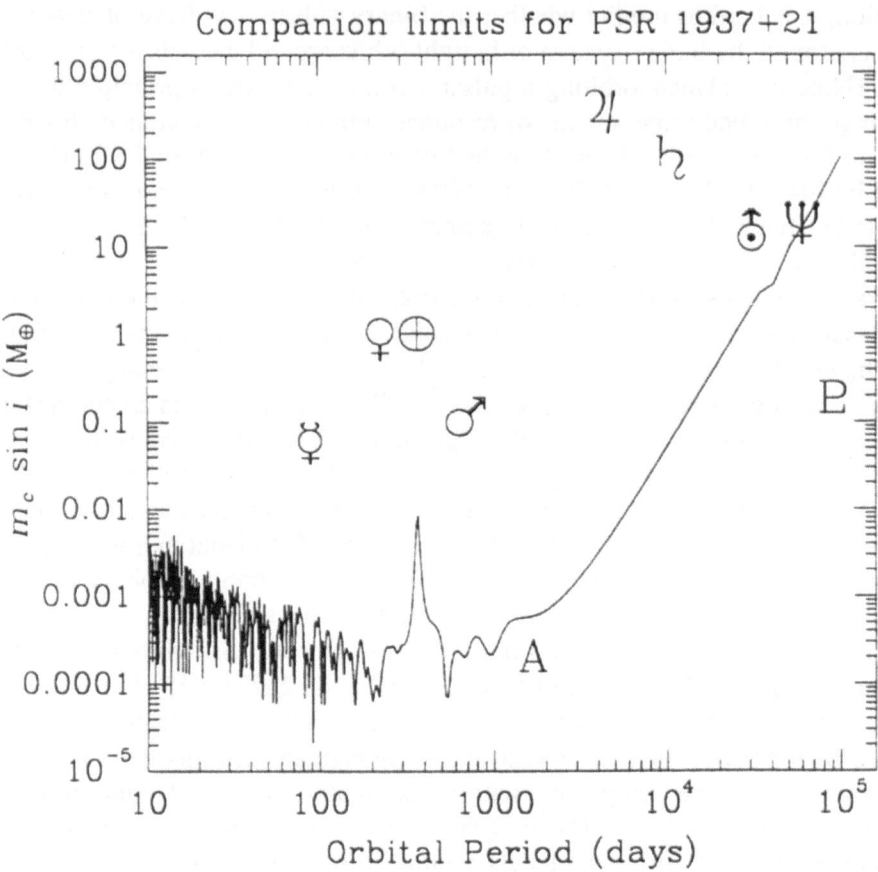

Fig. 6.   Limits on the mass of any planets around the millisecond pulsar PSR B1937+21 from 6.5 years of timing data (Thorsett and Phillips, 1992). The region of phase space above the indicated line is excluded. The planets of our solar system are indicated by their standard symbols, while the symbol 'A' indicates approximately the period and mass of the largest asteroids. A 1.4 $M_\odot$ neutron star mass is assumed. As discussed in the text, the loss of sensitivity near periods of 1 yr is due to the fit for the (unknown) pulsar position. The sensitivity roll-off at long periods is due to the fit for pulsar period and period derivative combined with the finite length of the data set.

Banit, M., Ruderman, M. and Shaham, J.: 1993, in: J.A. Phillips, S.E. Thorsett and S.R. Kulkarni (eds.), *Planets around Pulsars*, , Astron. Soc. Pac. Conf. Ser. Vol. 36, 167–172.

Banit, M., Ruderman, M., Shaham, J. and Applegate, J.H.: 1993, preprint.

Cochran, W.D. and Hatzes, A.P.: 1993, in: J.A. Phillips, S.E. Thorsett and S.R. Kulkarni (eds.), *Planets around Pulsars*, Astron. Soc. Pac. Conf. Ser. Vol. 36, 267–273.

Cordes, J.M. 1993, in: J.A. Phillips, S.E. Thorsett and S.R. Kulkarni (eds.), *Planets around Pulsars*, Astron. Soc. Pac. Conf. Ser. Vol. 36, 43–60.

Cordes, J.M. and Downs, G.S.: 1985, *Astrophys. J. Supp. Series*, **59**, 343–382.

Demiański, M. and Prószyński, M.: 1979, *Nature*, **282**, 383–385.

Dolginov, A.Z. and Stepinski, T.F. 1993, in: J.A. Phillips, S.E. Thorsett and S.R. Kulkarni (eds.), *Planets around Pulsars*, Astron. Soc. Pac. Conf. Ser. Vol. 36, 61–71.

Fabian, A.C. and Podsiadlowski, P.: 1991, *Nature*, **353**, 801.

Field, R.G., Roger, R.D. and Rybicki, G.B.: 1992, preprint.
Fruchter, A.S. *et al.* : 1990, *Astrophys. J.*, **351**, 642–650.
Gil, J.A. and Jessner, A.: 1993, in J.A. Phillips, S.E. Thorsett and S.R. Kulkarni (eds.), *Planets around Pulsars*, Astron. Soc. Pac. Conf. Ser. Vol. 36, 71–79.
Hills, J.G.: 1983, *Astrophys. J.*, **267**, 322–333.
Kaspi, V.M. and Wolszczan, A.: 1993, in: J.A. Phillips, S.E. Thorsett and S.R. Kulkarni (eds.), *Planets around Pulsars*, Astron. Soc. Pac. Conf. Ser. Vol. 36, 81–85.
Krolik, J.H.: 1991, *Nature*, **353**, 829–831.
Lamb, D.Q. and Lamb, F.K.: 1976, *Astrophys. J.*, **204**, 168–186.
Levy, E.H. 1993, in: J.A. Phillips, S.E. Thorsett and S.R. Kulkarni (eds.), *Planets around Pulsars*, Astron. Soc. Pac. Conf. Ser. Vol. 36, 181–195.
Lin, D.N.C., Woosley, S.E. and Bodenheimer, P.H.: 1991, *Nature*, **353**, 827–829.
Lissauer, J.J. and Stewart, G.R.: 1993, in: J.A. Phillips, S.E. Thorsett and S.R. Kulkarni (eds.), *Planets around Pulsars*, Astron. Soc. Pac. Conf. Ser. Vol. 36, 217–233.
Lyne, A.G. and Bailes, M.: 1992, *Nature*, **355**, 213.
Lyne, A.G. *et al.* : 1990, *Nature*, **347**, 650–652.
Lyne, A.G. and Smith, F.G.: 1990, *Pulsar Astronomy*, (Cambridge: Cambridge University Press).
Malhotra, R. 1993a, in: *Planets around Pulsars*, Astron. Soc. Pac. Conf. Ser. Vol. 36, 89–106.
Malhotra, R.: 1993b, preprint.
Malhotra, R., Black, D., Eck, A. and Jackson, A.: 1992, *Nature*, **356**, 583–585.
Manchester, R.N. and Taylor, J.H.: 1977, *Pulsars*, (San Francisco: Freeman).
Nice, D.J.: 1992. PhD thesis, Princeton University.
Nice, D.J. and Thorsett, S.E.: 1993, in: J.A. Phillips, S.E. Thorsett and S.R. Kulkarni (eds.), *Planets around Pulsars*, Astron. Soc. Pac. Conf. Ser. Vol. 36, 289–293.
Phillips, J.A., Thorsett, S.E. and Kulkarni, S.R.: 1993, *Planets around Pulsars*, (San Francisco: Astron. Soc. Pac.). Conf. Ser. Vol. 36.
Phinney, E.S. and Hansen, B.M.S.: 1993, in: J.A. Phillips, S.E. Thorsett and S.R. Kulkarni (eds.), *Planets around Pulsars*, Astron. Soc. Pac. Conf. Ser. Vol. 36, 371–390.
Podsiadlowski, P.: 1993, in: J.A. Phillips, S.E. Thorsett and S.R. Kulkarni (eds.), *Planets around Pulsars*, Astron. Soc. Pac. Conf. Ser. Vol. 36, 149–165.
Podsiadlowski, P., Pringle, J.E. and Rees, M.J.: 1991, *Nature*, **352**, 783–784.
Postnov, K.A. and Prokhorov, M.E.: 1992, *Astr. Astrophys.*, **258**, L17–L18.
Rasio, F.A., Nicholson, P.D., Shapiro, S.L. and Teukolsky, S.A.: 1992, *Nature*, **355**, 325–326.
Rasio, F.A., Nicholson, P.D., Shapiro, S.L. and Teukolsky, S.A.: 1993, in J.A. Phillips, S.E. Thorsett and S.R. Kulkarni (eds.), *Planets around Pulsars*, Astron. Soc. Pac. Conf. Ser. Vol. 36, 107–119.
Rasio, F.A., Shapiro, S.L. and Teukolsky, S.A.: 1992, *Astron. Astrophys.*, **256**, L35–L37.
Rees, M.J.: 1993, in: J.A. Phillips, S.E. Thorsett and S.R. Kulkarni (eds.), *Planets around Pulsars*, Astron. Soc. Pac. Conf. Ser. Vol. 36, 367–370.
Ruden, S.P.: 1993, in: J.A. Phillips, S.E. Thorsett and S.R. Kulkarni (eds.), *Planets around Pulsars*, Astron. Soc. Pac. Conf. Ser. Vol. 36, 107–119.
Ruderman, M. and Shaham, J.: 1983, *Nature*, **304**, 425–427.
Sigurdsson, S.: 1992, *Astrophys. J.*, **399**, L95–L97.
Sigurdsson, S.: 1993, in: J.A. Phillips, S.E. Thorsett and S.R. Kulkarni (eds.), *Planets around Pulsars*, Astron. Soc. Pac. Conf. Ser. Vol. 36, 173–179.
Stevens, I.R., Rees, M.J. and Podsiadlowski, P.: 1992, *Mon. Not. R. astr. Soc.*, **254**, P19–P22.
Stinebring, D.R., Ryba, M.F., Taylor, J.H. and Romani, R.W.: 1990, *Phys. Rev. Lett.*, **65**, 285–288.
Strom, S.E. and Edwards, S.: 1993, in: J.A. Phillips, S.E. Thorsett and S.R. Kulkarni (eds.), *Planets around Pulsars*, Astron. Soc. Pac. Conf. Ser. Vol. 36, 235–256.
Tavani, M. and Brookshaw, L.: 1992, *Nature*, **356**, 320–322.
Thorsett, S.E. and Phillips, J.A.: 1992, *Astrophys. J.*, **387**, L69–L71.
Thorsett, S.E., Phillips, J.A. and Cordes, J.M. 1993, in: J.A. Phillips, S.E. Thorsett and S.R. Kulkarni (eds.), *Planets around Pulsars*, Astron. Soc. Pac. Conf. Ser. Vol. 36, 31–39.
Tutukov, A.V.: 1991, *Sov. Astron.*, **35**, 415–417.
van den Heuvel, E.P.J. 1992, *Nature*, **356**, 668.

Wijers, R.A.M.J., van den Heuvel, E.P.J., van Kerkwijk, M.H. and Bhattacharya, D.: 1992, *Nature*, **355**, 593.

Wolszczan, A.: 1993, in: J.A. Phillips, S.E. Thorsett and S.R. Kulkarni (eds.), *Planets around Pulsars*, Astron. Soc. Pac. Conf. Ser. Vol. 36, 3.

Wolszczan, A. and Frail, D.A.: 1992, *Nature*, **355**, 145.

DETECTION OF ACCRETING CIRCUMSTELLAR GAS AROUND
WEAK EMISSION-LINE HERBIG AE/BE STARS *

C. A. GRADY**

Applied Research Corporation, Landover Maryland, USA

M. R. PÉREZ**

*Science Programs, Computer Sciences Corporation and IUE Observatory,
Greenbelt, Maryland, USA*

and

P. S. THÉ

*Astronomical Institute "Anton Pannekoek," University of Amsterdam,
Amsterdam, The Netherlands*

Abstract. We present archival and recent IUE high dispersion spectra of late B stars which reveal the presence of accreting gas with velocities as high as 350 km s^{-1}, collisional ionization of the accreting gas to temperatures above the stellar T_{eff}, and column densities intermediate between those observed toward classical Herbig Ae/Be stars and the nearby proto-planetary system β Pictoris. One of the stars, HD 176386, while lacking obvious optical signatures of youth, is a member of the R CrA star formation region, and with an inferred age of 2.8 Myr has not yet arrived on the zero-age main sequence (ZAMS). The other object, an isolated, field B star with pronounced IR excess due to warm, circumstellar dust, 51 Oph, exhibits only modest Hα emission. The combination of high velocity, accreting gas in systems with IR excesses due to circumstellar dust suggests that not only are these objects candidate proto-planetary systems, but that they may represent an extension to higher stellar masses of the weak-emission pre-main sequence (PMS) stars.

1. Introduction

Since the discovery of a large circumstellar disk surrounding the nearby A star, β Pictoris, some of the most important questions raised have been the age of the system, its relation to phenomena seen in pre-main sequence objects (PMS), and the time scale for the evolution of the circumstellar envelope to something resembling the β Pic system. Recent advances in the study of 2–10 $M_\odot$ PMS stars, the Herbig Ae/Be stars, suggest that these objects are surrounded by large, optically thick, viscously heated accretion disks (Hillenbrand *et al.*, 1992), which apparently become optically thin on time scales of a few $\times 10^5$ years (early B stars) to 1.5 Myr (late A stars) (Strom *et al.*, 1993).

Detailed studies of 2 Herbig Ae/Be systems, HR 5999 (Pérez *et al.*, 1993a,b) and HD 45677 (Grady *et al.*, 1992, 1993b), which are fortuitously oriented with the disk edge-on to our line of sight, demonstrate that the inner, optically thin cavities needed to fit the IR data are filled by optically thick, accreting gas with behavior similar to β Pic, although with substantially higher column densities in both the

---

* Paper presented at the Conference on *Planetary Systems: Formation, Evolution, and Detection* held 7–10 December, 1992 at CalTech, Pasadena, California, U.S.A.
** Guest Observer, IUE Observatory, NASA/GSFC

Astrophysics and Space Science **212**: 107–114, 1994.
© 1994 *Kluwer Academic Publishers*.

TABLE I

HD 176386 and 51 Oph.

Star	Spectral Type	$v \sin i$	V_{max}
HD 176386	B9.5 IV	180	300
51 Oph	B9.5 IVe	220	100-150:

high velocity gas, and in the material seen near the system radial velocity. These data suggest that the β Pic system is comparatively young.

Due to the lack of stellar age indicators for mid-A stars, the suggestion of comparative youth for the β Pic system cannot be directly tested with additional observations of that object. Thus, it becomes important to explore the evolution of proto-planetary systems as the mass accretion rates begin to decline and the star approaches the ZAMS, with stars of similar orientation to β Pic, and known ages. A natural place to search for such systems is in areas of on-going or comparatively recent star formation with known distances, permitting accurate estimation of the stellar luminosity, and hence its location in the HR diagram and thence the age of the star.

One such star, which from its luminosity is still slightly above the ZAMS, and is lacking in obvious optical indicators of a high accretion rate such as strong $(W_{H\alpha} \geq 5 \text{ Å})$ Hα emission, is HD 176386 (B9.5 IV, Grady et al., 1993a; $v \sin i = 180 \text{ km s}^{-1}$, Bibo et al., 1992, see Table I). With its location in the R CrA star formation region and its inferred age of 2.8 Myr, this object may be the higher mass analog of the weak-emission line T Tau stars (Walter, 1986), or if we adopt a parallel nomenclature, a weak-emission line Herbig Ae star. A very similar, but isolated, object of unknown age is 51 Oph (B9.5 Ve, $v \sin i = 220 \text{ km s}^{-1}$, Slettebak, 1982) which IR (Waters et al., 1988; Dougherty et al., 1991; Farjado-Acosta et al., 1992) and UV studies (Grady et al., 1991; Grady and Silvis, 1993) have strongly suggested is another field, proto-planetary system similar to β Pic.

2. Detection of Accreting Gas

Both HD 176386 and 51 Oph are sufficiently bright to permit high dispersion (R = 20,000) UV spectral observations with the IUE. As the brighter object, 51 Oph has been more extensively studied (Grady et al., 1990; Grady and Silvis, 1993) and is known to exhibit variable column densities of accreting gas, similar to β Pic. The more recently identified system, HD 176386, has been observed on 3 dates spanning 5 years, most recently in September 1992 (see Grady et al., 1993a

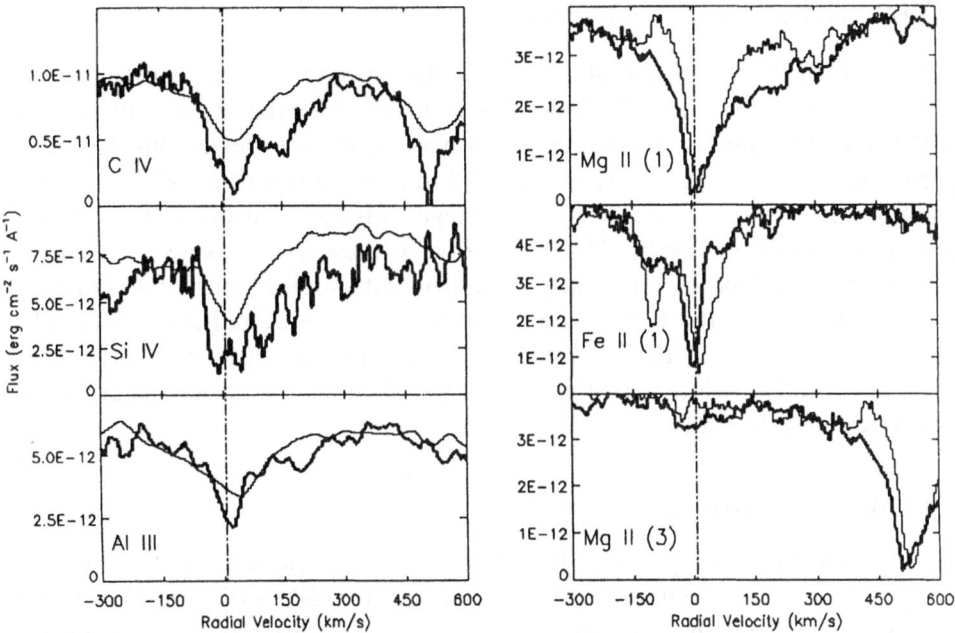

Fig. 1. Accreting gas toward HD 176386 (bold) and 51 Oph (light), in C IV 1548, Si IV 1393, Al III 1854, Mg II (1) 2795, Fe II (1) 2599, and Mg II (3) 2791 Å. The HD 176386 data are from September 23, 1992. The 51 Oph spectra have been chosen to exhibit the highest line-of-sight accretion and are from May 4, 1987, for C IV-Al III and March 12, 1989, for Mg II and Fe II).

for a more detailed discussion). The UV spectra of these proto-planetary system candidates share a number of features.

2.1. HIGH VELOCITY, ACCRETING GAS

Only modest indicators of accreting gas are visible in the lower oscillator strength transitions of singly ionized species such as Fe II, Si II, Al II, and C II. The Mg II (1) lines in both stars provide unambiguous evidence for the presence of excess absorption on the long wavelength side of the transitions extending to 90–100 and 220 km s^{-1} for 51 Oph and HD 176386, respectively. In the case of 51 Oph (Grady and Silvis, 1993) the larger number of IUE observations demonstrates that the accreting gas is variable in column density, much as has been observed for β Pic. Transitions of higher ionization stages, such as the resonance doublets of Al III, Si IV, and C IV show much more pronounced profile asymmetries to positive radial velocities, with absorption in 51 Oph followed to +100–150 km s^{-1} and for HD 176386 to 300–350 km s^{-1} (Figure 1). For HD 176386, the September 1992 data show the C IV column density, which traces the highest ionization plasma observed toward this star, peaking at +30 km s^{-1}, and there is a suggestion of a secondary maximum in the column density at 150 km s^{-1}.

2.2. COLLISIONAL IONIZATION OF THE ACCRETING GAS

Detection of C IV and Si IV absorption in the spectra of very late B stars is unexpected, as is detection of absorption with the observed equivalent widths. At B9.5, no photospheric contribution to either species is expected, nor are main sequence stars of this spectral type expected to have sufficient FUV/EUV fluxes to photoionize carbon or silicon to these stages anywhere in the vicinity of the star (Bruhweiler *et al.*, 1989). Thus, detection of these species suggests that the circumstellar envelope of both stars is collisionally ionized to temperatures in excess of those expected from equilibrium with the stellar radiation field. The positive displacement of the largest apparent column densities of both species relative to the stellar velocities suggests that the collisional ionization occurs as the gas accretes onto the star.

2.3. COMPARISON WITH β PIC

The circumstellar envelope of β Pic displays similar signatures of collisional ionization, with the detection of Al III extending to +300–400 km s^{-1}, and the Al III apparent column density typically peaking near 60–80 km s^{-1} (heliocentric), a displacement of 40–60 km s^{-1} relative to the stellar radial velocity (Lagrange-Henri *et al.*, 1987). Collisional ionization of accreting gas toward the two edge-on Herbig Ae/Be stars studied in detail to date is also observed, again with the accreting gas being visible to progressively higher velocities as the ionization potential of the species increases. The available data suggest that collisional ionization of the accreting gas may be a characteristic feature of all massive, proto-planetary systems beginning at comparatively young ages ($1–5 \times 10^5$ years), and continuing for at least several Myr.

2.4. THE LOW VELOCITY GAS

The stars HD 176386, 51 Oph, and β Pic, as well as the classical Herbig Ae/Be stars, have UV and optical spectra with strong low velocity absorption features which are present in all available IUE observations, independent of the column density of high velocity gas. These features are present not only in the lower ionization species, but also in species indicating collisional ionization of the gas. Detection of such species at these velocities suggests that this material, like the high velocity gas is produced comparatively close to the star. In the case of the Herbig Ae/Be stars, particularly, HD 45677, the absorption is well-resolved in the IUE high dispersion data. Optically thin transitions from Fe II and Si II in that system exhibit the double-absorption profile previously seen in FUORs, and interpreted by Hartmann and Kenyon (1985) as detection of rotationally broadened absorption produced by gas in Keplerian orbit in an optically thick disk-photosphere, rather than in the more conventional spheroidal stellar atmosphere. In such a model, progressive broadening of the profile or increasing separation of the absorption peaks, if the ion is confined to a comparatively thin annulus, with increasing ionization of the gas is expected, and in fact, observed in HD 45677 (Grady *et al.*, 1992, 1993b).

Neither 51 Oph nor β Pic exhibit such resolved profiles at the resolution of the IUE in the singly ionized gas. However, higher resolution (R = 100,000) optical spectra have revealed a trend of progressively increasing profile width in the low velocity Na I and Ca II absorption (Hobbs *et al.*, 1985; Vidal-Madjar *et al.*, 1986) for β Pic. HST/GHRS observations of Mg I, Fe I, Mn II, and Fe II confirm this observation and suggest that the circumstellar disk is marginally spectrally resolved with a resolution $R = 100,000$ (Bruhweiler *et al.*, 1993).

Na I and Ca II data for 51 Oph, presented by Lagrange-Henri *et al.* (1990), show three well-separated components in Ca II at -37, -21, and -16 km s^{-1}. Similar, although narrower absorption is seen in Na I with a broad absorption profile with central sharp component at -22 km s^{-1}. It is tempting to identify the two flanking features in Ca II as gas in Keplerian orbit about that star, but confirmation will require similar resolution spectra of additional species, such as can be provided by HST GHRS data. The Al III, Si IV, and C IV profiles for 51 Oph, are substantially broader than the lower ionization gas profiles. In particular, Al III and Si IV have absorption profiles with rather broad, unsaturated absorption minima suggestive of similar features in the FUORs and the Herbig Ae/Be stars. If supported by high S/N and higher resolution UV spectra, these data suggest that the low velocity absorption in these field systems also originates within a few tens of stellar radii of the photosphere.

The HD 176386 low velocity absorption profiles, while blended on the long wavelength side with higher velocity, accreting gas, are similar to the 51 Oph profiles in exhibiting progressive broadening with increasing ionization potential. The velocity half-widths for the profiles are intermediate between those seen in β Pic and in the younger, Herbig Ae/Be stars.

3. Comparison of the IR Excesses

Waters, Coté, and Geballe (1988) first noted the striking IR excess from multi-temperature circumstellar dust for 51 Oph. Overall, the excess closely resembles the IR excesses seen in the Herbig Ae/Be stars as first noted by Hillenbrand *et al.* (1992) and is distinctly different from other field, classical Be stars which typically have modest IR excesses consistent with free-free emission. The turnover in 51 Oph's IR excess suggests the presence of an interior, optically thin cavity, much as is found in the younger, Herbig Ae/Be stars. The low inferred accretion rate ($< 10^{-7}$ $M_\odot$ yr^{-1}, from comparison with Figure 6 of Hillenbrand *et al.*, 1992) is consistent with the modest column density of high velocity, accreting gas observed with IUE.

HD 176386 presents a somewhat different picture in the IR. Despite a line-of-sight high velocity, accreting gas column density which is intermediate between the Herbig Ae/Be stars and 51 Oph or β Pic, there is no evidence for optically thick, warm circumstellar dust such as is seen toward 51 Oph and the majority of PMS Herbig Ae/Be stars. Instead, the IR excess, as presented in Bibo *et al.* (1992), shows an excess only beginning at 10 μm, and suggestive of contamination from surrounding nebulosity and the nearby binary, TY CrA (HD 176386 lies 59″ SW of TY CrA), with whom HD 176386 shares a common proper motion. It is interesting to note that TY CrA has a similar IR excess shortward of 25 μm (Hillenbrand *et al.*, 1992; Casey *et al.*, 1993). The lack of warm circumstellar dust associated with these stars caused Bibo *et al.* (1992), and more recently Casey *et al.* (1993) to conclude that the IR excess was primarily produced by nearby clouds, and not intimately associated with the stars. However, the large ratio of total to selective extinction for the line of sight to HD 176386, $R_V = 5.5 \pm 0.2$ (Bibo *et al.*, 1992), coupled with the unambiguous detection of high velocity, accreting gas similar to that observed toward β Pic, suggests instead that there must be a reservoir of circumstellar material around HD 176386. Recent high dispersion spectroscopic studies of TY CrA by Lagrange *et al.* (1993) suggest that this system also has circumstellar gas similar to that seen toward β Pic. The lack of a pronounced IR excess at short wavelengths for both systems suggests then that much of the reservoir has become optically thin. We can reconcile the IR excess data with the line-of-sight accreting gas observations by positing that the lack of a near- and mid-IR excess in HD 176386 compared to younger Herbig Be stars such as R CrA (Hillenbrand *et al.*, 1992; Graham 1992) is due to grain agglomeration into larger bodies which do not emit as efficiently, but are still capable of efficient accretion into the immediate vicinity of the star. If correct, the data imply that the chemical evolution of the proto-planetary disk in a B9.5 star is well underway by at least 2.8 Myr, and that at least some of the precursors to planetesimal formation are occurring.

4. Discussion

The UV and IR data for the two known weak-emission line Herbig Be or candidate proto-planetary disk systems have a number of implications for our understanding of the evolution of these disk systems as the star arrives on the ZAMS. If proto-stars of comparable mass initially possess circumstellar disks with similar properties, the respective IR excesses of 51 Oph and HD 176368 would suggest that HD 176386, with an age of 2.8 Myr, is the older of the two systems, with a grain population which has begun to agglomerate into larger bodies and hence has become optically thin. If correct, the comparative inferred youth for one isolated and reasonably nearby proto-planetary system candidate suggests that β Pic may also be a rather young system. Previous suggestions of youth for β Pic have met with the objection that it also is an isolated A star without surrounding nebulosity and lacking nearby PMS stars. However the identification of a number of Herbig Ae/Be stars, the presumed progenitors of A and B stars, which are apparently isolated or in very small associations (e.g. HD 163296, Thé *et al.*, 1985; 17 Lep, Polidan, 1992;

UX Ori, Grinin *et al.*, 1991) suggests that this may not be as serious an objection as previously thought. Alternatively, the HD 176386 and 51 Oph data may suggest that there is a wide range in either the initial circumstellar disk properties or that the clearing time varies considerably from system to system.

Determining which of these possibilities is correct will require identification of additional weak-emission line or post Herbig Ae/Be stars in associations covering a wider range in stellar properties and with age estimates from 1–10 Myr. Detection of high velocity, accreting gas toward one star with a ratio of total to selective extinction consistent with a circumstellar disk suggests that searches for accreting gas toward ostensibly Main Sequence stars in associations and clusters with large R values may yield additional proto-planetary candidates and should more stringently constrain the evolutionary time scale for the β Pic disk in its current state.

Acknowledgements

Partial support for this study was provided by NASA Contract NAS 5-32059 to the Applied Research Corporation, and by NASA Contract NAS 5-31841 to the Computer Sciences Corporation. Computing resources were provided by the IUE Regional Data Analysis Facility at NASA/GSFC. This study made use of the SIMBAD database operated at CDS, Strasbourg, France.

References

Bibo, E.A., Thé, P.S. and Dawanas, D.N.: 1992, *Astron. Astroph.* **260**, 293.
Bruhweiler, F.C., Grady, C.A. and Chiu, W.A.: 1989, *Astroph. J.* **340**, 1038.
Bruhweiler, F.C., Lyu, J.-H., Boggess, A., Grady, C.A. and Kondo, Y.: 1993, in preparation.
Casey, B.W., Mathieu, R., Suntzeff, N.B., Lee, C.-W. and Cardelli, J.A.: 1993, *Astron. J.* **105**, 2276.
Dougherty, S.M., Taylor, A.R. and Clark, T.A.: 1991, *Astron. J.* **102**, 1753.
Farjado-Acosta, S.B., Telesco, C.M. and Knacke, R.F.: 1992, *BAAS* **24**, 1151.
Grady, C.A., Bruhweiler, F.C., Cheng, K.-P., Chiu, W.A. and Kondo, Y.: 1991, *Astroph. J.* **367**, 296.
Grady, C.A. and Silvis, J.M.S.: 1993, *Astroph. J.* **402**, L61.
Grady, C.A., Pérez, M.R. and Thé, P.S.: 1993a, *Astron. Astroph.* **274**, 847.
Grady, C.A., Bjorkman, K.S., Pérez, M.R., Schulte-Ladbeck, R.E., de Winter, D. and Thé, P.S.: 1992, *BAAS* **24**, 1141.
Grady, C.A., Bjorkman, K.S., Pérez, M.R., Schulte-Ladbeck, R.E., de Winter, D. and Thé, P.S.: 1993b, *Astroph. J.* **415**, L39.
Graham, J.A.: 1992, *Proc. Astron. Soc. Pac.* **104**, 479.
Grinin, V.P., Kiselev, N.N., Minikulov, N.Kh., Chernova, G.P. and Voshchinnikov, N.V.: 1991, *Astrophys. and Space Sci.* **186**, 283.
Hartmann, L.A. and Kenyon, S.J.: 1985, *Astroph. J.* **299**, 467.
Hillenbrand, L.A., Strom, S.E., Vrba,F.J. and Keene, J.: 1992, *Astroph. J.* **397**, 613.
Hobbs, L.M., Vidal-Madjar, A., Ferlet, R., Albert, C.E. and Gry, C.: 1985, *Astroph. J.* **293**, L29.
Lagrange, A.M., Ferlet, R. and Vidal-Madjar, A.: 1987, *Astron. Astroph.* **173**, 289.
Lagrange, A.M., Bouvier, J. and Corporon, P.: 1993, *ESO Messenger* **No. 71**, 24.
Lagrange-Henri, A.M., Ferlet, R., Vidal-Madjar, A., Beust, H., Gry, C. and Lallement, R.: 1990, *Astron. Astroph.S* **85**, 1089.
Pérez, M.R., Grady, C.A. and Thé, P.S.: 1993a, *Astron. Astroph.* **274**, 381.
Pérez, M.R., Grady, C.A. and Thé, P.S.: 1993b, this volume.
Polidan, R.S.: 1992, private communication.

Slettebak, A.: 1982, *Astroph. J. S.* **50**, 55.
Strom, S.E. *et al.*: 1993, in: E.H. Levy and J.I. Lunine (eds.) *Protostars and Planets, III*, Univ. Arizona
    Press, p. 837.
Thé, P.S., Felenbok, P., Cuypers, H. and Tjin A Djie, H.R.E.: 1985, *Astron. Astroph.* **149**, 429.
Vidal-Madjar, A., Hobbs, L.M., Ferlet, R., Gry, C. and Albert, C.E.: 1986, *Astron. Astroph.* **167**, 325.
Walters, F.M.: 1986, *Astroph. J.* **306**, 573.
Waters, L.B.F.M, Coté, J. and Geballe, T.R.: 1988, *Astron. Astroph.* **203**, 348.

THE EVIDENCE FOR CLUMPY ACCRETION IN THE HERBIG AE STAR HR 5999 *

M.R. PÉREZ

Astronomy Programs-CSC, IUE Observatory,
NASA–Goddard Space Flight Center, Greenbelt, Maryland, USA

C.A. GRADY

Applied Research Corporation, Landover, Maryland, USA

and

P.S. THÉ

Astronomical Institute "Anton Pannekoek," University of Amsterdam,
Amsterdam, The Netherlands

Abstract. Analysis of IUE high- and low-dispersion spectra of the young Herbig Ae star HR 5999 (HD 144668) covering 1978–1992 has revealed dramatic changes in the Mg II h and k (2795.5, 2802.7 Å) emission profiles, changes in the column density and distribution in radial velocity of accreting gas, and flux in the Lyα, O I and C IV emission lines, which are correlated with the UV excess luminosity. We also observe variability in the spectral type inferred from the UV spectral energy distribution, ranging from A5 IV-III in high state to A7 III in the low state. The trend of earlier inferred spectral type with decreasing wavelength and with increasing UV continuum flux has previously been noted as a signature of accretion disks in lower mass pre-main sequence stars (PMS) and in systems undergoing FU Orionis-type outbursts. Our data represent the first detection of similar phenomena in an intermediate mass ($M \geq 2\ M_\odot$) PMS star. Recent IUE spectra show gas accreting toward the star with velocities as high as +300 km s^{-1}, much as is seen toward β Pic, and suggest that we also view this system through the debris disk. The absence of UV lines with the rotational broadening expected given the optical data (A7 IV, $v \sin\ i = 180 \pm 20$ km s^{-1}) for this system also suggests that most of the UV light originates in the disk, even in the low continuum state. The dramatic variability in the column density of accreting gas, consistent with clumpy accretion, such as has been observed toward β Pic, is a hallmark of accretion onto young stars, and is not restricted to the clearing phase, since detectable amounts of accretion are present for stars with $0.5 < t_{\mathrm{age}} < 2.8$ Myr. The implications for models of β Pic and similar systems are briefly discussed.

1. Introduction

Recent studies of intermediate-mass pre-main sequence (PMS) or Herbig Ae/Be (HAEBE) stars, probably the progenitors of systems like β Pic, have suggested that many of the IR, optical, and UV features of these systems are associated with large, viscously heated accretion disks with $M_{\mathrm{acc}} \geq 10^{-7}\ M_\odot$ yr^{-1} (Hillenbrand *et al.*, 1992; Blondel *et al.*, 1993). This represents a significant departure from previous models which assumed, based on the presence of emission lines of species commonly seen in late-type star chromospheres, that most of the optical and UV spectral characteristics of HAEBE stars were produced in a chromosphere (cf. Catala *et al.*, 1984).

---

* Paper presented at the Conference on *Planetary Systems: Formation, Evolution, and Detection* held 7–10 December, 1992 at CalTech, Pasadena, California, U.S.A.

The irregular variable star HR 5999 (HD 144668, V856 Sco, CPD −38° 6373) is one of the most studied intermediate-mass PMS stars at optical, UV and IR wavelengths. This star has shown quasi-periodic and random photometric and spectroscopic variabilities recently summarized by Tjin A Djie *et al.* (1989), Praderie *et al.* (1991) and Pérez, Webb and Thé (1992a). Three recent studies have suggested that many of the features of the circumstellar environment of HR 5999 can be accounted for by the presence of an optically thick accretion disk. Blondel *et al.* (1993) suggested the presence of such a disk as a result of their analysis of the extended Lyα emission around HR 5999, which is interpreted as formed by recombination of infalling matter onto the central source. Pérez *et al.* (1992a) have suggested instabilities in an accretion disk as the source of non-periodic photometric variability in the optical. More recently, Hillenbrand *et al.* (1992) have interpreted the infrared excess of HR 5999 in terms of a large, optically thick debris disk with a high accretion rate and a possible inner cavity.

Baade and Stahl (1989) extensively studied the photometric and spectroscopic variability of HR 5999; however, they barely mentioned that the Hα lines appeared in inverse Beals (1950) type III P Cygni profiles during the nights of June 7 and 8 in 1987. This immediately led us to monitoring the UV spectral changes starting on September 7, 1990, focusing on the shape of the Mg II lines (2800 Å), which are being formed in more extended regions than Hα (up to 50 stellar radii). Our first observation confirmed the prediction that Mg II lines in HR 5999 are a good tracer of the dynamics in extended shells and/or bipolar flows by appearing as an inverse type III P Cygni profile, similar to Hα, indicating the increase of accretion phenomena in the line of sight. A more detailed report of the observations and analysis discussed here is published elsewhere (Pérez *et al.* 1993).

2. Relevant Observational Features of HR 5999

Among the intermediate-mass PMS stars, HR 5999 is one of the best-studied objects; however, its photometric and spectroscopic behavior appears to be somewhat different from other objects in this class, such as AB Aur (for some researchers, this is the prototype of the HAEBE class), HD 163296, HD 104237, etc. Nevertheless, HR 5999 fits all three classification criteria of the PMS Ae and Be class defined by Herbig (1960), although it is not in his original list of 26 members. In Table I and subsequent paragraphs, we have described some of the observational parameters of this star, as best known at this time.

In the following list, we have summarized some of the relevant properties and observational features of this star.

1. Associated with the Lupus T3 star-forming cloud; embedded in nebulosity clearly visible on the POSS prints.

2. The ratio of total to selective extinction, $R_v = 5.8$ (Pérez *et al.*, 1992a) is consistent with dust in the line of sight with a particle size distribution dominated

TABLE I

HR 5999 observational parameters

Spectral Type	T_{eff} (K)	$\log g$	L ($L_\odot$)	$v \sin i$ (km s^{-1})	R ($R_\odot$)	distance (pc)	M ($M_\odot$)	t_{age} (yr)
A5-7 III-IVe	7,800	3.5–4.0	36	180 ± 20	3.2	140	2–2.5	5×10^5

by large grains and different significantly from that typical of the diffuse ISM ($R_v = 3.1$).

3. Variable intrinsic polarization in the optical ranging from 0.13–0.52% with a wavelength dependence consistent with dust (Thé et al., 1981).

4. Spectroscopic and photometric variabilities from 1975–1985 are summarized by Tjin A Djie et al. (1989) and Praderie et al. (1991). Optical absorption lines due to singly ionized elements are particularly variable. Based on the variability detected in HR 5999, a qualitative model involving a corotating and slowly expanding region, surrounded by slowly rotating layers with higher expansion velocities, is proposed by Praderie et al. (1991).

5. Visual photometric variability and color changes have been observed in the last 20 years. This star does not show photometric indications of the "blueing" or "color reversal" effect detected in other intermediate-mass PMS stars such as UX Ori, BF Ori, CQ Tau, etc. (Bibo and Thé, 1991); however, spectral *blue* and *red* outbursts were found in this dataset (Pérez et al., 1992a). Short-duration (8–10 days) enhancements in the luminosity of the star, randomly located in time, have been interpreted as evidence for accretion instabilities.

6. The photosphere is only visible at a few optical windows. The stellar projected rotation rate is inferred to be 180 ± 20 km s^{-1} based on broadening of Mg II 4481 Å. Photospheric absorption is not detectable at Na I D and Ca II lines (Lagrange-Henri al., 1990). Photospheric features are not seen in the UV from 2580–3000 Å (Pérez et al., 1993).

7. Blondel et al. (1989) find, in the first 7 years of IUE data, a tendency for Fe II and other absorption features to have smaller equivalent widths when the star is fainter.

8. Hα and Mg II have P Cygni type III (emission with central absorption reversal) profiles. Both emission profiles are variable and prior to 1986, published data indicate, $V < R$. Beginning in June 1987 (Baade and Stahl, 1989) and continuing through 1991, $V > R$. Recent optical and IUE spectra (August and September 1992) present $V < R$.

9. Hillenbrand *et al.* (1992) classify the IR excess for this star as consistent with an optically thick accretion disk extending from $6\ R_* < r < 23$ AU with an accretion rate of $1 \times 10^{-6}\ M_\odot\ \text{yr}^{-1}$.

10. Lyα emission peaking at 850 km s^{-1} and with a FWHM of 1240 km s^{-1} is seen at an epoch when the star is optically bright (Blondel *et al.*, 1993). If interpreted as originating in a bipolar flow, the inferred mass accretion rate is $6.8 \times 10^{-7}\ M_\odot\ \text{yr}^{-1}$.

3. Recent Ultraviolet Observations

After September 1990 several new IUE observations in high- and low-dispersion with both the long- (LW) and short-wavelength (SW) cameras have been carried out in order to continue the monitoring of this star started in 1978, during the early days of the IUE spacecraft. Some of the relevant results from the high-dispersion data have been reported by Blondel *et al.* (1989). We briefly summarize some of the results obtained from the archival and recent observations.

3.1. THE UV CONTINUUM VARIABILITY

A striking feature of HR 5999 in the optical is the pronounced photometric variability. As shown in Figure 1, the continuum exhibits significant variability down to at least 1450 Å. In the longer wavelength portion of the spectrum near the Mg II resonance doublet, the continuum luminosity can be grouped into a UV-bright state and a UV-faint state, with up to a magnitude difference in flux levels between the states. The UV bright state at 2815 Å has been observed only for V(FES) < 7.0 (this V magnitude has been calculated from the IUE Fine Error Sensor [FES] using the calibration by Pérez and Loomis, 1991). At 1630 Å, faint and bright states are observed, but the association with V(FES) is less apparent. UV-bright states are observed down to $V(\text{FES}) = 7.46$, while fainter spectra are observed over the same range at 2815 Å. The 1450 Å data exhibit significant variation which does not appear correlated with the continuum flux at longer wavelengths. In all spectra, even those obtained in the UV-faintest states observed to date, the IUE spectra provide convincing evidence for a continuum UV excess compared to similarly exposed A5-A7 standard star spectra down to Lyα.

3.2. CORRECTION FOR INTERSTELLAR AND CIRCUMSTELLAR EXTINCTION

It has been known for some time that the extinction law towards HR 5999 is quite anomalous (Thé and Tjin A Djie, 1978; Thé *et al.*, 1981; Hecht *et al.*, 1984), typical of other active star-forming regions, such as Orion. Preliminary estimates of the ratio of total to selective extinction, $R_v = A_v/E(B - V)$, accounting for both the foreground and circumstellar extinctions, were between 4 and 5. Our trial-and-error fit of the unreddened optical fluxes to the Kurucz (1991) models indicated an R_v value of $\sim$5.5–5.7. The latest R_v value based on Strömgren photometric data derived from 282 data points covering 7 years of monitoring is 5.8 (Pérez *et al.*,

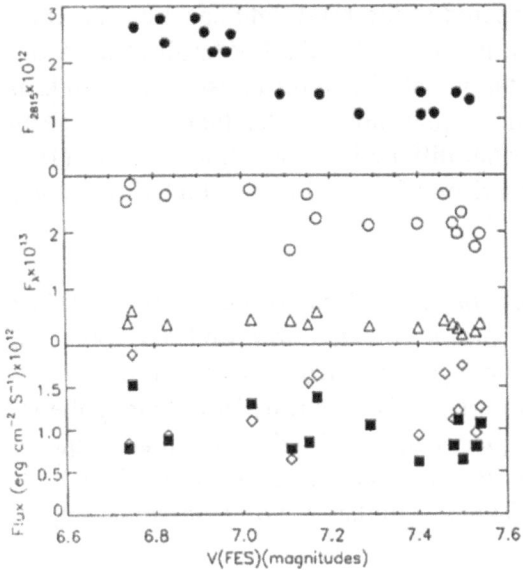

Fig. 1. UV photometric variability in HR 5999 of the observed fluxes as a function of the derived visual magnitude V(FES). Continuum variability at 2815 Å (filled circles), 1620 Å (open circles), 1450 Å (triangles) is shown. Net observed emission fluxes for C IV (filled squares) and O I (diamonds) are also indicated.

1992a). This value is intended to be an average since R_v becomes smaller when the star gets brighter as acknowledged by Thé et al. (1985) and Pérez et al., (1992a). By properly correcting for extinction the low-dispersion data, we found that even in the UV-faint state, the continuum flux mimics an A5 III at these wavelengths with detectable excess shortward of 1600 Å and longward of 2600 Å.

A trend of inferred spectral type shifting to earlier types with decreasing wavelength, larger UV excesses during episodes of high optical luminosity, and a flux distribution which is not consistent with a single-temperature stellar model, has previously been observed in lower mass pre-main sequence (PMS) stars, particularly in stars undergoing FU Orionis-type outbursts (Hartmann and Kenyon, 1985). These phenomena have been interpreted as evidence for the presence of emission from an optically thick accretion disk and boundary layer. Our data provide the first detection of similar phenomena in a more massive protostar. The extent of the UV excess, even in the faintest UV spectra obtained to date, is such that we would not expect to detect the stellar photosphere.

3.3. THE EMISSION LINES

The spectrum of HR 5999 shortward of 1600 Å is rich in emission lines from species commonly associated with chromospheres in late-type stars. Both O I (1320 Å) and C IV (1550 Å) exhibit significant variation in net emission fluxes and in the velocity range covered by the emission. The net fluxes and velocity widths are smaller in the

UV-faint spectra compared to the UV-bright spectra. The UV-faint spectra show much less scatter than is observed in the UV-bright data. A distinctive feature of the C IV emission is the variability in the emission centroid. In general, the C IV emission characteristics show more scatter than O I, suggesting at least partial formation in a somewhat different region. Blondel *et al.* (1993) reached similar conclusions in a comparison of O I and Lyα in high dispersion IUE SW spectra.

3.3.1. Mg II emission

It is known that in the shape of the Mg II lines in bright HAEBE stars there is a smooth progression of profiles ranging from well-developed type III P Cyg (Pérez *et al.*, 1992b) for the youngest to narrow absorption lines for the stars with more main sequence characteristics. For the truly PMS among the HAEBE sample, the Mg II profiles are unique when compared with B main sequence and Be classical stars and are generally the signature of accelerating winds in their extended disks (between 315 and 495 km s^{-1} in the case of the extreme object AB Aur [Praderie *et al.*, 1986]).

The Mg II profiles in HR 5999 are similar in shape when compared with the youngest members in the HAEBE class, but they are also substantially narrower. It is believed that this is due to the viewing stellar geometry, nearly edge-on, and due to the fact that a part of the Mg II lines are probably produced in the bipolar outflows. The suggestion that this star is seen almost edge-on comes from its observed high rotational velocity, $v \sin i = 180 \pm 20$ km s^{-1} ([Tjin A Djie *et al.*, 1989], ~66% of its breakup velocity), its large infrared excesses ($E(V - L) = 3.2$ mag), which in turn should produce high UV extinction ($1.2 \leq A_{2800} \leq 2.2$ mag for $R_v = 5.8$, according to Pérez *et al.*, 1992a); however, the Mg II, C IV, Si IV emission lines, for example, appear strong and relatively unaffected by changes in extinction, indicating that the emission volume generating these infrared excesses is not isotropically distributed and they are likely to be confined to a disk.

All of the HR 5999 Mg II profiles obtained through 1985 have type III P Cygni profiles, despite being obtained at epochs of both high UV continuum flux (the 1979 data), and low continuum flux (all intervening spectra). When IUE observations resumed in September 1990, the character of the Mg II profile had changed drastically to type III *inverse* P Cygni profiles closely resembling the Hα profiles of Baade and Stahl (1989) (see Figure 2). Subsequent spectra obtained between 1991 and March 1992 also exhibit inverse P Cygni profiles, although not as extreme as the 1990 September observations'. The Mg II emission was found to be more prominent relative to the continuum in spectra obtained during the low UV continuum state; the broad emission base extends to 300–350 km s^{-1}. However, our data suggest net emission is systematically higher in the high continuum state.

3.3.2. Singly ionized iron-peak species

The UV spectrum of HR 5999 is rich in absorption features to the ground configurations and excited levels of singly ionized iron-peak elements (Blondel *et al.*,

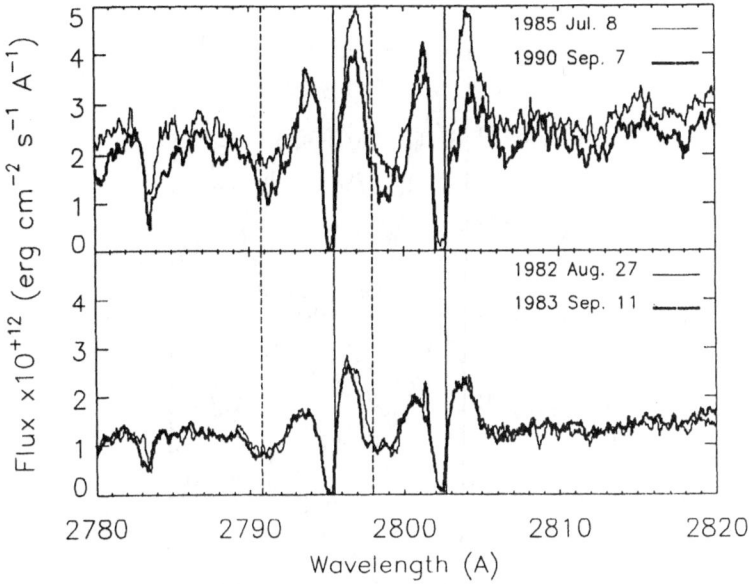

Fig. 2. Representative 2800 Å Mg II profiles for HR 5999. The upper panel shows two spectra obtained during the UV-bright state (LWP 6365, faint line; LWP 18717, bold line). Two spectra obtained during the UV-faint state are shown in the bottom panel (LWR 14026, faint line, and LWR 16769, bold line).

1989), and provides an opportunity to test our suggestion that the variability in the Mg II resonance profiles is due to variable column densities of accreting gas.

As noted by Blondel *et al.* (1989), strong absorption is present in the profiles of all of the lines of Fe II UV(1, 62, 63, 64, 68). The profiles from the UV(1) transitions are saturated over the velocity range -50 to 0 km s^{-1}. Longward of this saturated absorption the IUE spectra obtained prior to 1990 show absorption extending to approximately 300 km s^{-1}. From 1990 onward increased absorption from 0–300 km s^{-1}, and especially pronounced from 0–200 km s^{-1} is visible in all of the Fe II UV (1) lines (see Figure 3).

4. Comparison With β Pictoris

The singly ionized metal absorption profiles seen since 1990, with large column densities at low velocities and smaller column densities at progressively higher, positive velocities, are strongly reminiscent of the accreting gas profiles seen toward the proto-planetary system candidate β Pic. Both systems have circumstellar absorption profiles characterized by large column densities near the stellar radial velocity, with decreasing absorption extending to high positive velocities in at least some spectra. Both systems exhibit high-density, high velocity gas of variable column density, absorption components in some spectra which are particularly

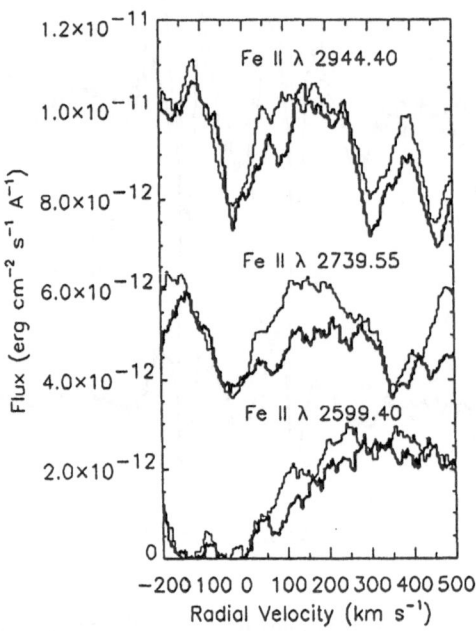

Fig. 3. Circumstellar Fe II absorption from 1985 July 8 (LWP 6365) and 1990 September (LWP 18717). The lines from top to bottom, 2944.4 Å (UV78), 2739.55 Å (UV 63) 2599.40 Å (UV1), both demonstrate the variability of the line-of-sight mass accretion rate during the UV-bright state and provide evidence for accreting gas with velocities as large as +300 km s^{-1}.

prominent in the low velocity portion of the flow, and a trend for the FWHM of the components to increase with increasing radial velocity (see Boggess *et al.*, 1991; Lagrange-Henri *et al.*, 1992). The relative prominence of the accreting gas relative to the material closer to the inferred systemic velocity increases with increasing excitation potential, as does the suggestion of discrete absorption components, which are particularly prominent in the 1985 July 8 and 1990 September data. Similarly, the outflowing material observed on 1983 September 11 (LWR 16769) also becomes more prominent with increasing excitation potential. Similar behavior is observed in the accreting and outflowing gas toward β Pic (Bruhweiler *et al.*, 1991, 1994; Boggess *et al.*, 1991).

In the case of β Pic, higher S/N and resolution HST GHRS data have demonstrated that the excitation temperature of the accreting gas exceeds the stellar T_{eff}, and that the components have excitation temperatures above the temperature of the accreting gas at nearby velocities (Bruhweiler *et al.*, 1994). The IUE data for HR 5999 suggest that the accreting gas in this, much younger system ($t_{age} \approx 5 \times 10^5$ year) with a significantly higher mass accretion rate, exhibits the same phenomena. The variability of the C IV net emission fluxes, compared with O I, may indicate the presence of C IV in absorption in the line-of-sight, since the UV-bright spectra with low C IV fluxes correspond to epochs with enhanced Mg II and Fe II absorption. If this assertion is supported by higher resolution UV spectra, such as can be provided by the HST, it will provide confirmation of the presence of collisionally ionized gas in the HR 5999 system. Detection of transient, outflowing material which is high density, and potentially collisionally ionized is a further point of similarity which must be accounted for in any model for accretion phenomena onto young, intermediate mass stars.

5. Conclusions

The UV data demonstrate, for the first time, that the photometric brightening of HR 5999 in the optical and UV is associated with increasing column densities of gas visible to 300–350 km s^{-1}. Sufficiently large amounts of material are involved that the UV continuum luminosity increases by one magnitude at 2800 Å. The photometric variability has previously been noted as a signature of accretion instabilities. Accretion signatures in this star have been independently detected by Hillebrand et al. (1992) and Blondel et al. (1993). If correct, the UV data suggest that the high velocity gas is perturbed into the inner portions of the accretion disk and the boundary layer by the same instabilities. Temporal variability in the high velocity, accreting gas is seen in the IUE data, although the spectra are sufficiently separated in time that we cannot infer the time scale for changes at high velocity. Previously such variability, termed clumpy accretion, has been reported for the HAEBE star R CrA (Graham, 1992) and BF Ori (Welty et al., 1992), and is characteristic of the accretion activity in older systems such as β Pic (Lagrange-Henri et al., 1992), 51 Oph (Grady and Silvis, 1993), and HD 176386 (Grady et al., 1993). This certainly indicates that the time scale for the characteristic accretion signatures to occur is extended beyond 2.8 Myr (lower limit for age of 51 Oph and β Pic systems).

The optically thin cavity inferred from the IR excess flux distribution by Hillebrand et al. (1992), marks the location of the dust sublimation zone rather than a true void. Interior to this region, the UV data indicate the presence of optically thick gas. Progressive broadening of the line profiles with increasing ionization potential suggests that the majority of the gas seen in absorption near the system velocity are in predominantly Keplerian orbits, with small velocity components toward the star. Since this material is seen in all the UV data, as well as in the available optical spectra, the gas at the system velocity may correspond to the sublimation products of grains originally spiraling into the star via Poynting-Robertson drag. Exterior to this region, Thé and Molster (1993) have argued that this star is surrounded by proto-planetary clouds orbiting in the outer parts of the circumstellar disk, which are responsible for the optical variability.

Acknowledgements

Partial support for this study was provided by NASA Contract NAS 5-32059 to the Applied Research Corporation, and by NASA Contract NAS 5-31841 to Computer Sciences Corporation. Computing resources were provided by the IUE Data Analysis Center at NASA/Goddard Space Flight Center.

References

Baade, D. and Stahl, O.: 1989, *Astron. Astrophys.* **209**, 255.

Beals, C.: 1950, *Publ. Dom. Astrophys. Obs.*, **9**, 1.

Bibo, E.A. and Thé, P.S.: 1991, *Astron. Astrophys. Suppl.* **89**, 319.

Blondel, P.F.C., Tjin A Djie, H.R.E. and Thé, P.S.: 1989, *Astron. Astrophys.* **80**, 115.

Blondel, P.F.C., Talavera, A. and Tjin A Djie, H.R.E.: 1993, *Astron. Astrophys.* **268**, 624.

Boggess, A., Bruhweiler, F.C., Grady, C.A., Ebbets, D.C., Kondo, Y., Trafton, L.M., Brandt, J.C. and Heap, S.R.: 1991, *Ap J* **377**, L49.

Bruhweiler, F.C., Kondo, Y. and Grady, C.A.: 1991, *Ap J* **371**, L27.

Bruhweiler, F.C., Lyu, H.-S., Boggess, A., Kondo, Y. and Grady, C.A.: 1994, (in preparation).

Catala, C., Kunasz, P.B. and Praderie, F.: 1984, *Astron. Astrophys.* **134**, 402.

Grady, C.A. and Silvis, J.M.S.: 1993, *Ap J*, **402**, L61.

Grady, C.A., Pérez, M.R. and Thé, P.S.: 1993 (this proceedings).

Graham, J.A.: 1992, *Pub. A. S. Pac.* **104**, 479.

Hartmann, L. and Kenyon, S.J.: 1985, *Ap J*, **299**, 462.

Hecht, J.H., Holm, A.V., Ake III, T.B., Imhoff, C.L., Oliversen, N.A. and Sonneborn, G.: 1984, in *Future of Ultraviolet Astronomy based on Six years of IUE Research*, NASA Conference Publication, No. 2349, 318.

Herbig, G.H.: 1960, *Astrophys. J. Suppl.* **4**, 337.

Hillenbrand, L.A., Strom, S.E., Vrba, F.J. and Keene, J.: 1992, *Ap J*, **397**, 613.

Kurucz, R.L.: 1991, in: A.G. Davis Philip, A.R. Upgren, K.A. Janes (eds.), *Precision Photometry: Astrophysics of the Galaxy*, (Schenectady, N.Y.: L. Davis Press).

Lagrange-Henri, A.M., Ferlet, R., Vidal-Madjar, A., Beust, H., Gry, C. and Lallement, R.: 1990, *Astron. Astrophys. Suppl.* **85**, 1089.

Lagrange-Henri, A.M., Gosset, E., Beust, H., Ferlet, R. and Vidal-Madjar, A.: 1992, *Astron. Astrophys.* **264**, 637.

Pérez, M.R., Webb, J. and Thé, P.S.: 1992a, *Astron. Astrophys.* **257**, 209.

Pérez, M.R., Imhoff, C.L. and Thé, P.S.: 1992b, *Bull. A.A.S.*, **23**, No. 4, 1374.

Pérez, M.R. and Loomis, C.: 1991, Record of the IUE Three Agency Coordination Meeting (NASA/ESA/SERC), Nov. 19–21, 1991 at GSFC, F-13.

Pérez, M.R., Grady, C.A. and Thé, P.S.: 1993, *Astron. Astrophys.* **274**, 381.

Praderie, F., Simon, T., Catala, C. and Boesgaard, A.M.: 1986, *Ap J*, **303**, 311.

Praderie, F., Catala, C., Czarny, J., Thé, P.S. and Tjin A Djie, H.R.E.: 1991, *Astron. Astrophys.* **89**, 91.

Thé, P.S. and Tjin A Djie, H.R.E.: 1978, *Astron. Astrophys.* **62**, 439.

Thé, P.S. and Molster, F.J.: 1993 (this proceedings).

Thé, P.S., Tjin A Djie, H.R.E., Bakker, R.S., Bastiaansen, P.S., Burger, M., Cassatella, A., Fredga, K., Gahm, G.F., Liseau, R., Smyth, M.J., Viotti, R., Wamsteker, W. and Zeuge, W.: 1981, *Astron. Astrophys. Suppl.* **44**, 451.

Thé, P.S., Tjin A Djie, H.R.E., Brown, A., Catala, C., Doazan, V., Linsky, J.L., Mewe, R., Praderie, F., Talavera, A. and Zwaan, C.: 1985, *Irish Astron. J.*, **17**, 79.

Thé, P.S. *et al.*: 1981, *Astron. Astrophys. Suppl.* **44**, 451.

Tjin A Djie, H.R.E., Thé, P.S., Andersen, J., Nordstrom, B., Finkenzeller, U. and Jankovics, I.: 1989, *Astron. Astrophys. Suppl.* **78**, 1.

Welty, A.D., Barden, S.C., Huenemoerder, D.P. and Ramsey, L.W.: 1992, *Astron. J.* **103**, 1673.

PROTOPLANETARY DUST CLOUDS IN DISKS OF HERBIG Ae/Be STARS *

P.S. THÉ and F.J. MOLSTER

*Astronomical Institute "Anton Pannekoek", University of Amsterdam,
Amsterdam, The Netherlands*

Abstract. The very large brightness decrease of late-type Herbig Ae/Be stars is believed to be caused by obscuring dust clouds orbiting in the outer parts of their circumstellar disks. The distances of the dust clouds to the central stars have been estimated using the wavelength at maximum flux of the excess near-IR radiation, Wien's displacement law, and a formula derived by Rowan-Robinson (1980). The critical masses of these clouds were calculated employing Chandrasekhar's (1943) formula. The minimum size of the dust grains in the obscuring clouds was estimated using Aumann *et al.*'s (1984) formula they had applied to the star α Lyr. However, it can be about ten times smaller if the dust grains are situated at the back of the cloud. The average size of these grains has been determined by assuming a size distribution similar to that in the asteroidal belt (Dohnanyi, 1969) and in the interstellar medium (Mathis *et al.*, 1977). Their number density was determined by means of the extinction power of the dust cloud at the V pass-band. The results of our calculations show that above parameters are similar to those in our solar system. Therefore, we believe that most probably (a) the formation of planetesimals in the circumstellar disks of Herbig Ae/Be stars is on-going; and (b) the obscuring clouds will, in the long run, become planet-like objects.

1. Introduction

During the last five years evidence accumulated that Herbig Ae/Be stars (HAEBESs), like T Tauri stars, also possess circumstellar disks. Their brightness Algol-type variability is believed to be caused by dust clouds orbiting in the outer regions of the disks. The reddening of these stars when they become fainter, the colour-reversal or blueing effect observed by many authors (Bibo and Thé, 1990; and references therein), the anti-correlation of brightness variations and changes in degree of polarization (Grinin *et al.*, 1991), the distribution of linear polarization in the wide surrounding of several HAEBESs (Bastien and Ménard, 1988), the excess radiation at long wavelengths of HAEBESs discussed by Strom *et al.* (1990), and their measurements of the energy-flux at 1.3 mm, are all signs of the presence of protoplanetary disks around Herbig's objects.

The reddening when the stars dim, the blueing effect, and the anti-correlation between brightness decrease and increase in linear polarization, are most probably caused by the obscuration of the central star by dust clouds, containing grains of minimum size smaller than 1 μm, moving in the outer regions of the circumstellar disk. If this is indeed the case it is important to know more about the physical properties of the obscuring clouds. In this paper the results of a study of the properties of several HAEBESs with large brightness variations, inspired by a paper published by Wenzel *et al.* (1971), are presented.

---

* Paper presented at the Conference on *Planetary Systems: Formation, Evolution, and Detection* held 7–10 December, 1992 at CalTech, Pasadena, California, U.S.A.

TABLE I

Data of several HAEBE central stars with large variable brightness (HIC: HIPPARCOS Input Catalogue).

Name	HIC	RA 1950	DEC 1950	SP.T.	Ref	Range in V	Ref
UX Ori	23602	05 02 00.6	-03 51 20	A2 IIIe	1	9.6 - 12.3	2
NX Pup	35488	07 17 56.5	-44 29 34	A0 IIIe	1	9.3 - 11.0	2
HR5999	79080	16 05 12.8	-38 58 23	A7 IIIe	4	6.7 - 8.5	2
KK Oph	-	17 07 00.7	-27 11 36	A5 Ve	5	10.3 - 12.0	3
R CrA	93449	18 58 31.3	-37 01 28	B8 Ve	6	9.4 - 13.8	3
T CrA	-	18 58 36.5	-37 02 10	F0 Ve	6	11.7 - 14.3	3

Name	M_* ($M_\odot$)	R_* ($R_\odot$)	L_* ($L_\odot$)	T_{eff} (K)	References
					(1) Tjin A Djie, Remijn, and Thé (1984)
UX Ori	2.8	3.3	65	9000	(2) Bibo and Thé (1991)
NX Pup	3.2	3.4	106	10100	(3) Herbig and Bell (1988)
HR5999	2.3	3.0	29	7700	(4) Thé et al. (1981)
KK Oph	1.9	1.9	14	8200	(5) de Winter and Thé
R CrA	3.5	3.1	180	12000	(6) Bibo, Thé, and Dawanas (1992)
T CrA	1.6	1.6	6.5	7200	(7) Bibo and Thé (1990)

2. Data of Several HAEBE Stars with Large Variable Brightness

We will apply the formulae derived below, to data of several HAEBESs, which show large ($> 1^m$) brightness variations. These data are listed in Table I. The ranges in brightness were taken preferably from publications in which the star has been observed during a long time interval, like that by Bibo and Thé (1991). In this study the Strömgren system has been used; we will assume that the range in y is equal to that in V. It should be noted also, that these ranges in brightness are supposed to be those which are only caused by the obscuring clouds. The influence of the extinction of the matter in the dust ring itself is excluded. The range in brightness variation of KK Oph and R CrA are given by Herbig and Bell (1988) in the blue photographic pass-band. In Table I we have transformed them into V using the colour indices $B - V$ given also by these authors.

The spectral types and luminosity classes of the 6 programme stars in Table I play an important role in the derivation of the other physical parameters. Sufficient discussions were given by the different authors for adopting the final MK spectra of these stars. The emission lines of the Balmer series are the cause for preventing spectral classification in the normal way. Therefore, one should then use, for example, the photometric method, in which the influence of the emission lines in the photometry is corrected as well as possible using low dispersion spectra. A method suggested by Jaschek and Jaschek (1987) can also be employed. In this statistical method one uses the relation between the emission in the Balmer series and the spectral type.

The luminosity class can be estimated using the determination of the luminosity, L_*, of the star when its approximate distance is known, the magnitude at maximum brightness is measured, and the value of the extinction $A_V = R_V E(B - V)$ is calculated, when $E(B - V)$ from photometry, and R_V, using the Kurucz spectral energy fitting procedure explained by Steenman and Thé (1989) are known. From a comparison with the tables of Schmidt-Kaler (1982) in the Landolt-Börnstein catalogue, one can then decide whether the star is a main sequence object or a giant.

In order to be able to compare the results of the programme stars the following procedure for the adoption of L_*, R_*, and M_*, all expressed in solar units, has been used for all stars. The L_* is taken from the tables of Schmidt-Kaler mentioned above. The stellar radius is calculated from

$$\frac{R_*}{R_\odot} = \left(\frac{L_*}{L_\odot}\right)^{1/2} \left(\frac{T_*}{T_\odot}\right)^{-2} \tag{1}$$

in which $T_\odot = 5800\ K$. For the M_* we use the mass-luminosity relation derived by Harris et al. (1963),

$$\log \frac{M_*}{M_\odot} = 0.46 - 0.10\,M_{bol} \tag{2}$$

which is applicable for $M_{bol} < 7.5$, in which M_{bol} is again taken from the tables of Schmidt-Kaler.

3. Temperature and Location of the Dust Ring

As mentioned previously, evidences have been accumulating that the circumstellar material of HAEBESs is distributed in a disk. The ratios of the symmetry axis have been estimated to be about 0.2 to 0.5 (Bastien and Ménard, 1990). The gaseous component of the circumstellar matter is located everywhere in the disk. In the outer regions of the disk, where the temperature is low enough, the dust grains are revolving around the central star in Keplerian orbits, which are concentrated in a dust ring. The dust particles can form clouds of different sizes, which remain stable in spite of the gravitational influence of the central star. Such an ensemble of dust grains is presumably located in the densest inner parts of the dust ring. The average temperature of the dust ring, T_d, can be estimated using the displacement law of Wien, $\lambda_{max} T_d = $ constant, in which λ_{max} is the wavelength at the maximum

TABLE II

Physical parameters of the dust ring and the dust cloud.

Name	T_d (K)	Ref	r_d (AU)	M_{cl} ($M_\oplus$)	θ_{cl} (km s^{-1})	P (day)
UX Ori	1490	7	1.4	3.7	42	225
NX Pup	1360	1	2.4	1.0	34	764
HR5999	1400	4	1.0	6.3	45	243
KK Oph	1250	5	1.0	1.3	41	266
R CrA	1370	6	3.3	0.3	29	1237
T CrA	1310	6	0.5	4.4	27	185

of excess infrared radiation. This excess radiation is due to re-emission by the dust grains which absorb radiation from the central star. For our programme stars the dust temperatures have been determined by several authors. They are listed in Table II. Using these temperatures, and assuming that the dust grains behave like black bodies, it is then possible to estimate the distance between the dust ring and the central star using a formula derived by Rowan-Robinson (1980),

$$r_d = R_* \left(\frac{T_*}{T_d}\right)^{2.5} \tag{3}$$

in which the values of R_* and T_* are taken from Table I. The results of the calculations for the different stars in our sample are given in Table II.

4. The Critical Mass of a Dust Cloud Orbiting Around the Central Star

When the assumption has been made that a dust cloud of certain mass is moving around the central star, the question can be raised what its critical mass should be in order that it will not be disrupted by the gravitational force of the central object. The dust in such an orbiting cloud is assumed to be in the form of grains, of different sizes. The motions in such a cloud of grains can be devided into a motion of the center of gravity of the cloud and the motions of the grains with respect to their center of gravity. If we now suppose that the center of gravity moves in an orbit in the gravitational field of force of the central star we have actually a similar situation as the case of a cluster of stars moving in the gravitational field of the galaxy, which, for instance, has been analysed by Chandrasekhar (1943).

What is actually the disrupting mechanism of such an ensemble of grains? Due to the variation of the gravitational potential over the spatial extent of the cloud a so-called tidal field will be produced. The tidal force that is associated with this field tends to disrupt the cloud. In case the cloud is too loose, the tidal field will succeed, whereas when the cloud is dense enough it will remain dynamically stable. The simplest case we are going to consider is that, in which the gravitational center of the cloud is moving in a circular orbit around the central star. It has been shown by Chandrasekhar (1943) that then the critical density of the cloud should be

$$\rho_{cr} = \frac{4}{\pi G \beta'_1} A(A - B) \tag{4}$$

in which G is the gravitational constant, β'_1 is a constant which has been derived by Mineur (1939) to be equal to 1.33, and A and B are the well known constants of Oort. For a circular orbit the orbital velocity is given by

$$\theta_{cl} = \left(\frac{GM_*}{r_d}\right)^{1/2} \tag{5}$$

in which r_d is the distance of the centre of gravity of the cloud to the central star, which we assume to be equal to the inner radius of the dust ring (see Section 3). It can be shown that

$$A = \frac{3}{4} \frac{(GM_*)^{1/2}}{r_d^{3/2}} \tag{6}$$

and

$$A - B = \frac{(GM_*)^{1/2}}{r_d^{3/2}} \tag{7}$$

in which M_* is the mass of the central star. With these values of A and A − B the critical density becomes (with $\beta'_1 = 1.33$)

$$\rho_{cr} = \frac{3}{\pi} \frac{M_*}{\beta'_1 r_d^3} = 0.72 \frac{M_*}{r_d^3} \tag{8}$$

and its critical mass

$$M_{cr} = 3.01 R_{cl}^3 \frac{M_*}{r_d^3} \tag{9}$$

in which R_{cl} is the radius of the cloud, which is assumed to be spherically shaped. From published observations of different authors there are indications that the dust cloud is probably elongated, but in our estimates we will assume that the cloud is spherically symmetric. When it is possible to make an estimate of the extent of the cloud it will thus also be possible to estimate the critical mass of the cloud; the mass of the central object, M_*, can be obtained from its luminosity (Table I).

5. The Dimension of the Dust Clouds

From many studies of the variability of HAEBESs it was found that at brightness minima the light curves are usually not flattened (see Bibo and Thé, 1991). This means that the angular diameter of the cloud must be limited if the obscuration is central. In Case I we have that

$$\alpha_{cl} = \alpha_* \tag{10}$$

in which α_{cl} and α_* are the angular diameters of the cloud and the central star, respectively. This Case I cannot only occur when the circumstellar disk is seen edge on, but also when the disk is somewhat inclined (see discussion in Section 6). A sharp brightness minimum can also be caused when the obscuration is partial. In Case II we have that

$$\alpha_{cl} > \alpha_* \tag{11}$$

There are also observations made especially for the study of the duration of a brightness minimum. These observations sometimes show that such a minimum can last several days or weeks. In Case III we believe that the cloud is elongated.

Going back to linear dimensions of the cloud and the central star we can write for Case I

$$R_{cl} = \left(1 - \frac{r_d}{D}\right) R_* \tag{12}$$

in which D is the distance of the central star from the earth. The ratio r_d/D is always very small. For our further calculations we will therefore assume that $R_{cl} = R_*$. In Case I we thus have that

$$M_{cr} = 3.01 \, R_*^3 \, \frac{M_*}{r_d^3} \tag{13}$$

For Case II we are not able to estimate the size R_{cl} of the dust cloud. We only know that its size cannot be too large, such as was found for the dust cloud associated with the star WW Vul by Voshchinnikov and Grinin (1992), because then the effect of differential rotation will destroy the cloud within a few times of orbital revolution around the star. Therefore, for Case II the critical mass as given in (9) is a lower limit. The same applies for Case III. In our further estimates we shall assume that M_{cr} is representative for the mass of the dust cloud, M_{cl}.

In Table II we have listed values of M_{cl} for several well studied HAEBESs. It is obvious that also in stellar objects having masses in the range of 2 to 10 times $M_\odot$, we can expect circumstellar dust clouds having masses like those of the planets in our solar system. Therefore, we believe that these dust clouds are progenitors of planet-like objects circulating around HAEBESs. The orbital velocity of the dust clouds calculated from (5) and the orbital period around the central star

$$P = \frac{2\pi r_d}{\theta_{cl}} \tag{14}$$

are also given in Table II.

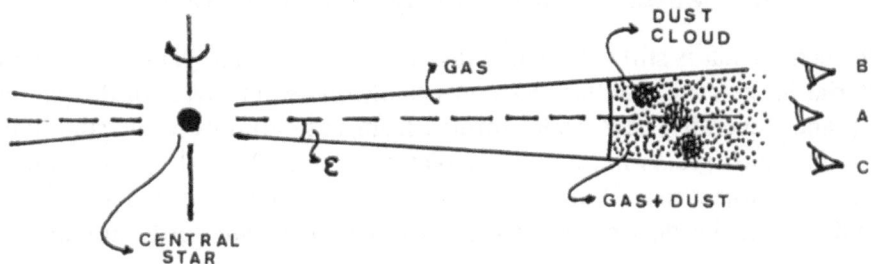

Fig. 1. A schematic intersection of the model of the gas and dust circumstellar disk and the meridional plane. The disk becomes wider outwards with half-width ϵ.

6. A Model of the Circumstellar Disk

At this stage it is important to explain how we can imagine the structure of the circumstellar disk of HAEBESs. It is well known that not all stars of this type are variable, and if variable only part of them show violent brightness changes, sometimes up to 2 to 3 magnitudes below their normal brightness. From this latter category we can have an idea how we should construct a model of the circumstellar disks of HAEBESs, which we believe will also be applicable as a model for the stars of the other categories. As can be seen from the lightcurves of several voilently variable HAEBESs published by Bibo and Thé (1991), the variations are completely irregular, usually with deep minima of short duration. These minima occur quite often, giving rise to the idea that we are not dealing with only one dust cloud orbiting around the central star, but with many clouds, of which the powers of extinction are different; in other words these clouds have different optical thicknesses.

Following Zickgraf (1992) who had proposed a model for disks of B[e]-stars, we can construct a model for the circumstellar disks of HAEBESs, of which the intersection with a meridional plane is given in Figure 1. In this simplified model we propose that the disk consists of an inner gaseous part which is dust-free, except for a number of dust grains which are spiraling into the central star due to the Poynting-Robertson effect, and an outer ring consisting of gas and dust grains, in which clouds of dust particles are orbiting around the central star. The projections of these clouds on the intersecting meridional plane are indicated in Figure 1 as darkened circles.

The inner boundary of the dust ring is determined by the temperature of the central star. The extent of the dust ring above and below the equatorial plane depends on the distance of the inner boundary to the central star, and the angular width of the circumstellar disk. In the model of the disk we have described above we will always be able to see a total or a partial eclipse of the central star by a dust cloud, when we are looking edge-on (Case A, Figure 1) or in an angle not larger than the halfwidth, ϵ, of the disk (Cases B and C).

7. The Size and Space Density of the Dust Grains

Since the central star is still very young (age between 10^5 - 10^6 yr) we believe that the material in the circumstellar disk consists of, firstly, left overs of the parental material, and, secondly, new matter formed from gas outflows of the star, which condense into dust grains in the outer cooler regions of the disk. This combined disk matter is subject to all kinds of external forces which tend to determine the minimum size of the dust grains. Those external forces are partly gravitational (Poynting-Robertson effect and the binding gravitational forces between the grains in the cloud) and partly radiation (evaporation and pressure).

7.1. THE MINIMUM SIZE

The minimum size of the dust particles in a circumstellar environment can be estimated by the balance of gravitational forces acting on the dust grains and the radiation pressure of the central star. If we assume that the efficiency of radiation pressure is equal to unity the minimum radius of a single dust grain is given by

$$a_{min} = 0.6 \frac{L_*}{\rho M_*} \tag{15}$$

In this formula, (see Aumann et al., 1984; Martin, 1978, p. 138) a_{min} is expressed in microns, the stellar luminosity L_* and stellar mass M_* expressed in solar units and the density of the dust grains ρ in g cm^{-3}. For example, adopting $\rho = 2$ g cm^{-3} (Aumann et al., 1984), for the HAEBE star UX Ori, with $L_* = 65 L_\odot$ and $M_* = 2.8 M_\odot$ from Table I, the minimum size of the dust grains in its circumstellar ring is 7 μm. Grains having radii smaller than this value will be blown away by the radiation pressure of the central star. For the other HAEBESs of our sample minimum radii are given in Table II.

Note that the minimum radius given by formula (15) is independent of the distance of the dust grain from the central star. This is because both the gravitational force and the radiation pressure decrease with r_d^2. It is possible that larger dust particles due to the Poynting-Robertson effect are slowly spiraling towards the central star, but as soon as these particles, because of evaporation, become small enough, they will be blown back outwards (see Martin, 1978, p. 139) by the radiation pressure of the central star.

From polarizational measurements of so-called "isolated" Herbig stars at deep minimum brightness by Grinin et al. (1992), it has been found that the light of the central star is highly polarized, meaning that there must exist in the dust ring a large amount of scattering dust grains of size an order of magnitude smaller than as given by (15). How such small grains can exist in this harsh surrounding is something that should be investigated.

7.2. THE EFFECTS OF SHIELDING AND EXTRA GRAVITATION ON THE MINIMUM SIZE

The dust grains with minumum size given by (15) which we have discussed above can be found in the general field of the circumstellar disk. Dust grains located at the back side of a dust cloud (towards the observer) can have much smaller minimum sizes due to the effects of shielding and of extra gravitational force by the cloud itself. In such a cloud the radiation of the central star is attenuated by the dust grains, so that the radiation pressure exerted on a single dust grain lying at the back of the dust cloud is reduced by a factor

$$f = 10^{-0.4\Delta m} \tag{16}$$

in which Δm is the difference in magnitude of the radiation of the star just before entering the dust cloud and when it arrives at the back of it.

A dust grain lying at the back of a dust cloud is pulled in the direction of the central star by the gravitational force of this star, and by the gravitational force of the ensemble of grains in the cloud. The ratio of these gravitational forces is given by

$$g \approx \frac{M_{cl}}{M_*} \left(\frac{r_d}{R_*} \right)^2 \tag{17}$$

because $r_d/R_* \gg 1$ and $R_{cl} \approx R_*$. The minimum size of such a dust grain is then

$$a'_{min} = 0.6 \frac{L_*}{\rho M_*} \frac{f}{1+g} \tag{18}$$

For the star UX Ori we can estimate with $\Delta m = 2.7^m$, $M_{cl} = 3.7 M_{\oplus}$, $r_d = 1.4$ AU and $R_{cl} \approx R_* = 3.3 R_{\odot} = 0.015$ AU, that $f \approx 0.08$ and $g \approx 0.03$. This means that the radiation pressure on the dust grain lying on the back of the dust cloud of UX Ori is reduced by a factor of 0.08 and the gravitational pull on this grain is increased by 3%. The minimum size of such a grain then becomes $a'_{min} = 0.5\ \mu m$, which means that due to the shielding effect and the extra gravitational attraction of the dust cloud a grain lying at the back of this cloud can be about 10 times smaller than when it is located outside the cloud. Note that the radius of such a grain ($a'_{min} \approx 5000$ A) is about the same size as the wavelength of blue/yellow light.

For the other stars in our sample the new minimum sizes are given in Table III.

7.3. THE SIZE DISTRIBUTION AND THE AVERAGE RADIUS

The size distribution of the dust grains in the obscuring cloud can generally be assumed to be

$$dN(a) = const \times a^{-q}\, da \tag{19}$$

in which $q = 3.5$ for asteroids of our solar system as derived by Dohnanyi (1969) and for the dust particles in the interstellar medium (Mathis *et al.*, 1977). We will apply the value $q = 3.5$ also for the size distribution of the dust grains in the protoplanetary dust cloud of Herbig stars.

TABLE III

Properties of the dust grains.

Name	a_{min} (μm)	a'_{min} (μm)	$\langle a \rangle$ (μm)	N (10^8 km^3)
UX Ori	7	0.5	12	17
NX Pup	10	2.0	17	5
HR5999	4	0.7	7	37
KK Oph	2	0.4	3	220
R CrA	15	0.3	25	6
T CrA	1	0.1	2	1600

The average radius $\langle a \rangle$ of the dust grains in the obscuring cloud can be estimated using the above assumed general power law

$$\langle a \rangle = \frac{\int_{a_{min}}^{a_{max}} a^{-q+1}\, da}{\int_{a_{min}}^{a_{max}} a^{-q}\, da} \tag{20}$$

When we assume that $a_{max} \gg a_{min}$ we obtain the relation

$$\langle a \rangle = \frac{q-1}{q-2} a_{min} \tag{21}$$

provided that $q > 2$. As mentioned above, we will take $q = 3.5$. For a_{min} we will use the values calculated using (15). The average sizes of the dust grains in the obscuring clouds of our sample of stars are listed in Table III.

7.4. THE SPACE DENSITY OF THE DUST GRAINS

The mass of the obscuring cloud can be estimated using

$$M_{cl} = \frac{16\,\pi^2}{9} R_{cl}^3\, \rho \langle a^3 \rangle N \tag{22}$$

In this formula ρ is the density of the dust grains and N is their space density.

Let us denote the extinction at the Johnson V passband during maximum obscuration as A_V. We then have that

$$A_V = 1.086\, \langle a^2 \rangle N D Q_e, \tag{23}$$

in which Q_e is the extinction efficiency of the dust particles, and D is the effective thickness of the cloud. For large particles $Q_e \approx 2$ (see Martin, 1978). Furthermore, for a spherical cloud $D = \frac{4}{3} R_{cl}$. From (24) and (25) we obtain using $R_{cl} \approx R_*$

$$N = 0.345 \frac{A_V}{R_*} \frac{q-3}{q-1} \frac{1}{a_{min}^2} \tag{24}$$

since

$$\langle a^2 \rangle = \frac{q-1}{q-3} a_{min}^2 \tag{25}$$

provided that $a_{max} \gg a_{min}$ and that $q > 3$. From this derivation we see that without knowing the exact value of a_{max} it is possible to derive the space density of the dust grains in the obscuring cloud, essentially using the extinction power of this cloud. The values of N for the different stars in our sample are listed in Table III.

8. Discussion and Conclusion

In order to have an idea about the properties of the obscuring clouds orbiting around a Herbig Ae/Be star, which show irregular Algol-type variations in brightness, we have estimated several of their parameters using simplified models of the circumstellar disk, the shape of the obscuring cloud, its extinction power, and the external forces acting on the cloud.

The results of our calculations depend strongly on two observational parameters: the spectral types and the distances of the programme stars. The spectral type of HAEBESs is usually difficult to determine spectroscopically, because most of the Balmer lines are in emission. One has to use other means, for instance, photometry with appropriate corrections for the influence of emission lines, and/or the method proposed by Jaschek and Jaschek (1987) using the emission lines of the Balmer series. The distances are usually estimated roughly by means of different methods. For this reason the luminosity of the objects is, in many cases, not very well known. However, this situation will be improved in the near future when the HIPPARCOS data of HAEBESs become available. In spite of these shortcomings we have made estimates of different physical properties of the dust clouds which are responsible for the very large irregular variations in brightness of HAEBESs; in this way we will have an approximate idea about the nature of these dust clouds. We especially want to know whether conditions exist for the possible formation of planet-like objects.

In this respect, using simple assumptions and approximate calculations, we have been able to show that the physical conditions of the obscuring dust clouds are similar to those found in our planetary system. We have studied six edge-on HAEBESs which show very large brightness decreases ($> 1^m$), ranging in spectra from B8 to F0; three of them are main sequence objects and the other three giants. Temperatures of the dust clouds were estimated to be lying between 1200 to 1500 K.

The distances of these clouds to the central stars ranges from 0.5 to 3.3 AU, and their masses are between 0.3 to 6.3 $M_{\oplus}$. These results indicate that formation of planet-like objects from these dust clouds is possible.

It is of interest also to know the approximate size of the dust grains in the obscuring clouds; these are the ingredients for planet formation. The minimum radius of these grains, determined using the effect of radiation pressure in comparison with the gravitational attraction of the central star, is found to be from 1 to 15 μm. However, smaller grains are most probably also present on the back of the obscuring dust cloud, due to the effects of shielding and of extra gravitational attraction. The existence of these small grains ($< 1\mu$m) can be expected from the fact that the blue light of the central star is scattered by these small grains, giving rise to the large increase of the linear polarization at deep brightness minima, such as observed by Grinin et al. (1991).

The average diameter of the dust grains estimated using their extinction power at visual wavelengths seems to lie between 2 to 25 μm. These are average sizes of grains like those planetesimals in the interplanetary space of our solar system. We have not been able to make an estimate of the maximum size of the rocks in the obscuring clouds. Such an estimate is actually very important, since the big rocks play an important role in the formation of planets.

From the above discussion about the sizes of the grains in the obscuring clouds, we are strengthened in our belief that in these clouds conditions for planet formation indeed exist.

References

Aumann, H.H., Gillett, F.C., Beichman, C.A., de Jong, T., Houck, J.R., Low, F.J., Neugebauer, G., Walker, R.G. and Wesselius, P.R.: 1984, *Astroph. J.* **278**, L23.
Bastian, P. and Ménard, F.: 1988, *Astroph. J.* **326**, 334.
Bastian, P. and Ménard, F.: 1990, *Astroph. J.* **364**, 232.
Bibo, E.A. and Thé, P.S.: 1990, *Astron. Astroph.* **134**, 273.
Bibo, E.A. and Thé, P.S.: 1991, *Astron. Astroph.* **89**, 319.
Bibo, E.A., Thé, P.S. and Dawanas, D.N.: 1992, *Astron. Astroph.* **260**, 293.
Chandrasekhar, S.: 1943, *Principles of Stellar Dynamics*, Dover Publ., New York, p. 220.
Grinin V.P., Kiselev, N.N., Minikulov, N. Kh., Chernova, G.P. and Voshchinnikov, N.V.: 1991, *Astroph. Space Sci.* **186**, 283.
Dohnanyi, J.S.: 1969, *J. Geophys. Res.* **74**, 2531.
Harris, D.L., Strand, K.Aa. and Worley, C.E.: 1963, Strand K.Aa (ed.), *Basic Astronomical Data, Stars and Stellar Systems* III, Chicago.
Herbig. G.H. and Bell, K.R.: 1988, *Lick Obs. Bull.* **No. 1111.**
Jaschek, C. and Jaschek, M.: 1987, *The Classification of Stars*, Cambridge Univ. Pres, Cambridge.
Martin, P.G.: 1978, *Cosmic Dust, Its Impact on Astronomy*, Oxford Univ. Press, Oxford.
Mathis, J., Rumpl, W. and Nordsieck, K.H.: 1977, *Astroph. J.* **217**, 425.
Mineur, H., 1939, *Ann. Astrophys.* **2**, 1.
Rowan-Robinson, M.: 1980, *Astroph. J. Suppl.* **44**, 403.
Schmidt-Kaler, Th.: 1982, in: *Landolt-Börnstein Catalogue*, Vol. 2b.
Steenman, H. and Thé, P.S.: 1989, *Astroph. Space Sci.* **159**, 189.
Strom, S.E., Keene, J., Edwards, S., Hillenbrand, L., Strom, K., Gauvin, L. and Condon, G.: 1990, *Angular Momentum Evolution of Young Stars*, NATO Advanced Research Workshop, in press.

Thé, P.S., Tjin A Djie, H.R.E., Bakker, R., Bastiaansen, P., Burger, M., Cassatella, A., Fredga, K., Gahm, G., Liseau, R., Smyth, M.J., Viotti, R., Wamsteker, W. and Zeuge, W.: 1981, *Astron. Astroph.* **44**, 451.
Tjin A Djie, H.R.E., Remijn, L. and Thé, P.S.: 1984, *Astron. Astroph.* **134**, 273.
Voshchinnikov N.V. and Grinin, V.P.: 1992, *Astrophysics* **34**, 84.
Wenzel, W., Dorschner, J. and Friedemann, C.: 1971 *Astron. Nachr.* **292**, 221.
de Winter, D. and Thé, P.S.: 1990, *Astroph. Space Sci.* **166**, 99.
Zickgraf, F.-J.: 1992, in: L. Drissen, C. Leitherer and A. Nota (eds.), *Nonisotropic and Variable Outflows from Stars*, *ASP Conference Series* **22**, 75.

IDENTIFICATION OF A COLLAPSING PROTOSTAR *

NEAL J. EVANS II

Department of Astronomy, The University of Texas at Austin,
Austin, Texas, USA

SHUDONG ZHOU

Department of Astronomy, University of Illinois,
Urbana, Illinois, USA

and

CARSTEN KÖMPE and C. M. WALMSLEY

Max-Planck-Institut für Radioastronomie,
Bonn, Germany

Abstract. The globular molecular cloud B335 contains a single, deeply embedded, far-infrared source. Our recent observations of H_2CO and CS lines toward this source provide direct kinematic evidence for collapse. Both the intensity and detailed shape of the line profiles match those expected from inside-out collapse inside a radius of 0.036 pc. The collapse began about 1.5×10^5 years ago, similar to the onset of the outflow. The mass accretion rate is about 10 times the outflow rate, and about $0.4 \, M_\odot$ should have now accumulated in the star and disk. Because B335 rotates only very slowly, any disk would still be very small (about 3 AU). The accretion luminosity should be adequate to power the observed luminosity. Consequently, we believe that B335 is indeed a collapsing protostar.

1. Introduction

It is by now a commonplace that stars form in collapsing parts of molecular clouds. Theories of both star and planet formation rely on this fundamental picture. It is less well known outside the star formation community that direct observational evidence for collapse is almost entirely lacking. Numerous claims of collapse motions have been made, but most have encountered considerable skepticism. More importantly, none have applied to the collapse of a region likely to form a single protostar (see, e.g., Evans, 1991).

The overwhelming kinematic signature in most regions of star formation is not collapse, but outflow (Lada 1985; Bachiller and Gómez-González, 1993). The outflow is indicated by a variety of tracers, including wide wings on CO and other molecular lines, masers, and Herbig-Haro objects. In a spherical picture of star formation, such outflows would indicate that the collapse phase had already ended in almost every object studied, even those which seem young by other indications.

The ubiquity of evidence for outflow and the nearly total absence of evidence for collapse led Wynn–Williams (1982) to refer to a collapsing protostar as the "holy grail" of star formation studies. As was no doubt true of the legendary grail, there is no shortage of candidates discovered by infrared and submillimeter continuum observations. The problem is one of authentication. A candidate must pass the

---

* Paper presented at the Conference on *Planetary Systems: Formation, Evolution, and Detection* held 7–10 December, 1992 at CalTech, Pasadena, California, U.S.A.

Astrophysics and Space Science **212**: 139–145, 1994.
© 1994 *Kluwer Academic Publishers.*

following tests: first, it must have a kinematic signature of collapse; second, its luminosity must plausibly result from accretion, rather than any nuclear reactions. The latter requirement suggests a focus on objects early in the collapse phase.

During the last decade, three new developments have again raised the possibility of identifying a collapsing protostar. The first of these was inside-out collapse (Shu, 1977; Shu *et al.*, 1987). In their picture, clouds or regions forming stars of low mass are supported almost entirely by thermal pressure. In these conditions, they would first relax to a centrally condensed distribution ($n(r) \propto r^{-2}$) and then initiate collapse from the inside. A wave of infall propagates outward at the effective sound speed, and the density distribution inside the infall radius (r_{inf}) relaxes to $n(r) \propto r^{-1.5}$. The implications of this idea for collapse searches are several. Since the collapse occurs first in the inside and occurs at relatively low velocity, we can detect kinematic evidence of collapse in the early, protostellar phases only with the use of high spatial and spectral resolution. Because the collapse begins in the innermost, densest part of the cloud, it will be best revealed by molecular lines that require high density for excitation. These will tend to "see" through the static envelope and probe the collapse region. Finally, since this picture was developed for low-mass star formation, low-mass clouds or regions would be the best candidates. In particular, small, globular molecular clouds of a few solar masses may have only a single collapse center, simplifying the kinematic signature.

The second important development was the abandonment of spherical symmetry. In non-spherical geometries, collapse and outflow can coexist. Since many outflows are bipolar, it is natural to think of allowing collapse in the plane perpendicular to the outflow direction. Theoretically, non-spherical geometries in the innermost regions of the collapse are a natural consequence of rotation. In calculations of the collapse of a cloud with initially slow rotation, Terebey, Shu, and Casson (1984) found that the geometry becomes quite non-spherical and a disk is likely to form inside the centrifugal radius (where the infall speed equals the rotation speed). The outflow would then be perpendicular to the plane of the disk. Since the disk is the likely site of planet formation, it is very important to check this picture observationally. The relevance of this idea to searches for collapse is that we need not reject clouds with outflows, but we must use tracers which are not dominated by outflow.

With the perspective of the first two developments, it is easy to see why observational evidence of collapse has remained elusive. Almost all simulations of line profiles from collapsing clouds (e.g., Anglada *et al.*, 1987) and most searches for kinematic evidence concentrated on CO, a molecule which is abundant and easily excited. These properties meant that CO emission probed mostly the outer regions of clouds and that it was especially prominent in outflows. Clearly, simulations of line profiles and observational efforts needed to focus on different molecular lines.

The third important development grew out of this last realization. Zhou (1992) modeled the evolution of line profiles during an inside-out collapse, focusing on lines of molecules which require high densities for excitation. The primary result

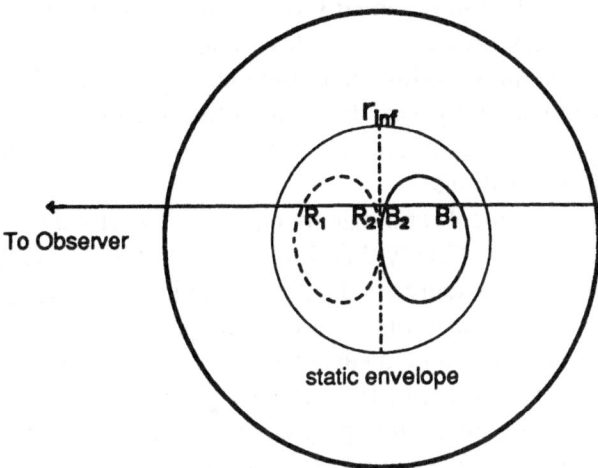

Fig. 1. A schematic diagram of a cloud experiencing inside-out collapse inside the inner circle (labeled r_{inf}). The ovals are the locus of points with the same velocity projected on the line of sight to the observer. For a density sensitive line and the density field of collapse, B_2 and R_2 produce stronger emission than B_1 and R_1. Since R_1 lies in front of R_2, and has the same projected velocity, R_1 obscures R_2, whereas the strong emission from B_2 is unobscured by B_1.

of that study is that the best line to employ for collapse searches is one requiring fairly high densities to excite, but which also has opacity of order 1. The modest opacity produces a distinctive kinematic signature of collapse, in which the profile appears self-absorbed (having a central minimum between two peaks), and the blue-shifted peak is brighter than the red-shifted peak. The self-absorption and the line width should decrease away from the center of the collapse or when observed with lower resolution. Finally, if several lines with different critical density are observed, the line width should increase with increasing critical density. These properties thus define a distinctive signature of inside-out collapse. In addition to predicting the general shapes of the lines, the inside-out collapse model, together with Zhou's radiative transport calculations, provides a prediction for the intensities and detailed shape of the lines, as a function of the time since collapse began, or equivalently the infall radius (r_{inf}), making the whole theory eminently testable.

2. The Candidate

Our candidate for collapse is an isolated, roundish globule called B335, located at a distance of 250 pc. It was probably discovered by Barnard (1927); at least it takes its name from his photographic atlas. The term globule was given to this class of objects by Bok and Reilly (1947), who also suggested that they may represent objects "... just preceding the formation of a star." Their suggestion was verified by Keene *et al.* (1983), who discovered a far-infrared source in B335, which they suggested "may prove to be ... low luminosity protostar." Two later papers presenting data on the submillimeter continuum emission (Gee *et al.*, 1985 and Chandler *et al.*, 1990) actually used the term protostar in the title, but both with question marks. The main source of the uncertainty was that outflow, not collapse, was indicated by CO observations.

B335 is in many ways an ideal candidate for collapse. The cloud is small and has very little turbulence, making it likely that Shu's picture of a thermally supported cloud is relevant. The infrared source is deeply embedded, being detected only at $\lambda > 60$ μm, indicative of an early phase in the collapse process. The outflow is nearly in the plane of the sky, oriented east–west, with an opening angle of about 45° (Hirano *et al.*, 1988; Cabrit *et al.*, 1988). These properties make it possible to avoid the outflow to some extent by mapping north–south.

Our work on this object began with a map of the 6-cm H_2CO line using the VLA (Zhou *et al.*, 1990). This line, a transition between two states of a K-doublet ($J_{K_{-1}K_1} = 1_{11} - 1_{10}$) appears in absorption against the cosmic background radiation because of a collisional pumping effect which cools its excitation temperature below 2.7 K (Townes and Cheung, 1969). Since the cosmic background temperature is extremely uniform, it provides a smoothly distributed background lamp, meaning that any structure in the observations arises in the cloud. We observed an apparent ring of absorption, with a hole centered near the infrared source and the location of peak emission in other molecular tracers. The explanation for this effect is that the collisional pump works optimally in a range of densities from about 10^3 to 10^4 cm^{-3}; above about 10^6 cm^{-3}, the collisions drive the line into emission, but there is a range of densities where the line has an excitation temperature close enough to that of the background radiation that neither emission nor absorption will be seen. We interpreted the ring as the effect of a density gradient in the source. Detailed modeling showed that a density gradient consistent with an inside-out collapse ($n(r) \propto r^{-\alpha}$, with $\alpha = 2.0$ outside a radius of about 0.03 pc and and 1.5 inside that radius) gave the best fit to the data. This model then predicted that the $\Delta J = 1$ transitions of H_2CO would appear in emission from the region of the "hole" in the absorption ring. Viewed with sufficient resolution, these lines might also show the kinematic signature of collapse.

3. The Evidence

To obtain the requisite resolution, the IRAM 30-m telescope was used to observe simultaneously two $\Delta J = 1$ lines of H_2CO: the $J_{K_{-1}K_1} = 2_{12} - 1_{11}$ line (140 GHz) and the $J_{K_{-1}K_1} = 3_{12} - 2_{11}$ line (225 GHz). The lines toward the peak of the map, coincident within uncertainties with the infrared source and a radio continuum source (Anglada *et al.*, 1992), match the predictions for an inside-out collapse remarkably well (Zhou *et al.*, 1993). The shapes of the lines provide strong evidence for collapse. Zhou *et al.* (1993) also observed three lines of CS

($J = 2 - 1, 3 - 2$, and $5 - 4$), and these lines also indicate collapse, although they are more affected by the outflow than are the H_2CO lines. Both molecules can be used to determine the parameters of the infall by varying the infall radius and molecular abundance to find the best fit to the observed profiles. The overall best fit is obtained for $r_{inf} = 0.036$ pc, with H_2CO favoring slightly smaller radii and CS favoring slightly larger radii. The resulting model profiles are compared with the observations in Figure 2 (note that the model profiles differ slightly from those in Zhou *et al.*, 1993, because we have corrected an error in the modeling program).

The best-fit abundance of H_2CO is 3.6×10^{-9} and the best-fit CS abundance is 3.2×10^{-9}, consistent with many other determinations, but much more constrained than previous measurements. The CS lines, especially the $3 - 2$ line, are more affected by the outflow than are the H_2CO lines, but spectra at positions to the north and south (perpendicular to the outflow axis) are relatively free of outflow emission and match predictions of the model well.

Zhou *et al.* (1993) considered alternative models for B335, including increased turbulence toward the center, rotation, spherical expansion, and outflow. None of these can explain the line profiles. The only alternative model which comes close to matching the observations requires a foreground cloud absorbing the emission from the background cloud, which contains the infrared source. This picture requires a very low velocity dispersion in the foreground gas ($\Delta v_f = 0.15$ km s^{-1}) for which there is no supporting evidence. In addition, the CS lines are not well-fitted in this model unless the central velocity of the background component shifts with J. The foreground absorption model is highly contrived and thus quite unlikely. While certainty is probably unattainable, simplicity certainly favors the conclusion that B335 is undergoing inside-out collapse, with density and velocity fields given by Shu *et al.* (1987).

4. Conclusions

Assuming that the collapse interpretation is correct, one can then use the inside-out collapse model to compute other quantities of interest. With $r_{inf} = 0.036$ pc and an effective sound speed (including a small turbulent contribution) of 0.23 km s^{-1}, the time since collapse began is 1.5×10^5 yr, similar to the (quite uncertain) age of the outflow, indicating that outflow may have begun very early in the collapse. The mass accretion rate would be $2.8 \times 10^{-6} M_\odot$ yr^{-1}, about 10 times the mass loss rate in the outflow. The total mass accumulated in the star and disk would be $0.4 M_\odot$, while the total reservoir from which material could eventually accrete is about $12 M_\odot$. The B335 cloud rotates only very slowly, with $\Omega = 1.4 \times 10^{-14}$ s^{-1} (Frerking *et al.*, 1987). Consequently, the centrifugal radius is only 3 AU. Deviations from spherical symmetry would thus occur on much smaller scales than our resolution and hence be negligible in our modeling.

Finally, we can ask whether B335 satisfies the other criterion for a collapsing protostar: luminosity derived from accretion. Given the mass accretion rate derived

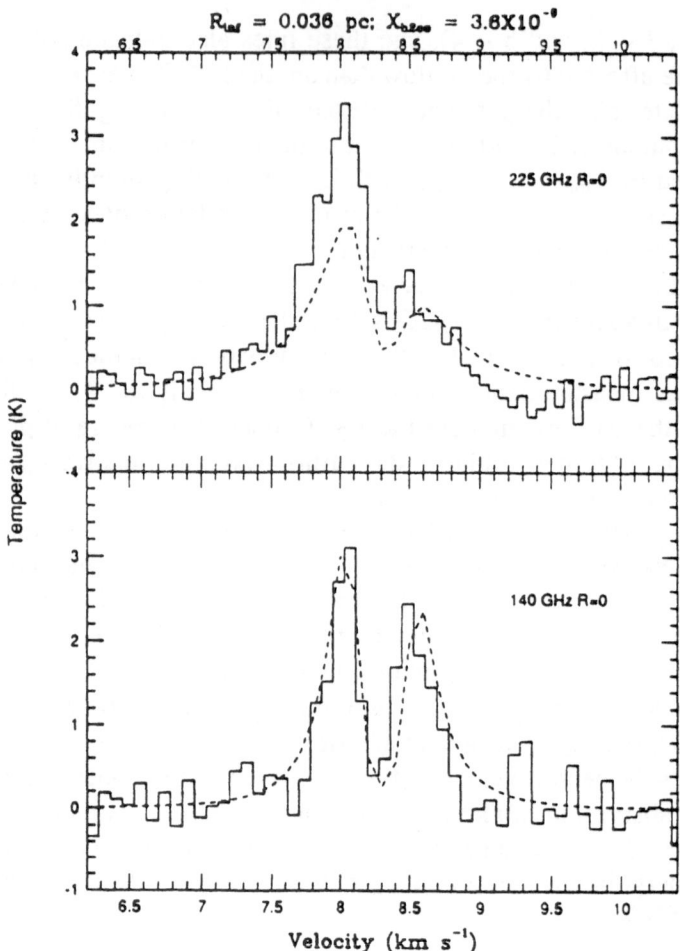

Fig. 2.   The observed spectra toward the center of B335 are shown as solid histograms. The dashed lines are model line profiles predicted by an inside-out collapse model with $r_{inf} = 0.036$ pc.

above, the accretion luminosity would equal the observed luminosity of 3 $L_\odot$ as long as the radius of the star is about 6 $R_\odot$, consistent with theoretical expectations (Stahler et al., 1980). All the facts are consistent with the interpretation of B335 as a collapsing protostar — the holy grail of star formation.

Acknowledgements

This work was supported by NSF Grants AST88-15801 and 90-17710, by NASA grant NAGW-2323, by a grant from the W. M. Keck Foundation, and by a grant from the Texas Advanced Research Program to the University of Texas.

References

Anglada, G. Rodríguez, L. F., Cantó, J., Estalella, R. and López, R.: 1987, *Astron. Astrophys.* **186**, 280.

Anglada, G. Rodríguez, L. F., Cantó, J., Estalella, R. and Torrelles, J.M.: 1992, *Astrophys. J.* **395**, 494.

Bachiller, R. and Gómez-González, J.: 1993, *Astron. Astrophys. Rev.* **3**, 257.

Barnard, E.E.: 1927, in: E.B. Frost and M. R. Calvert (eds.), *A Photographic Atlas of Selected Regions of the Milky Way*, Carnegie Institution of Washington.

Bok, B.J. and Reilly, E.F.: 1947, *Astrophys. J.* **105**, 255.

Cabrit, S., Goldsmith, P.F. and Snell, R.L.: 1988, *Astrophys. J.* **334**, 196.

Chandler, C.J., Gear, W.K., Sandell, G., Hayashi, S., Duncan, W.D., Griffin, M.J. and Hazell, A.S.: 1990, *Monthly Not. Roy. Astr. Soc.* **243**, 230.

Evans II, N.J.: 1991, Star formation: Observations. in: D.L. Lambert (ed.), *Frontiers of Stellar Evolution*, San Francisco, Astronomical Society of the Pacific, p. 45.

Frerking, M.A., Langer, W.D. and Wilson, R. W.: 1987, *Astrophys. J.* **313**, 320.

Gee, G., Griffen, M.J., Cunningham, C.T., Emerson, J.P., Ade, P.A.R. and Caroff, L.J.: 1985, *Monthly Not. Roy. Astr. Soc.* **215**, 15P.

Hirano, N., Kameya, O., Nakayama, M. and Takakubo, K.: 1988, *Astrophys. J. Letters* **327**, L69.

Keene, J. *et al.*: 1983, *Astrophys. J. Letters* **274**, L43.

Lada, C.J.: 1985, *Ann. Rev. Astron. Astrophys.* **23**, 267.

Shu, F.H.: 1977, *Astrophys. J.* **214**, 488.

Shu, F.H., Adams, F.C. and Lizano, S.: 1987, *Ann. Rev. Astron. Astrophys.* **25**, 23.

Stahler, S.W., Shu, F.H. and Taam, R.E.: 1980, *Astrophys. J.* **241**, 637.

Terebey, S., Shu, F.H. and Cassen, P.: 1984, *Astrophys. J.* **284**, 529.

Townes, C.H. and Cheung, A. C.: 1969, *Astrophys. J. Letters* **157**, L103.

Wynn-Williams, C.G.: 1982, *Ann. Rev. Astron. Astrophys.* **20**, 587.

Zhou, S.: 1992, *Astrophys. J.* **394**, 204.

Zhou, S., Evans II, N.J., Butner, H.M., Kutner, M.L., Leung, C. M. and Mundy, L.G.: 1990, *Astrophys. J.* **363**, 168.

Zhou, S., Evans II, N.J., Kömpe, C. and Walmsley, C.M.: 1993, *Astrophys. J.* 404, 232.

References

COMETARY-LIKE BODIES IN THE PROTOPLANETARY DISK

AROUND β PICTORIS *

H. BEUST

Service d'Astrophysique, DAPNIA,
Centre d'Etudes de Saclay, Gif-sur-Yvette, France

A. VIDAL-MADJAR, R. FERLET

Institut d'Astrophysique de Paris,
CNRS, Paris, France

and

A.M. LAGRANGE-HENRI

Groupe d'Astrophysique de Grenoble,
CERMO, Grenoble, France

Abstract. Repeated spectroscopic observations of β Pictoris have been performed since 1985 and revealed the presence, in many metallic lines like Ca II, Mg II, Fe II, ..., of strong sporadic circumstellar absorption, redshifted by tens to hundreds of km s^{-1} with respect to the star, and highly time-variable (time-scales of days or hours).

We have tentatively interpreted these variable events as the spectral signature of infalling cometary-like bodies, when evaporating in the vicinity of the star. This scenario has been furthermore theoretically studied, and we showed that it could indeed explain correctly the observations with their peculiar characteristics, like (1) the behavior difference between visible and UV lines, (2) the unusual line ratios, (3) the surprising presence of Al III absorption lines. Constraints deduced from both observational data and theoretical study allowed us to suggest that a planet within the disk could be responsible, by perturbations, of this high rate infall of small bodies towards the star ($>$ 100 per year).

1. Introduction

The dusty disk around the southern star β Pictoris has been the subject of intense investigations since its discovery (Smith and Terrile, 1984). It is yet regarded as the best candidate for an extra-solar planetary system in a still unknown evolutionary state. It is however expected to be less evolved than the solar system, since the age of β Pic is estimated to be $\sim 10^8$ years (Paresce, 1991), i.e. close to its ZAMS. The favourable edge-on orientation of the disk enabled the study of its gaseous counterpart by spectroscopic absorption. This has been done as early as late 1984 (Hobbs *et al.*, 1985, hereafter Paper I; Vidal-Madjar *et al.*, 1986, Paper II). (Figure 1) Some of the observed lines soon revealed a strange behavior. At the bottom of the Ca II K stellar line (observed with the CES at ESO, La Silla, Chile), apart from a central, stable, circumstellar absorption feature, variable absorption components have been sporadically detected (Ferlet *et al.*, 1987, Paper V; Hobbs et al., 1988, Paper VII; Beust *et al.*, 1991a, Paper XI; Lagrange-Henri *et al.*, 1992, Paper XIII). These additional features always appeared redshifted with respect to

* Paper presented at the Conference on *Planetary Systems: Formation, Evolution, and Detection* held 7–10 December, 1992 at CalTech, Pasadena, California, U.S.A.

Astrophysics and Space Science **212**: 147–157, 1994.
© 1994 *Kluwer Academic Publishers.*

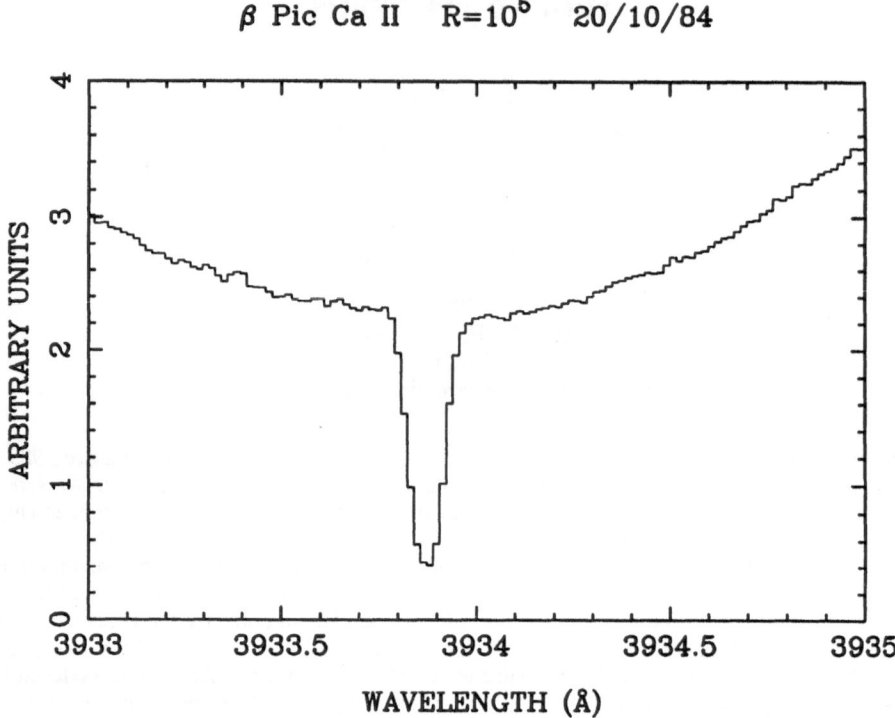

Fig. 1. High resolution spectrum towards β Pic on October 20, 1984. At the bottom of the rotationally broadened stellar line, a sharp and centered circumstellar absorption is detected.

the central one by a few tens of km s^{-1} (Figure 2). These spectral events could appear or disappear within less than one day (Papers V, XIII).

Meanwhile, the UV spectrum of β Pic was continuously monitored by the IUE satellite, recording frequent similar variations in lines of Mg II, Fe II, and Al III (Figure 3), with infall velocities towards the star reaching sometimes a few hundreds of km s^{-1} (Kondo and Bruhweiler, 1985; Lagrange *et al.*, 1986, Paper IV; Lagrange-Henri *et al.*, 1987, Paper VI; Lagrange-Henri *et al.*, 1989, Paper VIII).

All these spectral events were very surprising and needed to be interpreted.

2. The Characteristics of the Variations

The spectral events mentioned above have now been monitored for many years, and it is possible to draw a condensed list of their major characteristics.

First, they are very frequent and sporadic. The exact frequency of these events is difficult to evaluate, since it would require a continuous survey of the star, but from our sparse but frequent observations, it can be estimated to be ~ 100 per year. No evident periodicity has been detected in the apparition of these events which is absolutely sporadic. The bulk frequency seems to be changing over time scales

β Pic Ca II   R=10^5 11/01/86

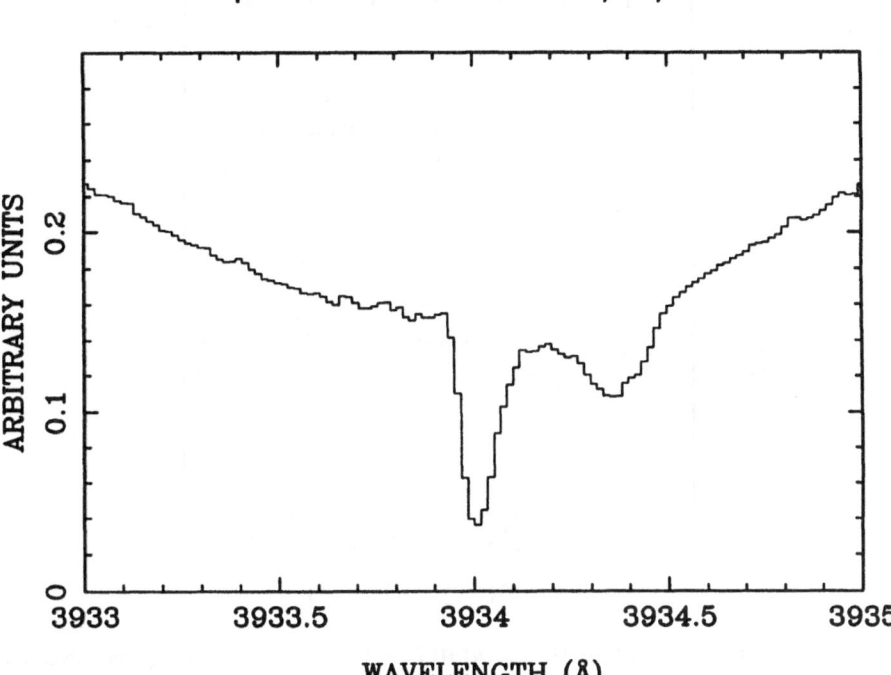

Fig. 2.   High resolution spectrum towards β Pic on January 11, 1986. Apart from the central and stable component, a redshifted one is clearly present at about 40 km s^{-1} redshift.

of years. This means there have been years with high activity and other ones with very low activity.

Second, the variations almost always appear in the red wing of the lines, meaning that we are observing infalling gas towards the star. The infall velocities themselves are subject to variations. This point is discussed below. Blueshifted features have been detected, but they are always very faint when compared to the main redshifted ones and not as obviously detectable (Paper XIII).

Finally, it must be stressed that these features are highly time variable, on time scales of days or even hours. This last point may be considered as their most striking one.

Apart from these major characteristics, a number of what should be called "puzzling points" need to be listed, meaning that every potential interpretation should take them into account.

It first became obvious after repeated observations that there is a behavior difference between the variations occurring in the visible lines of Ca II on the one hand, and those occurring in UV lines on the other hand. These differences mainly concern infall velocities and variation time scales. Redshifts in Ca II lines always remain in the range of few tens of km s^{-1} (< 50 km s^{-1}). On the contrary, the

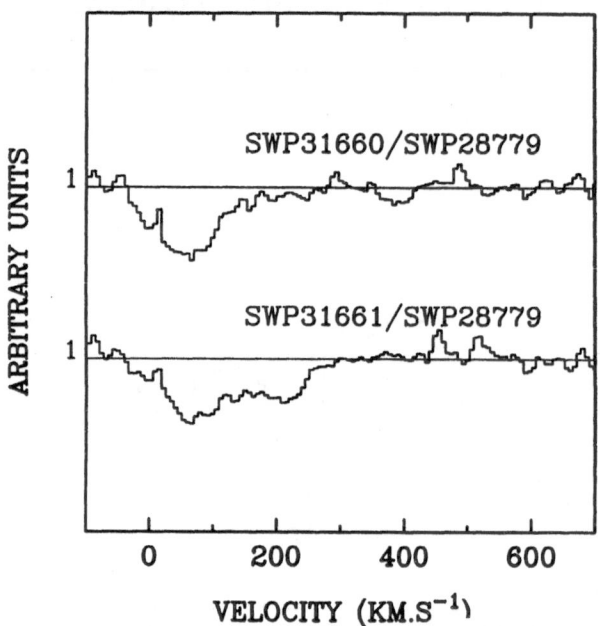

Fig. 3.   Two peculiar spectra of the 1854 Å Al III line. To make the time variation appear clearly, these spectra have been divided by a reference one. They reveal the apparition of a high velocity variation feature on a very short time scale: SWP31661 was recorded one hour after SWP31660.

infall velocities of UV features have been observed as high as 300 or 400 km s^{-1}, i.e. very close to the free fall velocity in the immediate vicinity of the star. It also appeared that UV events evolve more rapidly. Some of the high velocity features have been observed to appear or disappear in less than one hour (Papers VI, VIII). Ca II events never showed such an extreme behavior, their variation being much more of the order of the day.

When doublet lines are present (this is the case for Ca II H and K lines, and for the Al III doublet $\lambda\lambda1864, 1862$Å), it is possible to perform absolute saturation measurements of the lines, when dividing the two adjacent spectra. A ratio close to the oscillator strength ratio between the two doublet lines means that the observed features are optically thin, while a ratio close to unity indicates saturated lines. This procedure applied to the variable components (when present) revealed that they are always saturated (Papers VIII, XIII). However, all these lines (we mean the *variable* parts) were obviously far from reaching the zero intensity level.

Finally, the mere presence of Al III lines as mentioned above should be surprising. Al III ions are totally unable to be produced by stellar photoionization of Al II. β Pic is an A5V type star, and its flux at the Al II ionization wavelength (690 Å, i.e. extreme UV) is far too low to produce any sufficient quantity of Al III. Indeed, spectra of other A5V stars (e.g. 21 Vul) revealed *no* Al III absorption line.

3. Interpretation and Models

Understanding and interpreting such unusual observations appeared very early as an important problem, which had to be probably related to circumstellar material around β Pic.

The first idea one could think of is that we are observing a purely stellar phenomenon. However, this would require all stellar lines to be affected, not only the resonance lines or those arising from weakly excited levels. The observed behavior could also be related to an equatorial gas ring around the star which should be pulsating, meaning that β Pic is a so called "shell-star". β Pic had indeed been formerly classified as a shell star (Sletteback and Carpenter, 1983), and this does not have to be excluded. Nevertheless, the recorded variation cannot be interpreted that way, just because of the very short variation time scales and of the observed infall velocities, which are far above the usual pulsating velocities of shells, and much more like free fall velocity at a few stellar radii (hereafter R_*) from the star.

These basic characteristics concerning variation time scales and infall velocities led us to propose another interpretation (Papers VI, VIII). It is usually thought that cometary-like bodies or planetesimals (say, solid bodies in the kilometer range) were much more numerous in the early solar system than presently. Since β Pic is a much younger main sequence star than the Sun (Paresce, 1991), it should not be surprising to think that β Pic's system is somewhat close to that early stage. Therefore, we proposed that some of the suspected rocky bodies might (by perturbation) be set on star grazing orbits, and thus fall towards the star. These bodies, when approaching the star, should evaporate volatile material exactly like solar comets, and furthermore also release metallic material when coming into the immediate vicinity of the star ($\lesssim 100 R_*$). Thus the resulting ionic cloud surrounding the body may then generate the observed events when crossing the line of sight.

This scenario needed to be more precisely studied to check whether it was realistic or not. We thus carried out a theoretical study of the dynamics of metallic ions escaped from the parent body at the considered stellar distances (Beust *et al.*, 1989, Paper IX; Beust *et al.*, 1990, Paper X; Beust *et al.*, 1991a, Paper XI).

When released by the rocky body, evaporated dust is quickly ionized by the stellar flux, and the resulting ions are essentially subject to the following actions:

1. The gravity of the star. Its only effect is to make the ions follow their parent body on its path;
2. The stellar radiation pressure;
3. The collision with the surrounding medium which mainly consists of volatile material escaped from the same parent body.

The collisions can be computed as a dipole-induced action between charged ions and neutral atoms. The resulting effect can be described as a force opposed to the ion's motion relative to the neutral medium, proportional to the density of the medium and to the relative velocity (see Paper IX for details). Radiation pressure

TABLE I

Values of the ratio β *radiation pressure/gravity* for Ca II, Al III and Mg II (see Paper IX for the details of the calculation)

Element and Calculated β value
Ca II and 77.2
Al III and 7.76
Mg II and 4.45

results from photon absorption by ions at transition wavelengths, mainly those where we recorded spectral variations. Its intensity may be described as a constant ratio of β to gravity (both effects are $\propto r^2$). Since transition wavelengths are not the same, this ratio should vary among ions.

Table I shows that the ratio is always greater than unity, i.e. radiation pressure is larger than gravity. It also shows that radiation pressure is much more efficient on Ca II than on the other ions. This last point is due to the difference between β Pic's visible and UV fluxes.

The radiation pressure discrimination between Ca II and the other ions is in fact the basic reason explaining the behavior difference between visible (Ca II) and UV variation features. This appeared clearly through the dynamical simulations which followed this first study (Papers X, XI).

The simulation program has been designed to further investigate the proposed scenario. It first considers one infalling body on a given orbit which is intended to cross the line of sight. Since we are considering the part of these orbits located at stellar distances of about tens of R_* or less, and that the bodies are supposed to originate from much larger distances within the disk (typically few AUs), we can safely take the orbits as almost parabolic ones. The only orbital parameters are thus the perihelion q, and the angle Φ between the orbit axis and the line of sight.

Once the orbit is fixed, and once one is given parameters like vaporization rates of the comet (See Paper X for details), the motion of metallic ions escaped from the body is numerically computed and the spectral signature of the resulting cloud when crossing the line of sight is calculated.

This simulation gave excellent results. Figure 4 gives an example of what has been obtained with Ca II ions for a comet crossing the line of sight at $\simeq 22 \ R_*$. The simulated absorption feature is very similar to the one of Figure 2. Other simulations (Papers X, XI) with different input parameters have been able to reproduce faithfully a large variety of observed events, either visible or UV ones.

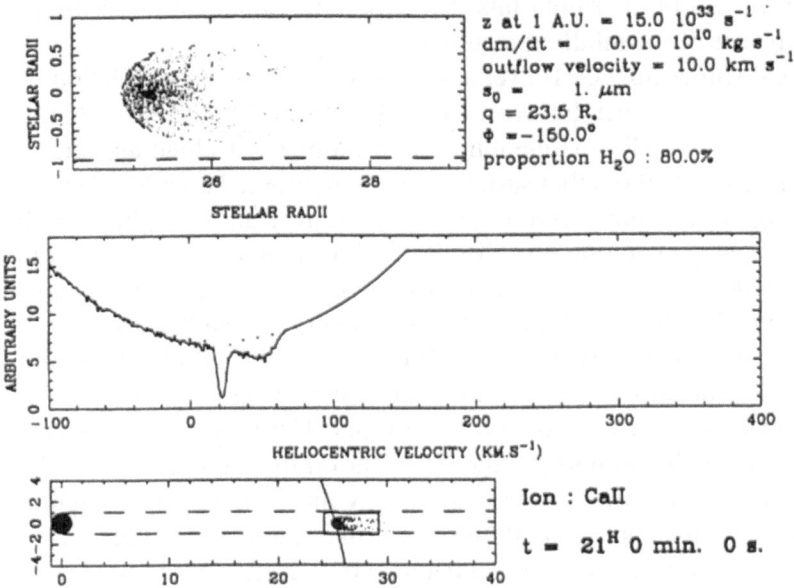

Fig. 4. Plot of the simulation reproducing the observations. The first plot shows a two-dimensional view of the Ca II cloud around the nucleus. Each single dot is one numerically followed ion. The second plot is the synthetic spectrum. The stellar rotationally broadened and the sharp central circumstellar absorptions are constant data. The third plot gives a view of the general situation, with the orbit, the star, and the line of sight (the two dashed lines). The box around the nucleus and the Ca II cloud represent the limits of the first plot. One can note here a deep additional absorption component.

It is interesting to look at the shape of the Ca II cloud around the nucleus. It is a parabolic-like cloud whose size is in fact entirely controlled by collisions and radiation pressure acting on Ca II ions. Once fresh ions are released by the nucleus, they found a dense medium where collisions dominate. They thus begin to follow this medium in its radial expansion around the nucleus. When reaching outer regions of the coma, the density of the surrounding medium drops, radiation pressure begins to overcome collisions, and ions begin to drift away from the coma in the anti-star direction. The net result is the formation of a cloud delimited by a parabolic-like surface.

A very important fact is that the size (i.e. its cross section) of the cloud is controlled by the balance between radiation pressure and collisions. If radiation pressure increases, it may overcome collisions earlier, and the resulting cloud is then smaller.

The size of the cloud is a determining factor for the resulting spectral signature. The fact that it is smaller than the stellar disk explains in fact why we observe saturated lines which do not reach the zero level. The lines are really saturated, but since the absorbing region is smaller than the stellar disk, it cannot absorb more than a given part of the stellar flux. Saturation measurements are a direct proof of the clumpiness of the infalling material.

Figure 4 shows how Ca II events may be reproduced by simulation. If we now consider the same comet, but crossing the line of sight much closer to the star (i.e. we take a much smaller perihelion), the resulting Ca II cloud appears very small once projected onto the stellar surface (see e.g. figures in Paper X). This is due to the r^{-2} scaling of radiation pressure. The closer to the star, the higher it is. If we take now UV events generating ions, Mg II or Al III, although the scaling of radiation pressure is the same, its effect is not as drastic, just because the β ratio is about ten times smaller. The consequence is that at very few stellar distances ($\lesssim 8R_*$), infalling comets may only generate UV events, since Ca II clouds are too small to generate deep absorption features. But these comets are the only ones which can produce high velocity ($\gtrsim 100$ km s^{-1}) events. These events are also expected to be those which should appear and disappear on the shortest time scales, simply because the crossing time of one given body through the line of sight is smaller when closer to the star. It finally turns out that the recorded behavior difference between visible and UV events is a natural consequence of our model (Paper X).

Recently, we carried out a hydrodynamical approach of the flow of metallic ions around cometary nuclei (Beust and Tagger, 1993). The basic idea was to calculate the delimiting surface of the ionic cloud as a shock, first to confirm the results of the numerical simulations, and second to have access to parameters along the shock surface such as temperature or velocity profiles. This work is still under way, but the first results seem to prove that the generated heat (i.e. the temperature profile of ions) is very high along the shock surface (a few times 10,000 K). This is mainly due to compression and to friction with the neutral medium as soon as ions begin their drift in the anti-star direction. The main result is that such temperatures are able to explain easily the collisional ionization of Al II into Al III in the absence of EUV stellar flux.

The conclusion we can draw now is that the proposed scenario is very well suited for explaining all observational characteristics of the spectral events. Moreover, all the puzzling points we mentioned in the previous section appear as natural consequences of the model.

4. Perspectives and Actual Status

4.1. ACTUAL PROBLEMS WITH THE MODEL

The scenario presented in the previous section seems to explain very well the recorded spectral events. However, all is still not clear and this study is not yet finished.

A first research field that now needs to be investigated concerns all that is related to the chemical composition of the bodies themselves. Observing a few ionic species is not enough to derive any valuable information about this. We would like to have a complete overview of all species involved in the variation process, and also have information concerning various ionization stages of ions. However, this is not always easy to do, since every ion does not have spectral lines in the observable part of the spectrum. Many metallic lines are in fact located in the UV spectrum. But the low resolution of IUE actually allows only the main ones to be used. Future HST/GHRS data will probably cure this observational problem. It is also interesting to seek spectral molecules' signatures related to our bodies. A first CO feature has been identified in the UV spectrum (Deleuil *et al.* 1993, Paper XV).

Another puzzling problem concerns the time duration of Ca II events. The variation time scale of these features is typically one day. That means that they are able to appear or disappear within that time, but some of them may last a few days before disappearing. This is somewhat in contradiction with the typical crossing time of one body through the line of sight at a few tens of stellar radii, which is ∼6 hours. That is why we are now thinking that the infalling bodies may not be isolated on their path, but probably fall in groups, on nearby orbits, producing long lasting events (Ferlet *et al.*, 1993, Paper XIV). This concept of "body showers" is reinforced by some recent observations showing apparently multi-component events (Paper XIII).

The origin of the central stable circumstellar component is not entirely clear. In a first model, Vidal-Madjar *et al.* (Paper II) claimed it may be due to grains spiraling towards the star by Poynting-Robertson effect releasing depleted Calcium when approaching the star (∼0.5 A.U.). The problem with this model is the stability of the component when Ca II is exposed to the stellar radiation pressure. An interaction with an eventual stellar magnetic field, where charged particles could be trapped, is possible.

4.2. IMPLICATIONS OF THE MODEL ON β PIC'S DISK

Apart from the above mentioned peculiar problems, the most interesting research field is the constraints such a model dynamically implies on the disk itself. Basically, one has to understand how solid bodies can be set on star-grazing orbits.

The most common idea is to imagine that planetary perturbations are responsible for that. Constraints deduced on the orbital parameters (essentially on the direction of the axis of orbits) of the infalling bodies (Paper X) from the simulations led us to propose that one of those objects could be sent to the star by the perturbing action of one planet.

This idea has been more precisely studied in a more recent study (Beust *et al.* 1991b, Paper XII). The major difficulty with this process is that the perturbation has to be very strong and precise enough to make the body loose almost all its angular momentum. This is absolutely not possible if the encounter occurs between a planet and an incoming object located on nearly circular orbits, just because their velocity difference is small with respect to their orbital velocity. That is why we had to

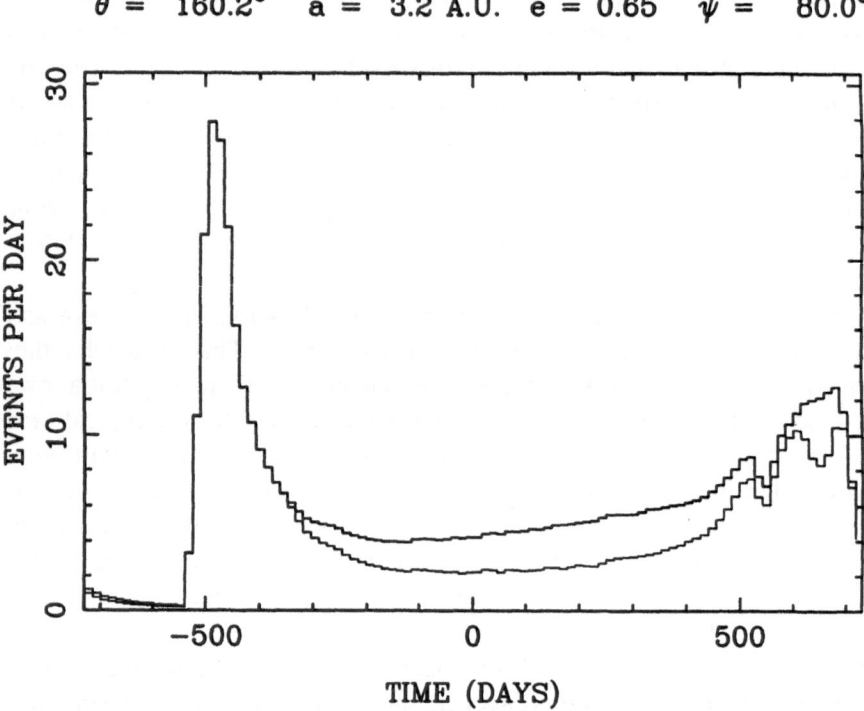

$\theta\ =\ 160.2°\quad a\ =\ 3.2\ \text{A.U.}\quad e\ =\ 0.65\quad \psi\ =\ 80.0°$

Fig. 5. Simulated evolution of event frequency over one orbital period of the planet in the elliptical orbit model, from perihelion to perihelion. Day 0 is aphelion. The frequency of generated events appears very sensitive to the position of the planet on its orbit.

suppose for this model that the suspected planet has an elliptical orbit to allow the perturbations to be efficient.

The model developed in Paper XII showed that, as soon as the eccentricity of the planet reaches values of 0.6–0.65, it is very easy to generate star-grazers from usual planetesimals. Comparison with our sample data gathered since 1985 led to heavy constraints on the orbital elements of the planet, such as the value and the direction of its semi-major axis. Constraints on the planet mass were much more difficult to obtain. The simulated activity (in terms of event frequency) of β Pic's spectral lines over one orbital period of the planet, resulting from the varying efficiency of the process at different positions on its path, moreover partially explained how the rate of spectral events may evolve on a time scale of years, as it seems to be observed (Figure 5).

However, these data need to be further confirmed, and the required high eccentricity seems to be a somewhat unrealistic assumption. Although it remains a possible model, alternative models are to be investigated. One possible way is to look for the combined action of several planets, by terms of repeated perturbations which make small bodies diffuse through the disk (see e.g. Gould, 1991). Another possible model is to look for secular planetary perturbations acting on comets with inclinations primarily $\simeq 90°$, which can make them become star-grazers. This last model was recently proposed as an origin for the Kreutz group of Sun-grazer comets (Bailey *et al.*, 1992).

5. Conclusion

The spectral events we have been continuously observing for many years, even if the process still needs to be further investigated, seem to be now well explained by the infalling evaporating bodies scenario. The future of β Pic's study will probably mainly involve the search for planets within its disk. Our study shows that the survey of spectral variations related to the pass through the line of sight of cometary-like bodies at very small stellar distances should provide heavy constraints on that subject and maybe help finding the suspected planets.

References

Bailey, M.E., Chambers, J.E. and Hahn, G.: 1992, *Astron. Astroph.* **257**, 315.

Beust, H., Lagrange-Henri, A.M., Vidal-Madjar, A. and Ferlet, R.: 1989, *Astron. Astroph.* **223**, 304 (Paper IX).

Beust, H., Lagrange-Henri, A.M., Vidal-Madjar, A. and Ferlet, R.: 1990, *Astron. Astroph.* **236**, 202 (Paper X).

Beust, H., Vidal-Madjar, A., Ferlet, R. and Lagrange-Henri, A.M.: 1991a, *Astron. Astroph.* **241**, 488 (Paper XI).

Beust, H., Vidal-Madjar, A. and Ferlet, R.: 1991b, *Astron. Astroph.* **247**, 505 (Paper XII).

Beust, H. and Tagger, M.: 1993, *Icarus*, (in press).

Deleuil, M., Gry, C., Lagrange-Henri, A.M., Vidal-Madjar, A., Beust, H., Ferlet, R., Moos, H.W., Livengood, T.A., Ziskin, D., Feldman, P.D. and McGrath, M.A.: 1993, *Astron. Astroph.* **267**, 187 (Paper XV).

Ferlet, R., Hobbs, L.M. and Vidal-Madjar, A.: 1987, *Astron. Astroph.* **185**, 267 (Paper V).

Ferlet, R., Lagrange-Henri, A.M., Beust, H., Vitry, R., Zimmerman, J.P., Martin, M., Char, S., Belmahdi, M., Clavier, J.P., Coupiac, P., Foing, B., Sevre, F. and Vidal-Madjar, A.: 1993, *Astron. Astroph.* **267**, 137 (Paper XIV).

Gould, A.: 1991, *Astroph. J.* **368**, 610.

Hobbs, L.M, Vidal-Madjar, A., Ferlet, R., Albert, C.E. and Gry, C.: 1985, *Astroph. J.* **293**, L29 (Paper I).

Hobbs, L.M, Lagrange-Henri, A.M., Ferlet, R., Vidal-Madjar, A. and Welty, D.E.: 1988, *Astroph. J.* **334**, L41 (Paper VII).

Kondo, Y. and Bruhweiler, F.C.: 1985, *Astroph. J.* **291**, L1.

Lagrange, A.M, Ferlet, R. and Vidal-Madjar, A.: 1987, *Astron. Astroph.* **173**, 289 (Paper IV).

Lagrange-Henri, A.M., Beust, H., Vidal-Madjar, A. and Ferlet, R.: 1989, *Astron. Astroph.* **215**, L5 (Paper VIII).

Lagrange-Henri, A.M., Vidal-Madjar, A. and Ferlet, R.: 1988, *Astron. Astroph.* **190**, 275 (Paper VI).

Lagrange-Henri A.-M., Gosset, E., Beust, H., Ferlet, R. and Vidal-Madjar, A.: 1992, *Astron. Astroph.* **264**, 637 (Paper XIII).

Paresce, F.: 1991, *Astron. Astroph.* **247**, L25.

Sletteback, A. and Carpenter, K.G.: 1983, *Astroph. J. Suppl.* **53**, 869.

Smith, B.A. and Terrile, R.J.: 1984, *Science* **226**, 1421.

Vidal-Madjar, A., Hobbs, L.M., Ferlet, R., Gry, C. and Albert, C.E.: 1986, *Astron. Astroph.* **167**, 325 (Paper II).

SYNTHETIC IMAGES OF PROTO-PLANETARY DISKS AROUND YOUNG STARS *

J. BOUVIER, F. MALBET and J.-L. MONIN

Laboratoire d'Astrophysique, Observatoire de Grenoble,
Université Joseph Fourier, Grenoble, France

Abstract. We present synthetic images of accretion disks around young stars computed from a model where the disk's vertical structure is solved assuming hydrostatic equilibrium. The disk's brightness results from three emission processes: (1) the reprocessing of stellar photons in the optically thick disk's regions; (2) the scattering of stellar photons in the optically thin parts of the disk; and (3) the thermal emission of the disk due to viscous energy dissipation during the accretion process.

We discuss the relative importance of these emission processes at wavelengths ranging from 1.2 to 20 μm.

1. Introduction

High-resolution imaging techniques, such as speckle interferometry or adaptive optics, very recently began to reveal the complex circumstellar environment of young stars. In particular, direct clues to the existence of possibly proto-planetary disks have been obtained for a few T Tauri stars, a class of very active pre-main sequence stars (e.g., Beckwith *et al.*, 1989; Monin *et al.*, 1989; Malbet *et al.*, 1993). While protostellar disks are expected to result from angular momentum conservation during the gravitational collapse of the parental cloud, the evidence for circumstellar disks around T Tauri stars was mostly indirect so far and derived from the analysis of the spectral and polarimetric properties of the stars (e.g., Edwards *et al.*, 1987, Bertout *et al.*, 1988, Ménard, 1989).

Sub-arcsecond resolution is required to image circumstellar disks around young stars since, at the distance of the nearest sites of star formation (150 pc), solar system-sized disks have an angular diameter of a few tenths of an arcsecond at most. The multi-aperture interferometers that will become available in the next 10 years (ESO VLTI, Keck Telescopes, etc.) are expected to bring considerable insight into the structure and physical properties of proto-planetary disks. In order to estimate the impact of such instruments onto our knowledge of circumstellar disks, we computed synthetic images of accretion disks thought to exist around T Tauri stars. We report here preliminary results of this work and describe the aspect of the disk seen at various wavelengths and inclination angles. These synthetic maps can readily be used as a guide to interpret the results of current high angular resolution imaging of young stars and their immediate surroundings.

---

* Paper presented at the Conference on *Planetary Systems: Formation, Evolution, and Detection* held 7–10 December, 1992 at CalTech, Pasadena, California, U.S.A.

TABLE I

Model parameters

Stellar Parameters	M_*	1 $M_\odot$
	R_*	2 $R_\odot$
	T_*	5000 K
Disk Parameters	R_{min}^{disk}	1.1 R_*
	R_{max}^{disk}	400 AU
	$\dot{M}_{acc}$	10^{-7} $M_\odot$ yr^{-1}
	α_{visc}	0.1
Observer Parameters	λ	$0.5 - 500 \, \mu m$
	i	0–90°

2. The Disk Model

As a starting point for the model, the radial dependence of temperature and density within the disk ($T(r)$, $\rho(r)$) is derived from the classical formulation of Lynden-Bell and Pringle (1974) using model parameters listed in Table I. We then assume that hydrostatic equilibrium holds in the vertical direction, and compute the disk's vertical structure, i.e., $T(r, z)$ and $\rho(r, z)$, by solving the equation of radiative transfer in the z-direction (Malbet, 1992). The resulting disk's structure is illustrated in Figure 1 where isocontours of the optical depth, temperature, and density are displayed.

Using the disk's physical structure derived above, we then consider three emission processes in order to compute the disk's intensity maps:

1. thermal emission from the viscous energy dissipation in the accretion disk,
2. thermal emission due to reprocessing of the stellar photons that are absorbed and re-emitted in the optically thick layers of the disk,
3. single scattering of stellar photons in the upper, optically thin layers of the disk's atmosphere.

The intensity resulting from thermal emission mechanisms (viscous energy dissipation and reprocessing) is given by:

$$I_\nu^t(r, z) = (1 - \varpi)B_\nu(T(r, z)) \tag{1}$$

where ϖ is the albedo of the disk's atmosphere, while the intensity due to scattered light is obtained from:

$$I_\nu^s(r, z) = \frac{\varpi}{4\pi} \frac{\pi R_*^2}{d^2} B_\nu(T_*) \, e^{-\tau_*(r, z)} \tag{2}$$

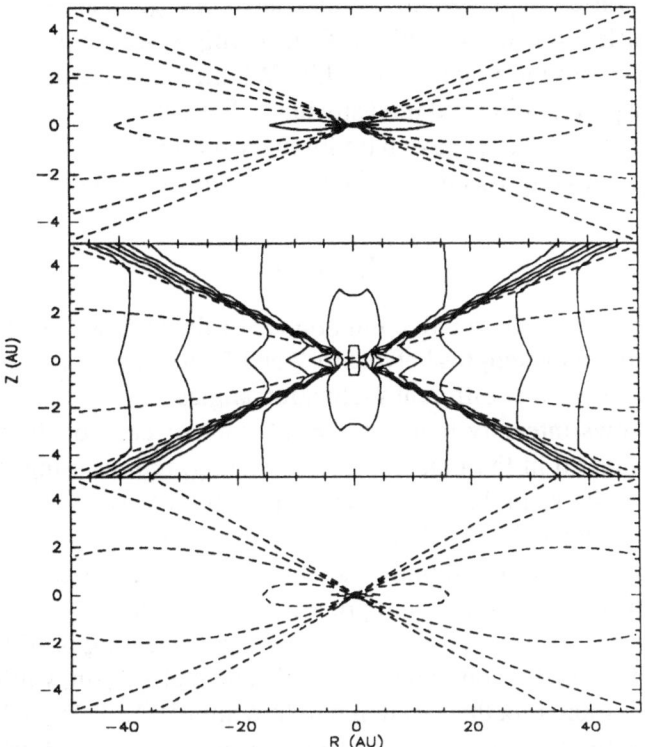

Fig. 1. Vertical structure of the inner 50 AU of a 400 AU circumstellar disk around a young low-mass star. *Upper panel*: optical depth contours ranging from $\tau = 10^{-4}$ to $\tau = 10$ by step of 10. Solid lines: optically thick regions; dashed lines: optically thin regions. *Middle panel*: temperature contours (solid lines) for $\log T = 0.9, 1.0, 1.2, 1.4, 1.6, 1.8, 2.0, 2.25$, and 2.5, and optical depth contours (dashed lines) for $\tau = 10^{-2}$ and $\tau = 10^{-4}$. Note the strong temperature gradient at the upper and lower edges of the disk, due to the reprocessing of stellar photons in the disk. *Lower panel*: density contours ranging from $\rho = 10^{-15}$ g cm^{-3} to $\rho = 10^{-11}$ g cm^{-3} by steps of 10.

where d is the distance between (r, z) and the star and τ_* is the optical depth from (r, z) to the star.

Finally, the emergent intensity on a line of sight $(x_{\text{obs}}, y_{\text{obs}})$ on the plane of the sky is computed as:

$$I_\nu(x_{\text{obs}}, y_{\text{obs}}) = \int_0^\infty \left[I_\nu^t(r, z) + I_\nu^s(r, z) \right] e^{-\tau_{\text{obs}}} \, d\tau_{\text{obs}} \tag{3}$$

where $d\tau_{\text{obs}} = \kappa \rho \, dz_{\text{obs}}$ is the optical depth towards the observer. Monochromatic opacities are interpolated from Wolfire and Cassinelli (1986) tables.

This model differs from previous, similar works in several ways. Bertout and Bouvier (1988) computed synthetic images of protoplanetary disks in the infrared for the Very Large Telescope Interferometer by taking into account only thermal emission and assuming flat disks. Lazareff *et al.* (1990) investigated only scattered emission in a series of parametrized disk/envelope geometries. Ménard (1992) computed polarization maps resulting from multiple scattering in dusty circumstellar envelopes around young stars while Whitney and Hartmann (1992, 1993) also performed polarization and intensity maps, but with a lower spatial resolution. Finally, Malbet *et al.* (1992) gave a preliminary report of the work presented here except that reprocessing was not yet included in the computations.

3. Results

This section illustrates the relative importance of the various disk's emission processes at different wavelengths between 1.2 and 20 μm, as well as the aspect of the circumstellar disk seen at different inclination angles.

Figure 2 shows intensity maps at $\lambda = 2.2$ μm for each of the three emission processes considered in the model: viscous heating, reprocessing, and scattering. Figure 3 shows the same but at $\lambda = 20$ μm. In both Figures 2 and 3, the disk's inclination is 60°. Note the widely different spatial scales in each panel of Figures 2 and 3. At 2.2 μm, scattering dominates on the large scale while thermal emission due to either viscous heating or reprocessing is limited to the innermost disk regions. At 20 μm, thermal emission due to viscous heating is still limited to the inner 10 AU of the disk, though more extended than at 2.2 μm, while reprocessing now contributes more flux than scattered light within 100 AU from the star. At both 2.2 and 20 μm, thermal emission due to reprocessing extends further away from the star than thermal emission due to viscous heating as a result of disk flaring. The dark lane seen in the disk's equatorial plane results from the enhanced opacity in the disk's densest layers.

Comparison between Figures 2 and 3 shows that scattering dominates at short wavelengths while thermal emission becomes important at long wavelengths. This effect is also illustrated in Figure 4 where intensity maps including the three emission processes are shown as a function of wavelength. At $\lambda = 1.2$ and 2.2 μm, scattering dominates on the large spatial scales. At $\lambda = 10$ μm, scattering is greatly reduced and thermal emission is not yet important, while at $\lambda = 20$ μm, thermal emission, and in particular reprocessing, becomes the major source of flux in the inner disk regions.

Finally, intensity maps at $\lambda = 2.2$ μm are shown for an inclination of 0°, 60°, 75°, and 90° in Figure 5. At this wavelength, single scattering is the main source of light over the whole disk structure. It is seen from the edge-on image that scattering occurs only in the upper and lower optically thin edges of the disk while no scattering occurs in the optically thick, equatorial disk's plane.

4. Conclusion

We have computed intensity maps of circumstellar, possibly proto-planetary, disks around young low-mass stars. The direct detection and imaging of these disks is

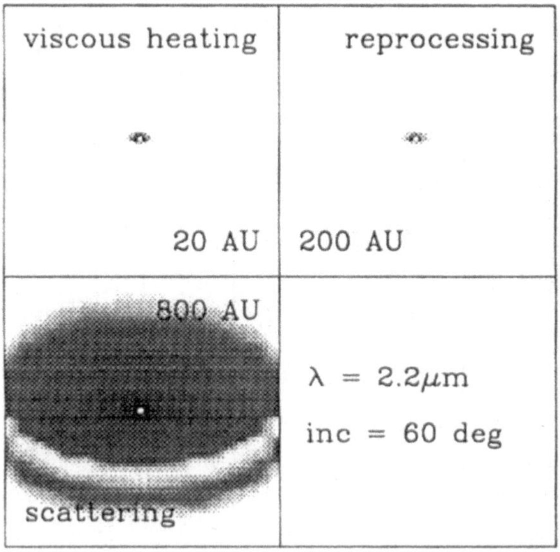

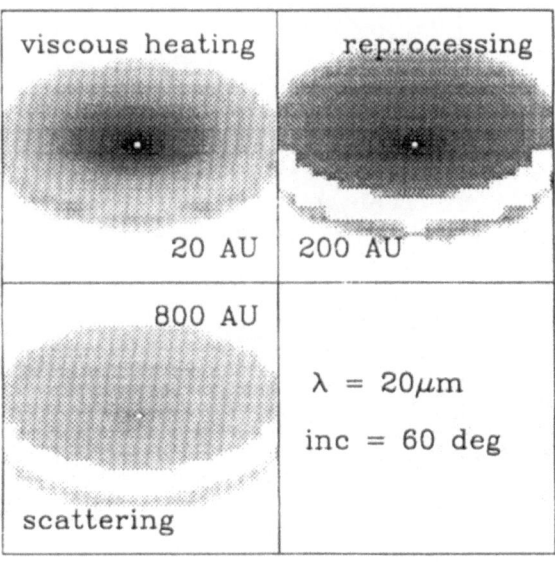

Fig. 2.   Thermal emission vs scattering at $\lambda = 2.2$ μm. Disk inclination is 60°. The gray scale spans 15 decades of intensity.

Fig. 3.   Same as Figure 2 at $\lambda = 20$ μm.

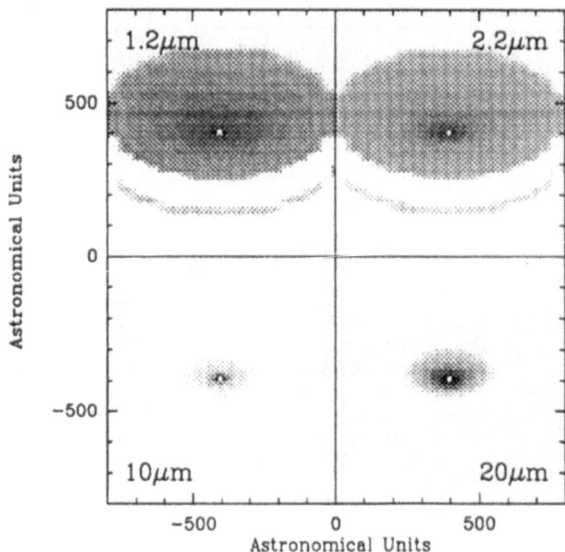

Fig. 4.   Wavelength dependence. Disk inclination is 60°. The gray scale spans 8 decades of intensity.

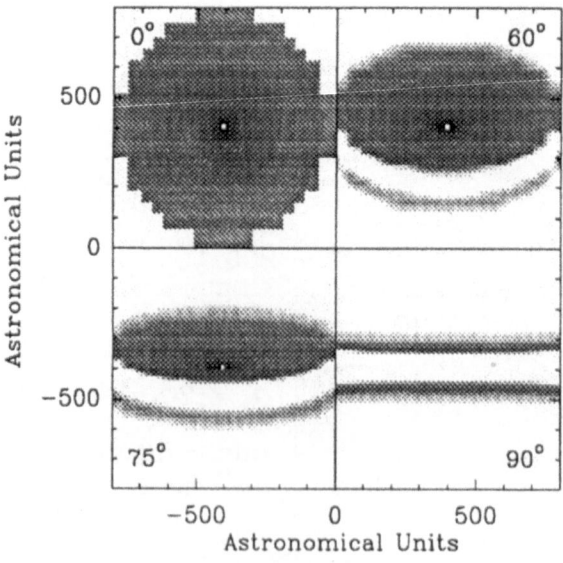

Fig. 5.   Dependence upon the inclination angle at $\lambda = 2.2$ μm. The gray scale spans 8 decades of intensity.

of crucial importance to understand how and how often planets might form around young solar-type stars. Our model shows that only scattering on the outer layers of the disk is expected to be detected at optical wavelengths with current high resolution imaging techniques. In contrast, thermal emission will be more easily detectable at mid- and far-infrared wavelengths ($\lambda \geq 20$ μm).

The primary constraint for detecting thermal emission is spatial resolution while a high dynamical range is mandatory to detect scattered light. More precisely, probing the structure of an accretion disk requires interferometers with a spatial resolution of at least 2 mas for an intensity ratio of 10^2 in the mid-infrared, i.e., baselines ranging from 100 m to 1 km. Hence, ground-based, space-based, and lunar-based interferometers appear as complementary tools that will provide both the range of baselines and the high dynamical range required to probe the detailed physical structure of circumstellar disks around young stars.

References

Beckwith, S.V.W., Sargent, A.I., Koresko, C.D. and Weintraub, D.A.: 1989, *Astrophys. J.* **343**, 393.

Bertout, C. and Bouvier, J.: 1988 , in F. Merkle, ed(s)., *Proc. NOAO-ESO Conf. on High-Resolution Imaging by Interferometry*, 69.

Edwards, S., Cabrit, S., Strom, S.E., Heyer, I., Strom, K.M. and Anderson, E.: 1987 , *Astrophys. J.* **321**, 473.

Lazareff, B., Pudritz, R.E. and Monin, J-L.: 1990, *Astrophys. J.* **358**, 170.

Lynden-Bell, D. and Pringle, J. E.: 1974, *Monthly Notes of Royal Astron. Soc.* **168**, 603.

Malbet, F.: 1992, Environnement Circumstellaire des Etoiles Jeunes, Ph.D. Thesis, Université of Paris VII.

Malbet, F., Monin, J.-L. and Bouvier, J.: 1992, *Proc. ESA Coll. Targets for Space-Based Interferometry*, Beaulieu, Oct. 92, (in press).

Malbet, F., Rigaut, F., Bertout, C. and Léna, P.: 1993, *Astron. Astrophys.*, submitted.

Ménard, F.: 1989, Etude de la polarisation causée par des grains dans les enveloppes circumstellaires denses, Ph.D. Thesis, Univ. of Montreal.

Ménard, F.: 1992, *Astrophys. J.*, submitted.

Monin, J-L., Pudritz, R.E., Lacombe, F. and Rouan, D.: 1989, *Astron. Astrophys.* **215**, L1.

Whitney, B.A. and Hartmann, L.: 1992, *Astrophys. J.* **395**, 729.

Whitney, B.A. and Hartmann, L.: 1993, *Astrophys. J.*, (in press).

Wolfire, M.G. and Cassinelli, J.P.: 1986, *Astrophys. J.* **310**, 207.

NEW OBSERVATIONS OF THE β PICTORIS CIRCUMSTELLAR DISK
WITH THE JHU ADAPTIVE OPTICS CORONAGRAPH *

M. CLAMPIN

Space Telescope Science Institute, Baltimore, Maryland, USA

and

D.A. GOLIMOWSKI and S.T. DURRANCE

The Johns Hopkins University, Baltimore, Maryland, USA

Abstract. We report new coronagraphic observations of the β Pictoris circumstellar disk with the JHU Adaptive Optics Coronagraph. We are able to image the inner region of the disk to a radius of 40 AU from the star, and find an abrupt change in the gradient of the disk's midplane surface brightness at $\sim$100 AU. Our results are consistent with recently published models which propose two disk components in order to unify optical, IR and IRAS data.

1. Introduction

Since the initial discovery by IRAS of a large excess in the far-IR spectrum of α Lyrae (Aumannn *et al.*, 1984), over 100 main-sequence stars have been found to possess similar far-IR excesses (Backman and Paresce, 1993). IR-excess is generally interpreted as thermal radiation from cool ($T \approx$ 50–150 K) disks of optically-thin, orbiting dust grains. Despite the large number of cases now known, only three of the objects have been spatially resolved by IRAS: α Lyrae, β Pictoris and α Piscis Austrinus. The observational properties of these three sources are generally interpreted in the context of planetary system formation (Gillett, 1986).

In 1984, optical observations of β Pictoris with a stellar coronagraph revealed an almost edge-on disk of scattered light around the star (Smith and Terrile, 1984). The disk extended from the edge of the occulting mask, at a radius of 6 arcsec (100 AU), out to a distance of $\approx$ 25 arcsec (450 AU), with a position angle of 30°. Smith and Terrile determined that the surface brightness of the disk decreased as $r^{-4.3}$, with a disk surface brightness of 16 mag arcsec^{-2} at 100 AU. The difference in color between the disk and β Pictoris was found to be neutral by Paresce and Burrows (1987), indicating a characteristic grain size of $\geq$ 1 μm. β Pictoris remains the only resolved IR-excess source with an optical counterpart.

An important result from the IRAS data is that the three resolved objects have relatively clear central regions (Backman and Gillett, 1987). The size of the clear regions cannot be constrained uniquely by the IRAS data alone, since it is dependent upon the disk model assumed. On the basis of IR and optical data, Artymowicz *et al.* (1989) derived an estimate of 5–15 AU for the half-clearing radius of the β Pictoris disk, in reasonable agreement with similar models by Diner and Appleby

---

* Paper presented at the Conference on *Planetary Systems: Formation, Evolution, and Detection* held 7–10 December, 1992 at CalTech, Pasadena, California, U.S.A.

Astrophysics and Space Science **212**: 167–172, 1994.

(1986). Interesting questions are raised, however, by the mechanism responsible for the removal of grains, since Poynting–Robertson radiation drag on grains would be expected to fill any cleared region. Gravitational perturbations, removal by radiation pressure forces and ice sublimation have been proposed as candidate mechanisms (Artymowicz *et al.*, 1989; Backman *et al.*, 1992; Beust *et al.*, 1991).

The size, composition and spatial distribution of the dust grains play a fundamental role in understanding grain removal mechanisms and the dynamics of the inner regions of the disk. Near-IR observations have recently provided additional information on the properties of the dust within the central 100 AU. On the basis of 10 μm and 20 μm imaging of the inner disk Telesco *et al.* (1988) proposed submicron grain sizes. Subsequently, IR spectroscopy by Telesco and Knacke (1990) revealed a 10 μm silicate emission feature emitted by grains with radii < 10 μm, and did not exclude the sub-micron sized particles previously proposed by Telesco *et al.* (1988). Sub-mm observations by Becklin and Zuckerman (1990) also support two populations of micron and sub-micron grain sizes. Backman *et al.* (1992) have recently reported 3.5–20μm photometry and find that their data, in combination with IRAS and optical imaging results, are best fit by a two-component-disk model with a boundary at $\approx$ 80 AU. The properties of this inner component can be modeled either by a change in the particle size or a change in the spatial gradient with respect to the outer component of the disk. The optical characteristics of the disk within 100 AU have remained difficult to probe until recently, due to the limitations of coronagraphic techniques. In an attempt to address the issue of the surface brightness profile of the disk within 100 AU, we have observed β Pictoris with the Johns Hopkins Adaptive Optics Coronagraph (AOC). The first results of this study were presented by Golimowski *et al.* (1992). In this paper we discuss the scattering properties of the inner 40–100 AU of the disk.

2. Observations

The observations of β Pictoris were conducted with the AOC on 23 and 24 December 1991 at the 2.54 m DuPont telescope at the Las Campanas Observatory, Chile. The AOC is unique in employing an image motion compensation system which improves the angular resolution of an observation by compensating for image motion due to atmospherically induced wavefront tilt and residual telescope guiding errors (Durrance and Clampin, 1989; Clampin *et al.*, 1991; Golimowski *et al.*, 1992). In this system, light from the star is reflected by the occulting mask onto a quadrant-CCD (Clampin *et al.*, 1990), which is used to monitor the changes in image position in real time. A 100 Hz tip/tilt mirror operates in a closed-loop with the quadrant-CCD to compensate for image position changes in real-time. Depending upon the ambient seeing conditions and telescope aperture, the AOC can improve image resolution by factors of 1.2–2.

For these observations a circular occulting mask of 2 arcsec in radius was used, corresponding to 32.8 AU at the distance of β Pictoris. The telescope's focal plane was reimaged by the AOC with a magnification of 6 onto a Tektronix 1024×1024 CCD, giving a plate scale of 0.037 arcsec/pixel. The CCD read noise was ≈ 8 e$^-$ RMS with a gain of 3.8 e$^-$/DN. A sequence of ten exposures totalling 2450 seconds was made of β Pictoris with an R band filter (OG370+KG3; Bessel, 1990), followed by similar exposure sequences of two control stars HR 2435 and κ Phoenicis. Two additional observations were made of β Pictoris with the AOC rotated by $90°$, in order to isolate instrumental artifacts in the processed images. A precise estimate of the sky background level was obtained for each object by offsetting to a field of dark sky and obtaining an additional exposure.

Following bias subtraction and flat fielding of each image, the data sets were combined by summing the individual frames previously shifted to a common point of reference. In order to enhance the contrast of the disk it was necessary to remove the contribution from the outer wings of the central occulted star, which appears in the images as a halo of light surrounding the occulting mask. Two different methods were attempted to remove the halo contribution. The first consisted of subtracting a registered and scaled control star, the second, a radial subtraction of a profile obtained by fitting the halo at position angles $90°$ from the disk. The second method provided a better subtraction of the halo since it could account for the slight differences between the seeing profiles of β Pictoris and the control star. The general agreement between the two methods validates the assumptions made in fitting the halo of β Pictoris directly. A detailed description of the data reduction and analysis procedures is presented by Golimowski *et al.* (1993).

3. Results

The R band image of the β Pictoris disk is shown in Figure 1. The superimposed circle shows a radius of 100 AU, the inner limit of previous data (Smith and Terrile, 1984; Paresce and Burrows, 1987). The disk can be traced from the edge of the coronagraphic mask (40 AU), out to 300 AU from the star. The very high signal to noise of these data allows us to analyze the surface brightness distributions of the NE and the SW extensions of the disk independently. Figure 2 shows the variation of disk midplane surface brightness in mag/arcsec^2 as a function of the radial distance from the star, for the NE and SW extensions of the disk, respectively. The displayed error bars represent the combination of three sources: photon statistics, residuals resulting from the subtraction of the fitted halo profile and a contribution due to sub-pixel misalignments in subtracting the seeing disk halo. It is clear from both surface brightness profiles that there is an unambiguous change in gradient within 100 AU. In the region 40–90 AU the surface brightness distributions for the NE and SW extensions can be fitted by a radial power law with indices:

NE extension $\nu = -2.377^{+0.788}_{-0.646}$

SW extension $\nu = -1.908^{+0.945}_{-0.823}$

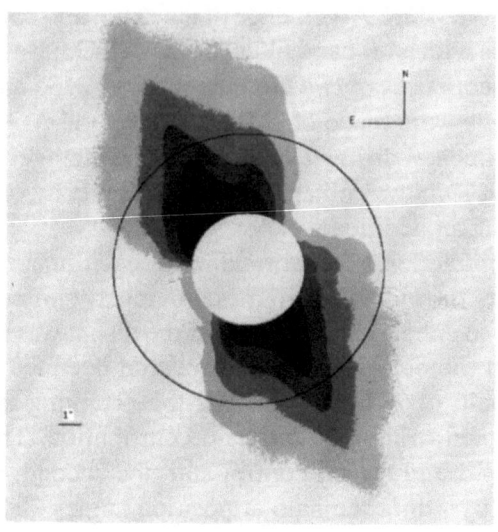

Fig. 1.   The R-band image of β Pictoris obtained with the JHU Adaptive Optics Coronagraph. The field size is 26" × 26". The occulting mask radius corresponds to 40 AU. A 100 AU radius is marked by the superimposed circle.

For the outer 100–300 AU region the corresponding indices are

NE extension $\nu = -3.508 \pm 0.003$
SW extension $\nu = -4.182 \pm 0.004$

4. Discussion

The possibility of a change in the index of the surface brightness distribution at around 100 AU has previously been suggested by Artymowicz *et al.* (1990). These new results show, for the first time, that a change in the index definitely occurs and, that it is even more pronounced than the function of r^{-3} suggested by Artymowicz *et al.* (1990). Another analysis of the inner region of the β Pictoris disk has recently been reported by Vidal-Madjar *et al.* (1992) using an anti-blooming CCD to image the disk in to 40 AU from the star. They find that for V, R and I band images the surface brightness distributions are fit by a single power law of index $r^{-3.6}$. A significant result of their study is the discovery of a large change in the surface brightness distribution within 100 AU for the B band data. Vidal-Madjar *et al.* attribute the change in surface brightness distribution in the B band to a change in albedo of the grains within 100 AU, ruling out a change in particle

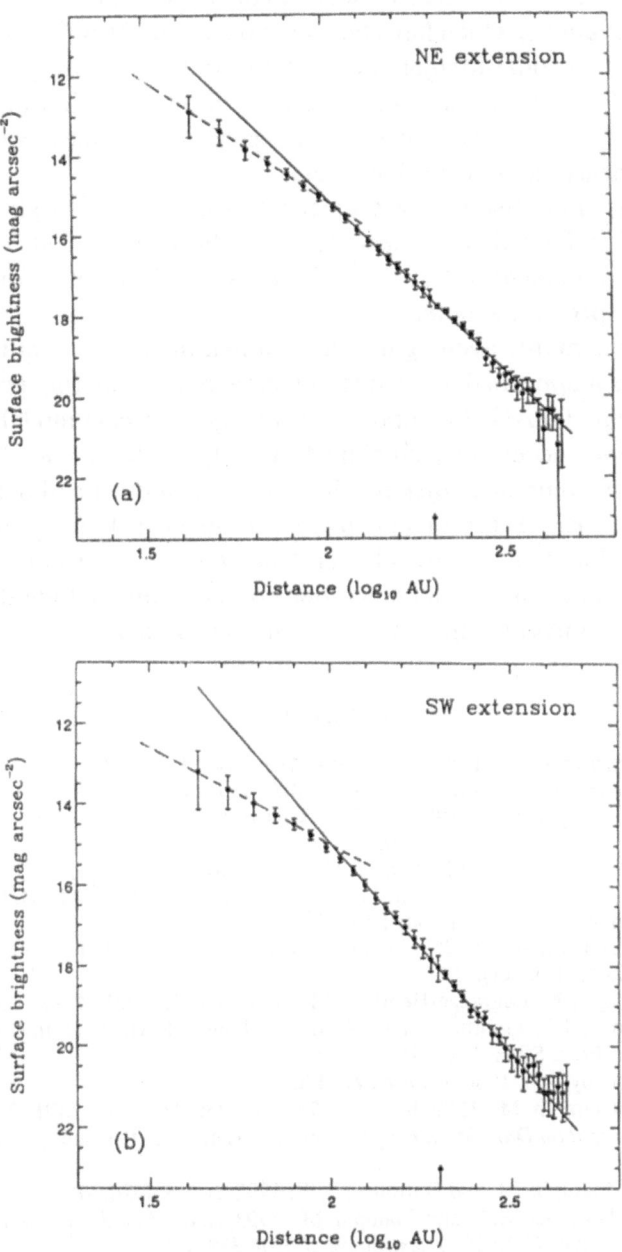

Fig. 2.   The midplane surface brightness of the NE extension of the disk (*top*) and the SW extension of the disk (*bottom*). The dashed and solid lines show the best fit to the 40–90 AU and 100–300 AU regions of the disk, respectively.

size on the basis that smaller grains should produce bluer disk colors. Our results do not support their R band measurements. However, it should be noted that our individual observations, which have typical exposure times of 250 s, have a higher signal-to-noise than Vidal-Madjar et al., who were limited to exposure times of a few seconds and were also unable to apodise the diffracted light from the telescope optics. These effects may combine to mask surface brightness variations at the level we have measured in our R band data.

Vidal-Madjar et al. also note asymmetries in the surface brightness gradients between the NE and SW disk extensions. We confirm that the indices for the outer and inner disk components differ from NE to SW, with the inner disk asymmetry inverted with respect to the outer.

Although more multiwavelength data of similar quality are required to confirm our observations it appears that there is a change in the scattering properties of the disk grains within 100 AU at all optical wavelengths. Recent models by Backman et al. (1992) have successfully combined optical, IRAS and IR observations by considering a two component disk in which there is an inner and outer component to the β Pictoris disk. These models predict an inner disk component extending from $\approx$ 1–80 AU in which there is lower density and either smaller particles or a less steep spatial gradient. Model predictions of the scattered light distribution are required to permit direct comparison with our optical data.

References

Artymowicz, P., Burrows, C. and Paresce, F.: 1989, *Astrophys. J.* **337**, 494.
Artymowicz, P., Paresce, F. and Burrows, C.: 1990, *Adv. Space Res.* **10**, 3.
Backman, D. E. and Gillett, F. C.: 1987, in *Cool Stars and Stellar Systems*, eds. J. Linsky and R. Stencel, (Berlin: Springer-Verlag), 340.
Backman, D. E., Gillett, F. C. and Witteborn, F. C.: 1992, *Astrophys. J.* **385**, 502.
Backman, D. E. and Paresce F. P.: 1993, in *Protostars and Planets III*, eds. E. H. Levy, J. I. Lunine and M. S. Matthews, Univ. of Arizona Press: Tucson, Az., in press.
Becklin, E. E. and Zuckerman, B.: 1990, in *Sub-Millimeter Astronomy*, G.D. Watt and A.S. Webster (eds.), (Netherlands: Kluwer), 147.
Beust, H., Vidal-Madjar, A., Lagrange-Henri, A. M. and Ferlet, R.: 1991, *A. Ap.* **236**, 202.
Clampin, M., Durrance, S. T., Golimowski, D. A. and Barkhouser, R. H.: 1991, in *Active and Adaptive Optical Systems*, Proc. SPIE. **1542**, 165.
Diner, D. J. and Appleby, J.F.: 1986, *Nature* **322**, 436.
Durrance, S. T. and Clampin, M.: 1989, in *Active Telescope Systems*, Proc. SPIE. **1114**, 97.
Gillett, F.: 1986, In *Light on Dark Matter*, Space Sci. Lib. 124, ed. F. P. Israel, (Dordrecht: D. Reidel Co.), 61.
Golimowski, D. A., Clampin, M. and Durrance, S. T.: 1992, *B.A.A.S.* **24**, 789.
Golimowski, D. A., Durrance, S. T. and Clampin, M.: 1993, *Astrophys. J.* **411**, L41.
Paresce, F. P. and Burrows, C. J.: 1987, *Astrophys. J. Lett.* **319**, L25.
Smith B. A. and Terrile, R. J.: 1993, *Science* **226**, 1421.
Telesco, C. M., Becklin, E. E., Wolstencroft R. D. and Decker, E.: 1988, *Nature* **335**, 51.
Telesco, C. M. and Knacke, R. F.: 1991, *Astrophys. J. Lett.* **372**, 29.
Vidal-Madjar, A. M. et al.: 1992, *ESO Messenger* **69**, 45.

INNER PART OBSERVATION OF THE β PICTORIS DISK *

R. FERLET, A. LECAVELIER DES ETANGS, G. PERRIN, A. VIDAL-MADJAR and
F. SÈVRE
Institut d'Astrophysique, CNRS, Paris, France

F. COLAS and J.E. ARLOT
Bureau des Longitudes, Paris, France

C. BUIL
CNES, Toulouse, France

H. BEUST
Service d'Astrophysique, DAPNIA, CEA, CEN-Saclay,
Gif-sur-Yvette, France

A.M. LAGRANGE-HENRI
Groupe d'Astrophysique de Grenoble, Univ. J. Fourier,
Grenoble, France

J. LECACHEUX
Observatoire de Paris, Section de Meudon, Meudon, France

and

M. DELEUIL, C. GRY
Laboratoire d'Astronomie Spatiale, CNRS,
Marseille, France

Abstract. The inner regions of the circumstellar disk around β Pictoris might be dust-free due to a possible planet which may have cleared up dust particles. We present a new observational technique based on the use of an anti-blooming CCD in order to directly image this zone. The structure of the disk is revealed down to 2 arcsec from the star (30 AU). We show that the β Pic disk brightness is neutral in V, R, and I_C, but drops down when going inward in B, possibly related to a change in the dust composition. Also, a slight disk asymmetry is present but inverted from the outer to the inner zones.

1. Introduction

Since the discovery with the IRAS satellite of a number of main sequence stars which show an infrared excess — in the only case of the southern A5 V star β Pictoris — a disk of dust surrounding the star has been directly imaged (Smith and Terrile, 1984). Viewed nearly edge-on from Earth, this favorable orientation has allowed the further detection in absorption of the gaseous counterpart of the disk (Hobbs *et al.*, 1985; Kondo and Bruhweiler, 1985; Vidal-Madjar *et al.*, 1986). Subsequent observations have emphasized the complex time variability of the circumstellar lines, both in the visible and the UV (Ferlet, Hobbs, and Vidal-Madjar, 1987; Lagrange, Ferlet, and Vidal-Madjar, 1987).

From an extensive spectroscopic data set on the β Pic proto-planetary system, we have proposed a model in which the sporadic redshifted events are the result of

---

* Paper presented at the Conference on *Planetary Systems: Formation, Evolution, and Detection* held 7–10 December, 1992 at CalTech, Pasadena, California, U.S.A.

Astrophysics and Space Science **212**: 173–180, 1994.
© 1994 *Kluwer Academic Publishers.*

the evaporation of solid comet-like bodies when grazing the star (Lagrange-Henri, Vidal-Madjar, and Ferlet, 1989; Beust *et al.*, 1989). Numerical simulations of this scenario are able to reproduce the data. We then assume that the numerous events seen could be due to the presence of a giant planet (or proto-planet) already formed in the β Pic disk which perturbs a lot of small, passing-by objects and throws some of them towards the star (see, e.g., Beust, Vidal-Madjar, and Ferlet, 1991 and references therein). It is thus plausible that the inner part of the disk has been cleared up.

From their coronographic study, Smith and Terrile (1984) have shown that an $r^{-4.3}$ power law is well representing the disk dust distribution. However, they were limited by a mask of 6 arcsec in radius (100 AU). By extrapolating their power law to less than 30 AU, they deduced that a too strong extinction of the direct starlight through the edge-on disk would result. Therefore, they claimed that a cleared up region was probably present within 30 AU from the star. Later on, Artymowicz, Burrows, and Paresce (1989), coupling new coronographic observations (but still limited to more than 6 arcsec) with the IRAS data, confirmed the radial power law, although slightly less steep, $r^{-3.6}$, and concluded again that a 5 to 15 AU cleared inner region is compatible with the observations. Simultaneously, Telesco *et al.* (1988) completed 10 and 20 μm ground-based observations with 5 arcsec resolution, and deduced that the inner disk is relatively free of dust, possibly up to 50 AU (3 arcsec). More recently, Telesco and Knacke (1991) might have detected a spectral signature between 8 and 12 μm related to silicates, only within 3 arcsec from β Pic.

Clearly, the existence of a dust-free zone in the inner regions of the β Pictoris disk is plausible. Observations still closer to the star are thus crucial to more strongly constrain the models.

2. Observations with an Anti-Blooming CCD

The intrinsic difficulty associated with classical coronographic observations is due to the diffuse light around the mask which is extremely sensitive to the star position behind the mask, a location difficult to control without any adaptative optic device. In all the published data, the limit is around 6 arcsec from the star; otherwise, at greater distances, results are very good.

On the other hand, an anti-blooming CCD allows the direct observation of the disk next to the star without any coronograph. We used a Thomson THX 7852 CCD (Buil, 1989), which 221×145 pix matrix contains an anti-blooming drain every two lines of pixels, arranged in such a way that electrons flowing from saturated pixels are caught to avoid contamination of adjacent pixels. This technique was well developed for planetary studies, and led to the detection of very faint satellites ($V \sim 16$) nearby giant planets (Colas and Arlot, 1991; Colas and Buil, 1992). The size of the stellar image is thus simply defined by the atmospheric seeing and the exposure time. Nevertheless, the diffuse light produced by the cleanliness of the telescope mirrors and by the diffraction pattern due to the secondary spider still remains.

The images recorded with such a CCD have merely to be centered on the stellar image in order to either add them and improve the S/N ratio by selecting sharpest images, or correct for diffuse light within the telescope by subtracting the properly centered image of a template star, which was here the nearby star α Pictoris. Due to the drain zones, the sensitive pixel size is 19×30 μm. After transforming these shapes into squares, the actual size projected on the sky is 0.363×0.363 arcsec.

The observations were performed in October 1991, at the 2.2-m telescope at ESO La Silla (Chile), with a typical seeing between 1 and 2 arcsec. To probe the β Pic disk at different distances, series of images were taken with exposure times ranging from 0.5 to 300 sec (to detect the disk as far as possible and secure an overlap with previous coronographic images). This has been done with the four standard filters: B (440 nm), V (550 nm), R (700 nm) and I_C (800 nm). More than 450 CCD images were recorded, and an average of at least 10 was gathered for each filter and exposure time. Additional exposures were recorded with different angular positions of the bonette in order to test the effect of the CCD orientation relative to the β Pic disk. No changes were detected.

To remove the strong diffraction spikes, the subtraction of the template star image, taken almost simultaneously and with the same instrument setting, must be done after a proper scaling which is not a simple linear one. The radial variation of the correction factor was evaluated assuming that the diffuse light (away from the obvious spikes) presents a circular symmetry.

3. Results

The final results of the data analysis are shown in Figure 1 for β Pictoris and Figure 2 for another template star having no disk. Comparing these two figures in which the same correction was applied, the β Pic disk detection is obvious. The central part of the images, which appears darker and flat, is the saturated area, larger and larger as the exposure time increases. The disk extends in the North-East and South-West directions, with a tilt angle of $31.5 \pm 0.5°$ with respect to the North-South direction. It is clear also that the correction process is unable to perfectly eliminate the brightest spikes, because of the time fluctuations between the β Pic image and the template star one. However, this is a minor problem in evaluating the disk brightness, thanks to its angular separation with these spikes. The bar in Figures 1 and 2 is 6 arcsec long, which represents 100 AU at the β Pic distance of 16 pc. The shortest exposures allow a precise evaluation of the disk brightness down to 1.8 arcsec, i.e. down to less than 30 AU.

By evaluating the total signal perpendicularly to the disk plane, we have reconstructed the disk brightness for the four filters (Figure 3), from three series of images with different exposure times. A good approximation of the errors is given by the luminosity of the secondary spike quadrant which is disk free.

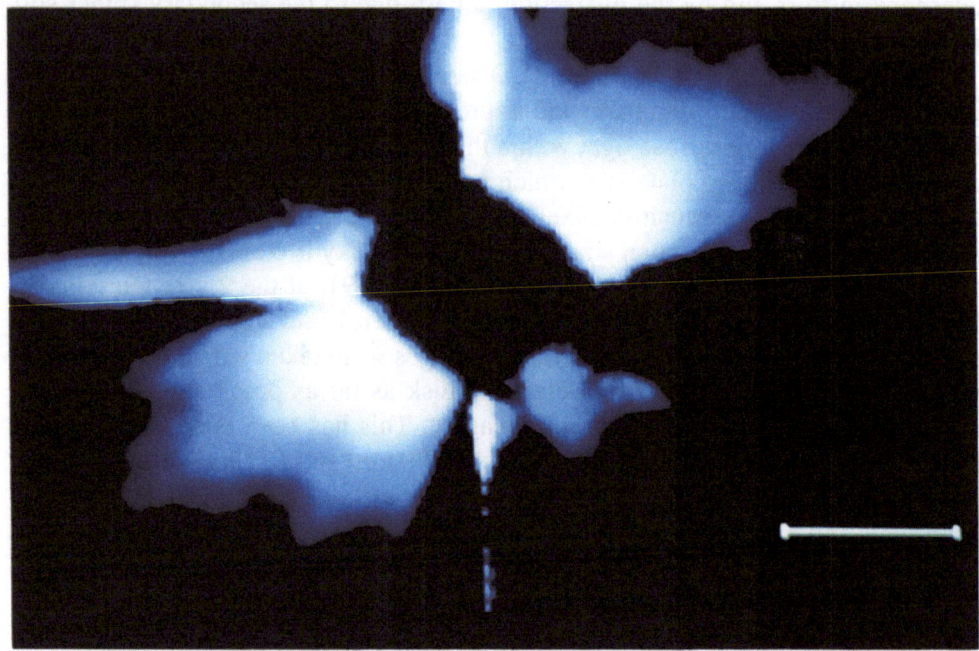

Fig. 1.   The β Pictoris disk image in the V filter with a 60-s exposure time. It is corrected for the diffuse light through the use of a template star. The bright spikes due to the secondary spider cannot be completely removed, but do not contaminate disk brightness evaluations due to their angular separation. The bar represents 6 arcsec in the sky.

In V, R, and I_C filters, the light distribution is quite well fitted by an $r^{-3.6}$ power law, in excellent agreement with Artymowicz *et al.* (1989). There is no extinction down to 30 AU, as it would be the case if an inner dust-free zone did exist at such a distance from the star.

A striking difference arise in the B filter. Although the power law found at more than 6 arcsec from the star matches extremely well the one for the other filters, at shorter distances, there is a clear drop, by at least a factor of 4 when moving from $\sim 7.5'$ to $2.5'$. However, even closer to the star, the luminosity seems to increase again, as it does for the other filters. These disk colors are shown in Figure 4 after normalization with the stellar colors themselves. In the outer disk we thus confirm the previous coronographic results and the flat albedo of the dust particles, demonstrating their probable large size ($> 1~\mu$m).

But in the inner disk, our results imply that the dust albedo is decreasing. In effect, a change in the grain size alone cannot explain the variation seen in B, because larger grains should induce no color changes while smaller grains should favor the blue. Therefore, it seems that the nature of the grains is changing when

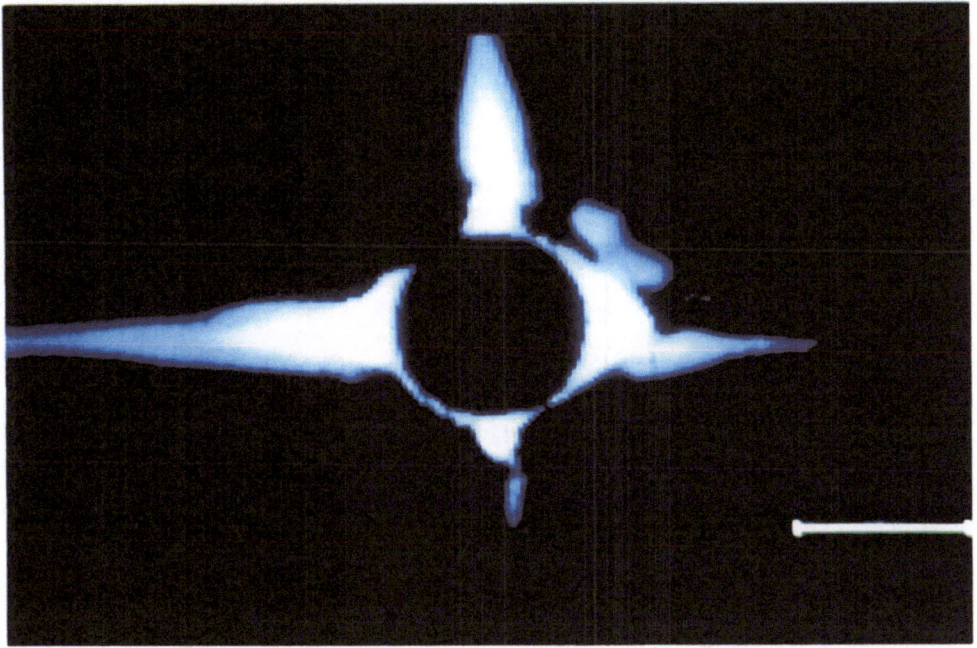

Fig. 2. Same as Figure 1, but for another star (with a similar light level) which shows obviously no circumstellar disk. The exposure time is corrected with respect to the β Pic brightness.

moving inward. From experimental albedos, a rather similar behavior has been found in icy materials more and more dusty (e.g., Gaffey and McCord, 1979).

This might be the first direct observation of a different situation prevailing in the inner regions of the β Pic disk, namely at less than about 80 AU (5 arcsec). This is in agreement with the analysis of Backman *et al.* (1992), in which a single component disk model is inconsistent with the combined optical and infrared data. Models with two structural components better match the available observations, with a transition around 80 AU: in the outer disk, water ice particles could be stable, whereas in the inner disk, ice sublimation is so rapid that particles must be refractory, the transition temperature being in the range of 90–140 K. In a theoretical model of circumstellar dust and planets formation, Nakano (1988) also considered a discontinuity in the disk due to ice sublimation.

Finally, as previously shown by Smith and Terrile (1987), we found also a slight asymmetry in the β Pic disk. The statistical brightness ratio between the two sides of the disk is rather constant, in favor of the South-West extension within about 7′ from the star, and in favor of the North-East extension outward about 8′ as detected by Smith and Terrile (1987). There is clearly a sharp zone of inversion close to 7′, which might again be a signature of a planetary formation process taking place

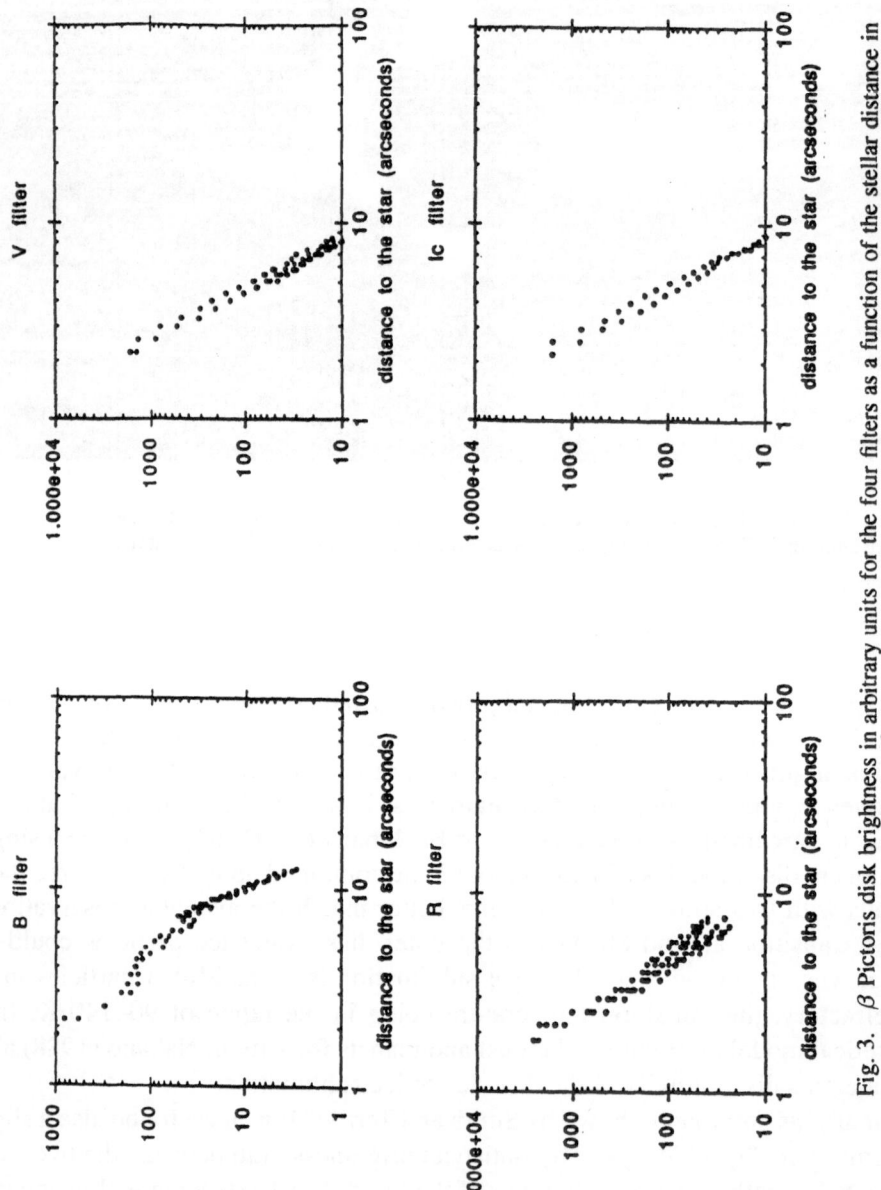

Fig. 3. β Pictoris disk brightness in arbitrary units for the four filters as a function of the stellar distance in arcsec. In V, R and I_C filters, an $r^{-3.6}$ power law is found. In B, a clear departure from this law is detected at less than 5 arcsec (75 AU).

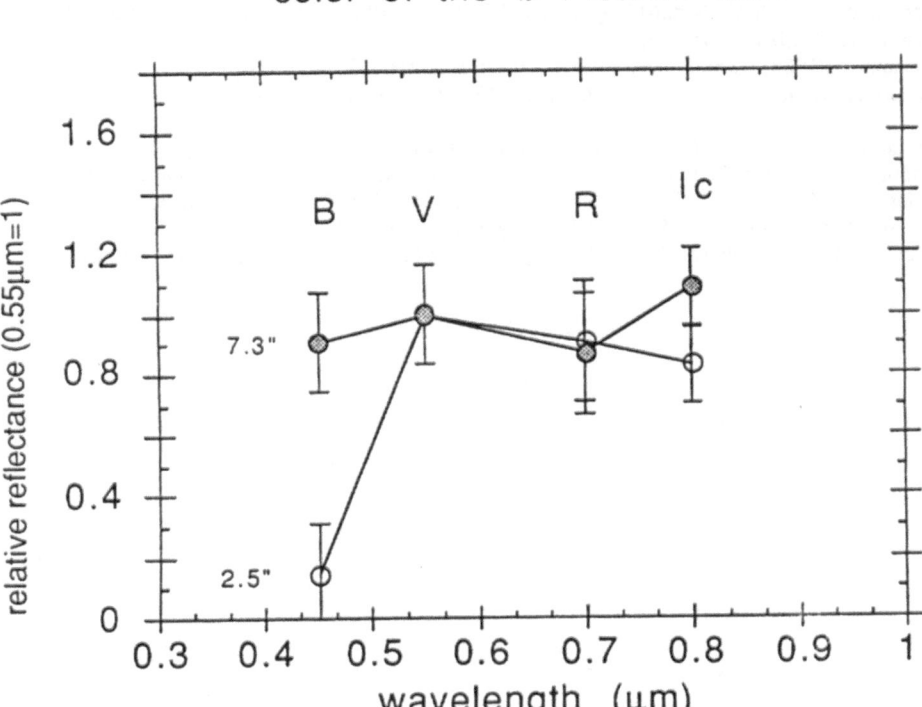

Fig. 4. The disk color is given at two different locations from the star, in the four filters (after normalization with the stellar light). Neutral colors are seen at large distances, while there is a clear drop in the *B* filter in the inner regions.

within the β Pic disk. This possibly could be linked to large asymmetry structures imposed to the dust particles, either during the planetary formation process (see, e.g., Lissauer, 1987), or by already existing planets (Sicardy, 1991; Scholl, Roques, and Sicardy, 1992).

References

Artymowicz, P., Burrows, C. and Paresce, F.: 1989, *Astrophys. J.* **337**, 494.
Backman, D.E., Gillett, F.C. and Witteborn, F.C.: 1992, *Astrophys. J.* **385**, 670.
Beust, H., Lagrange-Henri, A.M., Vidal-Madjar, A. and Ferlet, R.: 1989, *Astron. Astrophys.* **223**, 304.
Buil, C.: 1989, in: S.A.P. (eds.), *Astronomie CCD*, Toulouse.
Colas, F. and Arlot, J.E.: 1991, *Astron. Astrophys.* **252**, 402.
Colas, F. and Buil, C.: 1992, *Astron. Astrophys.* **262**, L13.
Ferlet, R., Hobbs, L.M. and Vidal-Madjar, A.: 1987, *Astron. Astrophys.* **185**, 267.
Gaffey, M. and McCord, T.B.: 1979, in : T. Gehrels, ed(s)., *Asteroids*, 688.

Hobbs, L.M., Vidal-Madjar, A., Ferlet, R., Albert, C. and Gry, C.: 1985, *Astrophys. J.* **293**, L29.
Kondo, Y. and Bruhweiler, F.C.: 1985, *Astrophys. J.* **291**, L1.
Lagrange, A.M., Ferlet, R. and Vidal-Madjar, A.: 1987, *Astron. Astrophys.* **173**, 289.
Lagrange-Henri, A.M., Vidal-Madjar, A. and Ferlet, R.: 1989, *Astron. Astrophys.* **190**, 275.
Lissauer, J.J.: 1987, *Icarus* **69**, 249.
Nakano, T.: 1988, *Monthly Notices Roy. Astron. Soc.* **230**, 551.
Scholl, H., Roques, F. and Sicardy, B.: 1992, *Celestial Mechanics* in press.
Sicardy, B.: 1991, *Icarus* **89**, 197.
Smith, B.A. and Terrile, R.J.: 1984, *Science* **226**, 1421.
Smith, B.A. and Terrile, R.J.: 1987, *Bulletin of the American Astronomical Society* **19**, 829.
Telesco, C.M., Becklin, E.E., Wolstencroft, R.D. and Decher, R.: 1988, *Nature* **335**, 51.
Telesco, C.M. and Knacke, R.F.: 1991, *Astrophys. J.* **372**, L 29.
Vidal-Madjar, A., Hobbs, L.M., Ferlet, R., Gry, C. and Albert, C.: 1986, *Astron. Astrophys.* **167**, 325.

THE DETECTION AND STUDY OF PRE-PLANETARY DISKS *

A.I. SARGENT

California Institute of Technology,
Pasadena, California, USA

and

S.V.W. BECKWITH

Max Planck Institut für Astronomie,
Heidelberg, Germany

Abstract. A variety of evidence suggests that at least 50% of low-mass stars are surrounded by disks of the gas and dust similar to the nebula that surrounded the Sun before the formation of the planets. The properties of these disks may bear strongly on the way in which planetary systems form and evolve. As a result of major instrumental developments over the last decade, it is now possible to detect and study the circumstellar environments of very young, solar-type stars in some detail, and to compare the results with theoretical models of the early solar system. For example, millimeter-wave aperture synthesis imaging provides a direct means of studying in detail the morphology, temperature and density distributions, velocity field and chemical constituents in the outer disks, while high resolution, near infrared spectroscopy probes the inner, warmer parts; the emergence of gaps in the disks, possibly reflecting the formation of planets, may be reflected in the variation of their dust continuum emission with wavelength. We review progress to date and discuss likely directions for future research.

1. Introduction

Stars like the Sun form in dense, 10^5 to 10^7 cm^{-3}, cool, 10–35 K concentrations in interstellar molecular clouds (e.g., Benson and Myers, 1989; Myers and Fuller, 1992). These slowly rotating cores, initially ranging in size from thousands of AU to tenths of a parsec, become gravitationally unstable and collapse to produce a central condensation, a proto-star, surrounded by an accretion disk, and embedded in an envelope of still-infalling gas and dust (cf. Shu *et al.*, 1987; 1993). Over a period of about 10^5 years, the proto-stellar core gradually increases in mass and a strong wind develops, breaking through the envelope at the rotational poles of the star-disk system in two oppositely-directed outflows. Ensuing dissipation of the obscuring material results in the emergence of a visible star with a circumstellar disk. Current theories of solar system formation suggest that the formation of planetesimals and eventually planets took place in such a disk (cf. Safronov, 1991).

As a result of recent instrumental advances, the circumstellar disks still associated with very young stars can now be observed. Many appear similar to the pre-solar nebula, the disk of gas and dust that surrounded the Sun in its infancy, 4.5×10^9 years ago, and from which the planets formed (cf. Lin and Papaloizou, 1985; Safronov, 1991). Studies of these pre-planetary disks may therefore supply clues to how the present solar system evolved. In addition, they provide a method

---

* Paper presented at the Conference on *Planetary Systems: Formation, Evolution, and Detection* held 7–10 December, 1992 at CalTech, Pasadena, California, U.S.A.

Astrophysics and Space Science **212**: 181–189, 1994.

of establishing how often other planetary systems occur. For the present, this cannot be determined by searching for mature planetary systems since many factors, notably the small surface area of planets and the enormous contrast in brightness between them and their central stars, prevent direct observation of planets around other stars.

2. Detecting Disks

We can estimate the properties of proto-planetary disks by extrapolating from the present solar system. Theory suggests that the mass of the young solar nebula was at least 0.01 $M_\odot$ (Lin and Papaloizou, 1985; Safronov, 1991). Co-planarity and co-revolution of the planets imply a highly-flattened disk with diameter of order 100 AU, the approximate diameter of Pluto's orbit. The velocity field should be Keplerian, of the form $v \propto r^{-0.5}$. Expected disk surface temperatures at about 1 AU from a young, solar-type star can be no more than a few $\times 100$ K, decreasing with increasing radius to a few $\times 10$ K. Disk observations are therefore best undertaken at the infrared and millimeter wavelengths where relatively cool dust grains preferentially emit. Spectral line emission from a wide variety of molecular species is also found at millimeter wavelengths, allowing important analyses of disk velocity structure.

Our understanding of pre-planetary disks has benefited enormously from technological developments over the last decade that have improved spectral and spatial resolution, as well as sensitivity in the infrared and millimeter-wave bands. Observations with the Infrared Astronomical Satellite (IRAS), launched in 1983 to survey the entire sky at wavelengths of 12, 25, 60 and 100 μm, have had a major impact. The high-sensitivity IRAS measurements demonstrated that as many as 30% of nearby A stars, including Vega, emit considerably more infrared radiation than is expected from the stellar photosphere alone (Aumann et al., 1984; Gillett, 1986; Backman and Gillett, 1987). Although these stars are almost as old as the Sun, typically a few $\times 10^8$ years, the "Vega phenomenon" is attributed to orbiting dust particles; optical coronagraph observations of one such object, β Pictoris, reveal that the dust lies in an edge-on disk with radius many hundreds of AU (Smith and Terrile, 1984). The considerable body of observational data that now exists suggests that 10^{-5} $M_\odot$ disks, perhaps the residue of planet formation, surround Vega-type stars (cf. Becklin and Zuckerman, 1990; Lagrange-Henri et al., 1990; Telesco and Knacke, 1991; Boggess et al., 1991; Chini et al., 1991; Backman and Paresce, 1993).

The IRAS observations of very young (3×10^5 to 3×10^7 years) solar-mass stars also show excess infrared emission that can be attributed to circumstellar disks (Adams et al., 1987; 1988; Strom et al., 1989). Many other features peculiar to these T Tauri stars, including enhanced infrared and ultraviolet fluxes, asymmetric emission line profiles, optical jets and bipolar molecular outflows, are also readily explained if disks of radius $\sim$ a few $\times$ 100 AU are present (cf. Appenzeller and

Mundt, 1989; Bertout, 1989; Edwards *et al.*, 1993). In the nearest star-forming clouds, at 140 pc distance, 100 AU corresponds to 0.7″. Confirmation of disk morphology and detailed investigations of disk properties therefore require imaging at arcsecond spatial resolution. Gross properties can, however, be derived by comparing lower resolution observations of the near-infrared to millimeter-wave dust continuum emission with those expected for star-disk systems (see Beckwith and Sargent, 1993a, and references therein). Using this technique, large numbers of stars can be searched quite rapidly for the dust continuum radiation typical of disks. Spectral energy distributions (SED's) based on IRAS flux measurements (Strom *et al.*, 1989) and IRAM 30-meter telescope observations at wavelength $\lambda =$ 1.3 mm (Beckwith *et al.*, 1990), imply that disks surround at least 50% of nearby T Tauri stars.

At wavelengths shortward of 100 μm, the dust emission is optically thick and the SED's give the disk radial temperature distributions, $T_0(r/r_0)^{-q}$; optically thin emission at sub-millimeter and millimeter wavelengths is sensitive to the disk mass, M_d. Beckwith *et al.* (1990) derive disk temperatures between 50 and 400 K at 1 AU from the star; the power law index of the temperature distribution, q, is between 0.5 and 0.75; outer radii are typically a few $\times$ 100 AU; masses, M_d, are between 0.003 and 3 $M_\odot$. These parameters are consistent with the presence of pre-planetary disks. Adams *et al.* (1990) obtain higher values of M_d for a smaller sample of T Tauri stars, using a lower adopted value of κ_ν, the mass opacity coefficient. Adopted values for κ_ν vary widely (cf. Beckwith and Sargent, 1991); within the uncertainties, M_d is in most cases greater than required to form the minimum mass solar nebula. These pre-planetary disks are very different from the 'debris' disks that surround the older, Vega-type stars; masses, for example, are orders of magnitude higher. Evolution between these extremes should, in principle, reflect the evolution of the solar system.

3. Investigating Disk Properties

High resolution millimeter-wavelength line and continuum observations can directly confirm the disk morphology inferred from low resolution surveys and establish if the gas is moving in Keplerian orbits, as would be expected for a reservoir of material from which planets form. If the disks are indeed proto-planetary in nature, more detailed studies of their physical and chemical characteristics are required for comparison with models of planetary system formation and evolution.

Millimeter-wave interferometers that can provide the required resolution began operation in the 1980's, and disk-like structures around a few very young objects have been mapped (Beckwith *et al.*, 1986; Mundy *et al.*, 1986; 1992; Sargent and Beckwith, 1987; 1991; Weintraub *et al.*, 1989; Sargent *et al.*, 1989; Walker *et al.*, 1990; Ohashi *et al.*, 1991; Simon and Guilloteau, 1992; Kawabe *et al.*, 1993; Koerner *et al.*, 1993a; 1993b; Sargent and Beckwith, 1993). Figures 1a and 1b are 2.7″ resolution images of the low velocity ^{13}CO. (1→0) emission from the

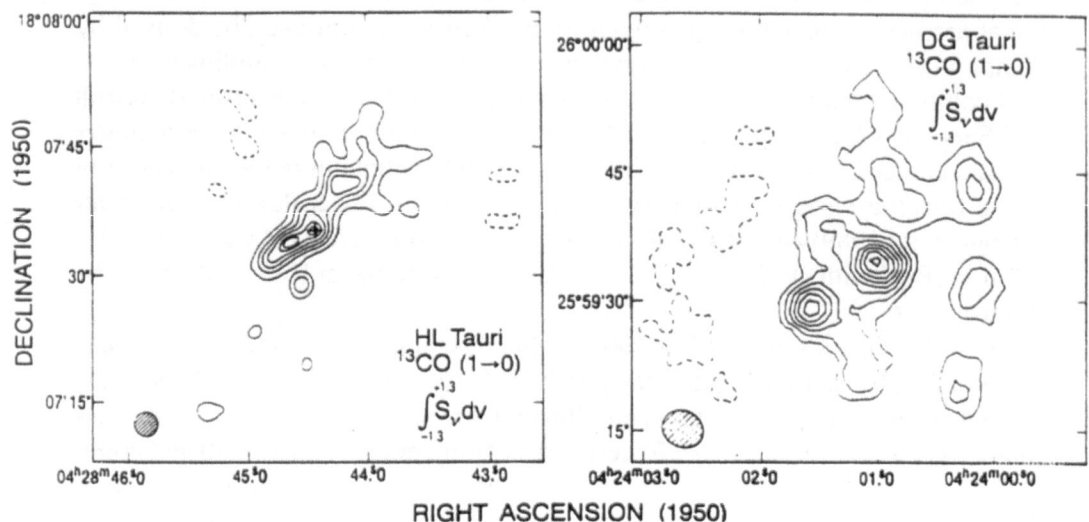

Fig. 1. Maps of the distributions of low-velocity ($v_{sys} \pm 1.3\,\mathrm{km\,s^{-1}}$.), ^{13}CO ($1\rightarrow 0$) emission around (a) HL Tauri and (b) DG Tauri. In each map a cross represents the position of the star. Contours begin at 60 mJy (2σ) and are spaced by 30 mJy. In maps that include contributions from the higher velocity gas, emission is unresolved and concentrated within the highest central contours shown above.

T Tauri stars HL Tauri and DG Tauri. For velocities, v(systemic) ± 1.3 km s^{-1}, both show elongated gaseous structures extending to radii of almost 2000 AU. In each case, high velocity gas, and also the $\lambda = 2.7$ mm dust continuum emission, is confined to an unresolved core, ≤ 200 AU in radius, centered on the star. Disk masses calculated from the ^{13}CO fluxes are $\sim 0.1\ M_\odot$, for HL Tau, and 0.04 to 0.05 $M_\odot$, for DG Tau, in good agreement with the values derived in the IRAM survey of 1.3 mm dust continuum emission (Beckwith *et al.*, 1990). Stellar ages are a few $\times\ 10^5$ years, as anticipated for stars supporting pre-planetary systems.

To date, evidence for Keplerian velocity structure in the disks associated with young stars is compelling but not unambiguous. A simple analysis of the velocity pattern in HL Tau suggests a circumstellar disk of gas is in Keplerian rotation about the star (Sargent and Beckwith, 1987; 1991). Studies of the velocity structure in DG Tau are not yet complete, but the fact that high velocity gas is concentrated in a core, while the low velocity material is extended as shown in Figure 1b, is consistent with gas in Keplerian orbits (Sargent and Beckwith, 1993).

A persistent problem is caused by contributions to the CO and ^{13}CO emission from the ambient clouds in which most of the T Tauri stars are found. Large-scale cloud structure ($\geq 15''$) is resolved out by the millimeter-wave interferometers, but smaller-scale features in the inhomogeneous cloud are detected as a result of the enhanced sensitivity. This makes it difficult to completely isolate the velocity structure of any disks. Aperture synthesis observations of molecular species that trace only denser gas should, in principle, alleviate the problem. The lower rotational transitions of CO are excited at densities an order of magnitude lower than required to produce CS ($2\rightarrow1$) emission. Imaging of the dense disks in the millimeter-wave spectral lines of the CS molecule should therefore discriminate against more diffuse ambient cloud features.

Interferometric CS ($2\rightarrow1$) observations of HL and DG Tau are reported by Ohashi *et al.* (1991). The CS ($2\rightarrow1$) emission around HL Tau has also been mapped by Blake *et al.* (1992). Extended structures with radii 2000 AU are observed, but in neither case is the gas centered on the stars; the displacements, $\sim 3 - 5''$, are significant. Imaging of the proto-type of the class, T Tauri, also indicates a dearth of CS ($2\rightarrow1$) molecular line emission at the stellar position (van Langevelde *et al.*, 1993). From excitation calculations that incorporate the results of IRAM searches for emission in higher rotational CS transitions, Blake *et al.* (1992) infer molecular hydrogen densities of $10^4 - 10^5$ cm^{-3}, and CS fractional abundances $\sim 10^{-8}$ for the gas around HL Tau. These values are typical for standard cloud cores in the Taurus region and imply that the CS observations are probing an outer envelope rather than the circumstellar disk. The models imply that CS is depleted by factors of 25–50 in the disk itself. An attractive possibility is that gas-phase CS is converted into other sulfur-bearing molecules, as in the inner part of the proto-solar nebula (Barshay and Lewis, 1976).

Modeling the millimeter-wave molecular line emission expected from proto-planetary disks for comparison with aperture synthesis observations of circumstellar gas is likely to be the most effective method of establishing the disk velocity fields. Calculations of the ^{13}CO. ($1\rightarrow0$) emission expected from a rotating disk have recently been carried out by Beckwith and Sargent (1993b). A comparison of synthetic aperture synthesis maps based on these calculations with ^{13}CO. images of the T Tauri star, GM Auriga, show very good agreement (Koerner *et al.*, 1993). It appears that a proto-planetary disk of mass 0.08 $M_\odot$ and radius ~ 200 AU orbits this 0.7 $M_\odot$ star, although higher spatial and spectral resolution maps are needed to remove all uncertainties from this interpretation.

Somewhat unexpectedly, the disk calculations of Beckwith and Sargent (1993) demonstrate that not only CO ($1\rightarrow0$) but also ^{13}CO ($1\rightarrow0$) emission is optically thick for disk masses $\geq 0.001 M_\odot$. This suggests that CO and ^{13}CO interferometric observations to date have reflected only disk temperature distributions and have provided no information about the density structure. Tracing the density variations in disks evidently requires observations in rarer isotopes, such as C^{18}O or C^{17}O, that are much less abundant, with optically thin emission. Preliminary HCO$^+$ maps of T Tauri indicate that, unlike CS, this species may prove a reliable density tracer (van Langevelde *et al.*, 1993).

Additional support for the presence of disks around very young stars is now being provided by observations of spectral line emission at near-infrared wavelengths. As described above, these observations probe the velocity field of the inner disk, close to the star. Echelle spectrographs capable of achieving the necessary high spectral resolution in this wavelength band have only recently begun operation. Model fits to observations of the 2.2 μm CO band-head emission from WL 16 by Chandler *et al.* (1993), for example, suggest an inner disk radius of 0.04 AU. Similar observations of DG Tau by these authors are more complicated to model but also indicate the presence of a disk.

4. Disk Evolution

Changing disk properties between the pre-planetary and debris extremes discussed above should reflect the evolution of the solar system. In fact, there is no clear indication of diminution of total disk mass over the stellar age range spanned by the Beckwith *et al.* (1990) survey, 2×10^5 – to 3×10^6 years. However, unexpectedly low 12 and 25 μm fluxes are noted for a number of stars and may indicate a form of evolution — depletion of dust in the inner regions of the disks. Such "clearing" could, in turn, presage the onset of planetary formation (Skrutskie *et al.*, 1989; Beckwith *et al.*, 1990; Strom *et al.*, 1993). Stars with inner holes tend to display other characteristics of disk evolution, such as decreased accretion activity in their disks (cf. Edwards *et al.*, 1993; Strom *et al.*, 1993).

With age 2×10^6 years, GM Aurigae, mentioned above, is among the oldest T Tauri stars in the samples searched for disks. Although the mass of the disk remains substantial, 0.08 $M_\odot$, the SED for GM Aur suggests clearing of dust within 0.36 AU of the star and there is little evidence of accretion (Beckwith *et al.*, 1990; Koerner *et al.*, 1993). Small-scale deviations of the near infrared fluxes from those implied by filled-disk models have been interpreted as signifying the existence of gaps in the GM Aur disk, perhaps the effect of forming planets (Marsh and Mahoney, 1992). Although depressed near- and mid-infrared flux values can be explained by the combined effects of dust grain opacity and vertical structure in the disks (Boss and Yorke, 1993), it seems likely that GM Aur is an example of a more evolved star-disk system.

Disk dissipation may also be affected by FU Orionis events. This phenomenon, a sudden brightening of a star by several magnitudes, with a subsequent decrease in visual intensity lasting for many years, is named for the object in which it was first observed (Herbig, 1977). Many of the properties of FUors can be explained by model in which an accretion disk surrounds a T Tauri star; the outburst is correlated with dramatic increase in the disk accretion rate (Hartmann and Kenyon, 1985). T Tauri stars may experience numerous FUor events in their early evolution, the disk material dissipating a little more with each outburst (cf. Hartmann *et al.*, 1993). It is not clear how this effect would impinge upon the formation of a planetary system, but it is notable that infrared and optical spectroscopic studies of at least one FUor, V1057 Cyg, suggest Keplerian rotation (Hartmann and Kenyon, 1987).

Further studies of FU Orionis objects are clearly in order. In addition, observations of disks around older stars, and of systems for which the SED's indicate less massive disks, are needed. These will extend studies of disk evolution and establish the evolutionary path of the early solar nebula.

5. The Future

With the launch of ESA's Infrared Space Observatory (ISO) and, in the more distant future, NASA's Space Infrared Telescope Facility (SIRTF), infrared surveys of the type pioneered by IRAS will continue. The higher resolution and sensitivity compared to IRAS will allow the determination of more accurate continuum SED's for a much larger sample of possible pre-planetary disks. A wide range of stellar ages and masses will be encompassed by these surveys. Inevitably, these will increase the pool of objects for mapping and should greatly enhance our knowledge of disk evolution.

Very few disks have been imaged to date. Because each of the millimeter-wave interferometers began operation with only a few telescopes, mapping has been a laborious and time-consuming process. All the arrays are currently expanding and upgrading equipment, and the VLA is being modified for operation at $\lambda = 7$ mm. As a result, interferometric surveys will be viable. These will enable the very sensitive searches for small-scale circumstellar material that are necessary to improve our understanding of disk dissipation, with obvious implications for solar system evolution. Such surveys will also clarify how many stars are likely to support forming solar systems.

The upgraded improved millimeter-wave arrays will also provide the very high spatial and spectral resolution that are vital to detailed studies of disk properties. Analysis of the velocity structure of the cool, outer disks will be readily undertaken. Studies of various molecular species will allow studies of the temperature, density and chemical variations across these disks. In addition, studies of the inner, warm regions of pre-planetary disks will advance as spectral resolution at infrared wavelengths continues to improve. It is anticipated that the two Keck telescopes and the European Southern Observatory's Very Large Telescope will employ a number of small out-rigger telescopes to carry out infrared interferometry and enable imaging of these warm disks. In terms of detecting and studying pre-planetary disks, we are on the threshold of a new scientific era.

Acknowledgements

The authors are grateful to G. Blake, J. Carlstrom, C. Chandler, D. Koerner, N. Scoville, E. van Dishoeck, and H. van Langevelde for communicating their results in advance of publication. A.I.S. thanks NSF for support through Grant AST-9016404 to the Owens Valley array, and NASA for support through the *Origins of the Solar System* program.

References

Adams, F. C., Emerson, J. P. and Fuller, G. A.: 1990, *Astroph. J.*, **357**, 606–620.

Adams, F. C., Lada, C. J. and Shu, F. H.: 1987, *Astroph. J.*, **312**, 788–806.

Adams, F. C., Lada, C. J. and Shu, F. H.: 1988, *Astroph. J.*, **326** 865–883.

Appenzeller, I. and Mundt, R.: 1989, *Astron. Astrophys. Rev.*, **1**, 291–334.

Aumann, H. H., Gillett, F. C., Beichman, C. A., de Jong, T., Houck, J. R., Low, F. J., Neugebauer, G., Walker, R. G. and Wesselius, P. R.: 1984, *Astroph. J.*, **278**, L23–L27.

Backman, D.E. and Gillett, F.C.: 1987, in: J.C. Linsky and R.E. Stencel (eds.), *Cool Stars, Stellar Systems and the Sun*, (Berlin: Springer-Verlag), 340–350.

Backman, D. E. and Paresce, F.: 1993, in: E.H. Levy, J.I. Lunine and M.S. Matthews (eds.), "Protostars and Planets III", (Tucson: Univ. of Arizona Press), in press.

Barshay, S.S., Lewis and J. S.: 1976, *Annu. Rev. Astron. Astrophys.*, **14**, 81–94.

Becklin, E. E. and Zuckerman, B.: 1990, in: G.D. Watt and A.D. Webster (eds.), "Submillimetre Astronomy" (Dordrecht: Kluwer), 147–153.

Beckwith, S. V. W. and Sargent, A. I.: 1991, *Astroph. J.* **381**, 250–258.

Beckwith, S. V. W. and Sargent, A. I.: 1993a, in: E.H. Levy, J.I. Lunine and M.S. Matthews (eds.), *Protostars and Planets III*, (Tucson: University of Arizona Press), in press.

Beckwith, S. V. W. and Sargent, A. I.: 1993b, *Astroph. J.*, **402**, 280–291.

Beckwith, S. V. W., Sargent, A. I., Chini, R. and Güsten, R.: 1990, *Astron. J.*, **99**, 924-945.

Beckwith, S. and Sargent, A. I., Scoville, N. Z., Masson, C. R., Zuckerman, B. and Phillips, T. G.: 1986, *Astroph. J.*, **309**, 755–761.

Benson, P. J. and Myers, P. C.: 1989, *Astroph. J. Suppl.*, **71**, 89–108.

Bertout, C.: 1989, *Annu. Rev. Astron. Astrophys.*, **27** 351–395.

Blake, G. A., van Dishoeck, E. F. and Sargent, A. I.: 1992, *Astroph. J. (Letters)*, **391**, L99–L103.

Boggess, A., Bruhweiler, F. C., Grady, C. A. and Ebbets, D. C.: 1991, *Astroph. J. (Letters)*, **377**, L49–L52.

Boss, A. P. and Yorke, H. W.: 1993, *Astroph. J.*, submitted.

Chandler, C. J., Carlstrom, J. E., Scoville, N. Z., Dent, W. R.F. and Geballe, T. R.: 1993, *Astroph. J. (Letters)* submitted.

Chini, R., Krugel, E., Shustov, B., Tutukov, A. and Kreysa, E.: 1991, *Astron. Astrophys.*, **252**, 220–228.

Edwards, S., Ray, T. and Mundt, R.: 1993, in: E. H. Levy, J.I. Lunine and M.S. Matthews (eds.), *Protostars and Planets III*, (Tucson: University of Arizona Press), in press.

Gillett, F. C.: 1985, in: F.P. Israel (ed.), *Light on Dark Matter*, (Dordrecht: Reidel), 61–69.

Hartmann, L. and Kenyon, S.: 1985, *Astroph. J.*, **299**, 462–478.

Hartmann, L. and Kenyon, S.: 1987, *Astroph. J.*, **322**, 393–398.

Hartmann, L., Kenyon, S. and Hartigan P.: 1993, in: E.H. Levy, J.I. Lunine and M.S. Matthews (eds.), *Protostars and Planets III*, (Tucson: University of Arizona Press), in press.

Herbig, G. H.: 1977, *Astroph. J.*, **217**, 693–715.

Koerner, D. K., Sargent, A. I. and Beckwith, S. V. W.: 1993a. *Icarus*, submitted.

Koerner, D. K., Sargent, A. I. and Beckwith, S. V. W.: 1993b. *Ap. J. (Letters)*, in press.

Lagrange-Henri, A. M., Ferlet, R., Vidal-Madjar, A., Beust, H. and Gry, C.: 1990, *Astron. Astrophys. Suppl.*, **85**, 1089–1100.

Lin, D.N.C. and Papaloizou, J.: 1985, in: D.C. Black and M. S. Matthews (eds.), *Protostars and Planets II*, (Tucson: University of Arizona Press), 981–1072.

Marsh, K. A. and Mahoney, M. J.: 1992, *Astroph. J. (Letters)*, **395**, L115–L118.
Mundy, L. G., Wilking, B. A. and Myers, S. T.: 1986, *Astroph. J. (Letters)*, **311**, L75–L79.
Mundy, L. G., Wootten, H. A., Wilking, B. A., Blake, G. A. and Sargent, A. I.: 1992, *Astroph. J.*, **385**, 306–313.
Myers, P. and Fuller, G. A.: 1992, *Astroph. J.*, **396**, 631–642.
Ohashi, N., Kawabe, R., Hayashi, M. and Ishiguro, M.: 1991, *Astron. J.*, **102**, 2054–2065.
Safronov, V. S.: 1991, *Icarus*, **94**, 260–271.
Sargent, A. I. and Beckwith, S. V. W.: 1987, *Astroph. J.*, **323**, 294–305.
Sargent, A. I. and Beckwith, S. V. W.: 1991, *Astroph. J. (Letters)*, **382**, L31–L35.
Sargent, A. I., Beckwith and S. V. W.: 1993, in *IAU Colloquium 140: Astronomy with Millimeter and Sub-millimeter Wave Interferometry*, (A.S.P. Conf. Series San Francisco: BookCrafters), in press.
Sargent, A. I., Beckwith, S. V. W., Keene, J. and Masson, C. R.: 1988, *Astroph. J.*, **333**, 936–942.
Shu, F. H., Adams, F. C. and Lizano, S.: 1987, *Annu. Rev. Astron. Astrophys.* **25**, 23–81.
Shu, F. H., Najita, J., Galli, D. and Ostriker, E.: 1993, in: E. H. Levy, J. I. Lunine and M. S. Matthews (eds.), *Protostars and Planets III*, (Tucson: Univ. of Arizona Press), in press.
Simon, M. and Guilloteau, S.: 1992, *Astroph. J. (Letters)*, **397**, L47–L49.
Skrutskie, M. F., Snell, R. L., Dutkevitch, D., Strom, S. E. and Schloerb, F. P.: 1991, *Astron. J.*, **102**, 1749–1752.
Smith, B. A. and Terrile, R. J.: 1984, *Science*, **226**, 1421–1424.
Strom, K. M., Strom, S. E., Edwards, S., Cabrit, S. and Skrutskie, M. F.: 1989, *Astron. J.*, **97**, 1451–1470.
Strom, S. E., Edwards, S. and Skrutskie, M. F.: 1993, in: E. H. Levy, J. I. Lunine and M. S. Matthews (eds.), *Protostars and Planets III*, (Tucson: University of Arizona Press), in press.
Telesco, C. M. and Knacke, R. F.: 1991, *Astroph. J. (Letters)*, **372**, L29–L31.
van Langevelde, H., van Dishoeck, E.F. and Blake, G. A.: 1993, in: M. Ishiguro and R. Kawabe (eds.), *IAU Colloquium 140, Astronomy with Millimeter and Sub-millimeter Wave Interferometry*, A. S. P. Conf. Series (San Francisco: BookCrafters), in press.
Walker, C. K., Carlstrom, J. E., Bieging, J., Lada, C. and Young, E.: 1990, *Astroph. J.*, **364**, 173–177.
Weintraub, D. A., Masson, C. R., Zuckerman, B.: 1989, *Astroph. J.*, **344**, 915–924.

MILLIMETER CONTINUUM MEASUREMENTS OF CIRCUMSTELLAR DUST AROUND VERY YOUNG LOW-MASS STARS *

S. TEREBEY

Infrared Processing and Analysis Center, Jet Propulsion Laboratory and California Institute of Technology,
Pasadena, California, USA

C.J. CHANDLER

Owens Valley Radio Observatory, California Institute of Technology,
Pasadena, California, USA

and

P. ANDRÉ

Service d'Astrophysique, Centre d'Etudes de Saclay,
Gif-sur-Yvette, France

Abstract. We investigate the question of disk formation during the protostar phase. We build on the results of Keene and Masson (1990) whose analysis of L1551 showed the millimeter continuum emission comes from both an unresolved circumstellar component, i.e., a disk and an extended cloud core. We model the dust continuum emission from the cloud core and show how it is important at 1.3 mm but negligible at 2.7 mm. Combining new 2.7 mm Owens Valley Interferometer data of IRAS-Dense cores with data from the literature we conclude that massive disks are also seen toward a number of other sources. However, 1.3 mm data from the IRAM 30 m telescope for a larger sample shows that massive disks are relatively rare, occurring around perhaps 5% of young embedded stars. This implies that either massive disks occur briefly during the embedded phase or that relatively few young stars form massive disks. At 1.3 mm the median flux of IRAS-Dense cores is nearly the same as T Tauri stars in the sample of Beckwith *et al.* (1990). We conclude that the typical disk mass during the embedded phase is nearly the same or less than the typical disk mass during the T Tauri phase.

1. Introduction

Continuum measurements at millimeter wavelengths of stellar heated dust have proven to be an important and successful way to detect circumstellar disks around young stellar objects. A number of studies of T Tauri stars (Weintraub *et al.*, 1989; Adams *et al.*, 1990; Beckwith *et al.*, 1990) have concluded that (1) circumstellar disks are common; (2) some disks are massive, meaning the disk mass is comparable with the stellar mass; but that (3) typical disks have masses like 0.01 $M_\odot$, the mass inferred for the primitive solar nebula.

It is clearly important to also study circumstellar disk properties around younger, likely protostellar objects in order to understand the formation and evolutionary history of disks. However, by their nature protostars are deeply embedded objects whose extended envelopes contribute to the continuum emission. High spatial

---

* Paper presented at the Conference on *Planetary Systems: Formation, Evolution, and Detection* held 7–10 December, 1992 at CalTech, Pasadena, California, U.S.A.

resolution measurements by interferometers are needed to discriminate between the envelope and disk emission.

Due to the difficulty of studying large samples with interferometers the best statistical information comes from single-dish work. Recent studies (e.g., André *et al.*, 1990; André and Montmerle, 1993) with moderate resolution (11″) show that overall, deeply embedded IR sources have nearly the same peak flux densities as T Tauri stars. This suggests a weak dependence of disk mass with age from 10^5 to 10^6 years.

However high-spatial resolution observations remain key to detecting circumstellar disks around embedded sources. Given expected disk sizes on the order of 100 AU ($< 1″$ in Taurus), it is difficult to actually resolve the disks, but interferometers can provide good upper limits to their sizes. Keene and Masson (1990) successfully used 2.7- and 1.3-mm interferometer data to distinguish between the envelope and disk emission and convincingly demonstrated the presence of a massive circumstellar disk in L1551.

Extending this approach, we have done extensive theoretical modeling of the expected continuum emission from a collapsing dense cloud core. We then compare the models with both interferometer and single-dish data for a sample of embedded stars. Interested readers are referred to Terebey *et al.* (1993) for a detailed description of this work.

2. Data

2.1. SOURCE SAMPLE

To select very young sources we define our sample to be deeply embedded infrared sources that are found near the peaks of dense gas emission in nearby molecular clouds (Myers and Benson, 1983; Beichman *et al.*, 1986; Myers *et al.*, 1987; Benson and Myers, 1989). The "IRAS-Dense cores" are thought to contain young, embedded low-mass stars with estimated ages of a few $\times 10^5$ years. The proximity of the embedded infrared sources to the peaks of the dense gas distribution indicates the stars are very young. Theoretical models suggest the IRAS-Dense cores are protostars surrounded by infalling envelopes of gas and dust (Terebey *et al.*, 1984, hereafter TSC; Shu *et al.*, 1987).

2.2. OBSERVATIONS

We mapped ten IRAS-Dense cores with the Owens Valley Millimeter Interferometer. The spatial resolution ranged from 3″ to 7″, corresponding to a linear scale of 300 to 2000 AU. Continuum emission was detected from six of the ten sources at 2.7 mm and seemed to be spatially unresolved for most sources.

To improve our statistics we obtained fluxes for 25 IRAS-Dense cores at the shorter wavelength of 1.3 mm with the IRAM 30-m, single-dish telescope. The ≈11″ beam of the 30 m should be sufficient to probe circumstellar structures in nearby star-forming regions. Continuum emission was detected from nearly all the sources.

2.3. COMPARISON OF IRAS-DENSE CORES WITH T TAURI STARS

For the single-beam measurements we do not know the relative contributions of extended envelope and circumstellar disk to the emission. However the total flux does provide an upper limit to the amount of mass in a circumstellar disk. With this in mind we compared the IRAS-Dense cores with the T Tauri sample of Beckwith *et al.* (1990).

The median peak flux density of the 25 IRAS-Dense cores is 80 mJy inside an 11″ beam at 1.3 mm, similar to but somewhat higher than the median flux density of the T Tauri stars. This suggests the circumstellar disk mass does not change substantially between 10^5 and 10^6 yr, the rough age span of the IRAS-Dense core to T Tauri phase. The 80 mJy flux density corresponds to a circumstellar mass of about 0.02 $M_\odot$. This typical circumstellar mass is somewhat higher than the 0.01 $M_\odot$ inferred for the primitive solar nebula, but is still much less than the mass of the young star.

This result has interesting implications for the formation and evolution of disks during the protostellar phase. Angular momentum considerations during the cloud-collapse phase suggest that much of the collapsing gas does not fall directly onto the star but first falls onto a disk (TSC). Therefore, disks are important because much of the mass that is eventually incorporated into the star, on the order of half a solar mass, is first processed through a circumstellar disk. On the other hand, the observations show that very few protostellar sources have disk masses this large. This means that infalling material does not accumulate in the disk during most of the protostar phase but instead is efficiently transported to the central protostar. Whatever the relevant physics turns out to be, the transport of mass and angular momentum in the disk occurs rapidly with respect to the 10^5 yr formation time scale, even for relatively low-mass disks.

3. Predicted Continuum Emission from the Collapsing Core

Since all the sources in our sample are thought to be embedded within a dense infalling envelope of gas and dust, there will be a contribution to the total continuum flux from this extended component. In order to evaluate the magnitude of the contribution in both our interferometer measurements and the single-dish data, we have constructed a model of the expected emission from a collapsing dense cloud core using the TSC collapse models.

3.1. BASIC EQUATIONS

We computed the contribution to the millimeter continuum intensity from thermal dust grains. At millimeter wavelengths the emission from the cloud core is well approximated by low optical depth, so that the emergent intensity profile I as a function of frequency ν is then simply given by

$$I_\nu\left(\omega\right) = \int_{-\infty}^{\infty} B_\nu\left(T_D\right) \rho\left(r\right) \kappa_\nu \, dl \tag{1}$$

Here ω is the impact parameter in the plane of the sky, B_ν is the Planck function, T_D is the dust temperature, ρ is the gas density at radius r from the center of the dense core, κ_ν is the specific gas opacity and l is the line-of-sight distance through the dense core. If the dust temperature is not too low, the Rayleigh-Jeans approximation ($B_\nu \propto T_D$) holds at millimeter wavelengths.

The intensity computation therefore requires three quantities $\rho(r)$, $T_D(r)$ and κ_ν to be specified. The density profile is provided by the collapse models and in the limit of small radius approaches a power-law with $\rho \propto r^{-3/2}$. The dust temperature depends weakly on the bolometric luminosity and has a shallow power-law profile (typically $T_D \propto r^{-0.4}$) over the spatial scales of interest. The opacity is the most uncertain quantity in the calculation; we follow the usual practice of extrapolating from far-infrared wavelengths using a power-law with index β between one and two.

4. Results

The model predicts most of the continuum emission that arises at these spatial resolutions arises in the inner, collapsing region of the cloud (hence the name 'infall' envelope). Furthermore, because of the power-law profile of the density, the infall envelope has a characteristic signature in which the emission always looks spatially resolved, with size roughly 1.5 times the beam, no matter what the observed beam size.

The model quantitatively predicts that the continuum emission from the infall envelope scales as $F \propto a_s^3 M^{-0.5} L^{0.2}$ where a_s is the sound speed, M is the central (star plus disk) mass, and L is the bolometric luminosity. For our sample of about 25 embedded stars the models with $M = 0.5$ or $1.0\ M_\odot$ can account for much of the observed single-dish emission at 1.3 mm. However the emission from a few sources is too strong to be easily explained by the standard parameters and at 2.7 mm the discrepancy extends to all observed sources.

Our models suggest one possible explanation since they predict that younger sources (having lower central masses) naturally have higher continuum fluxes. However inspection of the 2.7-mm interferometer maps reveals that only a few sources show spatially resolved structure, the characteristic signature expected of continuum emission from the infall envelope.

For most sources the continuum emission at 2.7 mm appears pointlike. This argues that the excess flux arises in a compact circumstellar disk rather than the infall envelope. This provides evidence that circumstellar disks are common around

IRAS-Dense cores, although they typically are less massive than the disk found around L1551 (Keene and Masson, 1990).

After throwing out sources from the literature to correct our sample for selection biases we find that 1/20 or 5% have very strong millimeter continuum fluxes. The statistics are admittedly poor but the results do agree with other studies that find strong continuum sources are relatively rare (André *et al.*, 1990; André and Montmerle, 1993). This implies that either massive disks occur briefly during the protostar phase or that relatively few young stars form massive disks.

Our models suggest that future observations of the dust continuum focusing on 2–3 mm interferometer data will be extremely useful in measuring disk fluxes and disk masses around protostars. It is unfortunate that our poor knowledge of the opacity at millimeter wavelengths limits our ability to infer disk masses, and irksome that the opacity may even change with time if grains grow rapidly in the protosolar nebula. Despite the difficulties, millimeter-continuum measurements provide an important way to probe the formation of disks during the early protostar phase.

Acknowledgements

This work was carried out in part at the Jet Propulsion Laboratory, California Institute of Technology, under a contract with the National Aeronautics and Space Administration. CJC acknowledges the support of a SERC/NATO Fellowship during the course of this work.

References

Adams, F.C., Emerson, J.P. and Fuller, G.A.: 1990, *Astrophysical Journal,* **357**, 606–620.
André, P. and Montmerle, T.: 1993, From T Tauri Stars to Protostars: Circumstellar Material and Young Stellar Objects in the ρ Ophiuchi Cloud, submitted to the *Astrophysical Journal.*
André, P., Montmerle, T., Feigelson, D. and Steppe, H.: 1990, *Astronomy and Astrophysics,* **240**, 321–330.
Beckwith, S.V.W., Sargent, A.I., Chini, R.S. and Gusten, R.: 1990, *Astronomical Journal,* **99**, 924–945.
Beichman, C.A., Myers, P.C., Emerson, J.P., Harris, S., Mathieu, R., Benson, P.J. and Jennings, R.E.: 1986, *Astrophysical Journal,* **307**, 337–349.
Benson, P.J. and Myers, P.C.: 1989, *Astrophysical Journal Supplement,* **71**, 89–108.
Keene, J. and Masson, C.R.: 1990, *Astrophysical Journal,* **355**, 635–644.
Myers, P.C. and Benson, P.J.: 1983, *Astrophysical Journal,* **266**, 309–320.
Myers, P.C., Fuller, G.A., Mathieu, R.D., Beichman, C.A., Benson, P.J., Schild, R.E. and Emerson, J.P.: 1987, *Astrophysical Journal,* **319**, 340–357.
Shu, F.H., Adams, F.C. and Lizano, S.: 1987, *Annual Reviews of Astronomy and Astrophysics,* **25**, 23–81.
Terebey, S., Chandler, C.J. and André, P.: 1993, *Astrophys. J.* **414**, 759–772.
Terebey, S., Shu, F.H. and Cassen, P.C.: 1984, *Astrophysical Journal,* **286**, 529–551 (TSC).
Weintraub, D.A., Sandell, G. and Duncan, W.D.: 1989, *Astrophysical Journal,* **340**, L69–L72.

THE SEARCH FOR PROTOSTARS USING

MILLIMETER/SUBMILLIMETER DUST EMISSION AS A TRACER*

P.G. MEZGER

Max-Planck-Institut für Radioastronomie, Bonn, Germany

1. Formation and Evolution of Protostars: An Overview

In the disk of our Galaxy, about four solar masses of interstellar matter (ISM) per year are transformed into stars. Observations combined with evolutionary models (see also, e.g., Shu *et al.*, 1987) suggest the following sequence of events for the formation of medium- and high-mass stars, which is illustrated by the Orion Giant Molecular Cloud (OMC; see Figure 1):

1. In an extended GMC of medium density (Figure 1a; $n_H \sim 10^2$ to 10^3 cm^{-3}) a filament or ridge structure of increased density (Figure 1b; $n_H \sim 10^4$ cm^{-3}) has developed which is usually investigated using transitions of molecules with high dipole moment (e.g., CS isotopes).

2. Elongated cloud cores form in this filament (Figure 1c) with densities $n_H \sim 10^5$ to 10^6 cm^{-3}, which should be gravitationally unstable according to the Jeans criterion but may be supported by frozen-in magnetic fields and (non-thermal) turbulence. These cores can be best investigated using transitions of molecules with high dipole moment (e.g., CS isotopes) or dust emission.

3. Rotating protostellar condensations with mean densities $\geq 10^8$ cm^{-3} form along the major axis of such an elongated core like beads on a string (see, e.g., Figures 3 and 6c). In most cases these condensations can only be observed through their dust emission.

4. They collapse in free-fall until a stellar core forms at mean densities of $n_H \sim 10^{11}$ to 10^{12} cm^{-3}.

5. A rotating disk forms as a consequence of conservation of angular momentum, through which the stellar core accretes additional matter.

6. Molecular outflow perpendicular to the accretion disk begins at a relatively early evolutionary stage — perhaps as early as the formation of a stellar core.

7. These pre-MS stars evolve to MS stars through the states of dust embedded star and T-Tauri star (medium-mass stars) or cocoon star (high-mass stars; see Figure 1c), respectively.

The observations presented in the following sections relate to the above evolutionary stages (1) through (4). Note, however, that — from the point of view of observations — little is known about the protostellar evolution of low-mass stars.

---

* Paper presented at the Conference on *Planetary Systems: Formation, Evolution, and Detection* held 7–10 December, 1992 at CalTech, Pasadena, California, U.S.A.

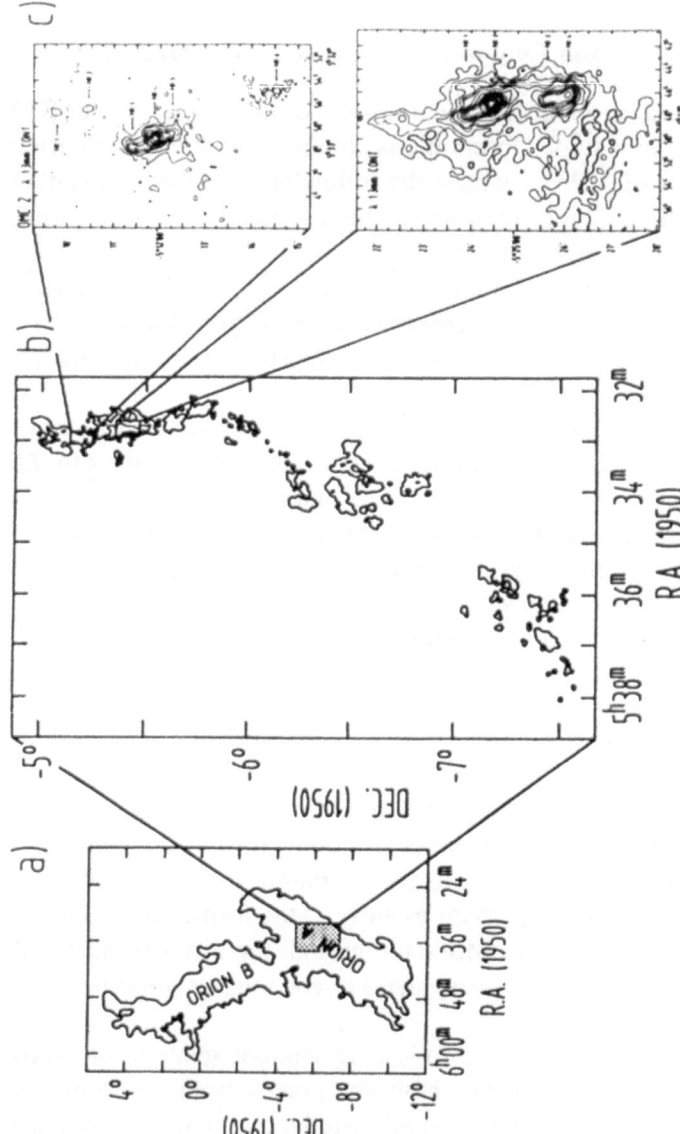

Fig. 1. Morphology of the massive star forming Orion GMC: (a) The GMC complex (which consists of the Orion A and B clouds) as outlined by CO mapping (Maddalena *et al.*, 1986). Each cloud has a mass of ~ 10^5 M$_\odot$. (b) The dense ridge in the Orion A cloud as outlined in CS(1–0) emission (Tatemutsu and Umemoto, 1992). It has a mass of ~ 10^4 M$_\odot$. (c) The dense cloud cores OMC1/2 outlined in λ1300 μm dust continuum (Mezger *et al.* 1990b) with masses of ~ 1.7×10^3 and 1.5×10^3 M$_\odot$.

2. Millimeter/Submillimeter Dust Emission: A New Tool for the Investigation of High Density Regions

High resolution interferometer observations have long shown evidence for chemical abundance variations due to the formation and evaporation of molecular ice mantles around dust grains, e.g., in the densest parts of OMC1 (Plambeck, 1987; for more recent observations see McMullin et al., 1993 and Zhou et al., 1991). At about the same time it was found that the massive disk surrounding the exciting star IRS4 of the bipolar outflow in S106 is only visible in single dish observations of dust emission, but not in molecular line emission observed with comparable angular resolution. It was also found that the high submillimeter optical depths were incompatible with NIR optical depths if normal grains (such as in the Mathis, Rumpl and Nordsieck, 1977, model) were responsible for both extinction and emission (Mezger et al., 1987). Based on these observations it was suggested that in high-density regions grains form molecule ice mantles and subsequently begin to coagulate, thus reducing the visual/NIR dust extinction, without changing the dust absorption cross section at millimeter/submillimeter wavelengths. Since then there is increasing evidence (see Section 4) that for hydrogen densities $n_H \leq 10^6$ cm^{-3}, dust emission and isotopic molecular line emission, such as $C^{18}O$ and $C^{34}S$, indeed yield similar hydrogen masses and column densities but that, at higher densities and low gas and dust temperatures (≤ 20 K), molecular transitions cease to be reliable tracers of hydrogen column densities.

The interpretation of dust emission in terms of optical depth, hydrogen column density and mass is rather straightforward. The surface brightness at the frequency ν of a dust layer of optical depth τ_ν and average dust temperature T_d is

$$I_\nu = S_\nu/\Omega_A = B_\nu(T_d)(1 - e^{-\tau_\nu}) \tag{1}$$

Here S_ν/Ω_A is the beam averaged intensity which is usually expressed in "flux density (S_ν/Jy) per solid beam angle (Ω_A/sr)" and

$$\tau_\nu(\lambda)/N_H = \sigma_\lambda^H = (Z/Z_\odot)b \begin{cases} 7 \times 10^{-21} \lambda_{\mu m}^{-2} & \lambda_{\mu m} \geq 100 \\ 7 \times 10^{-22} \lambda_{\mu m}^{-1.5} & 40 \leq \lambda_{\mu m} \leq 100 \end{cases} \tag{2}$$

is a parametrized representation of the FIR/submillimeter dust absorption cross section per H-atom σ_λ^H, with $N_H = N(H) + 2N(H_2)$ as the total hydrogen column density, $Z/Z_\odot$ as the relative metallicity, and b as a parameter which characterizes respectively dust in the diffuse ISM ($b = 1$), in molecular clouds of medium density ($b = 1.9$) and in very dense ($n_H \geq 10^6$ cm^{-3}) and cold ($T_d \leq 20$ K) condensations ($b = 3.4$). (For these and the following relations see Mezger et al., 1990b, Appendix A; and Mezger, 1990.)

Reliable hydrogen column density and mass determinations are only possible
at wavelengths where $\tau_\nu \ll 1$. In this case N_H and M_H are related to the observed
surface brightness S_ν/Ω_A and total integrated flux density $S_{\nu,\text{tot}}$, by

$$(N_H/\text{cm}^{-2}) = 1.93 \times 10^{15} \frac{(S_\nu/Jy)\,\lambda_{\mu m}^4}{(\theta_A/\text{arcsec})^2\,(Z/Z_\odot)bT}\,\frac{e^x - 1}{x} \qquad (3a)$$

and

$$(M_H/M_\odot) = 4.1 \times 10^{-10}\,\frac{(S_{\nu,\text{tot}}/Jy)\,\lambda_{\mu m}^4\,D_{kpc}^2}{(Z/Z_\odot)bT}\,\frac{e^x - 1}{x} \qquad (3b)$$

Here θ_A is the telescope FWHP, D_{kpc} is the distance of the source in kpc and
$x = 1.44 \times 10^4/\lambda_{\mu m}T$. Note that at $\lambda = 1300\ \mu m\ \tau_\nu \sim 1$ for $N_H \sim 10^{26}\ \text{cm}^{-2}$
and at $\lambda = 350\ \mu m$ for $N_H \sim 10^{25}\ \text{cm}^{-2}$.

How reliable are the dust cross sections given by Eq. (2)? Mezger *et al.* (1982)
and Mathis *et al.* (1983) extended the Mathis, Rumpl and Nordsieck (1977; MRN)
model to millimeter wavelengths and found reasonable agreement with the scarce
measurements of submillimeter absorption cross sections available at that time.
This model was further improved by Draine and Lee (1984) and Draine (1985)
based on a critical review of the optical characteristics of graphite and silicates. The
parameter value $b = 1$ in Eq. (2) describes their dust model. The larger values of b
agree with submillimeter observations by Hildebrand (1983) and Pajot *et al.* (1989;
$b = 1.9$) of medium density and by Rengarajan (1984; $b = 3.4$) of high density
molecular clouds. The dust model suggested by Rowan-Robinson (1986) uses a
$\sigma_\lambda^H \propto \nu$ opacity law at $\lambda > 30\ \mu m$ and thus has considerably higher submillimeter
dust absorption cross sections with equivalent b-values at *lambda* 1 mm as high
as ~ 30. More recently Mathis and Whiffen (1989) suggested an alternative dust
model, where small individual particles of silicates, amorphous carbon and graphite
are bonded together but include a substantial fraction of empty space.

Figure 2 (adapted from Draine, 1989) shows the dust absorption cross section
per H-atom of the four model computations listed. Also indicated in this diagram is
— as stippled area — σ_λ^H as given by Eq. (2) for the parameter range $1 \le b \le 3.4$;
$Z/Z_\odot = 1$. Only the Rowan-Robinson model cross sections lie — for $\lambda > 100\ \mu m$
— considerably above the regime of cross sections, recommended here by Eq. (2)
for the conversion of dust surface brightnesses into hydrogen column densities.
One argument against the high Rowan-Robinson cross sections is — at least for
dust in interstellar space — the low temperatures which they predict but which are
not in accordance with observations (see, e.g., Cox and Mezger, 1989).

Based on a comparison of cloud characteristics obtained under quite varying
physical conditions with molecular spectroscopy and dust emission I conclude that
in MCs of intermediate densities millimeter/submillimeter dust absorption cross
sections as given by Eq. (2) are correct within $\pm 50\%$. For densities $n_H \ge 10^6\ \text{cm}^{-3}$,
where I recommend the use of the parameter value $b = 3.4$, the uncertainties may be

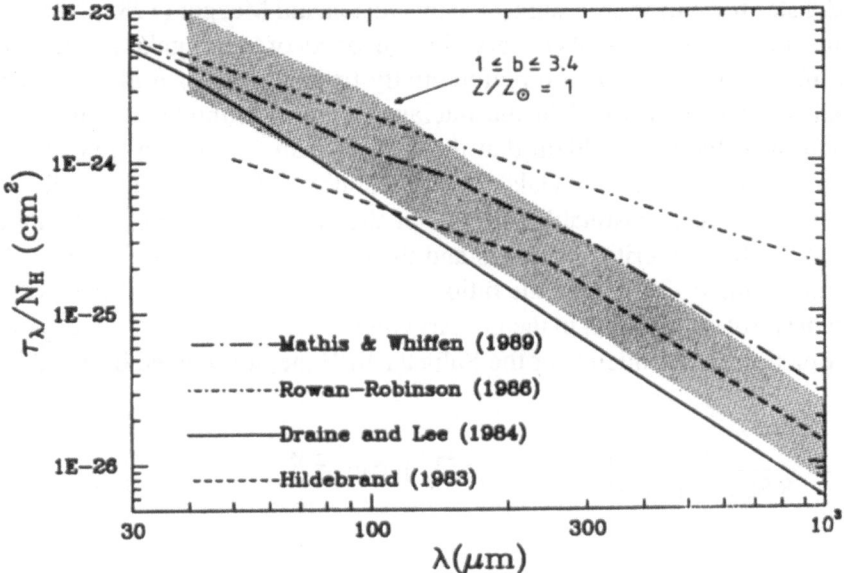

Fig. 2. Dust absorption cross sections predicted by four models (adapted from Draine, 1989) compared with the cross section approximation Eq. (2) for the range of the dust parameter b recommended in this paper (see text).

as large as 100%, but still much better than the uncertainties with which gas phase abundances of molecules are known. For further discussions of dust characteristics I refer to Mezger (1990).

3. Star Formation Rates and the Number of Detectable Isothermal Protostars

Isothermal protostars — as compared to pre-main-sequence stars (PMS stars) — appear to be extremely rare. How many stars in the mass range $(m, m + dm)$ are formed at a given time t in the galactic disk and in the solar vicinity? The answer gives the creation function

$$C(t, m)\,dm = \Psi(t)\,\Phi(m)\,dm \qquad (4)$$

where $\Psi(t)$ is the time-dependent star formation rate (SFR) and $\Phi(m)$ is the mass-dependent initial mass function (IMF) which is normalized to unity

$$\int_{m_l}^{m_u} \Psi(m)m\,dm = 1.$$

Here m_u, m_l are the upper and lower stellar mass limits. In the following discussion I use the Salpeter (1955) IMF with $m_u = 100\ M_\odot$ and $m_l = 0.1\ M_\odot$.

$$\Phi(m) = 0.172\,m^{-2.35}$$

Bimodal star formation was suggested by Güsten and Mezger (1983; GM83) to explain the abundance gradients observed in the disks of spiral galaxies. It means that high-mass stars $m \geq m_c$ are preferentially formed in the spiral arms, while stars of all masses are formed in the interarm regions. Bimodal star formation can in addition solve the problem that the high Lyman continuum (Lyc) photon production rate observed in the Galactic Disk cannot — for a continuous IMF — be reconciled with any reasonable estimate of the present-day SFR. In the model preferred by GM83 the critical mass is and the discontinuity in the bimodal IMF hence occurs at $m_c = 3M_\odot$, and the ratio — at the solar circle — of massive star formation in spiral arms to that in the interarm region is $\alpha = \Psi^{sa}_{m>m_c} / \Psi^{ia}_{m>m_c} = 1$. With the functional dependence of the Salpeter IMF the normalized bimodal IMF is

$$\Phi(m) = \frac{1}{5.824 - 4.45\beta}[(1 - \beta \, m^{-2.35}_{m>m_l} + \beta m^{-2.35}_{m>m_c}].$$ (5)

Here

$$\beta = \Phi^{sa}_{m>m_c}(\Phi^{sa}_{m>m_c} + \Phi^{ia}_{m>m_c})^{-1}$$

is the fraction of all massive stars formed in spiral arms and m extends to $m_u = 100 \, M_\odot$.

In Table I, I give the present-day bimodal SFR $\Psi(t_o)$ (i.e., the mass of interstellar matter converted per year into stars) and the corresponding bimodal creation rate

$$C^{bm}(t_{o,m})dm = \Psi^{bm}(t_o) \, \Phi^{bm}(m)dm$$

(i.e., the number of stars formed per year in a given mass range m, m + Δm) for the galactic disk (R = 1.7 to 8.5 kpc), the "solar bin" (R = 7.7 to 8.5 kpc) and an area of D $\leq$ 1 kpc around the sun. Input parameter for the derivation of the SFR is the Lyc-photon production rate derived from free-free emission. The procedure is outlined in GM83 and Mezger (1985) but the results are adjusted for $R_\odot$ = 8.5 kpc. The net effect of bimodal star formation is that the creation of massive stars m > m_c is increased relative to that of medium and low-mass stars.

The *total* number of isothermal protostars in the mass range m'–m'' given in Table II

$$N_{iso} = \tau_{iso} \int_{m''}^{m'} C^{bm}(m) \, dm$$

is obtained with C_m and τ_{iso}, the time the protostar spends in free-fall contraction. Here I substitute $\tau_{iso} \sim 4 \times 10^4$ yr, the estimated lifetime of medium mass protostars (see Section 5).

The number of *detectable* isothermal protostars depends on the detection limit of the bolometer in the mapping mode, for which I use the conservative estimate

TABLE I

Star formation rates

R/kpc	β	Φ^{bm} $M_\odot$ yr^{-1}	$\Phi^{bm}/M_\odot$ yr^{-1}			C^{bm}/stars yr^{-1}		
			0.1–0.8 $M_\odot$	0.8–3 $M_\odot$	3–100 $M_\odot$	0.1–0.8 $M_\odot$	0.8–3 $M_\odot$	3–100 $M_\odot$
1.7–8.5	0.70	3.8	1.37	0.49	1.94	6.6	0.35	0.23
7.7–8.5	0.50	0.23	0.10	0.04	0.09	0.50	0.03	0.01
$D < 1$ kpc	0.50	0.08	0.01	0.003	0.007	0.04	0.002	0.001

TABLE II

Detection rate of isothermal protostars

R/kpc	N_{iso} (total)			N_{iso} (detectable)		
	0.1–0.8 $M_\odot$	0.8–3 $M_\odot$	3–100 $M_\odot$	$\langle m \rangle = 0.3\ M_\odot$	1.6 $M_\odot$	20 $M_\odot$
1.7–8.5	2.6×10^5	1.4×10^4	9.2×10^3	280	74	1000
7.7–8.5	2.0×10^4	1.2×10^3	4.0×10^2			
$D < 1$ kpc	1.6×10^3	8.0×10^1	4.0×10^1	280	74	40

of $S_{1300\mu m} \sim 100$ mJy. Substitution in Eq. (3b) with $T_d \sim 20$ K yields a relation between the mass M_H of the isothermal protostar, for which I substitute — in the mass range m'–m'' — the IMF weighted mean stellar mass $\langle m \rangle$ and the distance D out to which it can be detected, provided the dust sphere is still optically thin at $\lambda\ 1300\ \mu m$. For $S_{1300}\nu = 100$ mJy, these distances are — for increasing protostellar masses — $D = 0.4$, 1.0 and 3.4 kpc. If we limit the search of protostars to the solar surroundings, the number of detected low- and intermediate-mass protostars should outweigh that of massive stars. An unbiased survey for protostars, however, will strongly favor the detection of high-mass stars.

4. The Case of NGC 2024: Molecular Spectroscopy Meets Its Limits

In 1988 Mezger *et al.*, in the cloud core associated with the compact H II region NGC 2024, detected (what they considered to be) the first isothermal protostars (Figure 3). The cloud core has a total mass of $M_H \sim 500\ M_\odot$, of which the

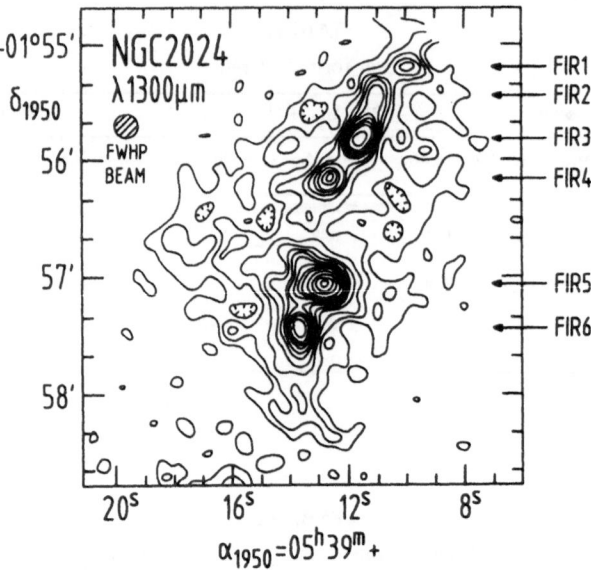

Fig. 3. The cloud core associated with the compact H II-region NGC 2024. One clearly recognizes its three components: Extended envelope ($M_H \sim 300\ M_\odot$), ridge ($\sim 140\ M_\odot$) and protostellar condensations ($\sim 70\ M_\odot$) (Mezger et al., 1988).

central ridge structure accounts for $\sim 140\ M_\odot$ and the protostellar condensations for another 70 $m_\odot$. This discovery was met with substantial skepticism, amongst other arguments also since the authors had suggested that molecules were depleted and therefore were no longer reliable tracers of hydrogen column densities (see, e.g., the paper by Schulz et al., 1991).

In two subsequent papers Mezger et al. (1992a) and Mauersberger et al. (1992) presented new observations: An $8''$, $\lambda\ 870\ \mu m$ dust emission MRT map which reveals a seventh condensation; and $C^{34}S$ and $C^{16}O$ MRT maps with angular resolutions of $16''$ and $12''$, which definitely show that molecular and dust emission peaks do not coincide (see Figure 4). Figure 5a shows the $\lambda\ 870\ \mu m$ map once more. The column densities derived from dust emission and from $C^{34}S$ and $C^{18}O$ transitions, respectively, along the ridge line indicated by the dashed curve are shown in Figure 5b. Note that in general only $C^{13}O$ is considered as a reliable tracer of N_H while $C^{34}S$ traces higher-density regions but its abundance is rather uncertain. It is obvious that in the case of the cloud core NGC 2024 $C^{18}O$ traces only the low-density envelope, while $C^{34}S$ traces in addition — at least qualitatively — the medium-density ridge. The protostellar condensations, however, are visible only in dust emission. NH_3 VLA maps made with an angular resolution of $\sim 3''$ indicate NH_3 clumps close to the dust condensations, but it is also found that this molecule must be heavily depleted, if the observed line emission should come from the dust condensations (Mauersberger et al.).

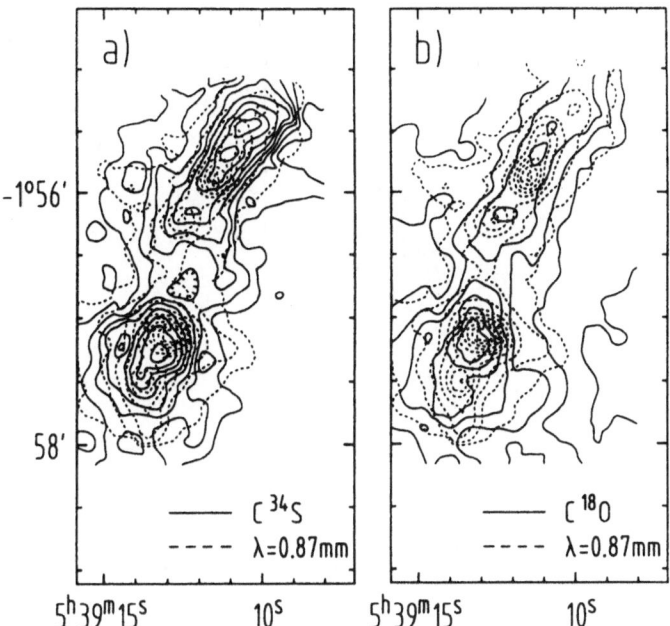

Fig. 4. (a) A $\lambda 870$ μm dust emission map convolved to an angular resolution of $11''$ (dashed contours) is superimposed on a velocity-integrated $C^{18}O$ map observed with similar angular resolution (solid contours). (b) Same as Figure 4a, but solid contours refer here to the velocity-integrated $C^{34}S$ map. Angular resolution is $16''$ (Mezger et al., 1992a).

Considering these facts the authors arrive at the following conclusions: Molecular transitions in NGC 2024 only trace the central ridge of the cloud core with a density of $n_H \sim 3 \times 10^6$ cm^{-3}, but not the protostellar condensations with average densities of $\sim 2.5 \times 10^8$ cm^{-3}. Gas and dust in the condensations FIR1, 2, 3 have about the same temperature, $T_d \sim T_g \sim 20$ K, while in the condensations FIR4, 5 (and perhaps in FIR6) the gas temperatures derived from ammonia are ≥ 40 K. The latter two condensations also show strong molecular outflow and may represent protostars evolved beyond the isothermal stage, where stellar core formation has begun.

One problem remained, however: The narrow widths of the NH$_3$ lines were not compatible with a virial mass as large as the mass derived from dust emission. To reconcile the narrow NH$_3$ lines with the large protostellar masses derived from dust emission Mauersberger et al. suggested that ammonia is present in only a small part of the dust condensations, and would have to be far from most of the mass in the dust condensation and not distributed symmetrically. Ongoing observations with the Plateau-de-Bure array (in preparation) of the $\sim \lambda$ 3-mm continuum and $C^{34}S$ emission towards FIR5 and 6 will further constrain the free parameters of any model of the NGC 2024 cloud core: They show a number of compact $C^{34}S$ sources

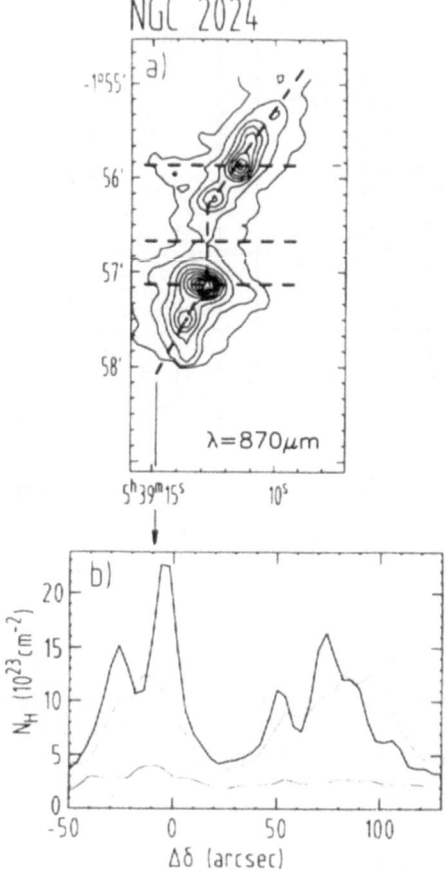

Fig. 5.   A λ 870 μm dust emission map smoothed to an angular resolution of $11''$. Dashed lines indicate sections approximately parallel and perpendicular to the elongated axis of the bar. Along these sections hydrogen column densities have been constructed from the λ 870 μm dust surface brightness (heavy solid curves), the $C^{34}S$ (dashed curves) and the $C^{18}O$ luminosities (thin solid curves; see text). Column densities along the ridge line are shown as Figure 5b. Offset coordinates refer to the position $\alpha_{1950} = 05^h.39^m.12^s.8$, $\delta_{1950} = -1°57'04''$ (Mezger *et al.*, 1992a).

in the $2''$ interferometer images, but those closest to FIR5 and 6 are clearly offset with respect to the dust emission peaks. The other CS clumps have no counterpart in dust emission.

The Orion B cloud (see Figure 1c), of which the cloud core NGC 2024 is part of, has a total mass of $\sim 10^5$ $M_\odot$. Lada *et al.* (1991) have surveyed in the CS(2-1) transition the densest cloud fragment with a mass of ~ 4000 $M_\odot$. They found five massive ($M_H > 200$ $M_\odot$) cloud cores associated with well-known massive star formation. However, the bulk of CS condensations in the mass range ~ 20 to 200 $M_\odot$ show no signs of star formation. We recently mapped the dust emission from this cloud with SEST (Launhard and Mezger, in preparation) and found the conclusions of Lada *et al.* in essence confirmed: four (out of five) massive cores are highly structured and may be actively involved in the formation of medium- to high-mass stars, while with a few — but very interesting — exceptions the lower-mass CS cores are not associated with compact dust emission sources. In summary: the gas in the Orion B cloud is clumped but not every clump actually forms stars.

5. Morphology of Low- and Medium-Mass Star Forming Clouds: The Ophiuchus Cloud

The nearby ($D \sim 160$ pc) Ophiuchus cloud complex extends for tens of parsecs. It is comprised of filamentary clouds of low density molecular gas (Figure 6a). In its westernmost part, however, this cloud contains a massive cloud fragment of size ~ 2 pc $\times 2$ pc, mass $M_H \sim 500$ *to* 600 $M_\odot$ and column densities of a few 10^{23} H-atoms cm^{-2} averaged over a $90''$-beam. In this fragment C^{18}O observations (Wilking and Lada, 1983; see also the review by Wilking, 1991) reveal the presence of a ridge of size $\sim 1 \times 2$ pc which extends from SE to NW (Figure 6b). In the following I refer to this fragment as the Rho Oph cloud.

This cloud appears to form low-mass stars over its whole extent. Wilking *et al.* (1989 and references therein) within an area of ~ 4.3 pc^2, have identified a total of 78 Young Stellar Objects (YSO) as members of a young cluster of low-mass stars (see Figure 5b). About one half of these YSOs appear to be dust-embedded stars and the other half T Tauri stars.

We have mapped with SEST and MRT the $\lambda 1300$ μm dust emission within the dense ridge outlined in C^{18}O (Mezger *et al.*, 1992b). The detection limit in the different fields ranges from ~ 0.2 to 0.04 $M_\odot$ per beam-area. We have detected altogether four cloud cores with masses ranging from $M_H \sim 1$ to 15 $M_\odot$ and typical densities of $n_H \sim 10^6$ cm^{-3}. Three of these cores were structureless (see, e.g., Oph B2 shown in Figure 6d); only the core SM1 (Figure 6c) shows four protostellar condensations with masses $M_H \sim 0.4$ to 3 $M_\odot$ and typical densities of $n_H \sim 10^7$ cm^{-3}. The submillimeter spectrum of SM1 is compatible with a dust temperature of $T_d \sim 15$ to 20 K (Ward-Thompson *et al.*, 1989; André *et al.*, 1993). The southernmost of these condensations, labeled FIR4 in Figure 6c, is associated with the ultra-compact HII region VLA 1623 which drives one of the youngest and best collimated CO bipolar outflows.

The active star-forming core SM1 may represent the low-to-medium mass pro-tostellar pendant of the high-mass star-forming core NGC2024. Its ridge-structure contains a total mass of ~ 15 $M_\odot$, of which one third has collapsed into isothermal protostars. As in the case of NGC2024 the cold high-density condensations FIR1–4 in SM1 are barely visible in molecular isotopic transitions (Mauersberger, private communication). Of the 80 YSOs identified in the Ophiuchus cloud as members of a young stellar cluster, one half are identified as T-Tauri stars with estimated lifetimes of $\sim 4 \times 10^5$ yr (Wilking and Lada, 1983). If the SFR in this cloud was

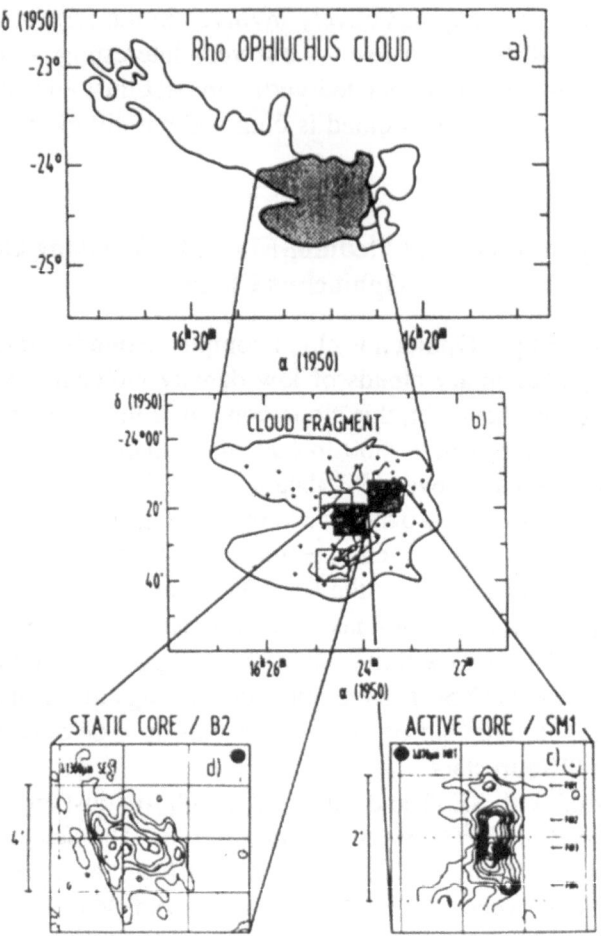

Fig. 6. Morphology of the low- and medium-mass star forming Ophiuchus cloud. (a) The low-density envelope (of mass $n_H \sim 3000\ M_\odot$) is shown as outlined in CO-emission. The densest part (shown stippled) with $M_H \sim 500\ M_\odot$ and $\langle n_H \rangle \sim 10^4$ cm^{-3} is shown in (b) together with the positions of 78 YSOs, observed at $\lambda 2.2\ \mu$m by Wilking *et al.* (1989) and identified as members of a young star cluster. (c) The active star forming cloud core SM1 ($M_H \sim 15\ M_\odot$), and (d) the static cloud core B2 ($M_H \sim 15\ M_\odot$) are shown as observed in $\lambda 870\ \mu$m (c) and $\lambda 1300\ \mu$m (d) dust emission (adapted from Mezger *et al.*, 1992b). Angular resolution is 8″(c) and 22″(d).

approximately constant over the past million years, and if statistics can be applied to the small number of four isothermal protostars, we derive a typical time scale for isothermal protostellar evolution of $\tau_{iso} \sim 4 \times 10^4$ yr, the value I used in Section 3 to estimate the number of detectable isothermal protostars. This time scale accounts for only $\sim 5\%$ of the total pre-MS evolution time of medium-mass stars and thus

TABLE III

Characteristics of W49 derived from its integrated spectrum

	T_d (K)	Mass ($M_\odot$)	N_H (cm^{-2})	$\langle n_H \rangle$ (cm^{-3})	L_{IR} ($L_\odot$)
(cold)	20	5.8×10^5	7.5×10^{23}	2.7×10^4	2.5×10^6
(warm)	50	4.4×10^4	7.5×10^{23}	9.8×10^4	1.3×10^7
(hot)	140	—	—	—	4.2×10^6

can explain the scarceness of isothermal protostars. It is also of interest to note that $\sim 4 \times 10^4$ yr corresponds to the free-fall time of a protostar which begins its contraction at a mean density $\sim 10^6$ cm^{-3}, as is observed in the static cores B1, B2 and F1.

6. Morphology of High-Mass Star Forming Clouds

In the theory of bimodal star formation (Güsten and Mezger, 1983; see also Section 3) one has spontaneous high-mass star formation in GMCs in the interarm region and induced high-mass star formation in spiral arms. Induced star formation results in clusters of very massive HII regions, of which W49 and W51 are well-known examples. W49A, at a distance of 14 kpc, shows one (or possibly several) warm components (Figure 7b) embedded in an extended but — compared to clouds in the interarm region — relatively compact cold GMC (Figure 7a). The well determined dust emission spectrum (Figure 7c) can be decomposed into three components (Sievers *et al.*, 1991). Source characteristics derived from this spectral decomposition using the opacities Eq. (2) are given in Table III.

Table III shows that warm dust accounts for $\sim 65\%$ of the total luminosity of $\sim 2 \times 10^7$ $L_\odot$ but for only $\sim 7\%$ of the total mass, while cold dust accounts for $\sim 13\%$ of the luminosity but for $\sim 93\%$ of the total hydrogen mass of $\sim 6 \times 10^5$ $M_\odot$. The 140 K dust emission, which accounts for $\sim 21\%$ of the luminosity, is unlikely to be produced by normal dust grains, but is rather emitted by very small particles, which are temporarily heated to relatively high temperatures by absorption of an energetic photon.

Note that the source parameters are volume averaged and thus column and volume densities are probably severely underestimated. There is mounting evidence of strong clumping in the interstellar gas in general and specifically of the gas in GMCs. For example, if the warm central component were surrounded by a dense layer of cold gas, the 50 K spectrum in Figure 7c would be truncated at the

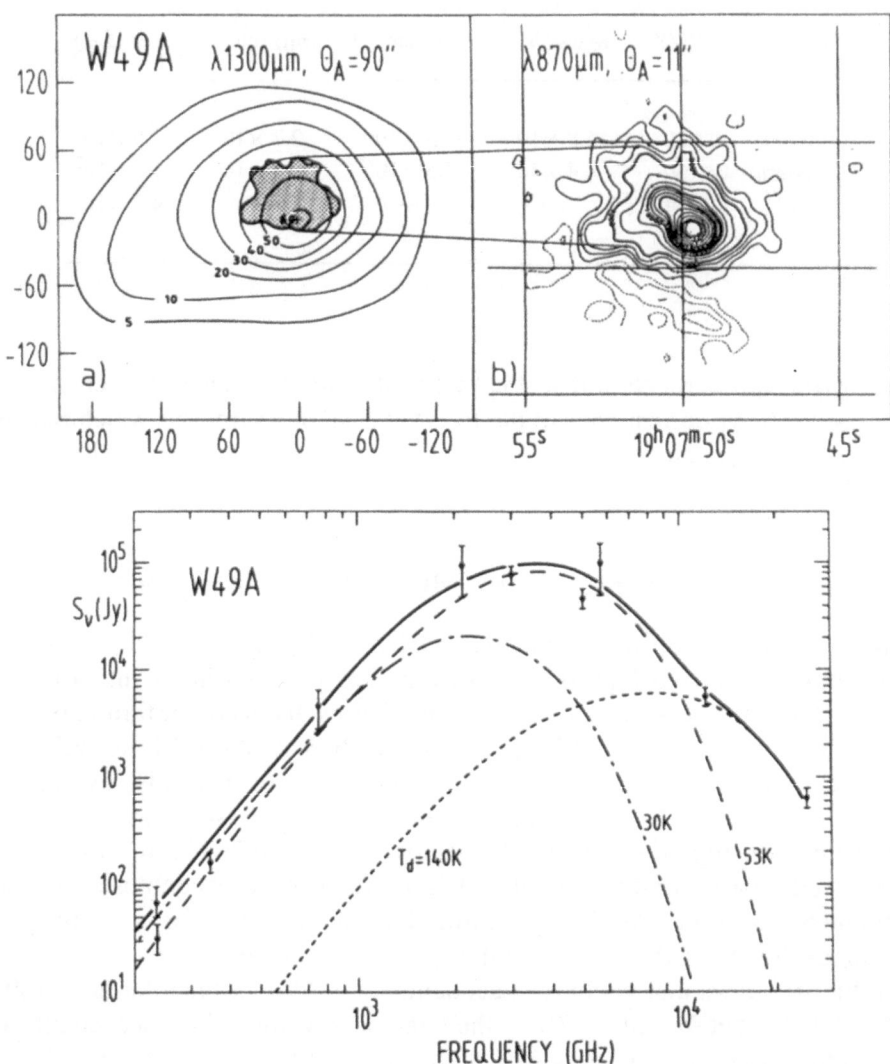

Fig. 7.   Morphology of the Giant HII/GMC complex W49. (a) The compact GMC observed in $\lambda 1300-\mu$m dust emission with an angular resolution of 90″. Most of the mass of $M_H \sim 6 \times 10^5\ M_\odot$ is associated with cold (~ 20 K) dust. (b) The central part of this cloud observed at 870 μm with an angular resolution of 8″. Its warm (~ 50 K) dust accounts for most of the luminosity of $L_{IR} \sim 2 \times 10^7\ L_\odot$. (c) The dust emission spectrum, decomposed into three components, using the dust opacities given by Eq. (2) (adapted from Sievers *et al.*, 1991).

wavelength where the cold dust layer becomes opaque. The very fact that the warm dust emission is so clearly discernible indicates that this cold dust layer must be very clumpy.

As discussed in Section 3 an unbiased $\lambda 1300$ μm continuum survey of the galactic plane should reveal three to four times more massive stars than medium- and low-mass stars. This, together with the fact that high-mass protostars are preferentially detected in the vicinity of well-known radio-HII regions has led to the situation that much more is known about high-mass star formation. Dust emission maps and spectra of star forming regions associated with SgrB2 (Gordon et al., 1993) and M17 (Tuffs et al., in preparation) will soon be published. Recently we finished a $\lambda 1300$ μm survey for compact GMCs which do not contain developed HII regions, but could be progenitors of high-mass star forming regions such as W49A and W51A. The very interesting results will yield further insight in the physical conditions prevailing at the onset of induced star formation (Mooney et al., in preparation).

High-mass star formation in the Orion A and B clouds (Figures 1 and 3) appear to be examples of spontaneous star formation in the interarm region. Contrary to induced star formation, where all stars appear to be formed simultaneously, spontaneous high-mass star formation is seen as a sequence of events spread over several lifetimes of O-stars.

7. Where Are the Low-Mass Protostars?

According to the estimated detection rates in Table II it should not be too difficult to detect low-mass isothermal protostars — provided this evolutionary phase exists at all for low-mass stars and provided one knows where to look for them.

Searches for dust emission from protostellar condensations in "Myers NH_3 cloud cores" and in globules were up to now not very successful (for a progress report see Mezger et al. 1990a). One finds similar gas masses as derived from NH_3 observations but there is no indication of a centrally peaked density distribution. More promising are the results of a recent investigation of Globular Filaments (GF) by Fiebig et al. (1992), who combined molecular spectroscopy with dust continuum observations. Multi-line excitation studies revealed the masses, density and temperature structure on the various scales. Figure 8 shows the high-density ridge of the filament GF9 in H_2CO absorption. Masses of the individual fragments which form this ridge are of order 10 $M_\odot$, with densities $\langle n_H \rangle \sim 10^3$ cm^{-3}. This ridge is surrounded by a low-density ($< n_H >$ of some hundred cm^{-3}) envelope, visible in HI and CO, which contains altogether ~ 500 $M_\odot$ of gas and dust. A search for higher-density cores in the ridge-fragments led to the identification of in total 10 compact, dense (10^5 cm^{-3}) solar-mass clumps (shown black in the grey-scale image Figure 8). The insert in Figure 8 shows clump #2 as seen in NH_3 λ 1.3 cm and dust λ 1.3 mm continuum emission. The latter, made with an angular resolution of $\sim 11''$, relates to a compact object of ~ 0.1 $M_\odot$, the NH_3

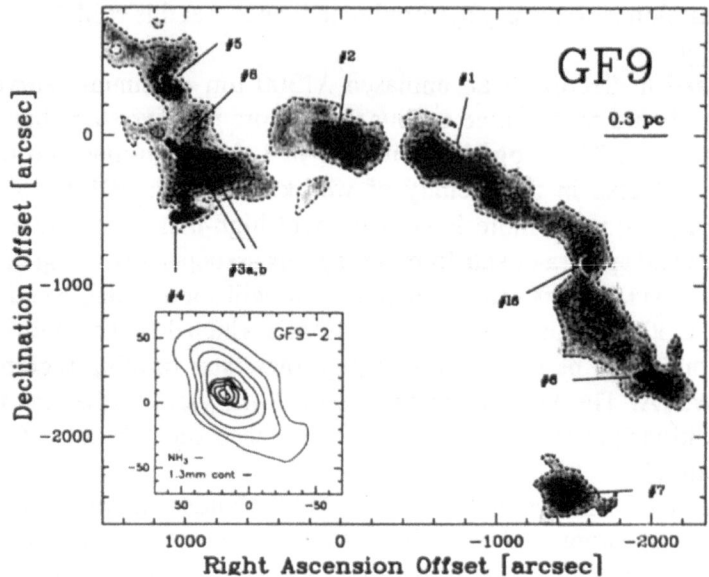

Fig. 8. The Globular Filament GF9 as seen in H_2CO absorption (greyscale). Embedded dense NH_3 (black areas) cores are superimposed. The distribution of NH_3 emission towards GF9-2, the suspected low-mass protostellar object, is shown in the insert. The central (accretion) core detected in $\lambda 1.3$-mm dust continuum is superimposed and shown with heavy contours.

emission, observed with a 40″ beam, relates to a more extended elongated envelope of $\sim 1\ M_\odot$.

Subsequent inspection of the IRAS PS-catalogue and a search for associated CO outflows revealed that out of these ten cores, four contain cool low-luminosity (of a few $0.1\ L_\odot$) IRAS sources, two of these bearing compact CO outflows. Their IR spectra clearly classify them as very young "Extreme Class I"-type (protostellar) objects whose luminosity is assumed to derive entirely from accretion of infalling material onto the central (prestellar) core. Kinematic evidence for ongoing collapse is found towards GF9-2 (see insert in Figure 8) where deep redshifted self-absorption observed in the CS and CO lines seems consistent with model predictions. A high-resolution study of the velocity field with the Plateau-de-Bure millimeter-interferometer is presently being carried out, from which a final confirmation about the evolutionary state of the object is expected.

What causes this filamentary cloud structure, which is seen on all scales wherever stars are formed? During past years the importance of the interplay between magnetic fields and interstellar matter and its effects on gas dynamics and quite specifically on the process of star formation has been more and more appreciated. Should magnetic fields be held responsible for the filamentary structure observed, for example, in GF9 (Figure 8)? The sobering fact is that Fiebig and Güsten, searching for the Zeemann effect in the λ 21-cm line emission from the atomic envelope surrounding the filamentary feature, obtained an upper limit of 4 μG — too low to have an impact on the dynamics of the gas.

Acknowledgements

I wish to thank R. Güsten, C.G.T. Haslam and R. Mauersberger for their comments and R.G. and R.M. for letting me quote their results on GF9 and NGC 2024 prior to publication. Most of the observations discussed in this paper are based on the bolometer development by the MPIfR bolometer group consisting of: E. Kreysa, H.P. Gemünd, C.G.T. Haslam, R. Lemke and A. Sievers.

References

André, P., Ward-Thompson, D. and Barsony, M.: 1993, *Astrophys. J.*, in press.
Cox, P. and Mezger, P.G.: 1989, *The Astronomy and Astrophys. Rev.* 1, 49.
Draine, B.T. and Lee, H.M.: 1984, *Astrophys. J.* 285, 89.
Draine, B.T.: 1985, *Astrophys. J. Suppl.* 57, 587.
Draine, B.T.: 1989, in *Proc. 22nd ESLAB Symp.*, "Infrared Spectroscopy in Astronomy", Salamanca, Spain, ESA SP-290, 93.
Fiebig, D., Güsten, R. and Ungerechts, H.: 1992, in preparation.
Gordon *et al.*: 1993, *Astron. Astrophys.*, (in press).
Güsten, R. and Mezger, P.G.: 1983, *Vistas in Astronomy* 26, 159, 89. (GM83)
Hildebrand, R.H.: 1983, *Quart. J. R. A. S.* 24, 267.
Lada, E.A., Bally, J. and Stark, A.A.: 1991, *Astrophys. J.* 368, 432.
Maddalena, R.J., Morris, M., Moscowitz, J. and Thaddeus, P.: 1986, *Astrophys. J.* 303, 375.
Mathis, J.S., Rumpl, W. and Nordsieck, K.H.: 1977, *Astrophys. J.* 217, 425.
Mathis, J.S., Mezger, P.G. and Panagia, N.: 1983, *Astron. Astrophys.* 128, 212.
Mathis, J.S. and Whiffen, G.: 1989, *Astrophys. J.* 341, 808.
McMullin, J.P., Mundy, L.G. and Blake, G.A.: 1993, *to appear in Astrophys. J.*, March 10, 1993.
Mauersberger, R., Wilson, T.L., Mezger, P.G., Gaume, R. and Johnston, K.J.: 1992, *Astron. Astrophys.* 256, 640.
Mezger, P.G.: 1985, in *Proc. IAU Symp. No. 116*, de Loore, Willis and Laskarides, ed(s)., 479.
Mezger, P.G.: 1990, in *Proc. of a symposium*, J. Krelowski and J. Papaj, ed(s)., "Physics and Composition of Interstellar Matter" 4–9 June 1990, Institut of Astronomy, Nicolaus Copernicus University, Torun.
Mezger, P.G., Mathis, J.S. and Panagia, N.: 1982, *Astron. Astrophys.* 105, 372.
Mezger, P.G., Chini, R., Kreysa, E. and Wink, J.E.: 1987, *Astron. Astrophys.* 182, 127.
Mezger, P.G., Chini, R., Kreysa, E. and Wink, J.E.: 1988, *Astron. Astrophys.* 191, 44.
Mezger, P.G., Sievers, A. and Zylka, R.: 1990a, in *Proc. IAU Symp. No. 147*, Falgarone, Boulanger and Duvert, ed(s)., 245.
Mezger, P.G., Wink, J.E. and Zylka, R.: 1990b, *Astron. Astrophys.* 228, 95.
Mezger, P.G., Sievers, A., Zylka, R., Haslam, C.G.T., Kreysa, E., Lemke, R., Mauersberger, R. and Wilson, T.L.W.: 1992a, *Astron. Astrophys.* 256, 631.
Mezger, P.G., Sievers, A., Zylka, R., Haslam, C.G.T., Kreysa, E. and Lemke, R.: 1992b, *Astron. Astrophys.* 265, 743.
Pajot, F., Gispert, R., Lamarre, J.M., Pomerantz, M.A., Puget, J.-L. and Serra, G.: 1989, *Astron. Astrophys.* 223, 107.
Plambeck, R.L.: 1987, "Galactic and Extragalactic Star Formation", R. Pudritz and M. Fich, ed(s)., NATO ASI Series, 253.

Rengarajan, T.N.: 1984, *Astron. Astrophys.* **140**, 213.

Rowan-Robinson, M.: 1986, *Monthly Not. Roy. Astr. Soc.* **219**, 737.

Salpeter, E.E.: 1955, *Astrophys. J.* **121**, 161.

Shu, F.H., Adams, F.C. and Lizano, S.: 1987, *Ann. Rev. Astron. Astrophys.* **25**, 23.

Schulz, A., Güsten, R., Zylka, R. and Serabyn, E.: 1991, *Astron. Astrophys.* **246**, 570.

Sievers, A.W., Mezger, P.G., Gordon, M.A., Kreysa, E., Haslam, C.G.T. and Lemke, R.: 1991, *Astron. Astrophys.* **251**, 231.

Tatematsu, K. and Umemoto, T.: 1991, in *Proc. of workshop on "Young Star Clusters and Early Stellar Evolution"*, "Molecular cloud cores in the Orion A Cloud", held in Vulcano, Italy.

Ward-Thompson, D., Robson, E.I., Whittet, D.C.B., Gordon, M.A. and Duncan, W.D.: 1989, *Monthly Not. Roy. Astr. Soc.* **241**, 119.

Wilking, B.A. and Lada, C.J.: 1983, *Astrophys. J.* **274**, 698.

Wilking, B.A., Lada, C.J. and Young, T.E.: 1989, *Astrophys. J.* **340**, 823.

Wilking, B.A.: 1991, in *ESO Scientific Report No. 11*, B. Reipurth, ed., 159.

Zhou, Sh., Evans, II, N.J., Güsten, R., Mundy, L.G. and Kutner, M.: 1991, *Astrophys. J.* **372**, 518.

COLD DUST AROUND CHAMAELEON STARS *

TH. HENNING and E. THAMM

Max Planck Society,
Research group "Dust in Star-forming Regions",
Jena, Germany

Abstract. Here we present the results of 1.3 millimetre continuum measurements for intermediate-mass stars in the Chamaeleon system of dark clouds. The detected millimetre radiation is most probably thermal emission from cold circumstellar dust grains. The measured millimetre fluxes are combined with infrared observations to model the broad-band spectral energy distribution (SED). In this way the parameters of the emission regions are determined.

1. Introduction

The Chamaeleon dark cloud complex is a key region for understanding the formation of low- and intermediate-mass stars. The whole system is separated into the three dark clouds Cha I, II, and III (Hoffmeister, 1962; Schwartz, 1991. We should mention that Cha II was originally named Cha III by Hoffmeister whereas Cha III is Cha II in the Hoffmeister nomenclature).

It is important to note that the three regions show significant differences in their star-formation activity. X-ray (Feigelson and Kriss, 1989; Braun et al., 1993), Hα (Schwartz, 1977, 1991; Hartigan, 1993), NIR (McGregor et al., 1993), and IRAS (Prusti, 1992; Gauvin and Strom, 1992) data proved that both the Cha I and II clouds are active sites of star formation. In Cha III no YSOs have been detected yet because the gas density in this cloud is probably not sufficient to promote star formation. From the ratio of the number of embedded objects to the number of optically visible YSOs, Gauvin and Strom (1992) concluded that Cha II is in an earlier stage of evolution than Cha I.

There was a lot of discussion about the distance to Cha I. We will follow Whittet et al. (1987) in assuming 140 pc for Cha I. For Cha II, Franco (1991) estimated a distance of 180 pc.

Because of the proximity, the isolated location substantially out of the galactic plane ($b \approx -16°$), and the significant number of YSOs, the Cha I and II clouds are excellent target objects for the statistical investigation of the presence of cold circumstellar dust around YSOs. Henning et al. (1993) performed a 1.3 mm survey for circumstellar dust around 36 objects in Cha I and II and reached a total detection rate of 44%.

In this contribution, we will concentrate on the results for the young intermediate-mass stars of the Chamaeleon star-forming complex. The question of the general presence of dust disks around Herbig Ae/Be stars which are pre-main sequence

---

* Paper presented at the Conference on *Planetary Systems: Formation, Evolution, and Detection* held 7–10 December, 1992 at CalTech, Pasadena, California, U.S.A.

objects of intermediate mass and luminosity is still a matter of dispute (Berrilli
et al., 1992). The study by Hillenbrand *et al.* (1992) led to the conclusion that at
least some of the objects have optically thick disks. Model fitting to high spatial-
resolution data at 50 and 100 μm for a number of young intermediate-mass stars by
Natta *et al.* (1993) indicates the presence of a dusty envelope and a central energy
source composed of a star and a circumstellar disk.

2. Intermediate-Mass Stars

Cha I contains two B9 stars (HD 97048 and HD 97300) whereas Cha II has only
one star of intermediate luminosity (IRAS 12496-7650). HD 97048 belongs to the
class of Herbig Ae/Be stars whereas HD 97300 seems to be a ZAMS star. IRAS
12496-7650 is a deeply embedded Herbig Ae star (Hughes *et al.*, 1989, 1991),
probably in an earlier phase compared with the "naked" Herbig Be star HD 97048.
In addition, the isolated object HD 104237 is a Herbig Ae star most likely (Hu *et
al.*, 1989). The relevant data for the four target objects of intermediate mass are
collected in Table I.

TABLE I
Intermediate-mass stars

IRAS	Optical iden-tification	Region	RA[h min s] [1950]	DEC[° ' "] [1950]	Lada class	Comment
11066-7722	HD 97 048	I	11 06 39.6	−77 23 01	IID	Herbig Be star
11082-7620	HD 97 300	I	11 08 17.9	−76 20 30	IIID	ZAMS B9 star
11575-7754	HD 104 237	Isolated	11 57 33.5	−77 54 51	II	Herbig Ae star
12496-7650		II	12 49 38.0	−76 50 45	I	Embedded Herbig Ae star

3. Observations

The 1.3 mm continuum observations have been performed between November
17th and 19th, 1991 with the SEST 15 m telescope at La Silla, Chile, using the
^{3}He-cooled bolometer system developed by the Max-Planck-Institut für Radio-
astronomie (Kreysa, 1990). The beam size was 23″ at 1300 μm. The details of the
observations are given by Henning *et al.* (1993).

4. Results

In Table II, we present the results of our 1.3 mm measurements. The two Herbig
Ae/Be stars in Cha I and Cha II are the strongest sources of the whole 1.3 mm

survey performed by Henning *et al.* (1993). The ZAMS star HD 97300 could not be detected and only our 3σ limit is given.

TABLE II

Measured fluxes and circumstellar gas masses

IRAS name	Date	$S_{1.3mm}$[mJy]	σ[mJy]	M_g [$M_\odot$]	
				T_d=50 K	T_d=100 K
				$\kappa_{1.3}$=0.02 cm^2 g^{-1}	
11066-7722	17/11/91	451.5	33.7	$2.9\ 10^{-2}$	$1.4\ 10^{-2}$
11082-7620	17/11/91	<91.0	-	$<5.8\ 10^{-3}$	$<2.7\ 10^{-3}$
11575-7754	17/11/91	66.0	13.3	$4.2\ 10^{-3}$	$2.0\ 10^{-3}$
12496-7650	17/11/91	680.0	22.0	$7.2\ 10^{-2}$	$3.4\ 10^{-2}$

The total mass of dust particles can be estimated by assuming optically thin emission at 1.3 mm and an isothermal, uniformly distributed population of dust grains. The total gas mass M_g is obtained if we use the mass absorption coefficient per g of interstellar matter instead of the coefficient per g of dust.

$$M_g = S_{1.3mm}D^2/\kappa(1.3mm)B_{1.3mm}(T_d) \qquad (1)$$

Here D is the distance to the object, $B_{1.3mm}$ the Planck function at the dust temperature T_d, and κ the mass absorption coefficient per g of interstellar matter. We should note that this expression gives only a crude estimate of the gas mass because the temperature and density distribution existing in a dusty disk or a spherical envelope are not taken into account. For $\kappa(1.3$ mm), we adopt the same value 0.02 cm^2 g^{-1} as Beckwith *et al.* (1990) have applied. This κ value is an order of magnitude higher than the value applied for the diffuse interstellar medium. To use Equation (1), we have to specify not only κ but also the temperature. Mass-averaged temperatures obtained from detailed modelling of the SED of embedded objects range from 50 to 100 K. We use the extreme values for the derivation of the dust masses which are not very sensitive to the exact value of the assumed temperature (Rayleigh-Jeans limit). The gas masses obtained in this way range from about 5×10^{-2} to 2×10^{-3} $M_\odot$ for the assumed high value of the opacity (see Table II).

We combined the measured millimetre fluxes with NIR observations and IRAS data (Hu *et al.*, 1988; Assendorp *et al.*, 1990; Prusti, 1992; Gauvin and Strom, 1992) and modelled the broadband SEDs by both an exact spherically-symmetric radiative transfer model including scattering and properties of different dust populations and a model for geometrically thin disks with parametrized temperature and density distributions. A detailed description of the models is given by Henning *et al.* (1993).

For the spherically-symmetric transfer computations, we used three different dust models: a model with silicate and graphite grains with the optical constants of

TABLE III

Results of disk models

Source IRAS name	$\kappa_\nu \sim \nu^\beta$; $\beta = 1$ $\kappa_{1.3} = 0.0054$ cm^2 g^{-1}		Draine; $\beta = 2$ $\kappa_{1.3} = 0.0024$ cm^2 g^{-1}		$T_d = 50$ K $\kappa_{1.3} = 0.0054$ cm^2 g^{-1}	$T_d = 100$ K
	$L_* [L_\odot]$	$M_g [M_\odot]$	$L_* [L_\odot]$	$M_g [M_\odot]$	$M_g [M_\odot]$	
11066-7722	35	$4 \ 10^{-2}$	35	$1 \ 10^{-1}$	$1.1 \ 10^{-1}$	$5.2 \ 10^{-2}$
11082-7620	40	$1 \ 10^{-2}$	40	$4 \ 10^{-2}$	$< 2.1 \ 10^{-2}$	$< 1.0 \ 10^{-2}$
11575-7754	35	$1 \ 10^{-2}$	30	$2 \ 10^{-1}$	$1.6 \ 10^{-2}$	$7.4 \ 10^{-3}$
12496-7650	35	$1 \ 10^{-1}$	35	$4 \ 10^{-1}$	$2.7 \ 10^{-1}$	$1.3 \ 10^{-1}$

Draine (1985) and a MRN size distribution, the same dust model but with silicate core-dirty ice mantle grains instead of pure silicate grains (see Preibisch *et al.*, 1993) as well as a model with porous silicate grains instead of compact silicate grains (vacuum fraction of 40%). Although all of these spherical models could fit the NIR and MIR energy distribution, they resulted in far too low 1.3 mm fluxes if compared with the measured values.

In the disk models we used two different dust models, again the Draine model mentioned above which is characterized by an exponent of the FIR dust emissivity of $\beta = 2$ and $\kappa(1.3\text{mm}) = 0.0024$ cm^2 g^{-1} and a power law $\kappa \propto \nu^\beta$ with $\beta=1$ and $\kappa(1.3\text{mm}) = 0.0054$ cm^2 g^{-1}. These disk models resulted in a general good fit of the whole SEDs (see Figure 1). However, a perfect fit could not be reached in case of HD 97048 and IRAS 12496-7650. For the embedded object, we obtain too much stellar flux in the near-infrared due to the assumption of a geometrically thin disk. For this object, clearly a second component (probably an extended envelope) is needed. However, we did not try to fit our data with a combination of an envelope

TABLE IV

Disk parameters (constant density)

Source IRAS name	$\kappa_\nu \sim \nu^\beta$; $\beta = 1$			Draine; $\beta = 2$		
	r_1 [cm]	r_2 [cm]	Σ_1 [g cm^{-2}]	r_1 [cm]	r_2 [cm]	Σ_1 [g cm^{-2}]
11066-7722	$7.7 \ 10^{10}$	$7.5 \ 10^{14}$	0.50	$7.7 \ 10^{10}$	$9.0 \ 10^{14}$	1.00
11082-7620	$2.3 \ 10^{10}$	$1.2 \ 10^{15}$	0.05	$2.3 \ 10^{10}$	$5.2 \ 10^{14}$	0.80
11575-7754	$2.3 \ 10^{12}$	$2.5 \ 10^{14}$	1.00	$2.3 \ 10^{12}$	$1.3 \ 10^{14}$	60.00
12496-7650	$2.0 \ 10^{12}$	$9.0 \ 10^{14}$	1.00	$2.0 \ 10^{12}$	$8.7 \ 10^{14}$	3.00

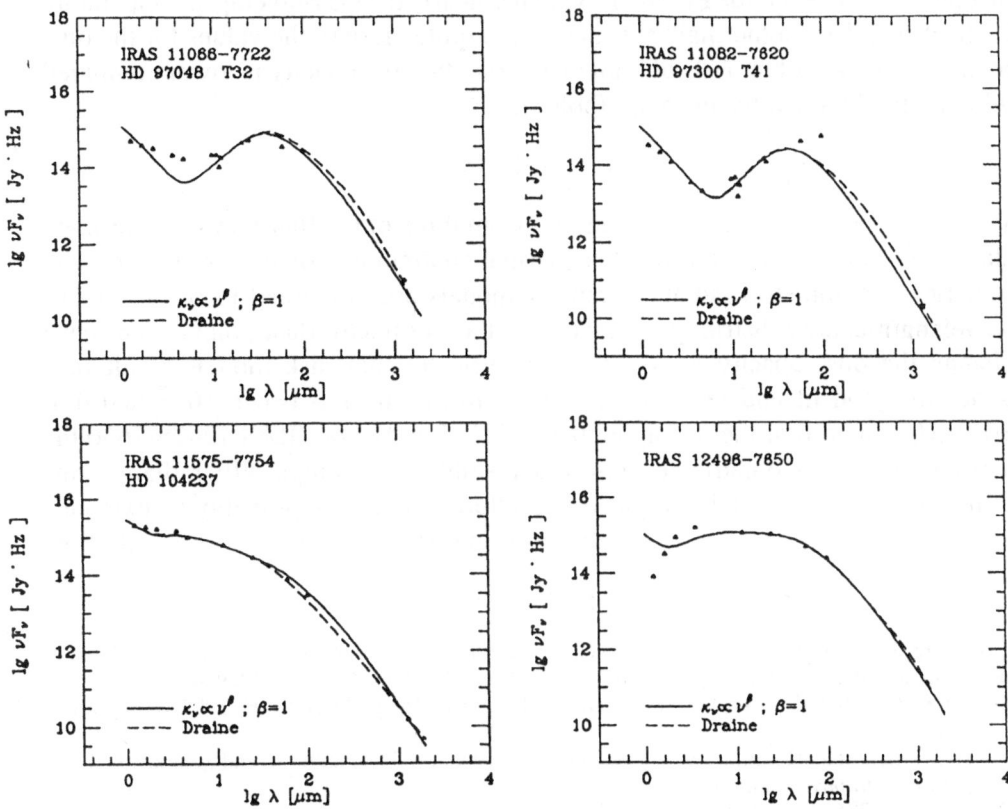

Fig. 1.   Comparison between measured fluxes (triangles) and disk models.

and a disk because of the lack of spatially resolved data and therefore a too large number of free parameters for such a fitting procedure.

The temperatures in the disks range from 1500 K at the inner boundary of the dust zone to about 50 K at the outer boundary. The mass-averaged temperatures have values of ~ 100 K. Most of the emission at 1.3 mm comes from the outermost and coolest parts of the disks.

The resulting parameters for the disk models are summarized in Table III and IV. The quantity L_* is the stellar luminosity, r_1 and r_2 denote the inner and outer radius of the disk, respectively. The parameter Σ_1 is the surface density at the inner radius. We should note that the surface density in our models is constant throughout the disk. From the surface density and the inner and outer radii of the disks, we calculated the gas masses of the disks. These mass values are given in column 3 and 5 of Table III. For the sake of comparison, we also give in Table III the "isothermal" gas masses computed with Equation (1), which are in general higher than the masses derived with the same opacity value and the disk parameters given in Table IV. However, for $T_d=100$ K, which is nearly the mass-averaged temperature in the disks, the "isothermal" gas masses are quite close to the values for the disk models given in column (3) of Table III. Note, that the opacity is lower compared with the Beckwith value used for Table II.

5. Conclusions

We were able to detect 1.3 mm emission from three of the four intermediate-mass stars. This emission is interpreted as thermal continuum radiation from cold dust. Spherically-symmetric radiative transfer models are not suited to fit the broad-wavelength energy distributions including the millimetre data points even if we change the dust opacity laws. The geometrically thin disk models result in a generally good fit and lead to gas masses in the disks between 10^{-2} and $4 \times 10^{-1} \, M_\odot$. In case of the deeply embedded Herbig Ae star IRAS 12496-7650 the assumption of a geometrically thin disk results in a too high NIR radiation due to the contribution of stellar light. Here, a third component probably an extended envelope in addition to the star and disk is needed to explain the energy distribution.

References

Assendorp, R., Wesselius, P.R., Whittet, D.C.B. and Prusti, T.: 1990, *Monthly Not. R.A.S.* **247**, 624.
Beckwith, S.V.W., Sargent, A.I., Chini, R.S. and Güsten, R.: 1990, *Astron. J.* **99**, 924.
Berrilli, F., Corciulo, G., Ingrosso, G., Lorenzetti, D., Nisini, B. and Strafella, F.: 1992, *Astrophys. J.* **398**, 254.
Braun, M., Henning, Th., Pfau, W., Zinnecker, H., Assendorp, R., Krautter, J. and Schmitt, J.: 1993, *Astron. Astrophys.*, in prep.
Draine, B.T.: 1985, *Astrophys. J. Suppl.* **57**, 587.
Gauvin, L.S. and Strom, K.M.: 1992, *Astrophys. J.* **385**, 217.
Feigelson, E.D. and Kriss, G.A.: 1989, *Astrophys. J.* **338**, 262.
Franco, G.A.P.: 1991, *Astron. Astrophys.* **251**, 581.
Hartigan, P.: 1993, *Astron. J.*, in press.
Henning, Th., Pfau, W., Zinnecker, H. and Prusti, T.: 1993, *Astron. Astrophys.* **276**, 129.
Hillenbrand, L.A., Strom, S.E., Vrba, F.J. and Keene, J.: 1992, *Astrophys. J.* **397**, 613.
Hoffmeister, C.: 1962, *Zs. f. Astrophysik* **55**, 290.
Hu, J.Y., The, P.S. and de Winter, D.: 1988, *Astron. Astrophys.* **208**, 213.
Hughes, J.D., Emerson, J.P., Zinnecker, H. and Whitelock, P.A.: 1989, *Monthly Not. R.A.S.* **236**, 117.
Hughes, J.D., Hartigan, P., Graham, J.A., Emerson, J.P. and Marang, F.: 1991, *Astron. J.* **101**, 1013.
Kreysa, E.: 1990, in: N. Longdon and B. Kaldeich, ed(s)., *From Ground-Based to Space-Borne Astronomy*, ESA:Nordwijk, 265.
McGregor, P.: 1993, in prep.
Natta, A., Palla, F., Butner, H.M., Evans II, N.J. and Harvey, P.M.: 1993, *Astrophys. J.* **406**, 674.
Preibisch, Th., Ossenkopf, V., Yorke, H.W., Henning, Th.: 1993, *Astron. Astrophys.*, in press.
Prusti, T.: 1992, *PhD. Thesis: Infrared Studies of Low-Mass Star Formation*, University Groningen: Groningen.
Schwartz, R.D.: 1977, *Astrophys. J. Suppl. Ser.* **35**, 161.
Schwartz, R.D.: 1991, in: B. Reipurth, ed(s)., *Low Mass Star Formation in Southern Dark Clouds*, ESO:Garching, 93.
Whittet, D.C.B., Kirrane, T.M., Kilkenny, D., Oates, A.P., Watson, F.G. and King, D.J.: 1987, *Monthly Not. R.A.S.* **224**, 497.

CLOUDY CIRCUMSTELLAR DUST SHELLS AROUND YOUNG
VARIABLE STARS *

C. FRIEDEMANN, H.-G. REIMANN and J. GÜRTLER

Astrophysikalisches Institut und Universitäts-Sternwarte
Friedrich-Schiller-Universität Jena, Jena, Germany

Abstract. A number of variable stars of the Orion population has been identified with IRAS point sources by us. This finding supports the conclusion that the prominent Algol-like minima in the lightcurves of these stars originate from obscurations by dust clouds in a circumstellar shell. The discussion of the existing *UBVR* data leads to the remarkable conclusion that the extinction properties of the grain populations contained in individual dust clouds moving in one and the same circumstellar shell are quite different.

From the multicolour photometric data of the different Algol-like minima we derived individual values of the reddening parameter $R = A_V/E(B - V)$. It covers a remarkable wide range of values from that one typical of the interstellar extinction law up to 7. In the case of SV Cep one of the grain populations produces a virtually neutral extinction. The large values of R speak in favour of larger than normal (interstellar) dust grains, which may have grown by coagulation processes. The cloudy circumstellar dust shell provides a natural explanation for the observed infrared excess. The properties derived from the optical light variations are fully compatible with the properties deduced from the infrared radiation. The irregularity of the light variations indicates that many clouds are involved and may sometimes superimpose themselves.

1. Introduction

SV Cephei (= BD+73°1031), WW Vulpeculae (=BD+20°4136), and RZ Piscium have been attributed to the Orion Population of pre-main-sequence stars (Herbig and Bell, 1988). According to spectrographic observations the spectral type of SV Cep can vary between A0 and A3 on time scales of days. Our own *uvby* observations resulted in a spectral type of B9 (Friedemann *et al.*, 1992). The spectral type of WW Vul is A0 (Friedemann *et al.*, 1993) and that of RZ Psc is K0 IV.

SV Cep, WW Vul, and RZ Psc have been observed frequently since the discovery of their variability. The lightcurves of these variables are similar and seem to consist of different components. Longer lasting quasi-periodic fadings of the brightness are superimposed by short-scale (10–50 days) Algol-like minima. Their amplitudes vary in the interval 0.3 to 2 mag. Discrete Fourier analyses of the lightcurves revealed no dominant periodicities.

Important for the explanation of the aperiodically occurring Algol-like minima in the case of SV Cep is the finding of Wenzel (1969) based on his *UBV* data that the deep minima show nearly neutral light changes. This fact led Wenzel to the proposal that dust clouds orbiting the star may cause these minima. This hypothesis has been worked out quantitatively by Wenzel *et al.* (1971), Zaytseva

---

* Paper presented at the Conference on *Planetary Systems: Formation, Evolution, and Detection* held 7–10 December, 1992 at CalTech, Pasadena, California, U.S.A.

and Chugainov (1984), Voshchinnikov (1989), Voshchinnikov and Grinin (1991), and Friedemann et al. (1992).

More elucidating for the explanation of the variability of SV Cep were the UBVR observations carried out by Kardopolov et al. (1985a, b). Also in the case of WW Vul these UBVR observations show that the Algol-like minima originate from individual circumstellar dust clouds.

The dust-cloud hypothesis is supported by the finding that two of the variables coincide with IRAS point sources: SV Cep = IRAS 22205+7325 and WW Vul = IRAS 19238+2106. The observed infrared excesses may be explained by stellar radiation thermalized by the dust grains in the circumstellar shells. At the position of RZ Psc no IRAS source could be found. But this finding does not contradict the existence of a circumstellar dust shell around this variable.

2. Properties of the Optical Variations

We have collected the existing photometric data for SV Cep and WW Vul, covering a time interval of about 100 years. Discrete Fourier analyses resulted in power spectra which show stochastic light variations only. We infer from these analyses that all "cyclic" variations claimed earlier by some authors (Wenzel, 1969; Kardopolov et al., 1985) are the result of the discussion of observations from relatively short time intervals. The lack of any periodicities of the Algol-like minima supports the assumption that a relative large number of dust clouds orbiting the stars must be responsible for them.

We discussed in detail the photometric properties of 18 Algol-like minima of WW Vul, 3 minima of SV Cep and 10 minima of RZ Psc by means of colour-magnitude as well as two-colour diagrams. Figures 1 to 3 present a typical example.

We infer from the V vs. $(V - R)$ and V vs. $(U - B)$ diagrams that the optical light variations can arise only to a small extent from variations of the effective temperature. Therefore, we conclude that the Algol-like light variations originate from individual circumstellar dust clouds orbiting the stars.

The wavelength dependence of the extinction by dust grains is usually described by the parameter $R = A_V/E(B - V)$. The standard value of this parameter for the average interstellar dust is $R = 3.2$. If we derive the parameter R for each Algol-like minimum we found values of R ranging from 3 to 8. The results of a comprehensive study of the Algol-like episodes of the light variations are collected in Figures 4 and 5 showing the relations R vs. $E(V - R)/E(B - V)$ and R vs. A_V. The distribution of the points shows clearly that the different minima were caused by dust clouds consisting of grain populations with different extinction properties.

The most probable reason for the greater R values than the interstellar mean is a deviating grain size distribution in the sense that a higher number of larger grains is present in circumstellar space. The higher density in the circumstellar environment should foster coagulation and accretion processes. As pointed out by Meakin and Donn (1988), these processes may be strengthened further by the fact that the

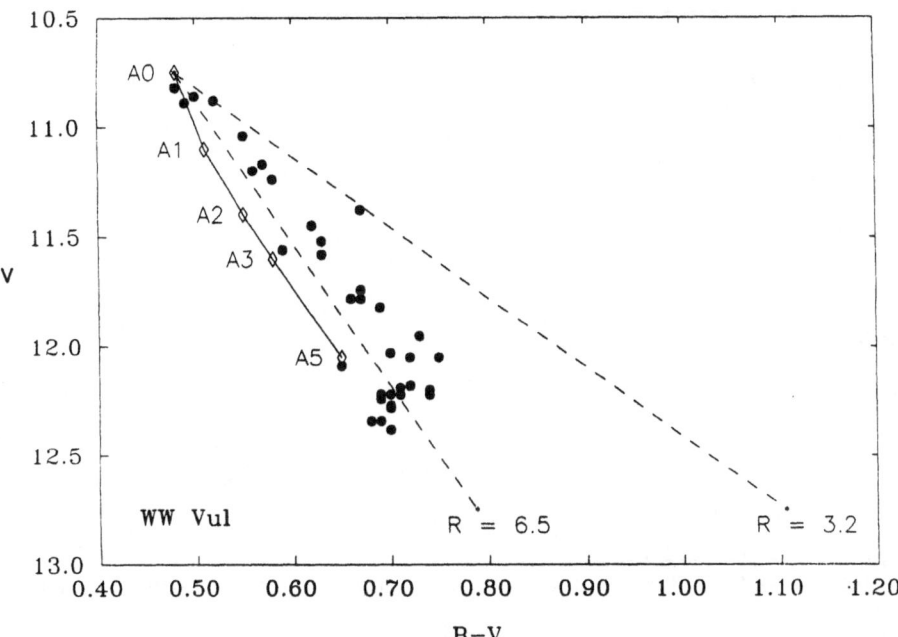

Fig. 1. Colour-magnitude diagrams for a deep minimum of WW Vul. The solid lines mark the star's path if it would change its effective temperature as indicated by the spectral types. The dashed lines correspond to the reddening lines with ratios $R = 3.2$ and 6.5, resp. (The photometric data are from Kardopolov and Filip'ev (1985)).

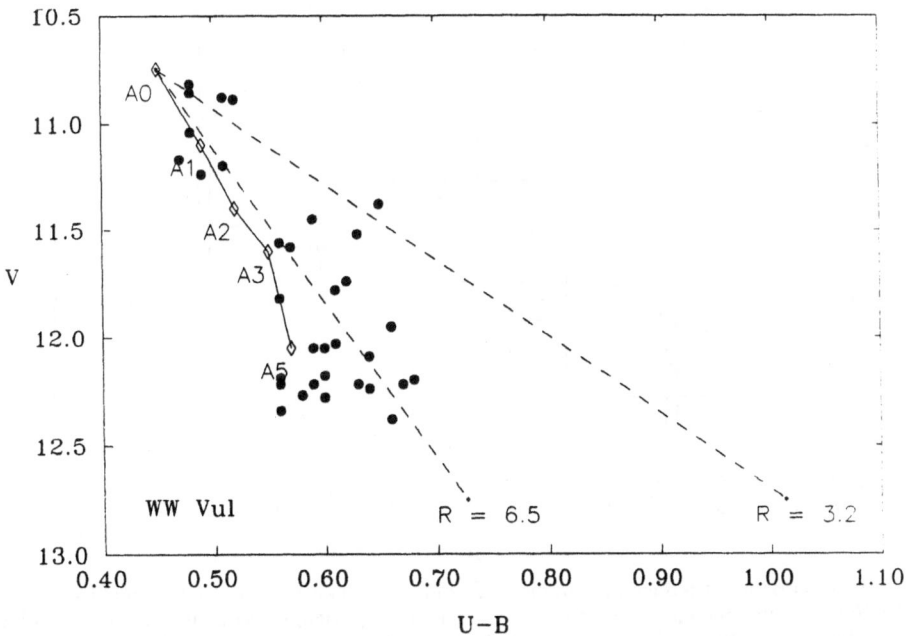

Fig. 2. Colour-magnitude diagram for a deep minimum. For explanation see Figure 1.

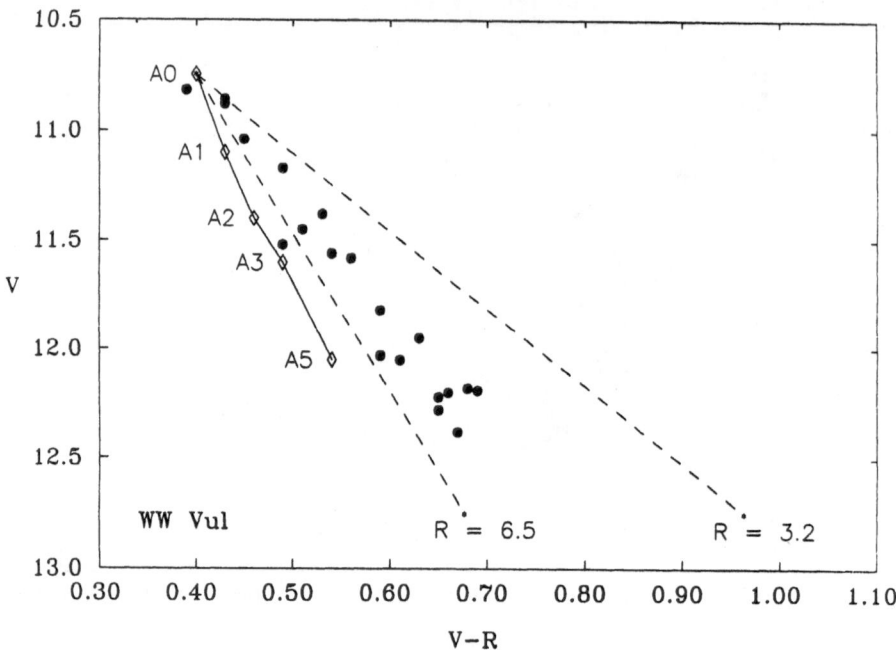

Fig. 3. Colour-magnitude diagram for a deep minimum. For explanation see Figure 1. Obviously, the variations of the spectral type cannot account for the minimum.

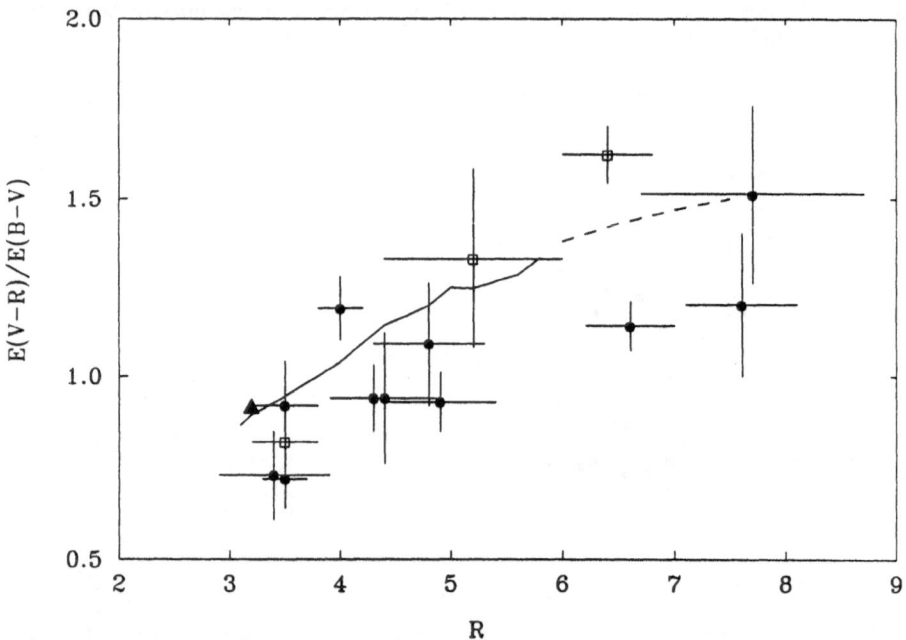

Fig. 4. Relation between reddening parameter R and colour excess ratio $E(V - R)/E(B - V)$ for WW Vul (full circles) and SV Cep (squares). The dashed triangle marks the standard value of the interstellar mean. The solid line (and the dashed part of it as an extrapolation from us) represent results reached by Steenman and Thé (1991) discussing anomalous extinction curves. Error bars indicate one standard deviation.

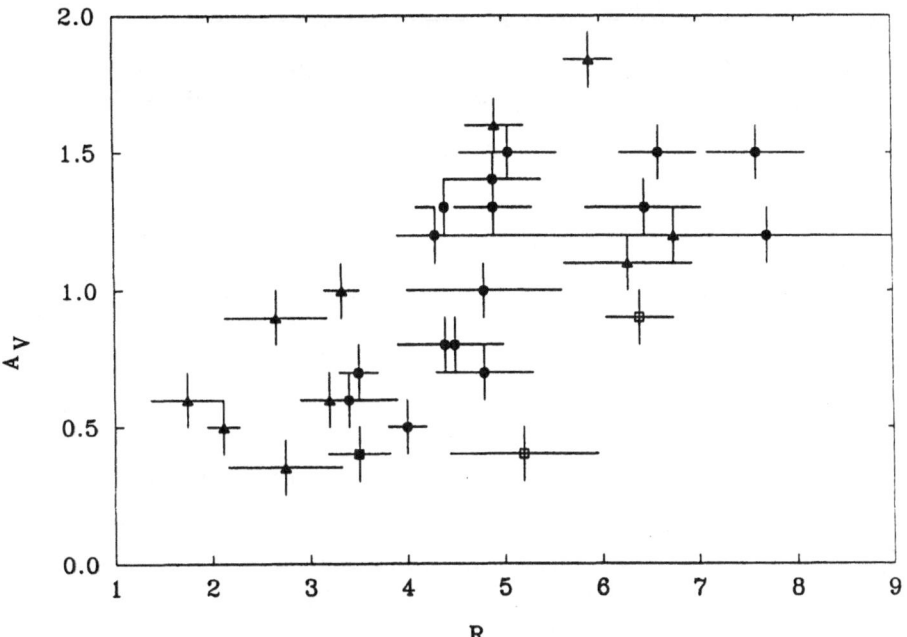

Fig. 5. Relation between R and A_V inferred from the observed amplitudes ΔV of individual Algol-like minima in the lightcurves of SV Cep (squares), WW Vul (filled circles), and RZ Psc (filled triangles). Error bars indicate one standard deviation. In the case of WW Vul some minima with the same amplitude show different values of R. This is an additional hint for differing extinction properties of dust present in individual clouds.

formation of fluffy and fractal particles is favoured and such are coupled to the gas up to larger sizes than more compact grains would. The line in Figure 4 represents results reached by Steenman and Thé (1991) discussing anomalous extinction curves. They started with a Mathis–Rumpl–Nordsieck grain model and studied the effect if the lower and upper cut-off limits of the particle size distribution are varied. The line (dashed parts are extrapolations by us) indicates the variation of the reddening parameter R and the colour excess ratio $E(V - R)/E(B - V)$ as a function of increasing minimum particle size. Figure 4 shows a good agreement between the theoretical results and the observational data. That means that the individual clouds responsible for the various Algol-like minima may be characterized by different depletion of small grains.

    An alternative interpretation put forward by Voshchinnikov (1991) envisages the obscuring dust clouds to be embedded in a diffuse oblate scattering envelope. During normal light stellar radiation scattered by this envelope does not contribute very much to the observed flux. When the star is eclipsed by an obscuring cloud the contribution of the shell becomes larger. Since blue light is scattered more strongly, the fading star does not follow the normal reddening path. The combined scattering and extinction make the colour variation mimic a reddening law with larger R.

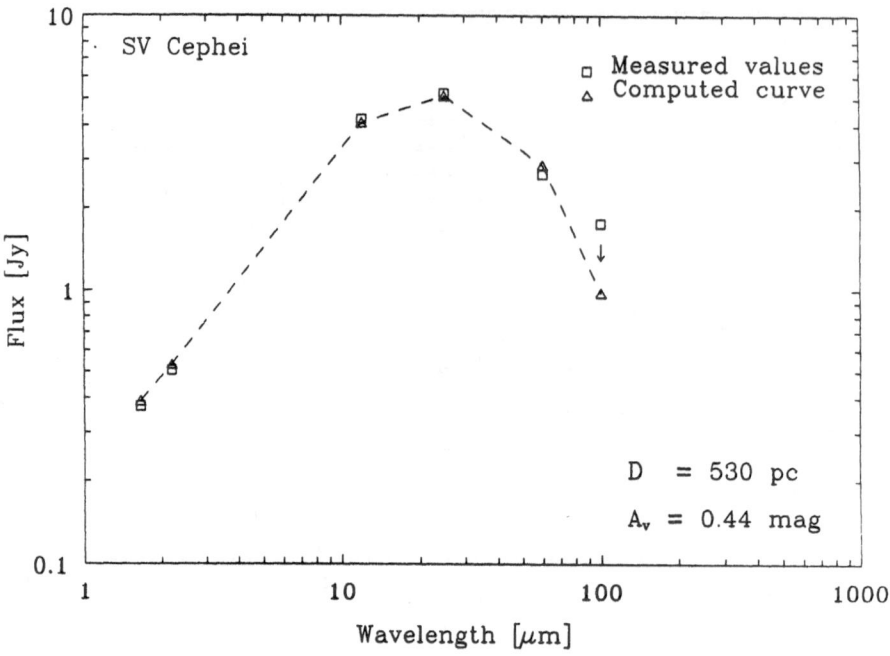

Fig. 6. Comparisons of the observed and calculated infrared fluxes. For the modelling a spherically symmetric dust shell with a power law density distribution was assumed. D is the distance of the star and A_V is the circumstellar extinction during normal light derived from photometric data. Infrared fluxes being upper limits are labelled by arrows.

Characteristic differences between both interpretations are expected only for very deep minima. Observing the linear polarization accompanying the scattering process will be another discriminating possibility.

After the observations have shown the existence of discrete circumstellar clouds, the question of their long-term stability arises. The discussion of this problem is beyond the scope of this paper, but the answer may be found within the framework of the paper by Wenzel *et al.* (1971).

3. Infrared Radiation from Circumstellar Dust Shells

SV Cep and WW Vul coincide with the IRAS point sources IRAS 22205+7325 and IRAS 19238+2106, respectively. Thermal emission by circumstellar dust provides a straightforward explanation of the infrared excess. Using the RTP code by Chini *et al.* (1986) we derived circumstellar model envelopes explaining the observed infrared excess radiation (Glass and Penston, 1974; IRAS data). The modelling assumes smooth spherically symmetric distributions of dust with interstellar properties. Good fits of the observed IR fluxes have been reached by density distributions following a power law with an exponent around -1.5. A total of some 10^{-6} solar masses of circumstellar dust is needed for the explanation of the IR radiation.

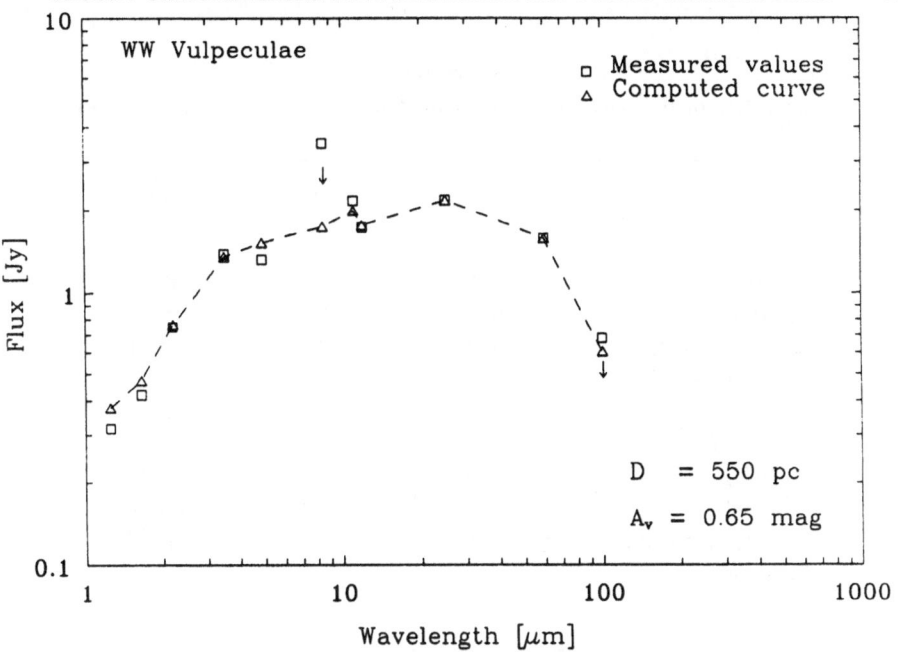

Fig. 7. Modelling of the IR flux. For explanation see Figure 6.

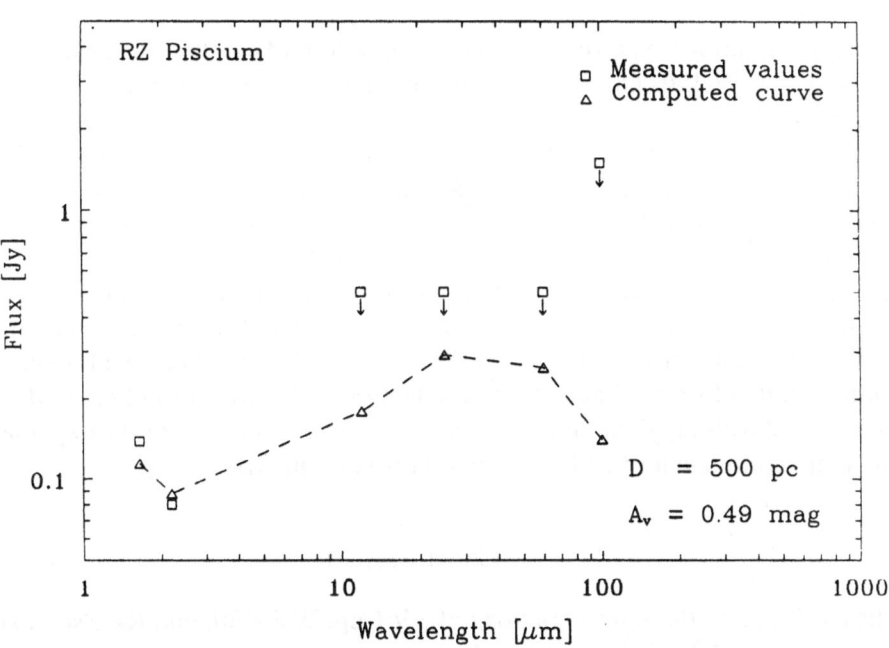

Fig. 8. Modelling of the IR flux. For explanation see Figure 6.

This amount of dust is compatible with the observed circumstellar extinction
(see Figures 6 to 8). For the separation of the circumstellar portion from the total

TABLE I

Parameters of the stars and the models of the spherical circumstellar dust shells.

Star	SV Cephei	WW Vulpeculae	RZ Piscium
Spectral type	B9 V	A0 V	K0 IV
T_{eff} (K)	10500	9520	5250
Distance D (pc)	530	550	500
$A_{V\text{circ}}$ (mag)	0.44	0.65	0.49
Boundaries of the shell			
r_{in} (AU)	13.4	2.7	4.0
r_{out} (AU)	$5.9\ 10^3$	$4.7\ 10^3$	$4.4\ 10^3$
Density distribution	$r^{-1.6}$	$r^{-1.45}$	$r^{-1.5}$
Total mass of the circumstellar dust ($M_\odot$)	$7.3\ 10^{-6}$	$6.3\ 10^{-6}$	$4.0\ 10^{-6}$

amount of extinction we carried out *uvby* observations of neighbouring stars around SV Cep and WW Vul. Our results indicate circumstellar extinction of about 0.5 mag at most.

We estimated the ratio of the number densities of the dust particles in an individual cloud and the surrounding. Adopting average values for the extinction in a single cloud and those in a cloud-free line of sight through the total dust shell we found a remarkably high ratio $\rho_{\text{cloud}}/\rho_{\text{shell}} \approx 10^4$–$10^6$.

For RZ Psc we calculated IR fluxes from a hypothetical circumstellar dust shell, too. They are fitted to the few existing NIR data. Presuming an analogous density gradient as for SV Cep and WW Vul we predict IR fluxes well below the detection limits of IRAS. Thus, the non-detection by IRAS does not contradict the assumption of a clumpy circumstellar dust shell around RZ Psc for the explanation of the main component of its light variation (see Figure 8).

4. Conclusions

We have discussed the light variations of SV Cep, WW Vul, and RZ Psc in some detail and reached the following conclusions:

1. SV Cep, WW Vul, and RZ Psc exhibit brightness variations with different time scales and amplitudes.
2. Intrinsic stellar changes can produce brightness variations up to 0.1 mag at best.

3. The prominent irregularly occurring Algol-like minima can be understood as a consequence of cloudy circumstellar dust shells.
4. The individual clouds have grain populations which may differ from that typical of the interstellar medium but also from each other. In most clouds the radii of the dust grains are larger than in interstellar space.
5. The circumstellar dust provides a consistent explanation of the observed IR radiation.
6. The scenario offers the hitherto unique opportunity to study localized dust particles with different properties in one and the same circumstellar environment.

Acknowledgements

The authors wish to thank the referee, Dr. B. Donn, for his valuable and constructive remarks. Part of this work was supported by a grant from the BMFT.

References

Chini, R., Krügel, E. and Kreysa, E.: 1986, *Astron. Astrophys.* **167**, 315.
Draine, B.T.: 1985, *Astrophys. J. Suppl.* **57**, 587.
Friedemann, C., Reimann, H.-G. and Gürtler, J.: 1992, *Astron. Astrophys.* **255**, 246.
Friedemann, C., Reimann, H.-G., Gürtler, J. and Toth, V.: 1993, *Astron. Astrophys.* **277**, 184.
Glass, I.S. and Penston, M.V.: 1974, *Monthly Notices Roy. Astron. Soc.* **167**, 237.
Herbig, G.H. and Bell, K.R.: 1988, *Lick Obs. Bull.* **No. 1111.**
Kardopolov, V.I. and Filip'ev, G.K.: 1985, *Peremen. Zvezd.* **22**, 122.
Mathis, J.S., Rumpl, W. and Nordsieck, K.H.: 1977, *Astrophys. J.* **217**, 425.
Meakin, P. and Donn, B.: 1988, *Astrophys. J.* **329**, L39.
Steenman, H. and Thé, P.S.: 1991, *Astrophys. Space Sci.* **184**, 9.
Voshchinnikov, N.V.: 1989, *Astrophys.* **30**, 509.
Voshchinnikov, N.V. and Grinin, V.P.: 1991, *Astrophys.* **34**, 181.
Wenzel, W.: 1969, *Mitt. Veränd. Sterne Sonneberg* **5**, 75.
Wenzel, W., Dorschner, J. and Friedemann, C.: 1971, *Astron. Nachr.* **292**, 221.
Zaytseva, G.V. and Chugainov, P.F.: 1984, *Astrophys.* **20**, 447.

10-μM IMAGES AND SPECTRA OF T TAURI STARS *

J.E. VAN CLEVE, T.L. HAYWARD, J.W. MILES, G.E. GULL, J. SCHOENWALD and
J.R. HOUCK

*Center for Radiophysics and Space Research, Cornell University,
Ithaca, New York, USA*

Abstract. T Tauri stars are young stars usually surrounded by dusty disks similar to the one from which we believe our own Solar System formed. Most T Tauri stars exhibit a broad emission or absorption band between 7.5 and 13.5 μm which is attributed to silicate grains in the circumstellar environment. We imaged three spatially resolved T Tauri binaries through a set of broadband filters which include the spectral region occupied by the silicate band. Two of these objects (T Tauri and Haro 6-10) are "infrared companion" systems in which one component is optically much fainter but contributes strongly in the infrared. Both infrared companions exhibit a deep silicate absorption which is not present in their primaries, indicating that they suffer very strong local extinction which may be due to an edge-on circumstellar disk or to a dense shell. We also took low resolution spectra of the silicate feature of two unresolved T Tauris to look for narrow features in the silicate band which would indicate the presence of specific minerals such as olivine. We observed GK Tau, for which Cohen and Witteborn (1985) reported a narrow emission feature at 9.7 μm, but do not find evidence for this feature, and conclude that it is either time-dependent or an artifact of absorption by telluric ozone.

1. T Tauri Stars and Planetary Systems

T Tauri stars are young stars of roughly solar mass or less, most of which show strong thermal infrared emission consistent with dusty disks. Although we do not know what fraction of these disks results in planetary systems, we can gain insight into the process of planet-forming by modeling the distribution and composition of the dust around these stars. In particular, we can look at the distribution of dust around binary stars and study dust composition using the structure of the 7.5 to 13.5 μm silicate feature of unresolved stars.

T Tauri binary stars with separations on the order of the diameter of the Solar System give insight into the question of planet formation in two ways. First, young binaries and planetary systems both have a primary orbited by a condensed object, a star in the first case and a Jovian planet in the second. The companion influences the subsequent disk structure by gravitationally torquing the disk material and accreting dust and gas onto itself (Ward, 1989). Second, the dissipation rate and optical depth of dusty disks in T Tauri binaries compared to single T Taus tells us whether binarity reduces the likelihood of planet formation. Simon *et al.* (1992) review the multiplicity of young stars in Taurus, based on K-band speckle and lunar occultation work, and show that about a third of the T Tau binaries have separations between 0.5″ and 3″, large enough to be resolved by modern IR cameras but close enough to insure the stars are physically bound. Ghez *et al.* (1991) present a detailed

---

* Paper presented at the Conference on *Planetary Systems: Formation, Evolution, and Detection* held 7–10 December, 1992 at CalTech, Pasadena, California, U.S.A.

study of T Tau itself and find that T Tau N shows silicate emission while T Tau S shows silicate absorption, demonstrating the need to spatially resolve the spectra of T Tauri binaries in order to understand their dusty disks.

The fine structure of the silicate feature, either in emission or absorption, can in principle be used to deduce the minerology of the dust in the disks in the region of terrestrial planet formation. The interpretation of spectra is complicated by the fact that the observed emission comes from an unknown distribution of grain sizes, and from a heterogeneous mixture of minerals. A convincing identification of a mineral would require near-IR and perhaps 18 to 24 μm spectra as well. Nonetheless, high S/N measurements at low ($R \approx 100$) resolution can be used to infer the presence of specific minerals in circumstellar disks, as reported in recent observations of β Pic (Fajardo et al., 1993). Cohen and Witteborn (1985) present possible evidence for such mineralogical features in a spectral survey of T Tauri stars. They find four stars with a relatively narrow ($R \equiv \lambda/\Delta\lambda \approx 20$) emission feature near 9.7 μm.

We used SpectroCam-10 (SC-10), a new mid-infrared imaging spectrometer built by Cornell (Hayward et al., 1993), to resolve the spatial distribution of flux in the binaries T Tau, Haro 6-10, and FV Tau, the first two of which are "infrared companion" systems in which one component is optically faint but contributes strongly to the infrared light. Our images show a deep absorption feature for the infrared component of both systems. We also examine the low resolution spectrum of one of the Cohen and Witteborn narrow-feature stars (GK Tau) which is known to be unresolved at the 50 milliarcsec level (Simon et al., 1992), and compare it to a neighboring T Tauri star which did not show the narrow feature. We do not find the 9.7 μm feature in either star.

2. Observations and Data Reduction

All the measurements reported here were taken with SC-10 on the 5-m Hale telescope. This instrument can switch between imaging at resolution $R \approx 10$, low resolution spectroscopy at $R \approx 100$, and moderate resolution spectroscopy at $R \approx 2000$. SC-10 uses a state of the art Si:As Blocked Impurity Band detector array to deliver the high spatial resolution and high sensitivity required for our T Tauri studies.

We imaged the T Tau binaries on 14 Nov 1992 using a 128 × 128 array, with 0.25″ pixels. The measured point spread function (PSF) on the night of observation had a full width at half maximum $\approx 0.7''$, at all wavelengths. For our measurements, we used filters with center wavelengths of 7.9, 10.3, and 12.5 μm, with 1 μm bandwidth. The 1σ point source sensitivities for our images are approximately 50 mJy for T Tau, 35 mJy for Haro 6-10, and 30 mJy for FV Tau. The images were taken while chopping the telescope secondary 20″ N-S at 10 Hz. After integrating for 5 sec the telescope nodded 20″ N-S for T Tau and FV Tau and 3″ E-W for Haro 6-10. Data were coadded during each 5 sec integration and saved to disk.

Because the separations of these T Tauri binaries are close to the resolution of the instrument and telescope, careful data reduction techniques are required to extract the brightness ratios of the components. The first step is to remove the positional offsets (1-2 pixels) between integrations due to imperfect telescope nodding and tracking. The integrations are mosaicked together using the centroid of the binary image in each integration as the alignment point, to produce the registered and cropped images of T Tau and Haro 6-10 shown in Figure 1. The individual integrations of the FV Tau data do not have sufficient S/N to use this technique, and the FV Tau integrations were simply added.

We then check the photometric quality of our results by obtaining the total system flux, with a 2.5″ square aperture centered on the intensity centroid of the system for source counts and the same aperture centered 3″ away for sky counts. For total flux calibration, we use images of β Peg and α Tau taken at roughly the same airmass and time as these observations. In this preliminary analysis, we neglect differential airmass and color corrections. An intercomparison of calibrator stars indicates that an accuracy of ±0.07 mag is obtained at this level of analysis. The results, shown in the Table I, are in good agreement with Ghez *et al.* (1991) for T Tau (Jan 1991) and with Whittet *et al.* (1988) for Haro 6-10 (1983-1986).

TABLE I

Total fluxes of T Tauri binaries.

System	λ μm	mag	F_ν Jy	$\log F_\lambda$ $(\mathrm{Wcm}^{-2}\mu\mathrm{m}^{-1})$	$\log \lambda F_\lambda$ (Wcm^{-2})
FV Tau	7.9	4.69	0.85	−7.39	−16.49
FV Tau	10.3	3.95	0.99	−7.55	−6.54
FV Tau	12.5	3.63	0.90	−7.76	−6.67
Haro 6-10	7.9	2.05	9.7	−6.33	−5.43
Haro 6-10	10.3	1.73	7.6	−6.66	−5.65
Haro 6-10	12.5	0.48	16.5	−6.50	−5.40
T Tau	7.9	0.92	27.5	−5.88	−4.98
T Tau	10.3	0.45	24.9	−6.15	−5.14
T Tau	12.5	−0.34	34.8	−6.17	−5.08

The next step in our analysis is to obtain the relative brightnesses of the component stars from the images of Figure 1. The S/N of these images is sufficiently high to apply a simple deconvolution technique which assumes that the source is two point sources with separation a and brightness ratio B. This method deconvolves the flux distribution along the axis of the binary by the distribution perpendicular to this axis. If the PSF is radially symmetric, these two parameters are sufficient to fit the model.

In reality, there is a small residual blurring ($\leq$ 1 pixel) in the images which is not removed by the centroid registry method. For example, the 10.3 μm image of

T Tau   Haro 6–10

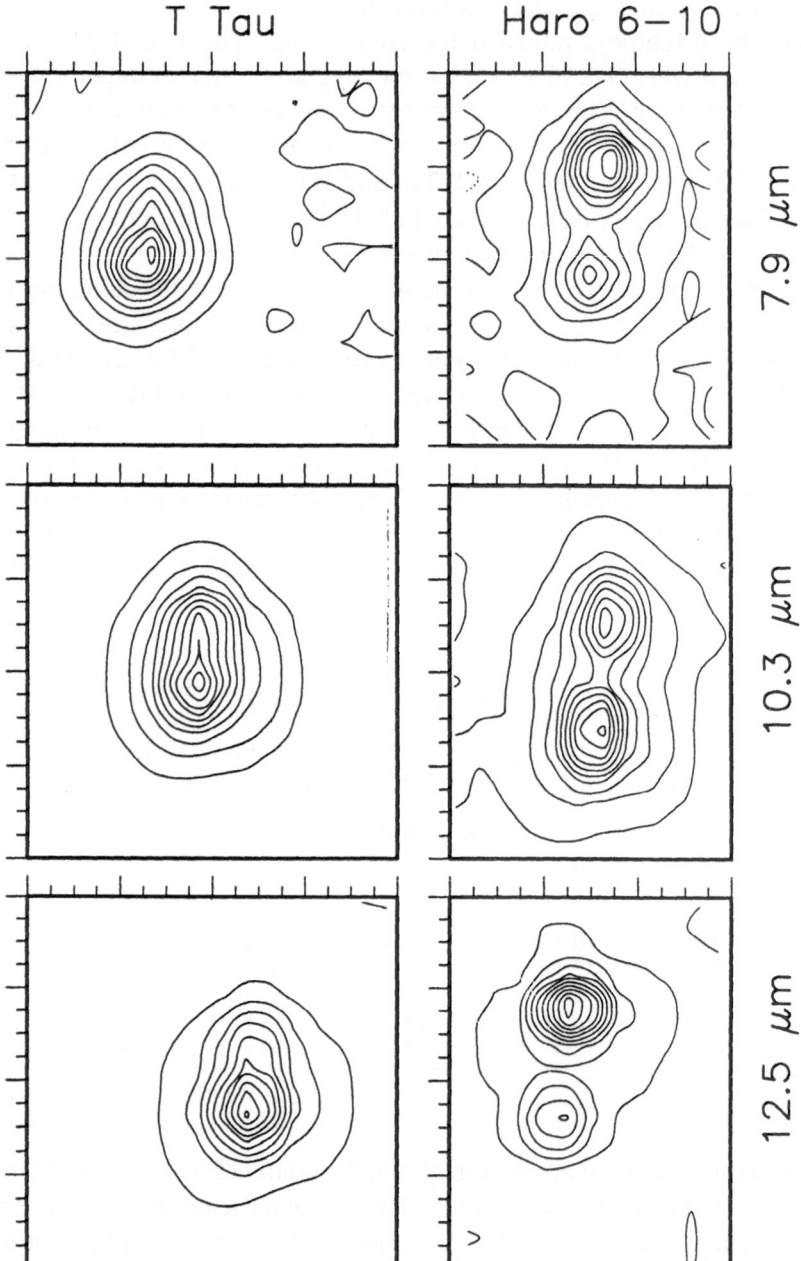

7.9 μm

10.3 μm

12.5 μm

Fig. 1. Contour plots of the flux of the T Tauri binaries T Tau and Haro 6-10. Each plot shows ten contour intervals between the highest and lowest values. Large ticks at 1.0″ intervals, small ticks at 0.25″ = 1 pixel intervals. North is up, E to the left.

Haro 6-10 (Figure 1) shows some elongation of the PSF in the N-S direction. This appears to be irreproducible – there is an E-W elongation in the 12.5 μm images of Haro 6-10, and the 10.3 μm α Tau image – so it is more likely due to telescope motion during an integration than aberrations in the telescope or instrument. The temperature profile required for emission of 1 pixel extent could not be maintained by reprocessing starlight from these low-mass stars and would require an unusually self-luminous disk around each component. Since we cannot distinguish extended emission from image blurring, we maintain the more conservative hypothesis that the image contains only two point sources. To better fit the data to a two point-source model, we add an additional parameter which is the difference in the width of the PSF between the N-S and E-W directions. The best fit parameters for the two point source model with blurring are shown in Table II.

TABLE II

Spectral energy distribution of individual components of the binaries T Tau and Haro 6-10. Uncertainties in a are .05″. Units of λF_λ are Wcm^{-2}.

System	λ (μm)	arcsec a	N/S flux ratio B	(Wcm^{-2}) log λF_λ N	log λF_λ S
Haro 6-10	7.9	1.15	1.8 ±.2	−15.63	−15.88
Haro 6-10	10.3	1.20	0.71 ±.1	−16.03	−15.88
Haro 6-10	12.5	1.20	2.20 ±.2	−15.56	−15.91
T Tau	7.9	0.70	0.42 ±.1	−15.51	−15.13
T Tau	10.3	0.70	0.73 ±.1	−15.51	−15.38
T Tau	12.5	0.70	0.45 ±.1	−15.59	−15.24

There are several ways of checking these results. First, the separation of the components a is in good agreement with speckle and lunar occultation measurements at shorter wavelengths (Ghez *et al.*, 1991; Leinert and Haas, 1989). Second, the Haro 6-10 binary is wide enough to compare the results of the two point source deconvolution method to small aperture photometry. We find agreement in B of ±10%, which is about as good as we can expect for the aperture photometry at this separation and pixel size. Third, we created synthetic binary images from copies of the 10.3 μm α Tau image, with $a = 0.75''$ and $B = 0.6, 0.7,$ and 0.8. The synthetic binary axis is along the elongated (0.5 pixels) direction of the α Tau image. The $B = 0.7$ contour plot is the best match to the 10.3 μm T Tau data, which agrees with the result in Table II.

The GK and GI Tau spectra were taken on 24 Jan 1992 using a 10×50 array with 0.5″ by 0.05 μm pixels and a 2″ slit. Outside the ozone feature, the per pixel S/N is ≈ 9 for GI Tau and 20 for GK Tau. At the maximum ozone absorption

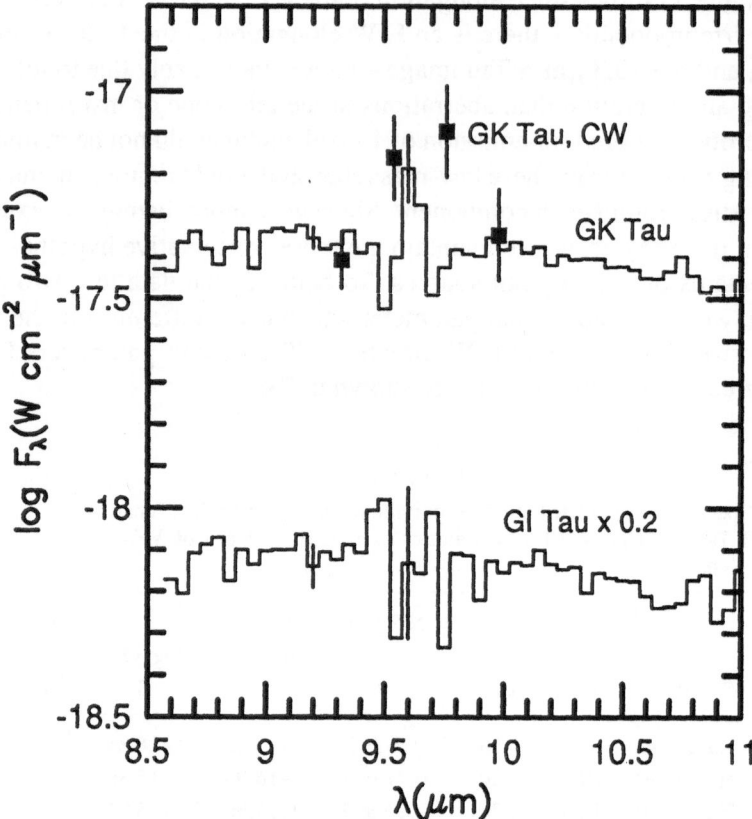

Fig. 2. Low-resolution spectra of GK Tau and GI Tau from 8.5 to 11 μm. Typical error bars outside (9.2 μm) and inside (9.6 μm) the telluric ozone absorption are shown. When the spectrum of GK Tau is normalized to ours for $\lambda < 9.3$ μm and $\lambda > 10.0$ μm, there is a significant feature centered at 9.7 μm defined by the four displayed points. This feature does not appear in our data. Our data agree with those of Cohen and Witteborn for GI Tau over this spectral range.

at 9.6 μm, the S/N is 3 for GI Tau and 6 for GK Tau. Spectra of these stars and the calibrator α Tau were taken at the same nominal grating position. To obtain fluxes for the low-resolution spectra, we use the true spectrum of α Tau (Cohen *et al.*, 1992) rather than assuming it to be a blackbody at the effective photospheric temperature, but do not apply airmass corrections in the results presented in Figure 2. We used the emission feature of telluric ozone to spectrally register the star and its calibrator within 0.2 pixels.

3. Discussion

A comparison of the spectral energy distributions λF_λ of T Tau and Haro 6-10 (Figure 3) shows that the infrared companion in each system – T Tau S and Haro 6-10 N – has a deep silicate absorption feature, while the optical component, which is

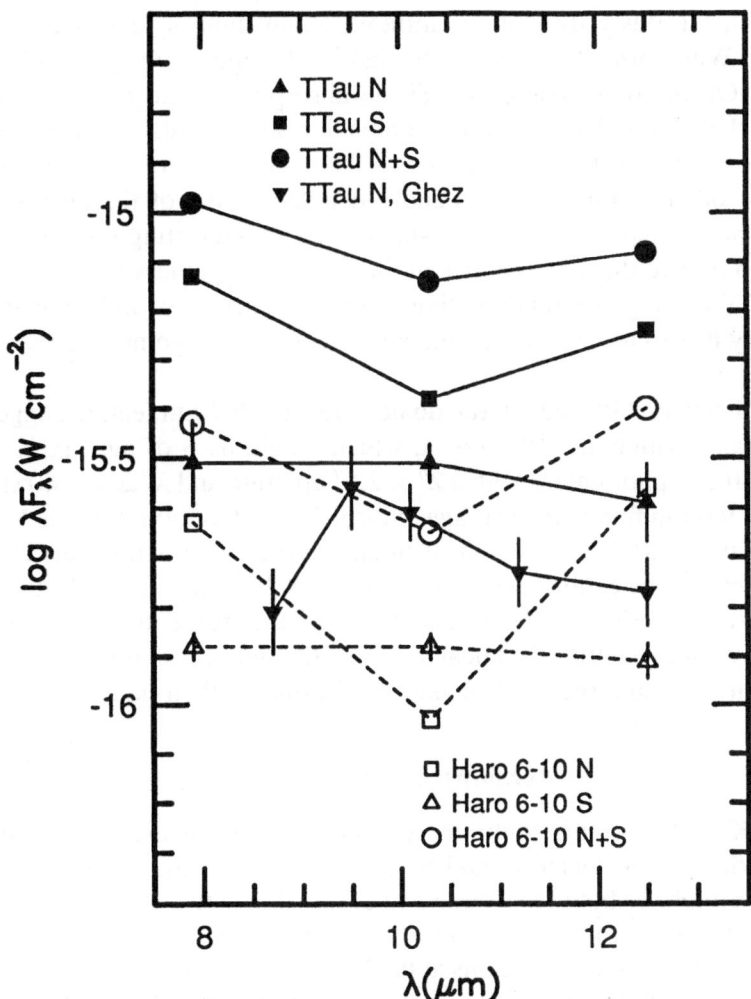

Fig. 3.   Spectral energy distribution λF_λ of the components of the T Tauri binaries T Tau and Haro 6-10 in 1 μm filter bands centered at 7.9, 10.3, and 12.5 μm, derived from the contour plots of Figure 1. Lines are guides to the eye, solid for T Tau and dashed for Haro 6-10. Data of Ghez *et al.* (1991) for T Tau N are also shown. Data for the optical component of each binary have error bars. In both T Tau and Haro 6-10, the infrared companion shows a deep absorption feature.

brighter at $\lambda < 2.2$ μm, shows an emission feature (T Tau N) or no silicate feature (Haro 6-10 S). While we find the same total flux as Ghez *et al.* (1991) for T Tau and agree that T Tau S shows a deep absorption feature, we do not find the large emission feature for T Tau N that they do, using a slit-scan technique. In particular, we need to investigate further why our 7.9 μm and Ghez *et al.*'s 8.7-μm fluxes for T Tau N differ to the degree they do.

The contrast between the components of each binary may be a disk geometry effect, in which the dusty disk of the infrared companion is seen nearly edge on (Cohen and Witteborn, 1985), while the disk of the optical component is seen more face-on. On the other hand, the silicate absorption could be caused by a residual dust shell around the infrared companion and its disk of warm dust. In neither case are the disks coplanar and in the same phase of their evolution. Observation of other T Tauri binaries and a detailed model of the dust around each component are required to decide which of these interesting possibilities is more likely. To reduce the uncertainties in the derived brightness ratios, we plan to reobserve T Tau using different nodding and readout schemes, and compare the observed binary flux to more general synthetic images of two point sources with a distorted PSF.

The spectral data on GK and GI Tau do not show the 9.7 μm feature suggested by the Cohen and Witteborn (1985) data, which would be 8 of our pixels wide (Figure 2). Synthetic photometry with a $2'' \times 2.5''$ aperture and $\lambda/\Delta\lambda = 40$ shows that the ratio of 9.6 μm flux to the average of the 9.35 and 9.85 μm fluxes is 1 ± 0.1 for both GK Tau and GI Tau, within the estimated airmass correction between the two observations and the calibrator star. This ratio is 1.7 for Cohen and Witteborn's GK Tau data. We conclude that GK Tau does not show a narrow solid-state emission feature indicative of crystalline dust near 9.7 μm, although we cannot rule out time variability in the 10 years since Cohen and Witteborn took their data.

Acknowledgements

We thank C. Koresko for the preliminary deconvolution of the binary images and for reviewing the manuscript, and S. Beckwith for pointing out that some T Tauri binaries might be both interesting and observable with our new instrument. We also thank the staff of Palomar Observatory for their support during these observations. The Cornell observations at the Palomar Observatory were made as part of a continuing collaborative agreement between Cornell and Caltech. Support for research at Cornell came from NASA.

References

Cohen, M., Walker, R.G. and Witteborn, F.C.: 1992, *Astron. J.* **104**, 2030.
Cohen, M. and Witteborn, F.C.: 1985, *Astroph. J.* **294**, 345.
Fajardo, S.B., Knacke, R.F., Telesco, C.M., Hackwell, J.A., Lynch, D.K. and Russell, R.W.: 1993, (these Conference Proceedings).
Ghez, A.M., Neugebauer, G., Gorham, P.W., Haniff, C.A., Kulkarni, S.R., Matthews, K., Koresko, C. and Beckwith, S.: 1991, *Astroph. J.* **102**, 2066.
Hayward, T L., *et al.*: 1993, in preparation.
Leinert, Ch. and Haas, M.: 1989, *Astroph. J.* **342**, 39L.
Simon, M., Chen, W P., Howell, R.R., Benson, J.A. and Slowik, D.: 1992, *Astroph. J.* **384**, 212.
Ward, W R.: 1989, *Astroph. J.* **345**, 99L.
Whittet, D.C B., Bode, M F., Longmore, A.J., Adamson, A.J., McFadzean, A.D., Aitken, D.K. and Roche, P.F.: 1988, *Monthly Not. Roy. Astr. Soc.* **233**, 321.

THE NOBEYAMA MILLIMETER ARRAY SURVEY FOR PROTOPLANETARY DISKS AROUND PROTOSTAR CANDIDATES AND T TAURI STARS IN TAURUS [*]

NAGAYOSHI OHASHI, RYOHEI KAWABE and MASATO ISHIGURO

Nobeyama Radio Observatory, National Astronomical Observatory, Minimimaki, Minamisaku, Nagano, Japan

and

MASAHIKO HAYASHI

Department of Astronomy, The University of Tokyo, Bunkyo-ku, Tokyo, Japan

Abstract. The Nobeyama Millimeter Array Survey for protoplanetary disks has been made for 19 protostellar IRAS sources in Taurus; 13 of them were optically invisible protostars and 6 were young T Tauri stars. We observed 98-GHz continuum and $CS(J = 2 - 1)$ line emissions simultaneously with spatial resolutions of $2\rlap{.}{''}8$–$8\rlap{.}{''}8$ (360–1,200 AU). The continuum emission was detected from 5 out of 6 T Tauri stars and 2 out of 13 protostar candidates: the emission was not spatially resolved and was consistent with being originated from compact circumstellar disks. Extended CS emission was detected around 2 T Tauri stars and 11 protostar candidates. There is a remarkable tendency for the detectability of the 98-GHz continuum emission to be small for protostar candidates. This tendency is explained if the mass of protoplanetary disks around protostars is not as large as that around T Tauri stars; the disk mass may increase with the increase of central stellar mass by dynamical accretion in the course of evolution from protostars to T Tauri stars.

1. Introduction

Recent observations have revealed that T Tauri stars often show evidence of compact and dense circumstellar dust disks (Beckwith *et al.*, 1990; Strom *et al.*, 1989; Adams *et al.*, 1990). Millimeter and submillimeter emissions have provided information about their mass and size, typically 10^{-3}–$1\ M_\odot$ and $\lesssim 100$ AU, respectively, which are similar to those of the "protosolar nebula" (Hayashi *et al.*, 1985); these disks are believed to be "protoplanetary disks".

Evolution of circumstellar disks as well as their physical properties is of great importance for understanding the mechanism of planetary system formation and, hence, is now a field of active investigation. For example, the decrease of near and mid-infrared excess emission with the age of T Tauri stars may suggest the dissipation of the inner region of disks (Strom *et al.*, 1989). It is likely that the formation of the disk proceeds in the stage prior to the T Tauri phase, namely in the embedded protostar phase. There were some observations of compact disks around embedded protostar candidates (e.g., L1551-IRS5, Keene *et al.*, 1990; IRAS 16293-2426, Mundy *et al.*, 1992). Because such observations of compact disks around embedded sources are limited in number, it is very important to make a survey of

---

[*] Paper presented at the Conference on *Planetary Systems: Formation, Evolution, and Detection* held 7–10 December, 1992 at CalTech, Pasadena, California, U.S.A.

Astrophysics and Space Science **212**: 239–250, 1994.

observations of disks around embedded sources in order to derive their physical properties and to understand the evolution of protoplanetary disks by comparing the results with those for T Tauri stars.

We have been carrying out survey observations for protoplanetary disks around protostar candidates and T Tauri stars with the Nobeyama Millimeter Array (NMA) in order to compare the total mass within $\lesssim$ 500 AU in radius between the two types of objects. The initial result of the survey was reported for 6 embedded sources and 5 young T Tauri stars (Ohashi *et al.*, 1991). We have increased the number of objects to 19 by now, and covered most of the embedded sources detected by IRAS toward the Taurus molecular cloud. In this article we report the latest results of the NMA survey.

2. Observations

Observations were made over three periods (Dec. 1989–May 1990, Dec. 1990–Mar. 1991, and Jan. 1992–May 1992) with NMA. The 98-GHz continuum and CS ($J = 2 - 1$) emission were simultaneously observed with spatial resolutions of $2\rlap{.}''8$–$8\rlap{.}''8$ (360–1,200 AU at a distance of 140 pc). Typical noise levels for the continuum and CS data were 4.0 mJy/beam and 130 mJy/beam, respectively.

We observed 19 low mass protostellar IRAS sources located in the Taurus molecular cloud complex. Table I shows the parameters of the observed sources. The NMA survey is focused on sources in the Taurus cloud because it is located close to the solar system (140 pc), has plenty of accumulated data in various wavelengths, and produces new stars with masses in a relatively narrow range 0.4–0.8 $M_{\odot}$. The last point is especially important when we discuss evolution of disks without the effect of mass variation among newly forming stars. The sample consists of 13 IRAS sources which are not visible on the Palomar Observatory Sky Survey and 6 IRAS sources which were identified as the youngest T Tauri stars. According to the studies for protostellar IRAS sources by Beichman *et al.* (1986), the optically invisible sources are in the younger evolutionary phase compared to the T Tauri stars; the invisible sources are probably accreting protostars. Ohashi (1991) cataloged the 13 invisible protostar candidates, and we made a complete survey for them. We selected the T Tauri stars from the catalog of Beckwith *et al.* (1990) under the criterion that the sources have large continuum flux not only at 1.3 mm but also at 100 μm.

3. Results

3.1. 98-GHz CONTINUUM EMISSION

Table II lists the results of continuum observations for all the observed sources. Figure 1 shows the maps of the 98-GHz continuum emission for all the sources with significant detection. Comparison of the 98-GHz flux with the spectral energy distribution from centimeter to submillimeter wavelengths suggests that the con-

TABLE I

Observed Sources with NMA

Observed Source	Ref.	Optical Appearence	$\log[F_{12}/F_{25}]^a$	F_{100}^b (Jy)	L_{bol}^c ($L_\odot$)	Ref.
L1489	1,2	invisible	−0.64	56	4.2	2
04108+2803	2	invisible	−0.65	11	0.74	2
04154+2823	2	invisible	−0.35	< −7.8	0.33	2
04169+2702		invisible	−0.84	17	1.3	15
04191+1523		invisible	< −0.71	14	0.48	
04239+2436	2	invisible	−0.61	16	1.4	2
04248+2612	2	invisible	−0.57	9.3	0.53	2
L1551-IRS5	3,4	invisible	−1.02	460	33	4
04295+2251	2	invisible	−0.48	7.4	0.64	2
04325+2402		invisible	< −0.92	22	0.85	
04361+2547	5	invisible	−1.02	35	3.7	15
04365+2535	1,2	invisible	−0.86	39	2.4	2
L1527	6	invisible	< −0.47	73	2.0	15
FS Tau	7,8	visible	−0.42	<11	1.7	16
T Tau	7	visible	−0.47	98	14	16
DG Tau	7,8	visible	−0.32	46	7.7	16
HL Tau	8,9	visible	−0.48	<460	7.3	16
GG Tau	7	visible	−0.11	5.2	2.3	16
DL Tau	7	visible	−0.14	<2.3	1.1	16

References: (1) Adams *et al.* 1987. (2) Myers *et al.* 1987. (3) Sargent *et al.* 1988. (4) Keene and Masson 1990. (5) Terebey *et al.* 1990. (6) Ladd *et al.* 1991. (7) Beckwith *et al.* 1990. (8) Adams *et al.* 1990. (9) Sargent and Beckwith 1987. (10) IRAS point sources catalog 1984. (11) Tamura *et al.* 1991. (12) Rodríguez *et al.* 1986. (13) Jones and Herbig 1979. (14) Rydgren *et al.* 1984. (15) Kenyon *et al.* (1990). (16) Cohen *et al.* (1989).

Notes: ^{a} infrared colors between 12 μmm and 25 μm
^{b} IRAS flux densities at 100 μm
^{c} far-infrared luminosities derived from IRAS data

tribution from free-free emission is at most 30% at 98 GHz without T Tau, so that most of the emission arises from dust grains. The typical beam size of NMA ($\sim$ 500 AU) could not spatially resolve the continuum emission, which is consistent with the thermal dust emission coming from compact dust disks whose sizes are $\lesssim$ 100 AU.

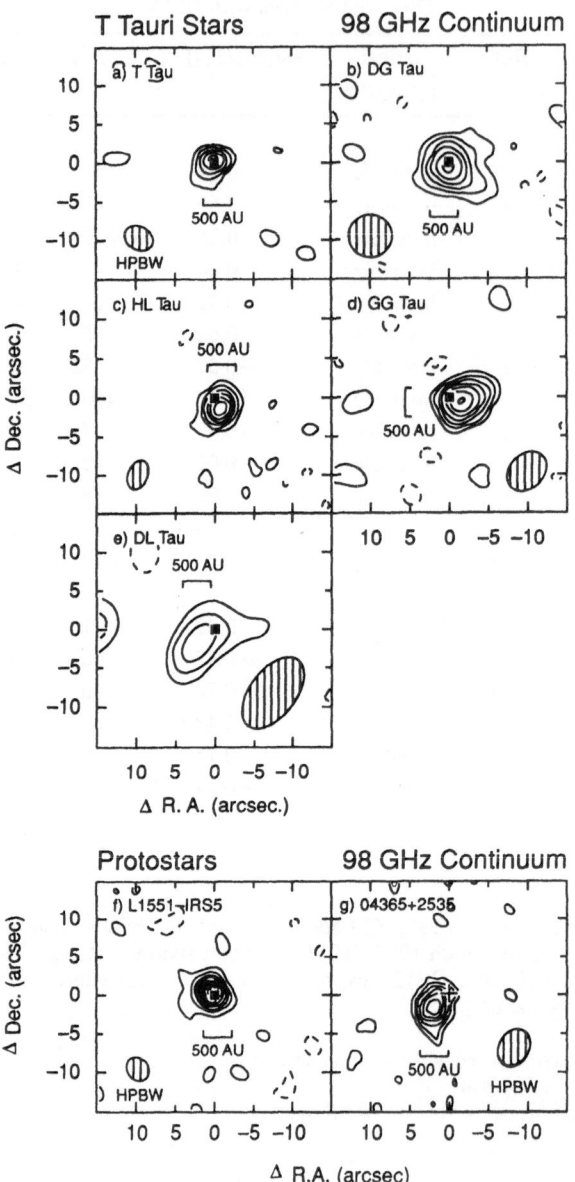

Fig. 1. The 98-GHz continuum maps. Hatched ellipses show the synthesized beam sizes in HPBW and their position angles. The contour levels are spaced 2σ, starting from 2σ for L1551–IRS5 and HL Tau. For the other sources, the contour spacing is 1σ, starting from 2σ. The filled square in Figure 1f marks the position of the 2-cm radio continuum source, and the cross in Figure 1g marks the position of the near-infrared source. The other filled squares show the optical positions of the T Tauri stars.

TABLE II
Observational Results of 98 GHz Continuum Emission

Observed Source	Optical Appearance	$F\nu$(mJy)[a]	Size (AU)[b]	M_d^c
L1489	invisible	<9.6		$<2.2 \times 10^{-2}$
04108+2803	invisible	<21		$<4.7 \times 10^{-2}$
04154+2823	invisible	<19		$<4.4 \times 10^{-2}$
04169+2702	invisible	<14		$<3.2 \times 10^{-2}$
04191+1523	invisible	<9.0		$<2.1 \times 10^{-2}$
04239+2436	invisible	<12		$<3.2 \times 10^{-2}$
04248+2612	invisible	<12		$<3.4 \times 10^{-2}$
04295+2251	invisible	<21		$<2.8 \times 10^{-2}$
L1551-IRS5	invisible	131 ± 4.0	<430 × 360	7.0×10^{-2}
04325+2402	invisible	<9.2		$<2.1 \times 10^{-2}$
04361+2547	invisible	<14		$<3.2 \times 10^{-2}$
04365+2535	invisible	30 ± 3.8	<940 × 700	6.8×10^{-2}
L1527	invisible	<11		$<2.5 \times 10^{-2}$
FS Tau	visible	<11		$<4.1 \times 10^{-2}$
T Tau	visible	56 ± 10	<550 × 420	1.0×10^{-1}
DG Tau	visible	57 ± 4.0	<800 × 770	1.0×10^{-1}
HL Tau	visible	74 ± 4.0	<530 × 420	1.5×10^{-1}
GG Tau	visible	41 ± 4.0	<780 × 670	1.6×10^{-1}
DL Tau	visible	23 ± 4.0	<1470 × 880	1.0×10^{-1}

Notes: [a] Upper limit values are 3σ and errors are 1σ.
[b] FWHM size
[c] in $M_\odot$

The last column in Table II shows the disk mass estimated from the 98-GHz continuum observations. Details of the mass estimation method are described in Ohashi *et al.* (1991). The masses of circumstellar disks around 5 T Tauri stars were 0.10 to 0.16 $M_\odot$ and for the 2 protostar candidates were ~ 0.07 $M_\odot$. The sources without detection at 98 GHz give upper limits to the disk mass of 0.02– 0.05 $M_\odot$. Hence the disks, if they exist, around 11 protostar candidates without continuum detection and around 1 T Tauri star FS Tau have masses smaller than ~ 0.03 $M_\odot$.

Comparing the results for protostar candidates with those for T Tauri stars, we note that the 98-GHz continuum emission was detected from 5 T Tauri stars out of 6 observed, whereas it was detected from 2 protostar candidates out of 13. It is natural that the continuum emission was detected from most of the observed T Tauri stars since we selected T Tauri stars with large fluxes at 100 μm and 1.3 mm. Contrary to the T Tauri stars, we detected continuum emission from only two embedded protostar candidates in spite of our selection for protostar candidates being also biased toward bright sources because we selected sources with significant detection in the IRAS point source catalog. This difference will be discussed in Section 4.1.

3.2. CS EMISSION

Table III shows the results for the CS ($J = 2 - 1$) emission. The CS emission was detected from 11 protostar candidates out of 13 and 2 T Tauri stars out of 6. In Figure 2 we present CS maps for 6 embedded sources (Figure 2a–2f) and for the 2 T Tauri stars HL Tau and DG Tau (Figure 2g and 2h). These maps clearly demonstrate that the CS emission is extended around the central sources with their typical size of $\sim$ 2,000 AU.

In most cases, distribution of the CS emission around the protostar candidates coincides with the position of the embedded stars within positional uncertainties. Because these sources are embedded in molecular cloud cores 10,000 AU in size and containing several solar masses (Benson and Myers, 1989), the CS emission distributed around the embedded sources is naturally understood to arise from dense inner regions of the extended protostellar envelopes. The optical depth of the CS emission must be large and we must be careful that the CS distribution may not follow the actual mass distribution of gas around the embedded sources, especially for those embedded sources with bad positional coincidence with the CS emission; the CS distribution might merely reflect the local perturbation of excitation temperature of CS.

The lower limits to the gas mass contained in the CS emitting area are listed in the last column of Table III. They are lower limits because we assumed optically thin emission which is not the actual case as mentioned above. The lower limit mass varies from 0.015 to 0.28 $M_{\odot}$, which means that the amount of gas within 2,000 AU around the embedded sources is sufficient to supply mass to the most massive circumstellar disks observed toward T Tauri stars in Taurus.

The detectability of the CS emission around T Tauri stars is small compared to the protostar candidates. This is a natural result of gas dissipation around the central star produced by stellar evolution from a protostar to a T Tauri star (Hayashi *et al.*, 1990). The CS distribution around HL Tau and DG Tau (Figures 2g and 2h) is different from that around some of the embedded sources; the CS emission avoids the stellar positions and tends to be smaller in size and weaker in intensity than the embedded sources. The CS emission around the T Tauri stars may hence arise from the remnant of gaseous envelopes, but not from circumstellar disks.

The non-detection of CS emission from compact disks associated with T Tauri stars is of great importance. It implies the following possibilities: (1) the CS gas in circumstellar disks is not abundant, or (2) the size of the CS emitting disks is so small that its emission is not detected due to beam dilution.

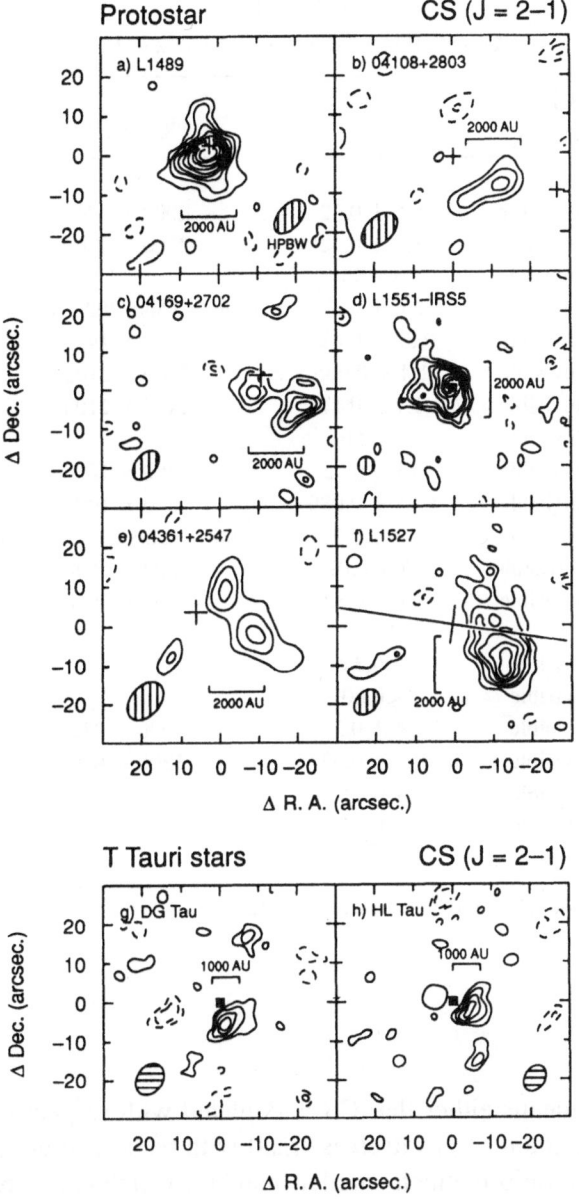

Fig. 2. The CS ($J = 2 - 1$) maps. Hatched ellipses show the synthesized beam sizes in HPBW and their position angles. The contour spacing is 1σ, starting from 2σ. The small crosses show the positions of near-infrared sources, and their sizes indicate the positional uncertainty of the infrared sources. The large cross in Figure 2f shows the position of the IRAS source, and its size indicates the positional uncertainty of the IRAS source. The filled square in Figure 2d shows the position of the 2-cm radio continuum source, and the squares in Figures 2g and 2h show optical positions of the T Tauri stars.

TABLE III

Observational Results of CS($J = 2 - 1$).

Observed Source	Optical Appearence	I_{CS}(Jy km s^{-1})^a	Size (AU)^b	M_{CS}^c
L1489	invisible	5.6 ± 0.32	2000×1700	1.1×10^{-1}
04108+2803	invisible	1.4 ± 0.32	2700×1200	2.6×10^{-2}
04154+2823	invisible	<0.44		
04169+2702	invisible	2.6 ± 0.32	$1600 \times 1500, 2000 \times 1200$	4.9×10^{-2}
04191+1523	invisible	0.80 ± 0.085	1400×1400	1.5×10^{-2}
04239+2436	invisible	1.4 ± 0.066	2400×2100	2.6×10^{-2}
04248+2612	invisible	1.7 ± 0.17	2400×2100	4.3×10^{-2}
04295+2251	invisible	<0.56		
L1551-IRS5	invisible	15 ± 0.32	2300×1600	2.8×10^{-1}
04325+2402	invisible	0.85 ± 0.080	1800×1400	1.6×10^{-1}
04361+2547	invisible	1.4 ± 0.32	$2100 \times 1500, 2100 \times 1500$	2.6×10^{-2}
04365+2535	vinvisible	2.5 ± 0.095	1600×1600	4.6×10^{-2}
L1527	invisible	11 ± 0.32	2600×2000	2.1×10^{-1}
FS Tau	visible	<0.34		
T Tau	visible	<0.60		
DG Tau	visible	1.4 ± 0.32	1600×730	2.6×10^{-2}
HL Tau	visible	1.5 ± 0.32	1200×850	2.8×10^{-2}
GG Tau	visible	<0.44		
DL Tau	visible	<0.37		

Notes: ^{a} Upper limit values are 3σ and errors are 1σ.
^{b} FWHM size
^{c} in $M_\odot$

The first case means either that CS is depleted with respect to the molecular hydrogen or the amount of gas itself is small with respect to the amount of dust; i.e., the gas to dust ratio is much smaller than 100 which is an ordinary assumed value for interstellar medium. The possibility of the depletion of gas phase CS was pointed out for the circumstellar disk around HL Tau by Blake *et al.* (1992). On the other hand the small abundance of gas components in circumstellar disks might also be suggested by the CO observations of GG Tau. Skrutskie *et al.* (1993) derived that the abundance of gas phase CO with respect to the total dust mass is much smaller than the ordinary interstellar value, suggesting that CO would also be depleted as CS in the disk of HL Tau. However, depletion of gas phase CO is difficult, because molecular CO has a very low condensation temperature (~ 12 K;

Nakano *et al.*, 1969), which requires very low disk temperatures unless indirect depletion of CO gas via other molecules, which are formed from CO in chemical reaction networks, is efficient. This might suggest that the entire molecular gas is less abundant in the circumstellar disks in HL Tau, DG Tau, and GG Tau, i.e., the gas to dust ratio would be much smaller than 100 as a result of gas dissipation.

The second case would give an upper limit of 70 AU (30 K/T_{disk}) to the radius of optically thick CS disks, which means that the size of CS disks is similar to or smaller than the size of dust disks. This means that the disk would have a very sharp cut-off at its outer boundary, although this may not be plausible because CS becomes optically thick at a column density of 10^{22} cm^{-2}, which is 2 orders of magnitude smaller than the column density for the dust emission to be optically thick.

4. Discussion

4.1. LOWER CONTINUUM DETECTABILITY TOWARD THE PROTOSTAR CANDIDATES

The results listed in Tables II and III showed that the 98-GHz continuum emission was detected from a larger number of T Tauri stars than embedded sources, while the CS ($J = 2 - 1$) emission has the opposite tendency. Although it is easy to understand the trend for the CS emission as argued in the previous section, it is not straightforward to understand the tendency for the continuum emission.

The high detectability of the 98-GHz continuum emission toward T Tauri stars may be a selection effect, because we selected T Tauri stars with strong continuum emission both at 1.3 mm and at 100 μm. The average luminosity for T Tauri stars is actually ~ 3 times larger than for protostars. We must hence take into account the continuum detectability for low luminosity T Tauri stars. When we evaluate the fraction of (age $< 10^{5.5}$ yr) T Tauri stars with disk masses larger than the NMA detection limit of 0.03 $M_\odot$ according to the catalog of Beckwith *et al.* (1990), there are 5 T Tauri stars out of 18. Thus if we observed less luminous T Tauri stars, we could have detected circumstellar disks at a rate of 5/18. This value is still larger than the detection rate of 2/13 toward protostellar sources, although this argument is not statistically significant. The low continuum detectability for protostar candidates compared to T Tauri stars is probably real.

4.2. FORMATION OF PROTOPLANETARY DISKS

If we assume that the low detectability of the continuum emission is real, it means that the circumstellar disk mass around T Tauri stars in Taurus is more massive than that around protostar candidates. On a larger scale, however, the total gravitationally bound mass is much larger for protostar candidates than for T Tauri stars, because protostar candidates are associated with molecular cloud cores with several solar masses (Benson, and Myers, 1989). This means that there is a significant difference in mass concentration between the two types of objects; protostar candidates are associated with large mass of 1–10 $M_\odot$ within $\sim 10,000$ AU in radius, while they are associated with relatively small masses $\lesssim 0.03\ M_\odot$ within 500 AU in radius compared to the T Tauri stars.

Because embedded protostar candidates are believed to be in an earlier stage of evolution than T Tauri stars, it is natural that this difference in mass concentration arises as a result of evolution from a protostar into a T Tauri star; i.e., the mass of a protoplanetary disk increases as a result of mass accretion from the protostellar envelope in the embedded phase, and the envelope mass decreases as a result of the accretion and of being blown off by molecular outflows. Such a scenario of disk formation can be easily understood on the basis of current theories of star formation (Stahler *et al.*, 1980) as below.

We consider a simple case where both the central star and disk steadily accrete mass under the condition that the disk accumulates matter at a rate of 10% of the total accreting rate from the envelope onto the central star and disk system. The physical base of this simple case is as follows: We consider a case where the central star accretes matter from its envelope through the disk. If the mass accumulation rate, which is the difference between the accretion rate from the envelope onto the disk and that from the disk onto the star, would exceed 10% of the total accretion rate, the disk mass should become larger than 10% of the central stellar mass, and the disk would eventually become gravitationally unstable. It would no longer evolve into an optically thick but geometrically thin disk in the coming T Tauri phase. Hence we consider that 10% is a reasonable upper limit to the mass accumulation rate for the accreting disk around a protostar.

For protostar candidates, most of their bolometric luminosity is due to accretion of infalling material. In the case that the infalling material reaches the vicinity of the boundary layer, all the released gravitational energy becomes thermal energy. Hence their bolometric luminosity is

$$L = \frac{GM\dot{M}}{R}$$

where M and R are stellar mass and radius, respectively, and $\dot{M}$ is the accretion rate onto the protostar (Stahler *et al.*, 1980). The final mass of the accreting protostar at the end of its embedded phase is $M = \dot{M}t$, where t is the duration of the embedded phase. The duration has been estimated to be $\sim 10^5$ yr from the number ratio of the protostar candidates to T Tauri stars whose typical age is $\sim 10^6$ yr. If we take the average bolometric luminosity of the protostar candidates without the continuum $\sim 1.5\ L_\odot$ as due to accretion, the mass accretion rate is

$$\dot{M} = \sqrt{\frac{LR}{Gt}} \sim 2 \times 10^{-6}\ M_\odot\ \mathrm{yr}^{-1}$$

This value shows good agreement with the theoretical one, i.e., $\dot{M} = c^3/G = 2 \times 10^{-6}\ M_\odot\ \mathrm{yr}^{-1}$, where $c = 0.2\ \mathrm{km\ s}^{-1}$ is the effective sound speed (Stahler *et al.*, 1980).

The accretion rate of $2 \times 10^{-6} \, M_\odot \, \text{yr}^{-1}$ gives the stellar mass of a protostar at the end of its embedded phase to be 0.2 $M_\odot$. Then the mass of the protoplanetary disk around it should be smaller than 0.02 $M_\odot$ in order for the disk to be gravitationally stable. This disk mass is comparable but is smaller than the mass of the NMA detection limit, so that there is no wonder we could not detect continuum emission toward protostar candidates at their final embedded stage.

The stellar mass of 0.2 $M_\odot$ at the end of its embedded phase is smaller than the typical stellar mass of T Tauri stars (0.5–0.8 $M_\odot$; Cohen and Kuhi, 1979). Those stars therefore need further accretion in the very young ($\lesssim 3 \times 10^5$ yr) T Tauri phase. Such accretion in the very early T Tauri phase is consistent with the high accretion luminosity for the youngest T Tauri stars. They sometimes have luminosity much larger than the embedded sources; the high luminosity gives the mass accretion rate of order $10^{-6} \, M_\odot \, \text{yr}^{-1}$, which can increase the stellar mass significantly in $\lesssim 3 \times 10^5$ yr. When the T Tauri star accretes sufficient mass, the circumstellar disk can exist stably around the central star with a detectable mass for NMA.

The above simple scenario of disk formation by steady accretion can explain our observational results. Recently, Nakamoto and Nakagawa (1993) investigated viscous evolution of protoplanetary disks, and found that the disk mass increases significantly in the late stage of infall (3–4×10^5 yr). Their numerical calculation is consistent with our observational results, and supports the scenario of disk formation in the course of protostellar evolution. Although we discussed on the basis of steady accretion, non-steady disk formation scenario is also possible to explain our results (see Ohashi *et al.*, 1991).

5. Summary

We have observed 13 protostar candidates and 6 young T Tauri stars in the 98-GHz continuum and CS ($J = 2 - 1$) emissions with the Nobeyama Millimeter Array, as part of the survey of protoplanetary disks. The main results are summarized as follows:

1. The 98-GHz continuum emission was detected from 5 T Tauri stars out of 6, and from 2 protostar candidates out of 13. The detected continuum emission was not resolved by the typical NMA beam size of 500 AU in radius, suggesting that the continuum emission arises from protoplanetary disks whose radius is $\lesssim 100$ AU.

2. The CS ($J = 2 - 1$) emission was detected from 11 protostar candidates out of 13, and from 2 T Tauri stars out of 6. The CS emission is spatially extended (1,000–2,000 AU), suggesting that it arises from the inner part of extended envelopes associated with the embedded protostars.

3. The continuum detectability is lower for the protostar candidates than for the T Tauri stars. This difference is due to the different degree of mass concentration; the mass distribution around T Tauri stars is concentrated around <500 AU

of the central sources as circumstellar disks, while protostar candidates show mass distribution extended more that 2,000 AU.
4. The difference in degree of mass concentration between two types of objects may be the result of their different evolutionary stages, i.e., masses of protoplanetary disks increase when protostar candidates evolve into T Tauri stars through dynamical accretion from their extended envelopes. A simple scenario of disk formation by steady accretion is consistent with our observational results.

Acknowledgements

We are grateful to Prof. A.P. Boss for his critical reading of our paper as a referee and to the Japanese Society for the Promotion of Science for their financial support of Nagayoshi Ohashi.

References

Adams, F.C., Lada, C.J. and Shu, F.H.: 1987, *Astroph. J.* **312**, 788–806.
Adams, D.C., Emerson, J.P. and Fuller G.A.: 1990, *Astroph. J.* **357**, 606.
Benson, P.J. and Myers, P.C.: 1989, *Astroph. J.S* **71**, 89–108.
Beichman, C.A., Myers, P.C., Emerson, J.P., Harris, S., Mathieu, R., Benson, P.J. and Jennings, R.E.: 1986, *Astroph. J.* **307**, 337–349.
Beckwith, S.V.W., Sargent, A.I., Chini, R.S. and Güsten, R.: 1990, *A J* **99**, 924–945.
Blake, G.A., van Dishoeck, E.F. and Sargent, A.I.: 1992, *Astroph. J. Lett.* **391**, L99–L103.
Cohen, M. and Kuhi, L.V.: 1979, *Astroph. J.S* **41**, 743–843.
Hayashi, M., Hasegawa, T., Ohashi, N., Sunada, K. and Fukui, Y.: 1991, in: A.D. Haschick and P.T.P. Ho, ed(s)., *Atoms, Ions and Molecules: New Results from Spectral Line Astrophysics*, the 3rd Haystack Conference, Astronomical Society of the Pacific, San Francisco, 223.
Hayashi, C., Nakazawa, K. and Nakagawa, Y.: 1985, in: D.C. Black and M.S. Matthews, ed(s)., *Protostars and Planets II*, University of Arizona Press, Tucson, Arizona, 1100.
Keene, J. and Masson, C.R.: 1990, *Astroph. J.* **355**, 635–644.
Mundy, L.G., Wootten, A., Wilking, B.A., Blake, G.A. and Sargent, A.I.: 1992, *Astroph. J.* **385**, 306–313.
Nakano, T., Ohyama, N. and Hayashi, C.: 1969, *Progress of Theoretical Physics* **39**, 1448.
Nakamoto, T. and Nakagawa, Y.: 1992, *Astroph. J.*, submitted.
Ohashi, N.: 1991, PhD Thesis
    University of Nagoya.
Ohashi, N., Kawabe, R., Hayashi, M. and Ishiguro, M.: 1991, *Astron. J.* **102**, 2054–2065.
Skrutskie, M.F., Snell, R.L., Strom, K.M., Strom, S.E., Edwards, S., Fukui, Y., Mizuno, A., Hayashi, M. and Ohashi, N.: 1993, *Astroph. J.*, in press.
Stahler, S.W., Shu, F.H. and Taam, R.E.: 1980, *Astroph. J.* **241**, 637–654.
Strom, K.M., Strom, S.E., Edwards, S., Cabrit, S. and Skrutskie, M.F.: 1989, *Astron. J.* **97**, 1451–1470.

HIGH-PRECISION VLBI ASTROMETRIC OBSERVATIONS

OF RADIO-EMITTING STARS FOR DETECTION OF

EXTRA-SOLAR PLANETS *

JEAN-FRANÇOIS LESTRADE, DAYTON L. JONES and ROBERT A. PRESTON

*Jet Propulsion Laboratory, California Institute of Technology,
Pasadena, California, USA*

and

ROBERT B. PHILLIPS

*Haystack Observatory, Massachusetts Institute of Technology,
Westford, Massachusetts, USA*

Abstract. The displacement of a radio-emitting star around the barycenter of a possible planetary system can be measured by astrometric very long baseline interferometry (VLBI) observations. We have observed the radio-emitting star σ^2 CrB at 8 epochs over 5 years by VLBI and fitted its 5 astrometric parameters to the observed coordinates. The post-fit coordinate residuals have an rms scatter of 0.22 milliarcseconds and show no systematic behavior. We use this result to set a limit on the presence of planets around σ^2 CrB and conclude that our present VLBI astrometric precision corresponds to the threshold to detect a Jupiter-like planet around this star. We also discuss the astrometric monitoring program of 11 radio-emitting stars that we are conducting for the Hipparcos space mission and its possible contribution to a long-term planet search program.

1. Introduction

Astrometric monitoring of the minute displacement of a star around the barycenter of the system that includes possible planetary companions has been an indirect method for detection of extra-solar planets for several decades at optical wavelength (van de Kamp and Lippincott, 1951). The motion of a single planet in a circular orbit around a star causes the star to undergo a reflexive circular motion around the star–planet barycenter. When projected on the sky, the orbit of the star appears as an ellipse with angular semimajor axis θ given by

$$\theta = \frac{m_p}{M_*} \frac{a}{d} \tag{1}$$

where θ is in arcsec when the semimajor axis a is in AU, the mass of the planet (m_p) and the mass of the star (M_*) are in solar masses, and the distance d is in pc. For example, observing the solar system from a distance of 10 pc, the presence of Jupiter would be revealed as a periodic circular displacement in the Sun's position, with an amplitude θ of 0.5 milliarcsecond (mas) and a period of 11.9 years. More generally, the astrometric accuracy required to detect giant planets is in the submilliarcsecond range, except for very nearby stars.

---

* Paper presented at the Conference on *Planetary Systems: Formation, Evolution, and Detection* held 7–10 December, 1992 at CalTech, Pasadena, California, U.S.A.

Astrophysics and Space Science **212**: 251–260, 1994.

Optical astrometry is generally limited to a precision of a few tens of milliarc-seconds (mas) but the best measurements are at the 1 mas level now (Gatewood *et al.*, 1992). Technical advances in very long baseline interferometry (VLBI) with the Mark III recording system (Rogers *et al.*, 1983) have provided sufficient sensitivity to reliably detect radio-emitting stars over the last few years. We have carried out VLBI measurements of the position of the radio star σ^2 CrB since 1987 and demonstrated that the level of precision is 0.2 mas during 5 years. This level of precision is not SNR-limited and could reach 20 microarcseconds if all systematics could be removed by an improved strategy of observations or data analysis. Such a high level of astrometric precision makes the VLBI technique a new tool for planetary searches.

2. Radio-Emitting Stars

There are about 400 stars that exhibit radio emission as compiled by Wend-ker (1987). About half of these stars exhibit thermal free-free emission from very large ionized circumstellar envelopes that are fully resolved by VLBI observations. The other stars exhibit non-thermal radio emission (gyrosynchrotron or possibly coherent emission mechanisms) with typical source size of a few mas or less that match the VLBI angular resolution. These non-thermal radio-emitting stars belong to a wide variety of physical classes, e.g., X-ray, RS CVn, Algol, dMe, FK Com, T Tauri. Many of these stars can be detected by the sensitive Very Large Array (VLA) in New Mexico, but are too weak to be detected by standard VLBI. Their radio flux density is only a few tens of mJy or less, i.e., 100–1000 times weaker than compact extragalactic radio sources usually observed by VLBI. Nonetheless, 30 stars can be detected by phase-referenced VLBI observations and this number should grow with future improvements of the technique.

We have selected 11 radio-emitting stars with non-thermal emission (7 RS CVn, 2 X-ray and 2 FK Com) for a high-accuracy VLBI astrometric monitoring program. The initial motivation of this program was to measure their radio positions and proper motions to make the future Hipparcos optical reference frame coincidental and at rest with respect to an extragalactic quasi-inertial reference system (Lestrade *et al.*, 1992). This is still our main motivation but the results and future developments of our ongoing program could contribute to the search for extra-solar planets. One of the selection criteria used to draw up the list of stars in our program was that they be of magnitude lower than 11th to match Hipparcos capability. In the future, this program could expand to monitor a larger sample of stars, especially with the use of the new Very Long Baseline Array (VLBA), which is the antenna array dedicated to VLBI and built by the National Radio Astronomy Observatory. A program has already started on radio-emitting red dwarfs (dMe) at Haystack Observatory since VLBI observations of these stars have shown that they have compact radio components at 8.4 GHz (Phillips *et al.*, 1988) and at 1.6 GHz (Benz and Alef, 1991).

3. Phase-Referenced VLBI Technique for High-Precision Astrometry of Weak Radio Objects

Phase-referenced observations are based on a strategy which allows high-accuracy differential astrometry because the prime observable used is the VLBI phase. A VLBI interferometer produces milliarcsecond fringe spacings on the sky and the phase of the complex visibility can measure the position of the radio source with an uncertainty corresponding to a small fraction of this fringe spacing. VLBI is an astronomical technique using an array of antennas (two or more) separated by baselines of a few thousands of kilometers which simultaneously observe the same radio source to record its continuum signal over a limited bandwidth, typically a few tens of MHz, on video magnetic tapes. After the observations, the tapes are shipped and the recorded signals of each pair of antennas are cross-correlated on a specialized processor. The observations are usually carried out at centimeter wavelengths (1 to 10 GHz) although recent observations at millimeter wavelengths have been successful.

The coherence of the radio signals recorded by antennas separated by long distance, and not electrically connected, is possible by locking the heterodyne reference frequency of the receiver at each site to a frequency standard like a hydrogen maser that has a stability ($\sim 10^{-14}$ s/s) and allows coherent cross-correlation over 10–15 minutes. Consequently, the observed radio source must have a flux density high enough to be detected over such an integration time.

This is the reason why VLBI is less sensitive than the VLA, which is a connected interferometer, and one can integrate data for several hours if high sensitivity is required. Of course, VLBI has a much finer angular resolution than the VLA, reaching < 1 mas on intercontinental baseline at centimeter wavelengths. At the VLBI processor, the cross-correlation of the recorded signals leads to the measurement of the amplitude and phase of the complex visibility induced by the source brightness distribution convolved with the beam of the antenna pair for the duration of each scan (a few minute integration period).

As mentioned above, the coherence time in standard VLBI is severely limited to less than ~ 15 minutes by non-linear instabilities in the independent frequency standards at the VLBI stations. When a radio source is so weak that it cannot be detected within this duration, one has to resort to the phase-referencing VLBI technique, which allows multiple scans to be combined in a single coherent integration period. A reference for the VLBI phase must be established by observing an angularly nearby strong extragalactic source alternately with the weak program source with a cycle time of a few minutes. Such a phase-referencing technique in VLBI allows increased sensitivity through use of much longer integration times (several hours) with minimum coherence loss. The phase-referencing VLBI technique as applied in our VLBI astrometric program is described in detail in Lestrade *et al.* (1990).

4. Results of a Series of VLBI Observations of the Star σ^2 CrB

σ^2 CrB is an RS CVn close binary whose orbital motion has a period of 1.1 day and a separation of 0.6 mas. Phase-referenced VLBI observations of σ^2 CrB were conducted at 8 epochs between May 1987 and August 1992. Observation dates, flux densities, and orbital phases are in Table I. At 5 GHz, our program used the VLBI array made of the following antennas: the Phased-VLA at the National Radio Astronomy Observatory, New Mexico (NRAO, NM); Bonn (Max-Planck-Institut, Germany); Medicina (Bologna, Italy); Greenbank (NRAO, West Virginia); Haystack Observatory (Massachusetts Institute of Technology, Massachusetts); Owens Valley Radio Observatory (OVRO), California Institute of Technology; and, at 8.4 GHz, the VLBI array was made of Goldstone (Deep Space Network, California); Hat Creek (Berkeley, California); VLA (NRAO, New Mexico); OVRO; Haystack, and newly commissioned VLBA radiotelescopes (USA). The total data integration times were between 5 and 8 hours at each epoch. The VLBI data acquisition system was the Mark III system (Rogers *et al.*, 1983) used in a mode to record a bandwidth of 28 MHz. The corresponding detection threshold is about 2 milliJansky (10σ). All the cross-correlation of the recorded signals was carried out at the Mark III Processor at Haystack Observatory (Whitney, 1988).

TABLE I

VLBI observations of σ^2 CrB at 8 epochs.

Obser. Date	Orbital phase (cycle)	Frequency (GHz)	Flux density (mJy)
87/05/26 04 UT	0.56	5.0	10
88/11/16 17 UT	0.93	5.0	28
89/04/13 06 UT	0.25	5.0	7
90/11/16 23 UT	0.37	5.0	3.8
91/04/12 10 UT	0.86	8.4	19.5
92/01/15 13 UT	0.88	8.4	4.6
92/06/08 04 UT	0.89	5.0	13
92/08/03 05 UT	0.06	8.4	8.3

The 5 astrometric parameters of σ^2 CrB (2 coordinates, 2 proper motion components and parallax) were estimated by a least-square fit with the 16 coordinates measured at the 8 epochs. Figure 1 shows the results of the fit. The uncertainties of the measured VLBI coordinates were set to 0.20 milliarcsec to make the reduced-χ^2 close to unity for the number of degree of freedom (11) in the fit. The rms of the post-fit coordinates residuals is 0.22 mas. With such an adjustment, the formal uncertainties for the 5 fitted parameters are 0.08 mas for the relative position between σ^2 CrB and the reference source 1611+343, 0.04 mas/year for the

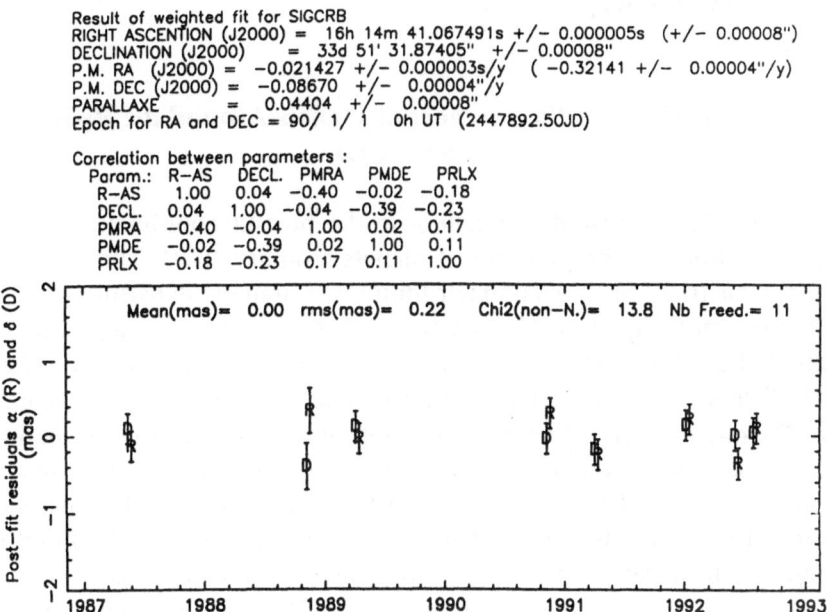

Fig. 1.   Result of the fit of the 5 astrometric parameters of σ^2 CrB adjusted to the coordinates measured by VLBI at 8 epochs.

proper motion and 0.08 mas for the trigonometric parallax. The correlation matrix indicates that the 5 parameters are well separated.

The number of degrees of freedom (11) is high enough to make the statistical significance of the formal uncertainties reliable. Various tests have been done for the robustness of the solution. One test has been to make two astrometric solutions, one with the first 4 epochs and one with the last 4 epochs of observations. The table below indicates the parameter differences between the two solutions.

Parameter	Differences
α	0.22 mas (1.2σ)
δ	0.19 mas (1σ)
μ_α	0.22 mas/y (0.8σ)
μ_δ	0.24 mas/y (0.9σ)
π	0.12 mas (0.5σ)

Also, the 5 astrometric parameters of σ^2 CrB determined by VLBI (see inset of Figure 1) all match, within uncertainties, the best but less precise optical determination by Requième and Mazurier (1987) (for the position and proper motion) and by Jenskin (1952) (for the parallax).

5. Implications for the Presence of Planets Around the Radio Star σ^2CrB

The lack of obvious sinusoidal signature in the post-fit coordinate residuals of Figure 1 sets a limit on the presence of planets around σ^2CrB. The rms of these post-fit residuals (0.22 mas) is an upper limit on systematic departure from linear motion for the star. Equation (1) can be used to exclude a range of planetary perturbations by taking $2\theta = 0.22$ mas, $M_* = 2.26\ M_\odot$ and d = 22.7 pc for σ^2CrB. The log-log representation of Equation (1) with these parameters is in Figure 2. The diagonal line of constant astrometric signature follows Equation (1), and all points above this line represent larger planetary perturbations. We assume that a full orbital period of the planet must be sampled during the total span of observations to separate the sinusoidal planetary signature from the fitted linear proper motion. In these conditions, the maximum semimajor axis a of a planet corresponds to the total observation span through Kepler's third law. This upper limit on a is 3.8 AU for our 5 years of observations and is the vertical dashed line in Figure 2. The shaded area indicates the parameter space (a, m_p) that is excluded by our observations for a possible planet. Note that for $a = 3.8$ AU, the mass m_p is 0.0014 $M_\odot$ and that this orbit radius $a = 3.8$ AU is very close to the lower boundary (4.0 AU) for the formation of giant planets in a planetary system controlled by the most luminous star F6 V in the close binary $\sigma^2\ CrB$. Finally, in Figure 2, we have also shown where a Jupiter-like planet would fall. Interestingly, the present accuracy of our VLBI measurement corresponds exactly to the detection threshold for a Jupiter-like planet around $\sigma^2\ CrB$ when 12 years of data are collected.

This interpretation is optimistic since Black and Scargle (1982) note that the fitted linear proper motion absorbs part of the planetary perturbation. These authors show that with observations that sample a single orbital period, the amplitude of the planetary perturbation is underestimated by as much as 47%. However, if the classical model (position, proper motion, and trigonometric parallax) is complemented by a sinusoidal function and the *a priori* values for the amplitude, period, and phase chosen to cover a large volume of the parameter space, no absorption of the planetary perturbation would occur and the 3 additional parameters could be fitted.

σ CrB is a triple system consisting of a visual pair with a G1V star (σ^1 CrB) separated by 140 AU (P=1000 years) from a spectroscopic binary (σ^2 CrB = F6V/F8V) whose separation is 6 $R_\odot$ (P=1.13 day). The two orbital planes are co-planar (Barden, 1985). The radio emission is identified with the spectroscopic binary classified as an RS CVn with two chromospherically active stars. The

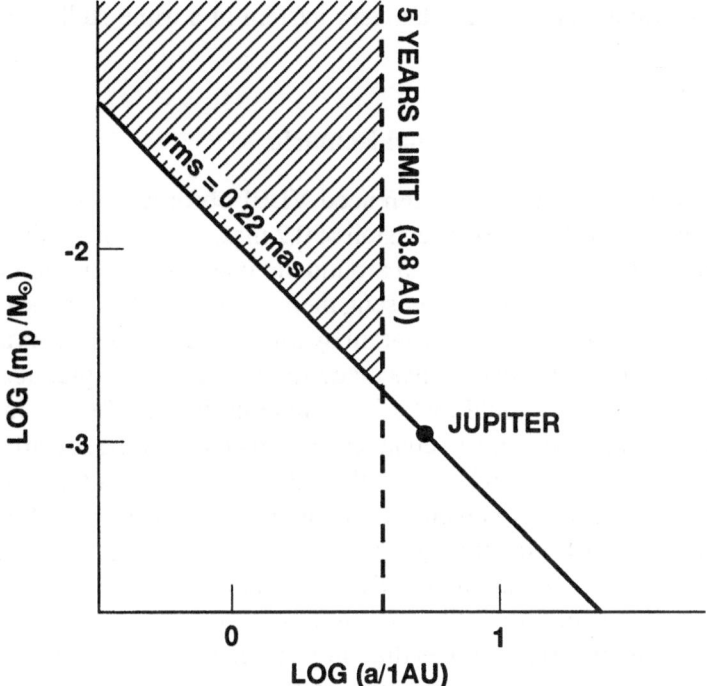

Fig. 2. Log-Log representation of Equation (1) for the rms of the post-fit coordinate residuals of σ^2 CrB. The shaded area is the parameter space (semimajor axis a, mass m_p) that is excluded by our observations for a planet around this star.

masses of these two stars, F6V and F8V, are 1.12 and 1.14 $M_\odot$ (Barden, 1985), respectively, and their sum, 2.26 $M_\odot$, has been used in the analysis above.

The location of the radio centroid within the spectroscopic binary is a crucial question. If the radio emission is associated with only one of the stars, then a total displacement of 1.2 mas correlated with the orbital phase of the system would be seen in our coordinates residuals since the observations were taken at various orbital phases of the spectroscopic system (see Table I). A precise ephemeris of orbital phases has been established for σ^2 CrB by Bakos (1984). The post-fit residuals might be dominated by this orbital motion and we are investigating this possibility with additional observation to uniformly cover the whole orbit. The stability of this radio centroid over time is an important question for a planetary search. There is no detailed model of the radio-emitting region that can be used to derive this stability. It must be determined observationally and the rms of our post-fit coordinate residuals (0.22 mas or 1 $R_\odot$ at σ^2 CrB) can also be interpreted as a measure of this stability.

Finally, another important theoretical question is to assess the possibility of the formation of a dynamically stable planetary system in a triple stellar system like σ CrB.

6. Final Remarks

We have demonstrated that phase-referenced VLBI observation of the radio star σ^2 CrB can achieve an astrometric precision of 0.2 mas. Interestingly, this precision corresponds to the level of perturbation around the linear proper motion of the star expected for a Jupiter-like planet. At present, there are about 30 radio stars that could be monitored for a planetary search program by astrometric VLBI observations. But the scarcity of VLBI observing time has made such a program unpractical till now. However, the new VLBA should make it possible in the near future. In addition, foreseen technical improvements to enhance the sensitivity of the recording system by the end of the decade should lengthen the list to about 60 candidate stars.

The theoretical precision for an interferometer is $\sigma_{\alpha,\delta} = 1/(2\pi)(1/\text{SNR})(\lambda/B)$ (Thompson *et al.*, 1986). For our observations, $\sigma_{\alpha,\delta}$ is 20 microarcseconds with $B = 3000$ km, $\lambda = 6$ or 3.6 cm and SNR > 15. Hence, the astrometric precision achieved for σ^2 CrB is not SNR-limited (signal-to-noise-ratio-limited). There are at least three systematic error sources that prevent reaching this ultimate precision of the observations: (1) the extrapolation of the reference source VLBI phase in switched observations to the time of the star observation, (2) the differential contribution of the atmosphere and ionosphere along the two lines of sight to the reference source and target star, and (3) the structures of the reference source and, possibly, of the star. The atmosphere and ionosphere propagation effects could be modelled by improved calibration and the structure of the reference source mapped with the data themselves. But the main error source, for the moment, is the extrapolation of the reference source VLBI phase over 2 to 3 minutes because of the switched observations. This can be eliminated entirely if the reference source and the star are angularly close enough to be in the main beam of the antennas, but this is a rare situation. A more general approach would be to use VLBI baselines with two antennas at each end for pointing continuously at the reference source and the star. The two antennas at each site should be phase-locked to the same frequency standard (hydrogen maser). The antenna pointing at the relatively strong reference source could be smaller. This prospect is not presently considered for the VLBA.

Table II summarizes the relevant information for the 11 radio-emitting stars of our Hipparcos astrometric program in order to compute the total sky displacement $2 \times \theta$ from Equation (1) expected for a Jupiter-like planet around these stars. The values calculated for $2 \times \theta$ in this table compare favorably to the potential SNR-limited astrometric precision of phase-referenced VLBI observations on intercontinental baselines. The last 2 stars in Table II (Hubble 4 and HDE283572 of the Taurus–Auriga dark clouds) are pre-main-sequence stars that are not part of our current program but have been detected on intercontinental VLBI baselines by Phillips *et al.* (1991). These two stars were part of a survey at 1.3 millimeters and were not detected, while others stars of the cloud were detected. The detections were interpreted as evidence for a dust-disk around the stars, i.e., proto-planetary material (Beckwith *et al.*, 1990). One can speculate that Hubble 4 and HDE283572 are more evolved and, possibly, that their initial dust-disks have already collapsed into planets.

TABLE II

The 11 radio-emitting stars of our VLBI program and the relevant information to compute the total sky displacement $2 \times \theta$ for a Jupiter-like planet around them.

Star	Class	Distance (pc)	Masses hot/cool (M_{sol})	Sp. Type	$2 \times \theta_{Jupiter}$ (μas)
LSI 61303	X-ray	2000			< 20
Algol	Algol	27	3.6/0.79	K0IV/B8V	80
UXARI	RS CVn	50	> 0.63/ > 0.71	G5V/K0IV	< 160
HR1099	RS CVn	36	1.1/1.4	G5IV/K1IV	100
HDE 283447	PMS (WTT)	160	1.74	K3	35
HR5110	RS CVn	53	1.5/0.8	F2V/G0IV	80
σ^2 CrB	RS CVn	21	1.12/1.14	F6V/G0V	210
Cyg X1	X-ray	2000			< 20
HD 199178	FK Com	140-90	3.2 ?	G5 III-IV	> 25
r AR Lac	RS CVn	47	> 1.3/ > 1.3	G2IV/K0IV	80
IM Peg	RS CVn	50	4 ?	K2III-II	25
Hubble 4	PMS (WTT)	160	0.5 - 2.0	K7	> 30
HDE283572	PMS (WTT)	160	0.5 - 2.0	G5	> 30

Acknowledgements

The research described in this report was carried out, in part, by the Jet Propulsion Laboratory, California Institute of Technology, under contract with the National Aeronautics and Space Administration.

References

Bakos: 1984, *Astron. J.* **89**, 1740.
Barden, S.C.: 1985, *Astrophys. J.* **295**, 162.
Beckwith, S.V.W., Sargent, A.I., Chini, R.S. and Gusten, R.: 1990, *A. J.* **99**, 924.
Benz, A.O. and Alef, W.: 1991, *Astron. and Astroph.* **252**, L19.

Black, D.C. and Scargle, J.D.: 1982, *Ap. J.* **263**, 854.

Gatewood, G., Stein, J., Joost Kiewiet de Jonge, T., Persinger, T. and Reiland, T.: 1992, *A. J.* **104**, 1237.

Jenskin, L.: 1952, *General Catalogue of Trigonometric Parallaxes.*

Lestrade, J.-F., Rogers, A.E.E., Whitney, A.R., Niell, A.E., Phillips, R.B. and Preston, R.A.: 1990, *Astron. J.* **99**, 1663.

Lestrade, J.-F., Phillips, R.B., Preston, R.A. and Gabuzda, D.C.: 1992, *Astron. Atroph.* **258**, 112.

Phillips, R.B., Lonsdale, C.J. and Feigelson, E.D.: 1991, *Ap. J.* **282**, 261.

Phillips, R.B., Hewitt, J.H., Corey, B.E., Cappallo, R.J., Lonsdale, C.J., Niell, A.E., Preston, R.A., Lestrade, J.-F. and Bookbinder, J.A.: 1988, *American Astronomical Society Bulletin of the* 173rd *Meeting,* held in Boston in January 1988.

Requième, Y. and Mazurier: 1987, *Astron. Astroph.* **89**, 311.

Rogers, A.E.E. *et al.*: 1983, *Science* **219**, 51.

Thompson, A.R., Moran, J.M. and Swenson, G.W.: 1986, *Interferometry and Synthesis in Radio Astronomy,* (Wiley, New York).

van de Kamp, P. and Lippincott, S.L.: 1951, *A. J.* **56**, 49.

Wendker, H.J.: 1987, *Astron. Astroph. Suppl.* **69**, 87.

Whitney, A.R.: 1988, in A.K. Babcock and G.A. Wilkins, ed(s)., *The Earth's Rotation and Reference Frames for Geodesy and Geodynamics, IAU Symposium 128,* (Reidel, Dordrecht), 429.

MILLIMETRE PHOTOMETRY AND INFRARED SPECTROSCOPY OF
VEGA-EXCESS STARS *

R.J. SYLVESTER and M.J. BARLOW

Department of Physics and Astronomy, University College London,
London, UK

and

C.J. SKINNER

Institute for Geophysics and Planetary Physics, Lawrence Livermore National Laboratory,
Livermore, California, USA

Abstract. JCMT millimetre-wave detections have been obtained for 11 Vega-excess stars having spectral types ranging from B9 to K5. UKIRT 10-μm spectra have been obtained for two of the sources, SAO 179815 and SAO 186777. The spectrum of SAO 179815 shows an unusually broad and diffuse silicate emission feature, whilst SAO 186777 shows the unidentified infrared (UIR) features, which are usually attributed to hydrocarbon vibrational modes. The mm photometry, along with optical, IRAS and near-IR photometry (much of the latter recently obtained by the authors), have been used to define the spectral energy distributions of the objects. A number of them show a 1–5 μm excess in addition to the longer wavelength excess. Values of the fractional excess luminosity, L_{dust}/L_*, have been derived from the spectral energy distributions; they exhibit a substantial range, from $\sim 10^{-5}$ up to almost 0.5, the theoretical maximum for a passive optically thick flared disc. Radiative transfer models have been constructed for several sources. One needs a well defined overall energy distribution, 10- and/or 20-μm spectra, and sub-mm and mm photometry in order to significantly constrain the model free parameters (disc density distribution, grain size power-law index, minimum and maximum grain radii).

1. Observations

A sample of Vega-excess stars taken mainly from the list of Walker and Wolstencroft (1988) was observed at 0.8 mm and 1.1 mm using the UKT14 bolometer at the James Clerk Maxwell Telescope in February and August 1992. Walker and Wolstencroft selected sources from the IRAS Point Source Catalog by requiring positional association with the SAO catalogue, a 60-/100-μm flux ratio similar to that of Vega and β Pic, and spatial extension in one or more of the IRAS bands. For the purposes of this paper, Vega-excess stars are defined as main-sequence stars with a far-infrared excess; however, luminosity classifications are not available for four of the 11 detected stars listed in Table I, so they should be treated with caution. Flux upper limits were obtained for a further 9 stars.

The majority of these stars lack published near-infrared photometry, which is important for defining the level of the photospheric continuum so one can determine the amount of longer wavelength excess emission. We have therefore obtained photometry in the J, H, K, L, L' and M bands for 9 Vega-excess candidates using UKT9 at UKIRT.

---

* Paper presented at the Conference on *Planetary Systems: Formation, Evolution, and Detection* held 7–10 December, 1992 at CalTech, Pasadena, California, U.S.A.

Astrophysics and Space Science **212**: 261–270, 1994.
© 1994 *Kluwer Academic Publishers.*

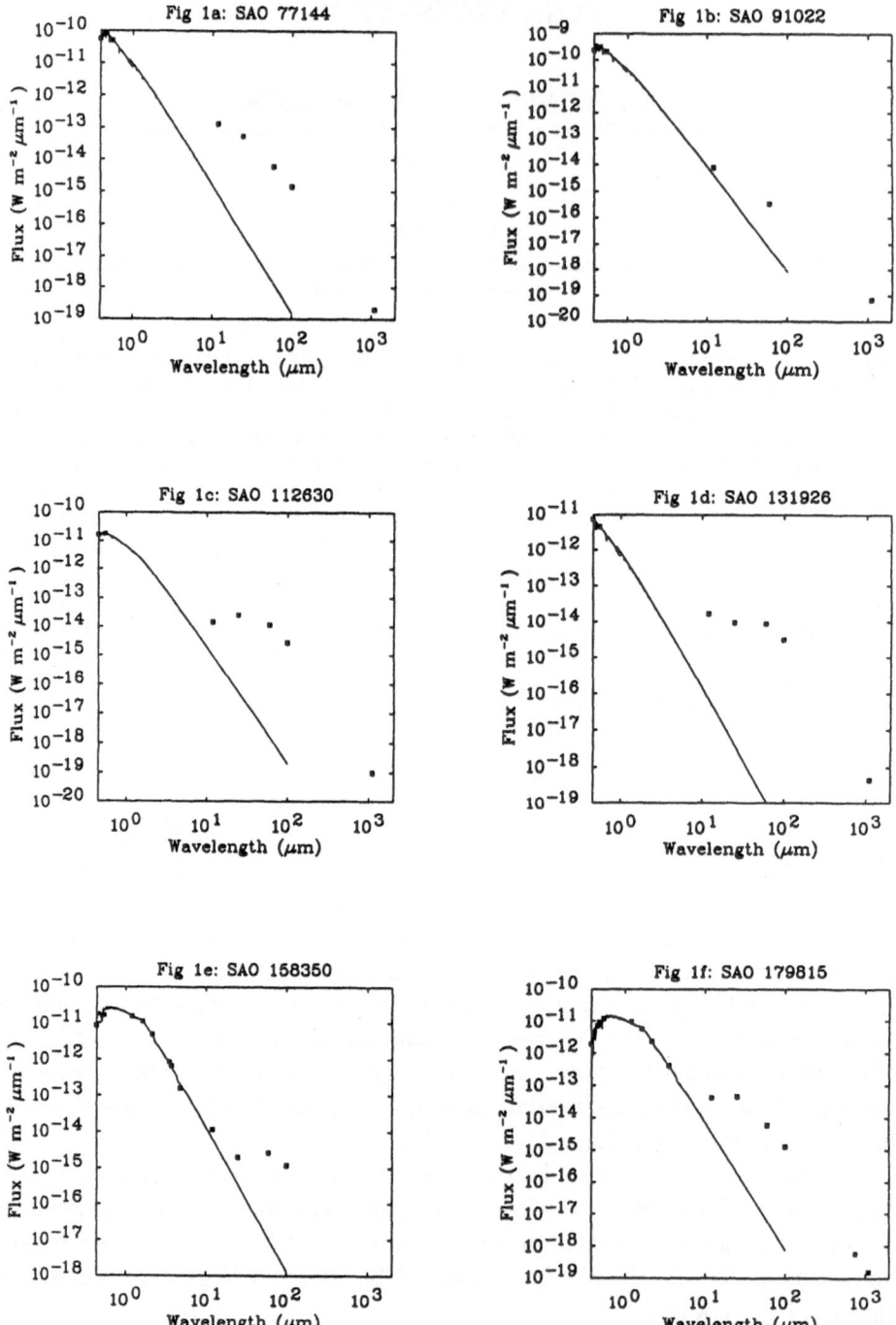

Fig. 1.   Optical, infrared and millimetre photometry of twelve Vega-excess stars. Filled squares: dereddened photometry; solid lines: Kurucz (1991) model atmospheres.

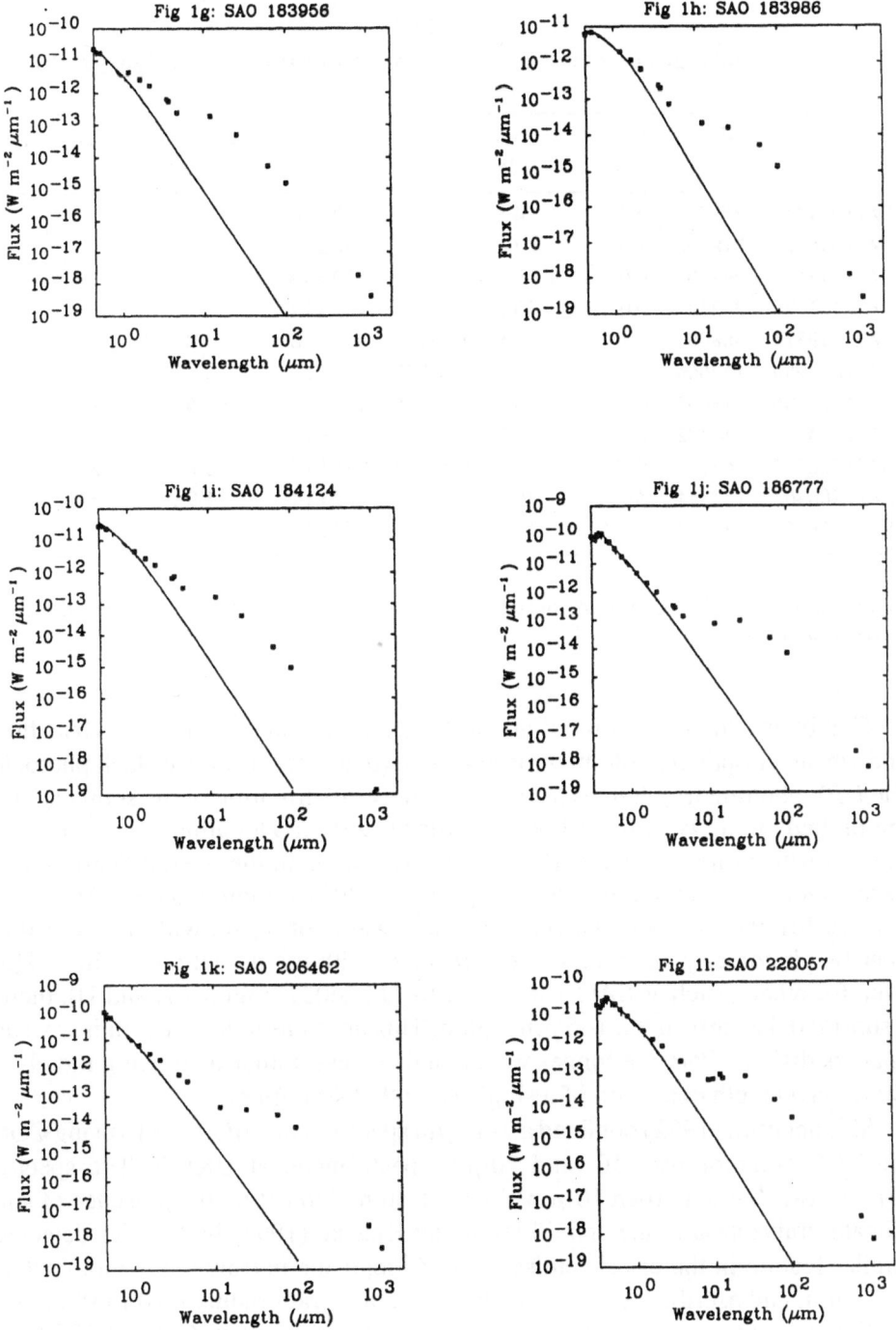

Fig. 1. *Continued.*

TABLE I

JCMT submillimetre fluxes for Vega-excess stars, obtained in February and August 1992.

Star	HD	Sp. type	F_{100} Jy	F_{800} mJy	F_{1100} mJy	$\alpha_{100-800}$	$\alpha_{800-1100}$	$\alpha_{100-1100}$
SAO 77144	35187	A2	4.8		80±10			1.7
SAO 91022	218396	A5V	<2.6		28±11			<1.9
SAO 112630	34700	G0	9.0		39±14			2.3
SAO 131926	34282	A0	10.8		183±17			1.7
SAO 179815	98800	K5V	4.5	111±12	63± 6	1.8	1.8	1.8
SAO 183956	142666	A8V	5.5	351±23	167±17	1.4	2.0	1.5
SAO 183986	143006	G5e	4.8	233±25	114±17	1.5	1.9	1.6
SAO 184124	144432	A9/F0V	3.2		72±12			1.6
SAO 186777	169142	B9V	23.4	610±50	271±19	1.8	2.6	1.9
SAO 206462	135344	A0V	25.7	570±21	209±14	1.9	2.7	2.0
SAO 226057	139614	A7V	13.9	608±26	264±16	1.5	2.3	1.7

Note: F_{100} is the IRAS PSC 100-μm flux density; F_{800} and F_{1100} are the JCMT 800-μm and 1100-μm fluxes respectively.

Combining the near-infrared and millimetre photometry with the IRAS fluxes and whatever optical photometry is available (many of the sources lack photoelectric *UBV* photometry) defines the spectral energy distribution of the sources. These are plotted, along with the relevant Kurucz (1991) model atmospheres in Figure 1, from which one can see that there is a large range in the overall distribution of excess emission. For some sources, e.g., SAO 179815 (Figure 1f) or SAO 158350 (Figure 1e), the excess emission begins longwards of 5 μm, with the optical and near-IR photometry showing close agreement with the model atmosphere. However, for others, such as SAO 186777, or SAO 226057 (Figures 1j and 1l), there is a substantial excess in the 1–5 μm region. This near-infrared excess in many cases appears distinct from the longer wavelength excess, rather than being merely the short-wavelength extension of a single smooth distribution.

Skinner *et al.* (1992) obtained a 7–13 μm spectrum of SAO 179815 using CGS3, the UCL common-user 10- and 20-μm spectrometer at UKIRT. The spectrum (Figure 2a) shows a broad 10-μm silicate feature, indicating the presence of small silicate grains around the star. Telesco and Knacke (1991) had earlier detected a similar feature in the excess emission of β Pic, using narrow-band photometry at $\approx$ 10 μm, and ascribed it to small silicate grains (grain radius <10 μm). Another CGS3 spectrum of a Vega-excess star was obtained by us in October 1992. This B9V star, SAO 186777, shows the unidentified infrared (UIR) bands in its spectrum at 8.7 and 11.3 μm (see Figure 2b), with the 7.8 μm feature also possibly present. These features are usually ascribed to hydrocarbon vibrational modes; this implies

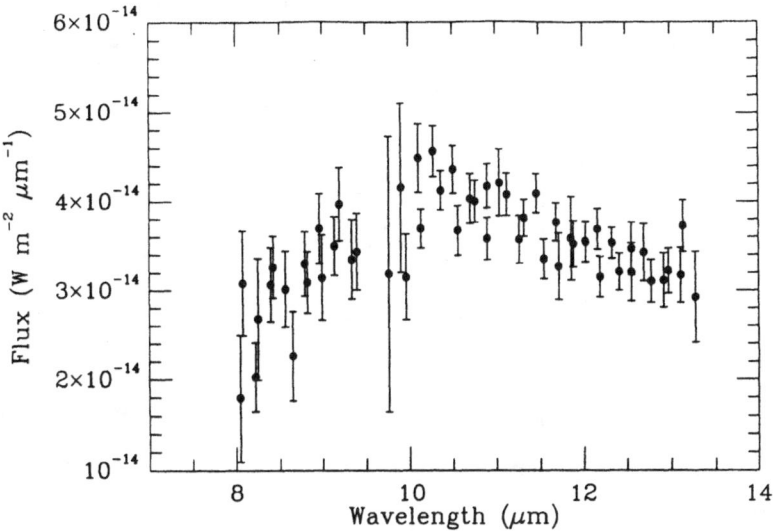

Fig. 2a. CGS3 spectrum of SAO 179815.

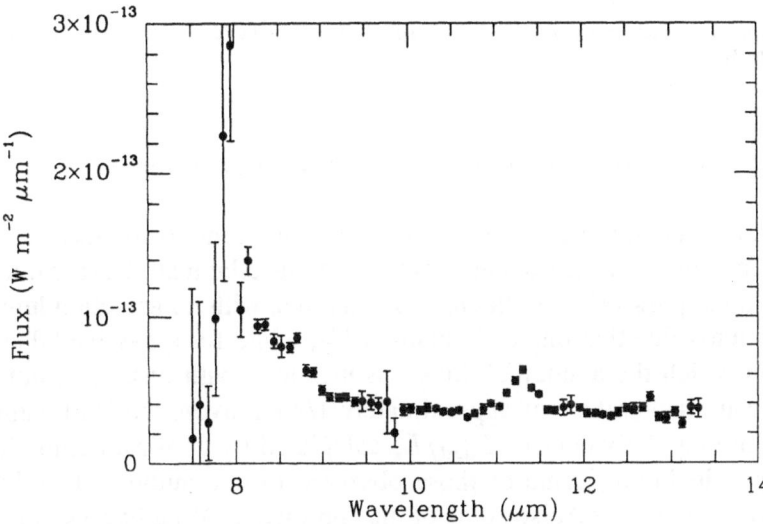

Fig. 2b. CGS3 spectrum of SAO 186777.

that there is a carbonaceous component in the dust around SAO 186777 similar to that found in Galactic reflection nebulae and HII regions.

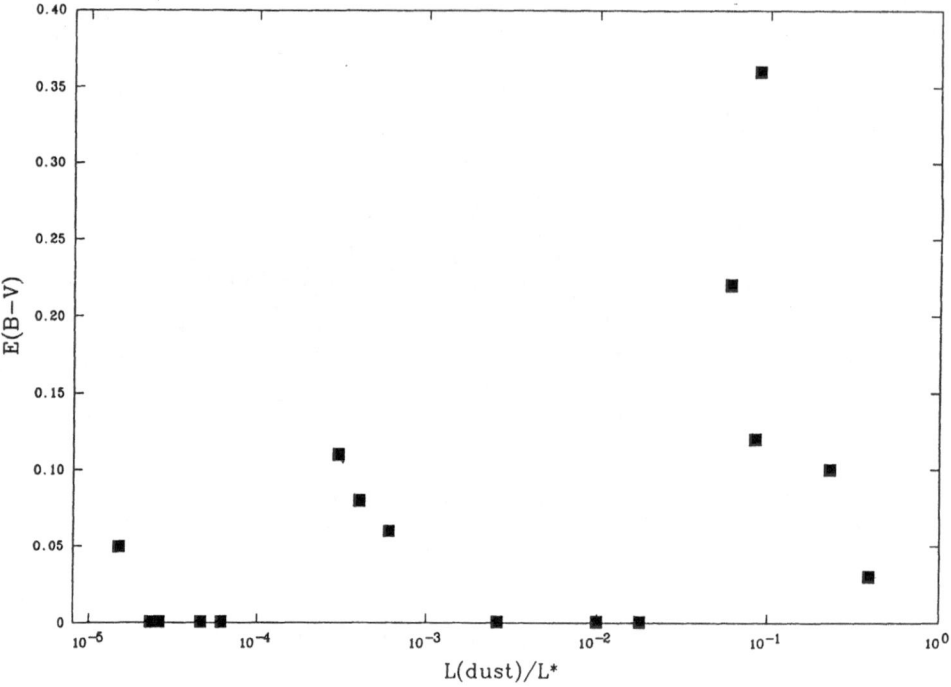

Fig. 3. Graph of the reddening, E($B - V$), versus the fractional excess luminosity, $L_{dust}/L_\star$, for the stars in Table II.

2. Dust Luminosities and Disc Optical Depths

The ratio of the energy radiated by dust to the stellar luminosity, $L_{dust}/L_\star$, gives a useful indication of the optical depth of the circumstellar material (assumed to be a disc). For a flat passively reradiating disc (i.e., one with no accretion luminosity), the maximum value that one can obtain is 1/4, while for a passive "flared" disc, i.e., one in which the azimuthal thickness increases with distance from the star, $L_{dust}/L_\star$ can reach values of approximately 1/2 (Kenyon and Hartmann, 1987). Table II presents the values of $L_{dust}/L_\star$ calculated by us for a number of Vega-excess stars, including some of those observed by the authors at millimetre or near-IR wavelengths, and also some of the "prototype" Vega-excess stars, such as α Lyr itself and β Pic. The values were obtained by subtracting an appropriate Kurucz model atmosphere, normalised to the dereddened optical photometry, from the dereddened overall flux distribution and then dividing the integrated infrared excess flux by the integrated flux of the normalised model atmosphere distribution. Where available, near-IR photometry, either obtained by the authors or taken from the literature, was included in the fit. Some of the resultant ratios are close to, or exceed, the "flat disc" maximum of 0.25 (Table II).

Table II also gives values of the reddening, $E(B - V)$, determined from photoelectric photometry available from the literature using the intrinsic colours of Schmidt-Kaler (1982). Figure 3 shows these values plotted against $L_{dust}/L_\star$ for the stars in Table II. The points lie in the region below a line going from the lower left to upper right of the diagram. This is what one might expect for a sample of nearby stars (i.e., ones with negligible interstellar extinction) with dust discs which are randomly distributed in inclination angle. Those stars which have very optically thin discs, and thus small values of $L_{dust}/L_\star$, would have little circumstellar extinction, even if viewed with the disc edge-on. Stars with a greater amount of dust in the disc, and so a large value of $L_{dust}/L_\star$, would have a wider range of $E(B - V)$ depending on the inclination angle — there would be little extinction if the disc was presented face-on, but considerably more, and hence a large $E(B - V)$, if the star was viewed through an edge-on disc.

Five of the stars for which we obtained JCMT data do not have photoelectric UBV photometry in the literature, making it impossible to obtain good estimates of $L_{dust}/L_\star$ and $E(B - V)$.

3. Radiative Transfer Modelling of Disc Properties

Our observations of Vega-excess stars are being modelled with an axisymmetric radiative transfer code. The disc is treated as being optically thin at present, hence there are no shadowing or self-heating effects taken into account. At present, only silicate grains are treated, but the code can incorporate other grain species too.

The temperature of the dust grains at any position in the disc is found by applying the condition of radiative equilibrium. The model calculates the flux from the object at 69 wavelengths, ranging from 0.3 μm to 2 mm. Up to fifteen grain sizes can be used, of radius $a = 0.005$ μm (50Å) to 5 cm.

The free parameters in the model are

1. γ—the index for the grain size distribution $n(a) \propto a^{-\gamma}$
2. β—the index for the dust density distribution $\rho(r) \propto r^{-\beta}$
3. The maximum and minimum grain sizes
4. The inner disc radius
5. The outer disc radius
6. The total dust mass

The parameters are varied to give the best agreement with the photometry and spectroscopy. Table III gives values for the best fit model parameters for two Vega-excess stars.

Figure 4a shows the results of modelling SAO 179815. The CGS3 spectrum, rescaled to the IRAS point, is included in the plot (small filled squares). The model fit (Table III) is very good, showing close agreement with all the photometry points. The only parameter not completely constrained is the maximum grain size, for which there is a lower limit, but no upper limit—longer wavelength millimetre photometry is needed.

TABLE II

The fractional IR excess luminosity L_{dust}/L_* for a number of Vega-excess stars.

Star	HD	Sp. Type	L_{dust}/L_*	$E(B-V)$	D (pc)
α Lyr	172167	A0V	2.3×10^{-5}	0.00	7.5*
β Pic	39060	A5V	2.6×10^{-3}	0.00	17*
ε Eri	22049	K2V	2.6×10^{-5}	0.00	3.3*
τ^1 Eri	17206	F5/F6V	4.6×10^{-5}	0.00	15*
ζ Lep	38678	A3V	$\leq 6.1 \times 10^{-5}$	0.00	25*
SAO 140789	141569	A0Ve	0.010	0.00	220
SAO 179815	98800	K5V	0.084	0.12	18
SAO 184124	144432	A9/F0V	0.23	0.10	108
SAO 186777	169142	B9V	0.087	0.36	230
SAO 208591	155826	G0V	0.018	0.00	25*
SAO 226057	131964	A7V	0.39	0.03	150
HR 890	18537	B7V	4×10^{-4}	0.08	130
SAO 77144	35187	A2	0.06	0.22	150
SAO 75532	16908	B3V	2×10^{-5}	0.05	160
SAO 111845	28375	B3V	3×10^{-4}	0.11	230
SAO 147886	9672	A1V	6×10^{-4}	0.06	76

Notes: Distances marked with * are derived from trigonometric parallaxes; the others by spectroscopic parallax. Extinction corrections are made assuming $A_V = 3.1 \times E(B-V)$. This may underestimate the extinction and overestimate the distances, since for circumstellar extinction by a grain size distribution including large and small grains, the value of R ($= E(B-V)/A_V$) may well be greater than 3.1.

The five last mentioned stars lack near-IR photometry; consequently the values of L_{dust}/L_* shown in the table could be underestimates of the true values.

TABLE III

Parameters for the models described in the text. (a) Model for SAO 179815. (b) Model for the excess of SAO 226057 from ~8 μm longwards. (c) A 'hot dust' model for the 1–5 μm excess of SAO 226057.

	Star	γ	β	R_{in} AU	R_{out} AU	a_{min} μm	a_{max} μm	M_{disc} $M_\odot$
(a)	SAO 179815	2.0	3.0	1.13	320	0.005	1000	9.0×10^{-9}
(b)	SAO 226057	1.0	2.0	8.00	6400	0.005	500	1.2×10^{-5}
(c)	SAO 226057	2.0	4.0	0.09	320	0.005	500	3.0×10^{-11}

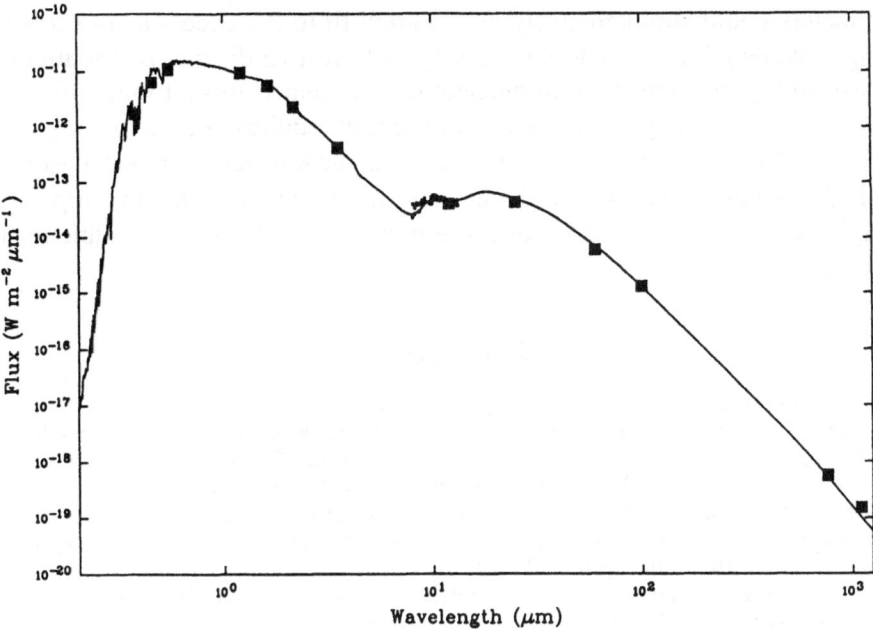

Fig. 4a. The best-fit model for SAO 179815. Large filled squares: dereddened photometric measurements; small filled squares: CGS3 spectrum normalised to the IRAS 12-μm point; solid line: model (a) from Table III.

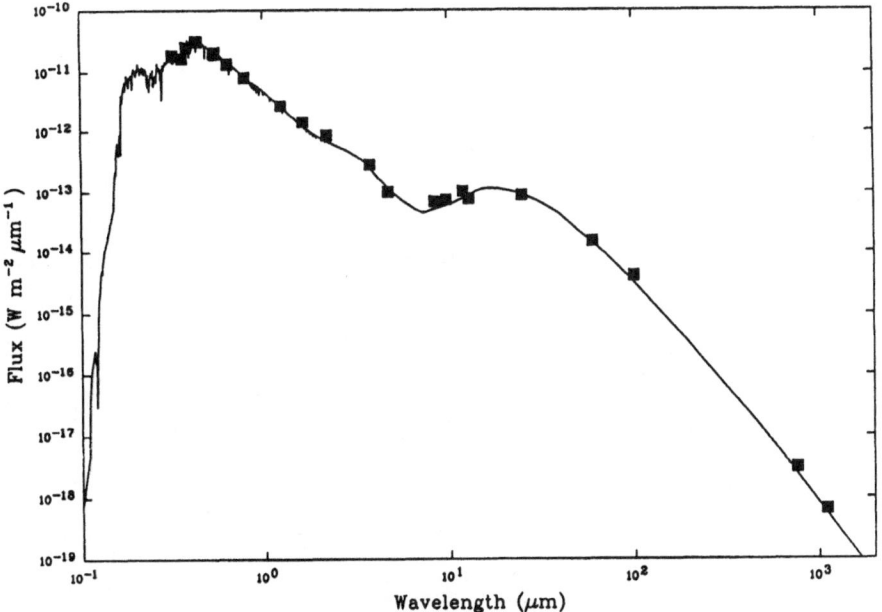

Fig. 4b. Model for SAO 226057. Filled squares: dereddened photometric measurements; solid line: sum of models (b) and (c) from Table III. Model (b) accounts for the excess emission from $\sim 8\mu$m longwards, while model (c) is required to fit the excess in the 1–5 μm region.

Figure 4b shows a model fit to SAO 226057. It was found that no single set of parameters could simultaneously give a good fit to the excess at 1–5 μm and the long-wavelength data. It was necessary to have two distinct distributions of grain-size and spatial density to fit the photometry from 1 μm–1.1 mm. Fitting the observed energy distribution in the 1–5 μm region requires grain temperatures of the order of 1000 K, about a factor of 5 higher than those needed to model the excess in the IRAS bands. Alternatively, the near-IR excess could be due to temperature spiking in very small grains, such as those hypothesised to be responsible for the UIR bands.

References

Kenyon, S.J. and Hartmann, L.: 1987, *Astrophys. J.* **323**, 714.

Kurucz, R.L.: 1991, in: A.G. Davis Philip, A.R. Upgren and K.A. Janes (eds.), Proceedings of the Workshop on *Precision Photometry: Astrophysics of the Galaxy*, held at Union College, Schenectady, New York, October 3–5, 1990, Schenectady: L. Davis Press), p. 27.

Schmidt-Kaler, Th.: 1982, in: K. Schaifers and H. H. Voigt (eds.), *Landolt-Börnstein, Numerical Data and Functional Relationships in Science and Technology, Group VI, Astronomy, Astrophysics and Space Research*, **2b**, Berlin: Springer-Verlag.

Skinner, C.J., Barlow, M.J. and Justtanont, K.: 1992, *Mon. Not. R. Astr. Soc.* **255**, 31P.

Telesco, C.M. and Knacke, R.F.: 1991, *Astrophys. J. Lett.* **372**, L29.

Walker, H.J. and Wolstencroft, R D.: 1988, *Publs Astr. Soc. Pacif.* **100**, 1509.

LONG, ACCURATE TIME SERIES MEASUREMENTS OF RADIAL VELOCITIES OF SOLAR-TYPE STARS *

R.S. MCMILLAN, T.L. MOORE, M.L. PERRY and P.H. SMITH

Lunar and Planetary Laboratory
University of Arizona, Tucson, Arizona, USA

Abstract. We have been measuring changes in the radial velocities (RV's) of solar-type stars to search for gravitational perturbations by planets. We transmit violet starlight through a Fabry-Perot etalon interferometer and sense changes in Doppler shift from changes in the fluxes of light on the slopes of stellar absorption lines. Our data now span 6 years. Our observations of the Sun showed earlier that both our technique and the profiles of solar photospheric violet absorption lines can be stable enough to reveal planetary perturbations. We now carry this validation to the spectra of other near-solar-type stars. Annual averages of our RV's of σ Draconis and β Virginis are stable to ± 6 m s^{-1}. The slope of our five-year series of RV's of ξ Bootis A is consistent with the star's well-determined visual astrometric orbit about ξ Bootis B. The Fabry-Perot technique of Doppler shift measurement is fully capable of detecting perturbations due to planets with masses and orbits similar to those of Jupiter.

1. Highlights of the Program to Date

McMillan *et al.* (1985, 1986, 1990) have described the instrument and technique, which is optimized for sensing small changes in radial velocity (RV) on long time scales. Our discovery of the 2-day pulsation of Arcturus (Smith *et al.*, 1987) was accurate to ± 20 m s^{-1} and was confirmed in detail by Cochran (1988). We discovered a 3-hour pulsation by β Gem (Smith and McMillan, 1987) and an RV drift due to a binary companion to ϵ Cygni A (McMillan *et al.*, 1992). McMillan and Smith (1987) strengthened Morbey and Griffin's (1987) refutation of the suspected binarity of η Cas A and McMillan *et al.* (1993) have found long-term stability better than ± 4 m s^{-1} in the apparent RV of the integrated disk of the Sun.

Table I shows the numbers and temporal extent of our observations. Since the phenomena we are seeking are very slow, each season's average must be accurate. We concentrate rigorously calibrated observations on a relatively small number of targets because most of our sources of error are random and can be reduced to very low values by repetition. The numbers of observations listed do not include some 400-odd observations used to calibrate the slopes of the spectra at the wavelengths of the transmitted interference orders. At the time of this writing, not all the observations have been reduced. However, we are now able to show reliable results for three more stars on our program of planet searching.

---

* Paper presented at the Conference on *Planetary Systems: Formation, Evolution, and Detection* held 7–10 December, 1992 at CalTech, Pasadena, California, U.S.A.

Astrophysics and Space Science **212**: 271–280, 1994.
© 1994 *Kluwer Academic Publishers*.

TABLE I

Quantity and Span of the Data, 1987 Jan 13–1993 Apr 30.[*]

HR	Name	B	Sp	#Obs	DT	Remarks
219A	Eta Cas A	4.0	F9 V	189	6	D & M const.
799	Theta Per A	4.6	F8 V	92	4	A & L const. D & M const.
937	Iota Per	4.7	G0 V	74	6	A & L const. CFHT. D & M const.
1084	Epsilon Eri	4.6	K2 V	151	5	CFHT var? Allegheny.
1729	Lambda Aur	5.3	G2IV–V	139	3	A & L const. D & M const.
2047	Chi[1] Ori	5.0	G0 V	151	6	SB (CFHT).
2943	Alpha CMi	0.8	F5 IV–V	142	5	CFHT.
2990	Beta Gem	2.1	K0 III	690	6	pulsations; CFHT, var.
3775	Theta UMa	3.6	F6 IV	148	6	M & G const.
4112	36 UMa A	5.4	F8 V	182	6	McCarthy prog. A & L const. D & M const. CFHT var?
4540	Beta Vir	4.2	F9 V	163	6	McCarthy prog. CFHT. Allegheny. A & L const.
4983	Beta Com	5.0	G0 V	164	6	CFHT const. D & M const.
5340	Alpha Boo	1.2	K2 III	3173	5	Irreg. puls.
5544	Xi Boo A	5.3	G8 V	168	6	CFHT var? Allegheny. Beavers const. D & M const.
5868	Lambda Ser	5.0	G0 V	130	5	M & G const. D & M const.
7462	Sigma Dra	5.5	K0 V	161	5	CFHT, const. Beavers "const."
7949	Epsilon Cyg A	3.5	K0 III	278	5	McMillan et al. 1992; drift
7957	Eta Cep	4.2	K0 IV	74	5	CFHT, const.
8969	Iota Psc	4.6	F7 V	58	5	A & L "const."
8974	Gamma Cep	4.2	K1 III	83	5	CFHT, puls., SB.
Sun	(via Moon)	"2"	G2 V	611	6	Integrated disk. McMillan et al. 1993; const. ± 8 m s^{-1}

[*]Notes: A & L: Abt and Levy (1976) RV survey. Allegheny: Gatewood's (1987) astrometric observing program at the Allegheny Observatory, University of Pittsburgh. "Beavers": Findings from the RV survey of Beavers et al. (1979). CFHT: Canada—France—Hawaii Telescope RV program of Campbell et al. (1988). (The current PI is Prof. Gordon Walker of UBC.) "const": No variation greater than observational errors was found by the stated investigator(s) or institution(s). "Drift": A drift of RV with time is observed. "DT": Time interval in years between the first and most recent observations. "D & M": RV study to ± 300 m s^{-1} by Duquennoy and Mayor (1991). McCarthy: The near-infrared speckle imaging programs of McCarthy et al. (1985). M & G: Morbey and Griffin's (1987) study of the reality of A & L's spectroscopic binary findings. "puls": A variation of RV due to stellar pulsation is observed. SB: Known spectroscopic binary. var: Variations of RV.

2. Results on Stars

2.1. PURPOSE

We have already validated the Fabry-Perot technique over a long time interval on sunlight reflected off the surface of the Moon (McMillan *et al.*, 1993). However, the lunar crater that we observe at full phase is bright and the excursion of its apparent RV is small, compared to stars on our program. In this paper we present similar validation of the long-term stability and sensitivity of our technique on three stars, as the next logical step in our campaign of exhaustive characterization of our method of detecting planets.

2.2. BETA VIRGINIS

Our RV series on β Virginis (F9 V; HR 4540) is illustrated in Figure 1. Prior to 1991, each observation was a 1-hr exposure. Since the input end of our optical fiber projects to a 3-arcsecond diameter on the sky and the drive and polar axis alignment of our telescope are very good, little guiding is necessary when observations are made near the meridian and the wind is light. A photomultiplier exposure meter monitors the zeroth order of light diffracted off our cross-disperser, providing the observer feedback in real time to the manual TV-assisted guiding. In 1991 we realized this star was bright enough that we could take 30-minute exposures with no significant degradation of precision, allowing more observations per season. Table II is a compact list of the observations averaged by season. All observations received equal weight in all the seasonal averages in this paper, and the listed Modified Julian Date (MJD) is the average of the observation times. As always, no celestial sources or iterative procedures were used to reduce data to a consistent (arbitrary) zero point. Although our interferometer provides absolute vacuum wavelengths, we have not attempted to determine absolute RV's of our target stars. Such an undertaking would, at best, be very labor intensive because of the need to identify many line species. Uncertainties in the wavelengths of absorption lines in the rest frame of the star would probably thwart any attempt to determine absolute RV's to any accuracy approaching the stability of our interferometer.

Observations of β Vir show a short-term standard deviation of ± 23 m s^{-1}. The standard deviation of the seasonal averages is ± 6.1 m s^{-1} and the slope of the series is $+0.47 \pm 1.1$ m s^{-1} yr^{-1}, comparable to our expectations of instrumental performance (McMillan *et al.*, 1993). Year-to-year variations by β Vir, if present, must be much less than ± 6 m s^{-1}.

The six-year series of 38 RV's measured by Campbell *et al.* (1988) does not overlap our data in the time domain, but it is appropriate to compare with our observations in a general way. Their slope of -3.9 ± 1.3 m s^{-1} yr^{-1} is not necessarily inconsistent with our flat series; a very slow RV curve could slope downward for 6 years and level off for 5 years. In any case, the "quietness" of our observations of β Vir demonstrates that our observations "average down" to a very consistent zero

TABLE II

Radial Velocity Averages for β Virginis, By Season.*

Date Interval	N	Av. MJD	Av. RV	± S.E.M.
87 Jan 22–87 Jun 02	19	46881.907	-21.3	5.8
87 Dec 28–88 May 25	19	47236.896	-4.6	5.3
89 Jan 17–89 Apr 20	18	47596.706	-19.3	3.8
89 Dec 20–90 May 14	19	47966.884	-14.4	6.2
90 Dec 03–91 May 30	29	48346.785	-10.7	3.8
92 Jan 14–92 Apr 22	34	48689.456	-15.6	4.0

*Notes:
"N" = Number of observations.
"RV" = Radial velocity, referred to a zero point arbitrarily
    assigned to one of the observations.
"S.E.M." = Standard error of the mean.
"MJD" = Modified Julian Date = Julian Date minus
    2 400 000.5.

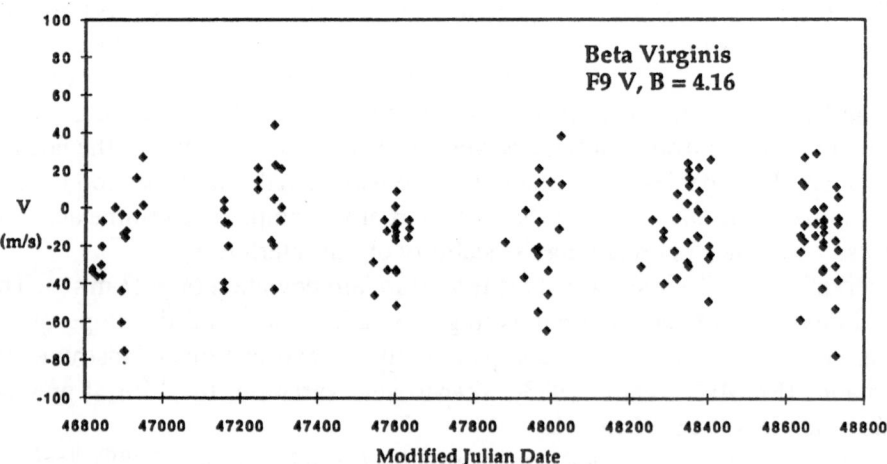

Fig. 1.   Radial Velocity Observations of β Virginis vs. Modified Julian Date. MJD = Julian Date minus 2 400 000.5. The error of a typical observation point is ±23 m s^{-1}. The velocity scale is referred arbitrarily to one of the observations.

TABLE III

Radial Velocity Averages for Sigma Draconis.

Date Interval	N	Av. MJD	Av. RV	± S.E.M.
87 Apr 15–87 Oct 09	28	46967.054	+3.9	4.6
88 May 05–88 Sep 26	18	47340.830	-1.8	3.7
89 Apr 20–89 Oct 20	29	47722.220	+5.3	4.5
90 Apr 12–90 Dec 05	16	48098.078	+0.1	4.7
91 Mar 31–91 Nov 20	40	48440.292	-9.4	3.4
92 Apr 14–92 Apr 22	5	48730.075	+3.3	13.0

point from year to year. It is extremely unlikely that changes in instrumental zero point and stellar variations would conspire to produce a spurious nondetection.

2.3. SIGMA DRACONIS

Our RV series on σ Draconis (K0 V; HR 7462) is illustrated in Figure 2 and the seasonal averages are listed in Table III. Individual observations were one-hour integrations and are good to ± 23 m s^{-1}. Since this is the faintest star on our program so far, some statistics are worth quoting. The total number of photometrically usable Fabry-Perot interference orders is 660, distributed over 18 echelle orders between 4250 and 4800 A. Of these, 191 orders have slopes with absolute values of at least 4% per (km s^{-1}) and 34 have slopes with absolute values of at least 12% per (km s^{-1}). The steepest slope is 32.6% per (km s^{-1}). Out of the 191 "RV-active" orders, those with slopes we have adequately calibrated number 158, with an intensity-weighted average absolute value of 7.94% per (km s^{-1}). Below we show there is a fortuitous astrophysical advantage to concentrating RV determinations on the steepest parts of the line profiles. Comparing these statistics with those we obtained in our short exposures on the bright lunar crater (McMillan *et al.*, 1993), we see that the later spectral type of σ Dra provides more and steeper slopes, partly compensating for its fainter magnitude.

In an average exposure on σ Dra approximately 113 of the 158 calibrated RV-active orders have adequate signal-to-noise ratios. An average "RV-active" interference order has approximately 60,000 electron-hole pairs of CCD signal accumulated in some tens of pixels. Allowing for all forms of detector noise and background noise, the typical signal-to-noise ratio of such an order is about 50:1.

Multi-year variations of σ Dra, if present, must be less than ± 5 m s^{-1}. Since this is the faintest star currently on our program, the stability of this series is a strong validation of our technique. Our slope of -2.4 ± 1.2 m s^{-1} yr^{-1} is consistent with Campbell *et al.*'s (1988) slope of -1.2 ± 1.4 m s^{-1} yr^{-1}. Periodogram analyses of

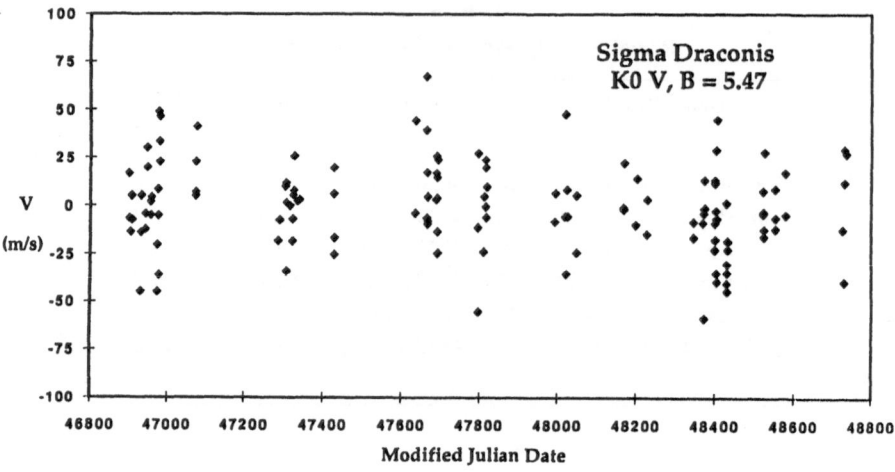

Fig. 2. Radial Velocity Observations of σ Draconis. A typical observation has an error of ± 23 m s^{-1}.

our observations of β Vir and σ Dra show no significant coherent variations within the accessible frequency domain.

High resolution profiles of red absorption lines of σ Dra obtained by Gray et al. (1992) showed no change in the velocity span of granulation greater than ± 3 m s^{-1} through the interval July 1984 – September 1992. However, those are not measurements of the Doppler shift of the star and are based on relative wavelengths of parts of line profiles that we avoid, so the stability observed by Gray et al. (1992) is at best a weak and indirect confirmation of our finding of RV stability.

2.4. XI BOOTIS A

Figure 3 and Table IV show our observations of ξ Bootis A (G8 V; HR 5544A). All observations were one hour in length and received equal weight. The slopes of all three data series are summarized in Table V. The slope of our series is consistent with Wielen's (1962) unusually reliable astrometric orbit of ξ Boo A about ξ Boo B; the seasonal averages scatter about the line ± 6.4 m s^{-1}, again consistent with expectations for this technique.

Toner and Gray (1988) discovered a periodic variation in the shapes of strong absorption lines in the red spectrum of this star. Figure 4 is a periodogram of our data including the frequency range appropriate to their 6.43-day period. One thousand periodograms of normally distributed artificial data with the same time sampling and variance as our observations showed peaks as high as 11.4 on the ordinate scale of Figure 4, so we regard the trace in Figure 4 as insignificant "grass."

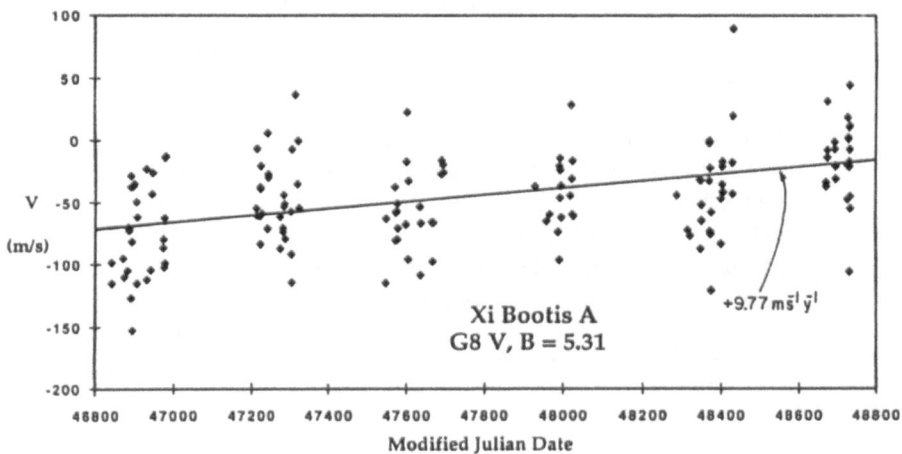

Fig. 3. Radial Velocity Observations of ξ Bootis A. A typical observation has an error of ±36 m s⁻¹.

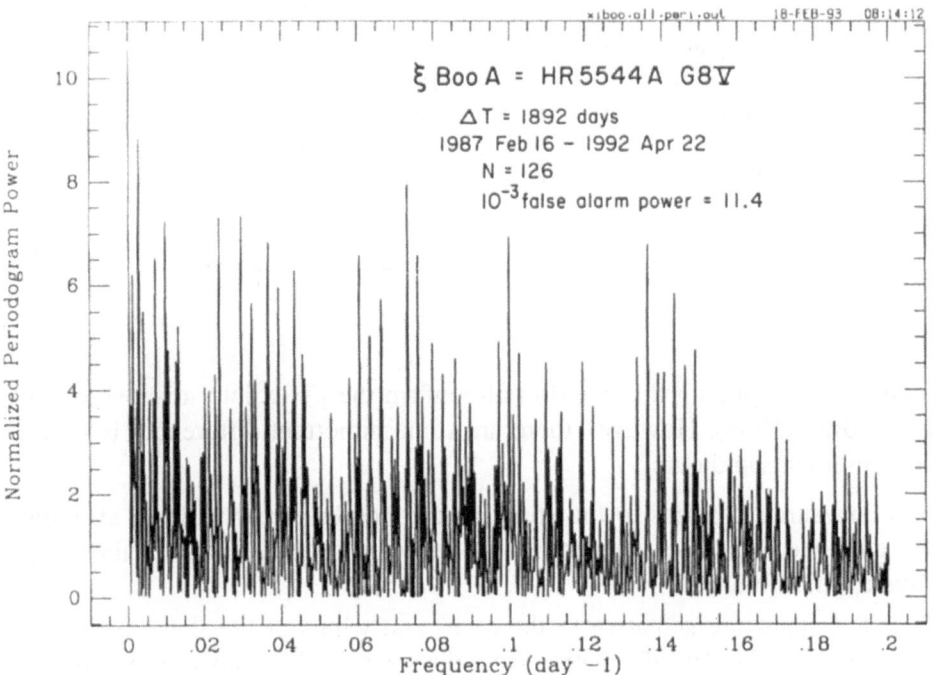

Fig. 4. Periodogram of the observations of ξ Bootis A.

TABLE IV

Radial Velocity Averages for ξ Bootis A.

Date Interval	N	Av. MJD	Av. RV	± S.E.M.
87 Feb 16–87 Jul 05	29	46918.431	-74.2	7.1
88 Feb 23–88 Jun 14	29	47267.994	-47.5	6.1
89 Jan 22–89 Jun 18	24	47617.663	-55.9	6.7
90 Feb 07–90 May 15	17	47995.604	-41.8	7.0
91 Apr 25–91 Jun 25	15	48402.388	-29.4	11.2
92 Apr 14–92 Apr 22	12	48731.171	-20.5	11.3

TABLE V

Line-of-Sight Accelerations.

Star	N	Span (days)	Av. RV $(m\,s^{-1})$	Slope $(m\,s^{-1}\,yr^{-1})$
Beta Vir	138	1917	-14.14 ±1.9	+0.47 ±1.1
Sigma Dra	136	1834	-0.95 ±1.9	-2.36 ±1.2
Xi Boo A	126	1892	-49.74 ±3.5	+9.77 ±2.1

It may seem surprising that we do not confirm the periodicity discovered by Toner and Gray (1988). However, there are some important differences between their measurements and ours:

1. They measured changes in the spans of the line bisectors with no absolute RV reference, while we measure RV's with respect to an absolute laboratory wavelength reference.
2. Their RV spans are sensitive to the cores and wings of the lines, while we derive most of the RV information from the steep flanks of absorption lines.
3. They deliberately selected strong lines while our technique samples a large number of unselected lines.
4. Their lines are in red light while ours are in the violet.

5. The data they used for period determination was in 1986, which does not overlap our series.

These differences allow several interpretations for our non-detection of their 60 m s^{-1} RV variation:

(a) The cores and wings of the lines could be shifting, leaving the steep flanks at rest. Dravins (1985) recommended the steep flanks of moderate-strength photospheric lines to minimize the effects of variations in convection on absolute RV work.

(b) The modulating effect of Toner and Gray's starpatch should be reduced by the greater limb darkening in violet light compared to red.

(c) Zirin (1988) describes greater granular contrast in blue light compared to that in the red. Granular contrast is greater in the blue and violet because on the blue side of the Planck curve the brightness varies exponentially with temperature, while in the red it varies linearly. Greater granular contrast means the violet line profiles are dominated by light from the hot, rising gases, so that changes in the relative luminosity of rising and falling streams should have less of an effect on line profiles in the violet.

A fourth possibility (d), that the starpatch has disappeared since 1986, seems unlikely because it was large and long-lived. Regardless of the relative importance of (a), (b), and (c), it appears RV measurements are best made on the steep flanks of average absorption lines in the violet.

3. Conclusions and Plans

Our instrument, calibration technique, and observing strategy show long-term stability well under $\pm 10 \text{ m s}^{-1}$. These demonstrations we believe to be original. The remaining weakness of our measurement approach is the comparatively large short-term random errors, which force us to commit many nights to a handful of targets. This noise originates in roughly equal amounts from photon statistics and detector noise.

We will continue to make an average of 24 observations per solar-type star per year. By the end of 1995 we will have a time series almost 9 years long for many of the targets on our original list. This should be long enough to make some statements about the presence or absence of Jupiter-class planets around that short list of stars.

We are designing a new RV instrument that will be systematically "quieter" and more photon-efficient than our present one. Better temperature control, a simpler tilt mechanism, faster optics, a state-of-the-art detector, and a larger telescope should bring many more target stars into our reach.

Acknowledgements

This work is funded by NASA grant NAGW-1283, NSF grant AST-9016572, and private donations to Prof. Tom Gehrels' Spacewatch Project. Telescope time is allocated by the Director of the Steward Observatory. We are grateful for the contributions of J.E. Frecker, T. Gehrels, R.L. James, W.J. Merline, J.V. Scotti, the late K.M. Serkowski, and M.S. Williams.

References

Abt, H.A. and Levy, S.G.: 1976, *Ap J Suppl.* **30**, 273.

Beavers, W.I., Eitter, J.J., Ketelsen, D.A. and Oesper, D.A.: 1979, *PASP* **91**, 698.

Campbell, B., Walker, G.A.H. and Yang, S.: 1988, *Ap J* **331**, 902.

Cochran, W.D.: 1988, *Astrophys. J.* **334**, 349.

Dravins, D.: 1985, in: A.G. Davis Philip and D.W. Latham, ed(s)., *IAU Colloq. 88, Stellar Radial Velocities*, Schenectady: L. Davis Press, 311.

Duquennoy, A. and Mayor, M.: 1991, *Astron. and Astrophys.* **248**, 485.

Gatewood, G.D.: 1987, *A J* **94**, 213.

Gray, D.F., Baliunas, S.L., Lockwood, G.W. and Skiff, B.A.: 1992, *Astrophys. J.* **400**, 681.

McCarthy, D.W., Probst, R.G. and Low, F.J.: 1985, *Astrophys. J. Lett.* **290**, L9.

McMillan, R.S., Moore, T.L., Perry, M.L. and Smith, P.H.: 1993, *Astrophys. J.* **403**, 801.

McMillan, R.S. and Smith, P.H.: 1987, *PASP* **99**, 849.

McMillan, R.S., Smith, P.H., Frecker, J.E., Merline, W.J., and Perry, M.L.: 1985, in: A.G. Davis Philip and D.W. Latham (eds.), *Stellar Radial Velocities*, IAU Colloq. 88, L. Davis Press, Schenectady, p. 63.

McMillan, R.S., Smith, P.H., Frecker, J.E., Merline, W.J. and Perry, M.L.: 1986, *Proc. SPIE* **627**, 2.

McMillan, R.S., Smith, P.H., Perry, M.L. and Moore, T.L.: 1992, *PASP* **104**, 1173.

McMillan, R.S., Smith, P.H., Perry, M.L., Moore, T.L., and Merline, W.J.: 1990, *Proc. SPIE* **1235**, 601.

Morbey, C.L. and Griffin, R.F.: 1987, *Ap J* **317**, 243.

Smith, P.H. and McMillan, R.S.: 1987, *The Impact of Very High Signal: Noise Spectroscopy on Stellar Physics*, IAU Symposium 132, D. Reidel Publ. Co., Dordrecht, p. 291.

Smith, P.H., McMillan, R.S. and Merline, W.J.: 1987, *Astrophys. J. Lett.* **317**, L79.

Toner, C.G. and Gray, D.F.: 1988, *Astrophys. J.* **334**, 1008.

Wielen, R.: 1962, *Astron. J.* **67**, 599.

Zirin, H.: 1988, *Astrophysics of the Sun*, 2nd ed., Cambridge University Press New York, pp. 33 and 122.

A HIGH-PRECISION RADIAL-VELOCITY SURVEY

FOR OTHER PLANETARY SYSTEMS *

WILLIAM D. COCHRAN and ARTIE P. HATZES

McDonald Observatory
The University of Texas at Austin
Austin, Texas, USA

Abstract. The precise measurement of variations in stellar radial velocities provides one of several promising methods of surveying a large sample of nearby solar type stars to detect planetary systems in orbit around them. The McDonald Observatory Planetary Search (MOPS) was started in 1987 September with the goal of detecting other nearby planetary systems. A stabilized I_2 gas absorption cell placed in front of the entrance slit to the McDonald Observatory 2.7 m telescope coudé spectrograph serves as the velocity metric. With this I_2 cell we can achieve radial velocity measurement precision better than $10\,\mathrm{m\,s^{-1}}$ in an individual measurement. At this level we can detect a Jupiter-like planet around a solar-type star, and have some hope of detecting Saturn-like planets in a long-term survey. The detectability of planets is ultimately limited by stellar pulsation modes and photospheric motions. Monthly MOPS observing runs allow us to obtain at least 5 independent observations per year of the 33 solar-type (F5–K7) stars on our observing list. We present representative results from the first five years of the survey.

1. Introduction

At present, the only confirmed example of a planetary system is the one surrounding our Sun. There is tantalizing evidence that other stars (and perhaps even pulsars) may possess planetary systems, but as yet no unambiguous evidence exists that planets orbit other stars. In order to understand fully the formation of our solar system, we must first find more than one example of a planetary system to study. It would be risky to base elaborate theories of the general problem of solar system formation on just the very limited sample of the solar system surrounding our Sun. We must determine whether the formation of planetary systems is a unique event, a rare phenomenon, or a common natural result of the process of star formation. By intercomparison of different planetary systems around stars of various masses, ages, and compositions, we can begin to develop a much deeper fundamental understanding of the process of planet formation. In this manner we can integrate our understanding of planet formation into the more general problem of star formation.

The McDonald Observatory Planetary Search (MOPS) program is designed to detect planetary systems in orbit around nearby solar-type stars. The program is based on a search for radial velocity variations in stars arising from the presence of planetary companions. The MOPS has been in operation since 1987 September, and now has sufficient time baseline that it is starting to produce extremely intriguing results. The survey is meeting all of its design goals on measurement

---

* Paper presented at the Conference on *Planetary Systems: Formation, Evolution, and Detection* held 7–10 December, 1992 at CalTech, Pasadena, California, U.S.A.

Astrophysics and Space Science **212**: 281–291, 1994.

precision, stability, and observational coverage of program stars. Details of the survey operations, reduction procedures, and sample results from the first five years of operation are presented in the following sections.

2. Observational Goals and Constraints

The detection of planetary systems around other stars is a difficult and challenging problem. We have adopted the radial velocity technique for the MOPS program because it allows Jovian-mass planets to be detected around solar-type stars using well proven ground-based astronomical techniques. While the radial velocity method may not be the most sensitive technique for all cases, we feel it will give scientifically interesting results in the shortest possible time.

Achieving the necessary long term radial velocity precision is an exacting task. The velocity of the Sun around the Jupiter-Sun barycenter averages $12.4 \, \mathrm{m \, s^{-1}}$. Lacking any better knowledge, we will consider our solar system to be typical and will set our observational goal as the detection of such systems around other stars. The full $12.4 \, \mathrm{m \, s^{-1}}$ velocity semi-amplitude is seen only in the plane of the orbit (fortunately the most probable orientation). To make an unambiguous identification of such a system over a large range of possible orbital inclinations, and to have any hope of unraveling the possible presence of more than one planet, we need to aim for a routine precision approaching $5 \, \mathrm{m \, s^{-1}}$ in an individual measurement.

What types of systems will we be able to detect with this precision? If we assume nearly circular orbits (as is the case with our solar system) the observed radial component of the velocity of the star around the barycenter is given by

$$V_\star = \frac{m_p \sin i}{M_\star + m_p} \sqrt{\frac{G(M_\star + m_p)}{a}} \tag{1}$$

where m_p is the planet mass, $M_\star$ is the mass of the star, a is the semi-major axis of the orbit, and i is the inclination of the orbital plane to the plane of the sky. The period of the orbit P is given by Kepler's third law as revised by Newton:

$$P^2 = \frac{4\pi^2 a^3}{G(M_\star + m_p)}. \tag{2}$$

If we combine these two equations, we find the observed stellar orbital velocity and the period are related by

$$V_\star = \left(\frac{2\pi G}{P}\right)^{1/3} \frac{m_p \sin i}{(M_\star + m_p)^{2/3}} \tag{3}$$

We see that the problem of planet detection requires not only high velocity precision, but also stability of the instrumental system over many years. Planetary systems with massive planets, or with Jupiter sized planets close to the star, would be considerably easier to detect than our own solar system. For example, a planet of

5 Jupiter masses at a distance of 0.45 AU from a 1 $M_\odot$ star would have an orbital period of about 0.3 years and a velocity semi-amplitude of about $210\,\mathrm{m\,s^{-1}}$. On the other hand, detection of an Earth-size planet at 1 AU would require precision of $9\,\mathrm{cm\,s^{-1}}$ to be maintained over a one year period.

Detection of brown dwarfs and very low mass stars would be considerably easier than detection of Jupiter sized planets. For instance, a 0.05 $M_\odot$ (50 M_J) object at 4 AU from its primary would have a velocity semi-amplitude of $700\,\mathrm{m\,s^{-1}}$ and a period of about 8 years. In a case such as this, the velocity precision is relaxed considerably with respect to the precision needed to detect a Jupiter-mass object, but the necessity remains for long-term stability of the instrument. This stability requirement is as difficult to achieve and as important as the required precision of an individual measurement.

3. Radial Velocity Measurement Technique

Classical techniques for measurement of stellar radial velocities with respect to a separate comparison source spectrum rarely achieve standard errors better than a few tenths of a kilometer per second. The principal limiting factor, as discussed by Griffin and Griffin (1973), is the difference in illumination of the spectrograph optics by the stellar and comparison sources. The stellar beam must traverse the terrestrial atmosphere and the telescope optics. The comparison source spectrum is generally taken at a different time (either before or after the stellar spectrum), has a different beam illumination pattern, and usually follows a significantly different path through the spectrograph. The Griffins suggested that these problems could be overcome if the wavelength comparison lines were imposed on the stellar spectrum before the light entered the spectrograph. They suggested the use of the telluric O_2 absorption band at 6300 Å as the reference spectrum. This method gives the obvious advantage that the comparison absorption spectrum is superimposed automatically on the stellar spectrum, and therefore both spectra must traverse exactly the same optical path. Thus, any instrumental wavelength shifts due to seeing, slit illumination, or spectrograph and detector drifts will affect both the stellar and the comparison spectrum equally. Very precise determination of the variations in stellar radial velocity are made by measuring the difference in the apparent Doppler shift of the stellar lines with respect to the reference lines. The disadvantages of the Griffin telluric line method are that one must observe in the spectral region of the telluric absorption band (which does not always have a high density of stellar lines), and that the precision is limited by the intrinsic variability of the telluric lines resulting from winds and variations in pressure and temperature. The Griffins claimed that this technique is capable of $10-20\,\mathrm{m\,s^{-1}}$ precision, and our extensive work in this area has shown this to be true (Cochran, 1988; Cochran *et al.*, 1991).

An obvious improvement to this technique is to pass the starlight through a gas absorption cell before the light enters the spectrograph. The absorption lines formed by the gas cell then serve as the fixed reference spectrum for the differential high precision measurement of variations in the stellar radial velocity. This circumvents the limitations imposed by the telluric atmosphere in providing the reference absorption spectrum. This technique was pioneered by Campbell and Walker (1979), who used HF as the absorbing gas. While this method has worked quite well, HF is a difficult gas to use even under the best of laboratory conditions.

An alternative is to use a more benign gas in a laboratory cell, like I_2 (Libbrecht, 1988; Marcy and Butler, 1992). An I_2 cell has several significant advantages over HF. I_2 has a strong electronic band ($B\ ^3\Pi_{0u}^+ - X\ ^1\Sigma_g^+$) in the 5000–6400 Å spectral region. The vapor pressure of I_2 is high enough (about 0.5 torr) to produce a significant absorption in a cell only 10–20 cm in length at room temperature. Moreover, I_2 vapor does not corrode the cell material, so a permanently sealed cell can easily be constructed. This cell then has a fixed number of I_2 molecules inside, and if the cell is operated at a temperature above that at which it was constructed, virtually all of the I_2 will be in the gas phase. I_2 has a rich spectrum of extremely narrow absorption lines. Pressure shifts of line frequencies are much smaller for I_2 than for HF. These factors all ensure that a sealed, temperature-regulated I_2 cell provides an extremely stable and compact metric for high-precision radial velocity measurement.

4. The McDonald Observatory Planet Search

4.1. PROGRAM HISTORY

The McDonald Observatory Planet Search program began operation in 1987 September. The survey began using the Griffins' telluric O_2 line method. Observations use the McDonald Observatory 2.7 m telescope coudé spectrograph, with its echelle grating and a Texas Instruments 800 × 800 CCD as a detector. This instrumental configuration gives a resolving power of about 220,000 at 6300 Å. The telluric line method gives us a routine radial velocity precision of about 15 m s^{-1} for stars down to about 5th magnitude. Perhaps the best indicator of both the random and systematic radial velocity measurement errors is the results from observations of an object thought to be constant in radial velocity: our Sun. In order to monitor overall system performance as a regular part of the MOPS, we have obtained observations of a standard spot on the lunar surface. This gives a measurement of disk-integrated sunlight in a manner approximating as closely as possible the observation of a star. The observed velocities were reduced to heliocentric velocities using the JPL DE200 planetary ephemeris. Figure 1a shows the results of the O_2 line lunar observations.

The rms scatter of the individual data points about the mean is ±11.9 m s^{-1}. There are, however, systematic patterns in the data which are revealed by Fourier analysis, indicating that the variance of the O_2 observations is dominated by an uncontrolled source of systematic error. The likely source for this systematic error

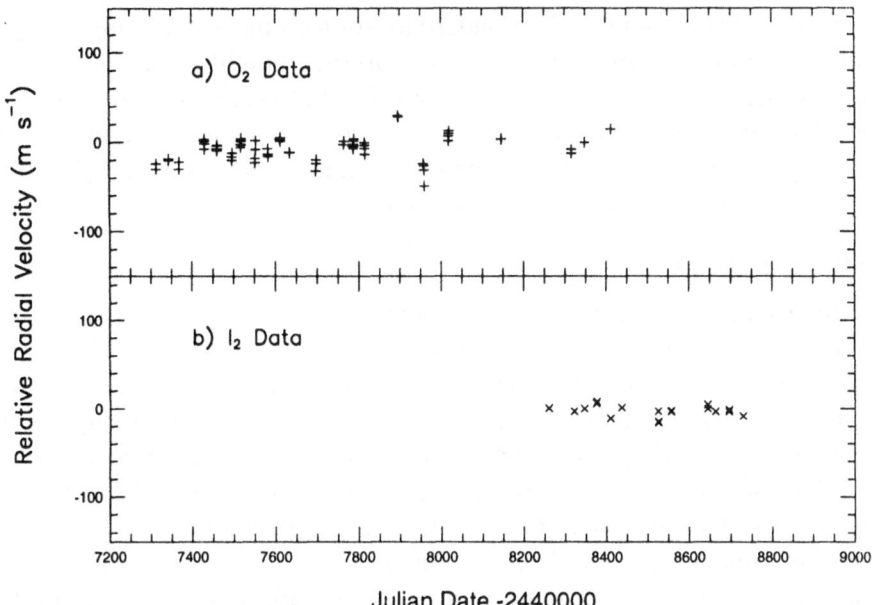

Fig. 1.   Radial velocity variations of a standard spot on the lunar surface as measured by the McDonald Observatory Planet Search. Panel a) shows results using the telluric O_2 line technique. The rms scatter of the individual observations around the mean is $11.9 \, m \, s^{-1}$. Panel b) shows the results obtained with the stabilized I_2 cell. The rms scatter of the I_2 data is $7.1 \, m \, s^{-1}$.

is the intrinsic velocity variability of the terrestrial atmosphere, i.e. seasonal wind patterns at McDonald Observatory.

While this level of precision was equal to that achieved by Campbell's HF cell (Campbell *et al.*, 1988) or McMillan's LPL instrument (McMillan *et al.*, 1990), we felt the need to eliminate this systematic error and improve our overall measurement precision. In 1990 October we installed a stabilized I_2 gas absorption cell as the velocity metric for the MOPS. The I_2 cell is a simple quartz tube 5 cm in diameter and 15 cm in length with sealed windows. A vacuum manifold apparatus is used to fill the cell with I_2 gas at its vapor pressure (about 0.5 torr) at room temperature. The cell is then sealed by a glassblower and removed from the manifold. The cell is thus *permanently sealed* with a fixed number of I_2 molecules inside. In use at the telescope, the cell is placed directly in front of the spectrograph entrance slit, with starlight passing through the cell. A temperature controller keeps the cell temperature constant at 50°C. This guarantees that essentially all of the I_2 will be in the gas phase, and that there will be no condensation on the cell walls or windows. Since I_2 does not react with the quartz cell or leak through it, this device produces a stable reference absorption line system. The use of this sealed stabilized I_2 cell removes potential problems with possible long-term drifts in the velocity metric. Our long-term stability relies on the stability of molecular physics, rather than on the thermal or mechanical stability of mechanical devices or the electronics which controls them. This increased level of stability is easily seen in Figure 1b, which shows MOPS observations of the standard location on the Moon using the I_2 cell. The rms scatter of the I_2 *individual observations* (not monthly or nightly averages) about their mean is $7.1 \, \mathrm{m \, s^{-1}}$. This improved level of precision demonstrates that we have indeed succeeded in controlling the major source of long-term systematic errors, and that we have met our design goal in radial velocity precision achieved in long-term operation of the MOPS program.

4.2. MOPS OPERATIONS

We decided to concentrate the survey on "solar-type" stars for a number of reasons. First, the only *known* planets in our galaxy orbit a G2V star. Second, it is extremely difficult to measure precise radial velocities for stars of spectral type F4 or earlier due to the stars' rapid rotation and paucity of deep photospheric absorption lines. We also limit the search to dwarfs and subgiants at this time because planets which are in orbits close enough to the star to be detected in a reasonable length of time will find themselves in or near the red giant envelope of an evolving star, where they will be quickly destroyed (Goldstein, 1987). The fate of gas giant planets in more distant orbits is uncertain. In addition, red giants are much more likely to show pulsation modes with radial velocity signals large enough to mask those signals due to planets. We started with the star lists used in the high precision radial velocity survey of Campbell, and then added other "interesting" stars which were bright enough and well placed for observations from McDonald Observatory. The observing list now consists of 33 F, G, and K stars of luminosity classes IV and V. These stars are listed in Table I.

There are no M dwarfs on our observing list simply because there are no satis-factory candidates bright enough to be observed with our high precision techniques. We strongly desire to have as much overlap as possible with Campbell's group and the LPL survey in the objects being observed. Any purported positive planetary detections will be of fundamental importance, and yet will be skeptically received. Therefore, it is essential to have several groups independently pursuing the goal of planet detection using different techniques, but having (at least initially) a large number of stars in common on their observing lists.

5. Technique

For our survey, we request one observing run of 3 nights each month. During each monthly observing run, we attempt to obtain at least one good set of observations of each available program star. Thus we strive to observe each star in 6−9 consecutive months each year, depending on the declination of the star. If sufficient good weather allows, we will often attempt to spend one night also observing with the telluric O_2 technique in order to maintain the same velocity zero-point between the O_2 and the I_2 data.

Figure 2a shows the spectrum of the I_2 cell from 5187 Å to 5197 Å. Figure 2b gives the spectrum of α Canis Minoris A in that same spectral region, and Fig-

TABLE I

List of Program Stars

HR	Star	Spec.	V	HR	Star	Spec.	V
219	η Cas	G0V	3.44	509	τ Cet	G8V	3.50
937	ι Per	G0V	4.04	963	α For	F8V	3.87
996	κ Cet	G5V	4.82	1084	ϵ Eri	K2V	3.73
1136	δ Eri	K0IV	3.54	1325	o^2 Eri	K1V	4.43
1543	π^3 Ori	F6V	3.19	2047	χ^1 Ori	G0V	4.41
2943	α CMi A	F5IV-V	0.38	3775	θ UMa	F6IV	3.17
4112	36 UMa	F8V	4.82	4540	β Vir	F8V	3.61
4983	β Com	G0V	4.28	–	HD114762	F9V	7.29
5019	61 Vir	G6V	4.75	5544	ξ Boo A	G8V	4.55
5933	γ Ser	F6V	3.85	6401	36 Oph B	K0V	5.33
6402	36 Oph A	K1V	5.29	6623	μ Her	G5IV	3.42
6752	70 Oph A	K0V	4.03	7462	σ Dra	K0V	4.68
7503	16 Cyg A	G1.5V	5.96	7504	16 Cyg B	G2.5V	6.20
7602	β Aql	G8IV	3.71	7948	γ^2 Del	K1IV	4.27
7957	η Cep	K0IV	3.43	8085	61 Cyg A	K5V	5.21
8086	61 Cyg B	K7V	6.03	8832	HR8832	K3V	5.56
8974	γ Cep	K1III-IV	3.21				

ure 2c shows the spectrum of α Canis Minoris A through the I_2 cell. This figure demonstrates that although the I_2 spectrum is dense and complex it does not totally obliterate the stellar spectrum with this size cell. The net throughput of starlight is quite high because the I_2 lines are extremely narrow and are unresolved by the spectrograph. From spectra such as shown in Figure 2, the variations in the shift of the stellar spectra with respect to the assumed constant I_2 spectrum may be measured, using a variant of the method of Fahlman and Glaspey (1973). These velocity shifts are then corrected to the solar system barycenter using the JPL DE200 planetary ephemeris. We note that we are attempting to measure only *relative velocity variations*, not absolute radial velocities. Therefore we do not care about the identifications of the I_2 or stellar lines, nor do we need to know the absolute wavelengths of the lines to high precision. This is an entirely differential measurement.

In using the radial velocity method to search for planetary systems around other stars, we must be sure that any signal we detect is due to orbital motion of the star, and not due to intrinsic stellar variability. Stellar radial and nonradial pulsations, as well as star spots rotating across the stellar disk or changes in the photospheric granulation pattern will all result in apparent radial velocity changes due to variations in spectral line profile shapes. We need to isolate these other sources of

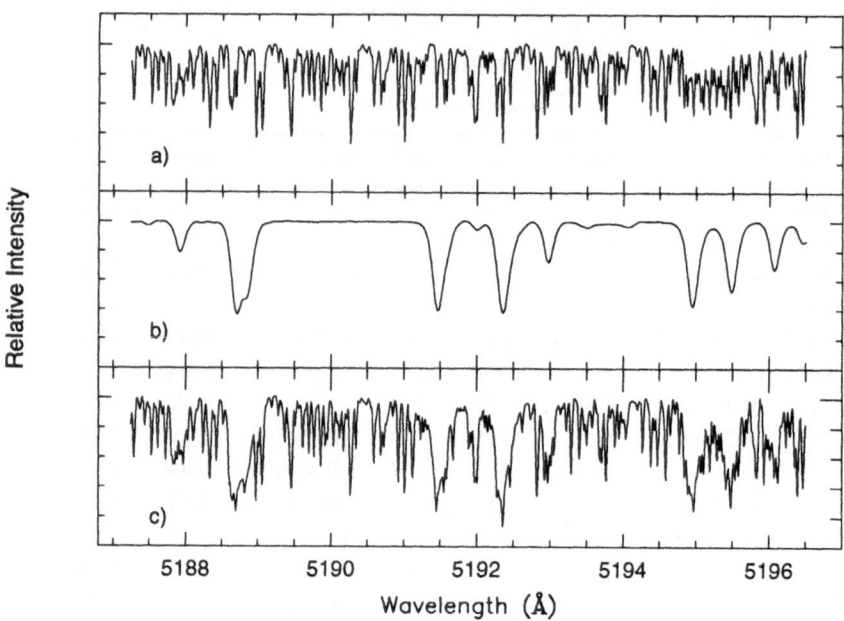

Fig. 2.   Sample spectra using an I$_2$ cell. Panel *(a)* shows the absorption spectrum of the I$_2$ cell (illuminated with white light) from 5187 Å to 5197 Å. Panel *(b)* gives the spectrum of α Canis Minoris A, without the I$_2$ cell, and panel *(c)* shows the spectrum of α Canis Minoris A taken with the cell in front of the entrance slit to the 2.7 m coudé spectrograph.

possible radial velocity variations from the desired signal due to orbital motion. Fortunately, this is easy, because only pure Doppler motion of the stellar center of mass will leave the line profiles constant. Our observing procedure routinely provides us with the data we need to separate orbital radial velocity variations from those due to intrinsic stellar sources. Our spectra are taken at $R = 220,000$, which is sufficient to resolve the stellar photospheric line profile shapes. Using the analysis techniques developed by Gray (1988), we can search for variations in the spectral line bisector shape. This spectral line shape information comes for free, as part of our observing process. This essential ability to measure directly the actual photospheric line profile shapes at high spectral resolution and high signal/noise gives us the unique advantage of easily being able to identify the Doppler origin of any detected radial velocity variation, and thus separate true detections of other solar systems from spurious stellar "noise" sources. In addition, high precision intermediate band and Ca II H and K line photometry of stars, coupled with spectroscopy of temperature sensitive photospheric lines, now permit stellar activity cycles of stars to be monitored on a routine basis, as was demonstrated for our survey star σ Draconis by Gray *et al.* (1992).

6. Results

The McDonald Observatory Planetary Search program has reached a level of maturity where it is starting to produce a wide variety of interesting results. Our orbital solution for the HD114762 system, with its possible brown dwarf secondary was given by Cochran *et al.* (1991). In this section we present additional results of the MOPS.

6.1. α CANIS MINORIS A

One of the brightest stars on our list, α Canis Minoris A (Procyon), shows significant night-to-night as well as month-to-month radial velocity variability. Our data for Procyon are shown in Figure 3. This variability of up to $\pm 50 \, \mathrm{m \, s^{-1}}$ is superimposed on a regular slope in radial velocities due to the orbit of Procyon and its distant white dwarf companion. Such variability is *not* shown by most other stars (both bright and faint) in the MOPS. Such radial velocity variability might result if Procyon were a marginal δ Scuti variable. The classical δ Scuti region is where the instability strip crosses the main sequence in the H-R diagram. This is nominally in the region of the late-A and early-F stars. However, this region has been defined by the precision of techniques (photometry and radial velocities) classically used to study these stars. As a F5 IV-V star, Procyon is the closest object in our survey to the δ Scuti region of the H-R diagram. Thus, it is not surprising that this star shows significant radial velocity variability. A few other stars of similar spectral type also seem to show this type of radial velocity variability. These include π^3 Orionis (F6V), θ Ursae Majoris (F6 IV), and γ Serpens (F6V). This type of variability is *not* seen in stars of later spectral type.

6.2. BROWN DWARFS ARE NOT COMMON

According to Equation 3, the presence of a brown dwarf companion to a star in a short period (< 20 years) orbit should reveal itself by large slopes in the measured radial velocities. While several stars (e.g. χ^1 Orionis, 70 Ophiuchi A, γ Cephei) appear to have unresolved *stellar* companions, there are no program stars which are yet showing radial velocity variations clearly in the range expected for brown dwarf companions. If there are any brown dwarf companions to stars in our survey list, they are either a) not now on the steep part of their radial velocity curves, b) in long period orbits, or c) in low inclination orbits.

6.3. POSSIBLE PLANETARY COMPANIONS?

As of 1992 December, none of the stars in the survey show convincing evidence for possible planetary companions. Examples of typical results are shown in Figure 4. Panel a) shows MOPS results for δ Eridani. To our level of precision, this star shows constant radial velocities. Panel b) shows results for μ Herculis. This star shows a clear decrease in radial velocities over the five year period for which we have data. Such a deceleration may well be caused by the presence of an unseen companion.

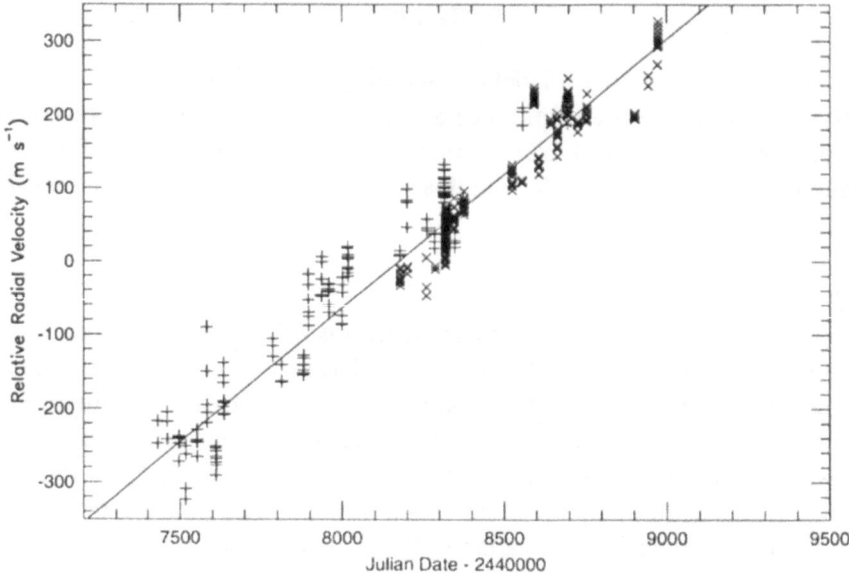

Fig. 3.   Measured radial velocity variations of α Canis Minoris A. Plus signs indicate data taken with the O_2 line technique, and the $\times$ symbols are data taken with the I_2 cell. The solid line is the variation expected from the orbit of the white dwarf companion. The scatter in the data is due to intrinsic variability of the star.

However, our data do not yet give any indication of the amplitude or period of such a variation. Therefore at this time, it is impossible to place any sort of meaningful estimate on the mass and semi-major axis of a companion. The observed variations could be caused by a planet, a brown dwarf, or a stellar companion. More data are necessary to determine the amplitude and period of the orbit, and thus the mass function of the system.

7. Conclusions

The McDonald Observatory Planetary Search has demonstrated that it is quite possible to achieve measurements of stellar radial velocity variations with a *long-term* precision of better than $10 \, \mathrm{m \, s^{-1}}$ in routine observations. This level of precision is capable of detecting Jovian planets in orbit around solar type stars. The MOPS program has been surveying a sample of 33 nearby F, G, and K stars since 1987 September to search for sub-stellar companion objects. While none of the stars (except for HD114762) yet show definitive evidence for low-mass companion objects, many stars do show secular changes in radial velocity. These variations will have to be followed for many more years to determine the period and amplitude (and thus the mass function) of the variation.

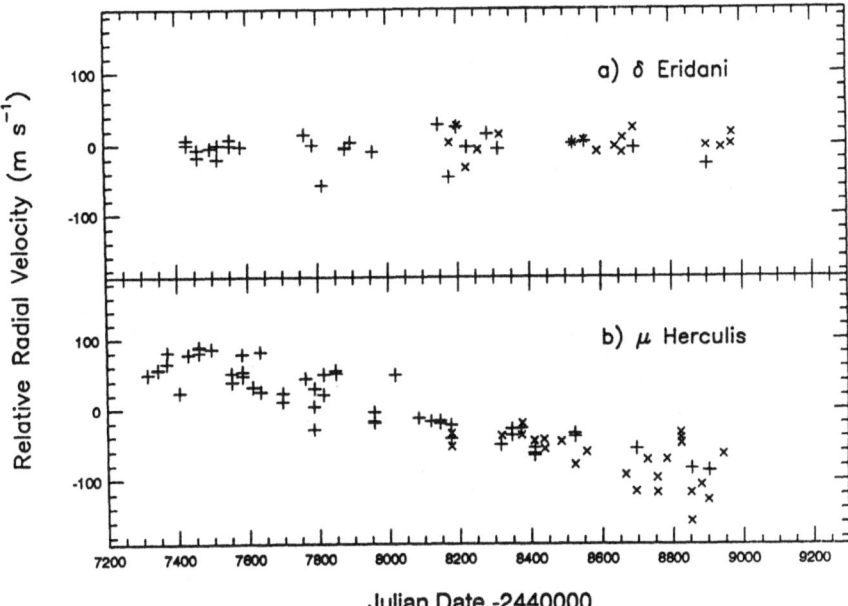

Fig. 4. Radial velocity variations of two typical stars in the McDonald Observatory Planet Search. For each star, +'s represent data taken with the O_2 technique, and ×'s represent data taken with the I_2 cell.

Acknowledgements

This work has been supported by NASA through grants NAGW-2302 and NAGW-2465.

References

Campbell, B. and Walker, G. A. H.: 1979, *Publ. Astron. Soc. Pacific* **91**, 540.

Campbell, B., Walker, G. A. H. and Yang, S.: 1988, *Astrophys. J.* **331**, 902.

Cochran, W. D.: 1988, *Astrophys. J.* **334**, 349.

Cochran W. D., Hatzes, A. P. and Hancock, T. J.: 1991, "Letters to the Editor," *Astrophys. J.*, **380**, L35.

Fahlman, G. G. and Glaspey, J. W.: 1973, "A Technique for Measuring Small Displacements in Digital Spectra" in *Astronomical Observations with Television Type Sensors*, eds., Glaspey, J. W. and Walker, G. A. H., Inst. of Astronomy and Space Science: Vancouver, 347.

Goldstein, J.: 1987, *Astron. Astrophys.* **178**, 283.

Gray, D. F.: 1988, *The Observation and Analysis of Stellar Photospheres*, Cambridge University Press: Cambridge, 417.

Gray D. F., Baliunas, S. L., Lockwood, G. W. and Skiff, B. A.: 1992, *Astrophys. J.* **400**, 681.

Griffin, R. and Griffin, R.: 1973, *Monthly Not. Roy. Astr. Soc.* **162**, 243.

Libbrecht, K. G.: 1988, "A Search for Radial Velocity Oscillations in Procyon" in I.A.U. Symposium 132: *The Impact of Very High S/N Spectroscopy on Stellar Physics*, eds., de Strobel, G. C. and Spite, M., Kluwer: Dordrecht, 83.

Marcy, G. W. and Butler, R. P.: 1992, *Publ. Astron. Soc. Pacific* **104**, 270.

McMillan, R. S., Smith, P. H., Perry, M. L., Moore, T. L. and Merline, W. J.: 1990, *Proceedings SPIE* **1235**, 601.

MULTIPLEX APPROACH TO THE PHOTOMETRIC DETECTION OF

PLANETS *

WILLIAM BORUCKI and DAVID KOCH

NASA Ames Research Center,
Moffett Field, California, USA

Abstract. If the hypothesis is correct that most solar-like stars have planetary systems and have planets in inner orbits, then approximately 1% of these stars should have planets with orbital planes close enough to our line of sight to show transits. To get a statistically significant estimate of the fraction of stars that have planets in inner orbits, it is necessary to monitor thousands of stars continuously for a period of several years. To accomplish this requires the use of a multi-channel photometer system. We present here several multi-channel methods that have been used for ground-based observations and a concept for applying multi-channel photometry to the detection of numerous Earth-sized planets.

1. Introduction

Current theories postulate that planets are a necessary consequence of the formation of a star. If this is correct, then most stars have planets and terrestrial-type planets are commonplace. If our solar system is typical of other stellar planetary systems, then the inner planets will be small and dense and the outer planets will be gas giants; approximately 1% of these stars should have planets in orbital planes close enough to our line of sight to show transits; a transit of an Earth-like planet should reduce the amplitude of the star light by about 0.01% for a period of about twelve hours; and the transit should reoccur with a period on the order of one Earth year (Borucki and Summers, 1984). To get a statistically significant estimate of the fraction of stars that have planets in inner orbits, it is necessary to monitor thousands of solar-like stars for several years. Although all the stars likely to show a transit will do so during the first year of observations, a second transit detection is necessary to determine the orbital period and to make a prediction for the third transit. Observation of a third transit with the predicted amplitude and time is necessary to confirm the detection. To accomplish this task in a period of several years, it is necessary to use a multi-channel photometer.

2. Existing Multiplex Systems

In the past decade, photometric and spectrometric systems have been constructed at most major observatories to simultaneously observe multiple targets to improve the precision of the measurements and to make more efficient use of the available telescope time. Geyer and Hoffmann (1975) review work with multi-channel photometers starting in 1918. They reference discussion of the use of two telescopes on

---

* Paper presented at the Conference on *Planetary Systems: Formation, Evolution, and Detection* held 7–10 December, 1992 at CalTech, Pasadena, California, U.S.A.

Astrophysics and Space Science **212**: 293–298, 1994.
© 1994 *Kluwer Academic Publishers.*

FIXED FIBER OPTIC FEED

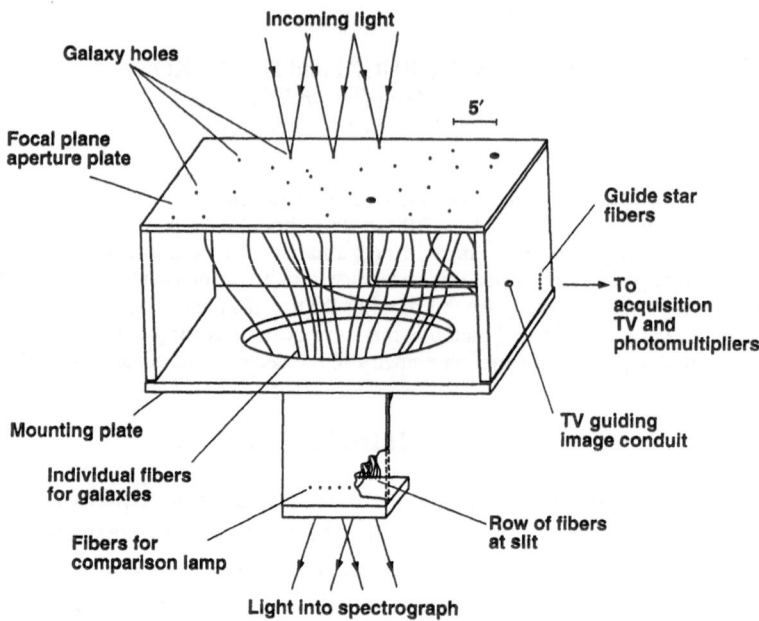

Fig. 1. The multiple object fiber optic spectroscope (Medusa). This system provides for simultaneous spectral recording of 37 galaxies (Hill *et al.*, 1982).

one mounting, switching light beams from each star, and the use of multiple detectors. In recent years, two-channel photometers have been used to reduce the effect of atmospheric extinction variations. De Biase *et al.* (1978) present calculations that show that their two-channel photometer can measure the brightness of stars five magnitudes fainter than a similar one-channel instrument with the same precision. Grauer and Bond (1981) demonstrated that their two-star photometer allows high-precision photometry to be obtained even under non-photometric conditions. Walker (1984) demonstrated a four-channel (three stars plus sky) photometer using fiber optics. Multiple channel photometers that observe the sky and a star at three or more wavelengths simultaneously have been constructed by Baum *et al.* (1959), Grønbech *et al.* (1976), De Biase *et al.* (1978), and Barwig *et al.* (1987). The fifteen-channel photometer constructed by Barwig *et al.* (1987) simultaneously monitors two stars and the sky at five wavelengths. With the exception of this photometer, however, the current photometers are too bulky to observe hundreds or thousands of stars in the focal plane of a telescope. More promising approaches are those used to make spectrometric measurements with optical fibers and the use of charged coupled devices (CCDs) to do synthetic aperture photometry.

MULTI-ACTUATOR SYSTEM

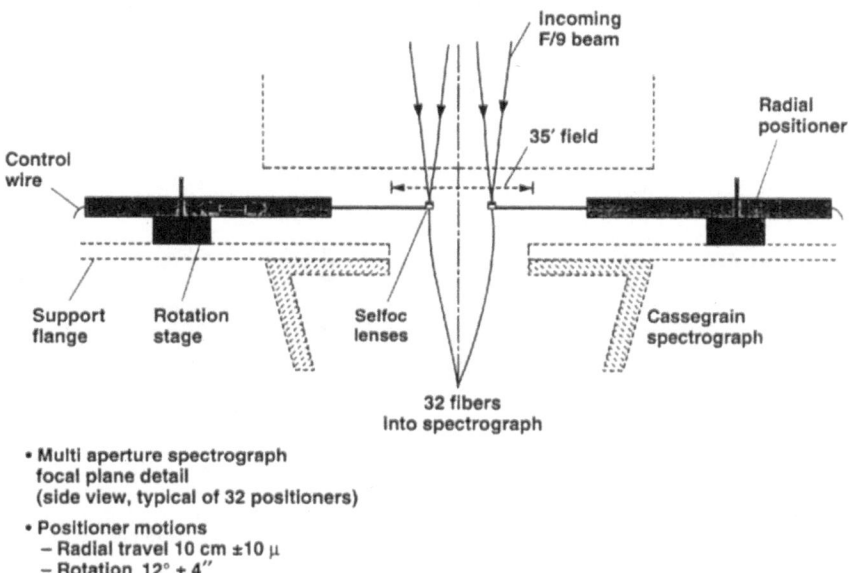

Fig. 2. A multi-aperture spectrograph with 32 dedicated fiber optic actuators. (Adapted from Hill *et al.*, 1982.)

2.1. FIXED FIBER OPTIC FEED SYSTEMS

Two different approaches have been used to make multi-channel spectrometric measurements at the focal plane; plug boards which are changed for each field of view (FOV) and robotic positioners. The plug board method (Hill *et al.* 1980, Gray 1983) uses many holes drilled into a template that is placed at the focal plane of the telescope, see Figure 1. A different pattern of holes is drilled to match the positions of each set of target objects. Gray (1983) used a plug board to observe 45 objects simultaneously in a 12 arcsecond FOV on the Anglo-Australian Telescope. The patterns for the star positions are obtained from measurements of positions recorded on photographic plates. With the use of computer-controlled machines, it is possible to drill thousands of apertures on a single template for a large field of view telescope. Although this concept is simple and has been used for many years, it has the disadvantage of being inflexible in the choice of targets once the aperture pattern has been chosen.

2.2. MOVABLE FIBER OPTIC FEEDS

If many fields of view are needed, then a robotic system that can position a large number of detectors and then quickly reposition them can be used. In one approach,

ROBOTICALLY POSITIONED DETECTORS

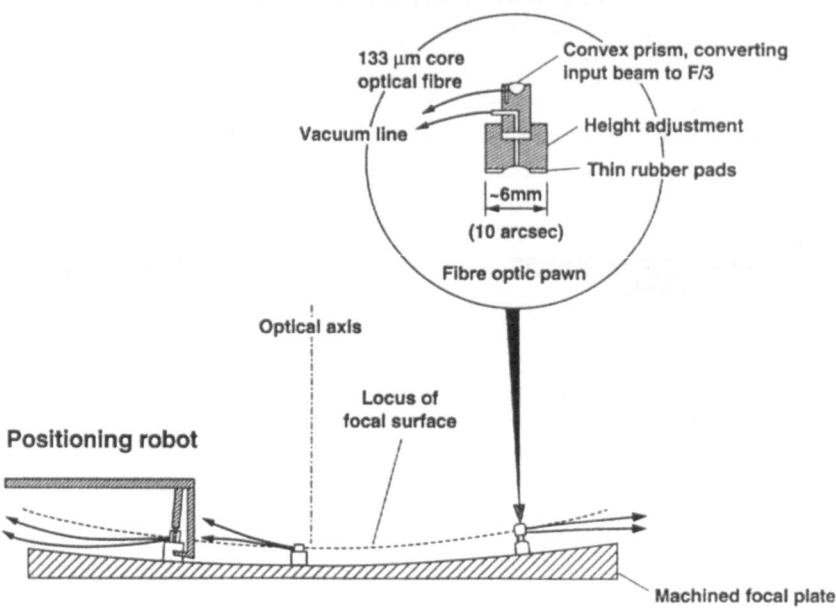

Fig. 3. An automated multiple object spectrometer (FASTAR), which uses a single robot system for positioning 50 pneumatically held fiber optic feeds. (Adapted from Lund, 1984.)

32 dedicated robotic micro-positioners are used to manipulate each of 32 fiber optic feeds (Hill *et al.*, 1982). See Figure 2. Another approach consists of a single micro-positioner used to relocate all of the fiber optic feeds. These are held in place pneumatically (Lund, 1984). See Figure 3. A robotic positioner to be used at the 3-meter Shane telescope at Lick Observatory has been designed to position 125 fibers and to accomplish a reconfiguration of these fibers in approximately five minutes (Craig *et al.*, to be submitted to PASP). Positioner systems to provide optical beams to simultaneously observe more than 400 objects are being constructed. For telescopes with a large plate-scale, even more targets should be available. The disadvantage of this approach is the complexity of the motions.

2.3. MULTIPLEXING WITH DISCRETE DETECTORS

Both the plug board and robotic positioner approaches can be used with small discrete detectors to monitor large numbers of stars. Laboratory photometers based on silicon diodes have already demonstrated the necessary precision to detect Earth-size planets around solar-like stars (Young *et al.*, 1992). Silicon detectors that can make high precision photometric measurements are available with diameters less

DISCOVERY SPACE FOR ASTROMETRIC, RADIAL VELOCITY AND PHOTOMETRIC METHODS

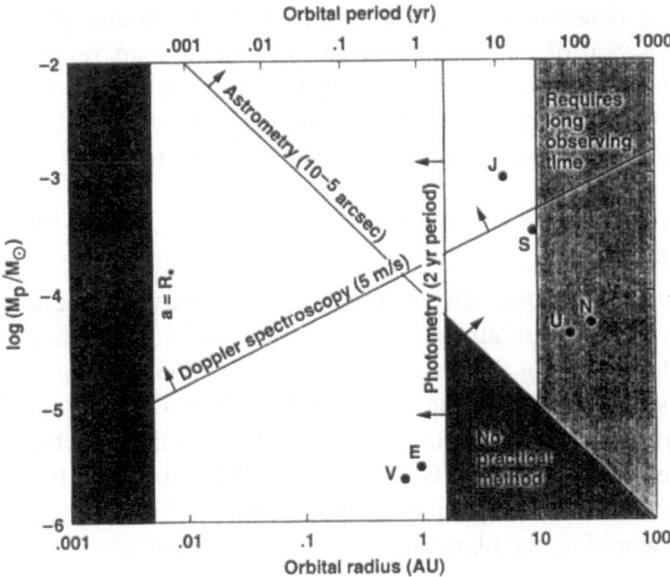

Fig. 4.   Comparison of discovery space for multiplex photometry, astrometry and Doppler spectroscopy. The discovery space for photometry is for Earth-size or larger planets around solar-like stars. That for astrometry assumes the stars to be within 10 parsecs.

than 1 mm. Even when the detectors are inside protective housings, it is possible to mount or move hundreds of detectors in the focal plane of a suitable telescope.

2.4. MULTIPLEXING WITH CHARGED COUPLED DEVICES

Recently CCD arrays have become available that allow simultaneous photometric measurements on many stars. Gilliland *et al.* (1991), observed a group of stars near 13th magnitude and obtained a precision of 8 parts per 10,000. Their precision was limited by the atmosphere rather than by the intrinsic precision of the CCD detector. Buffington (1990, 1991) made both laboratory measurements and an analysis of the noise sources of a CCD array and found a frame-to-frame precision of nearly 1 part in 10,000. If averaging many frames together can increase the precision by a factor of three, then CCDs will be useful for detecting transits by planets as small as the Earth or Venus.

3. Applying Multiplexing to Planet Detection

Given the limitations imposed by ground-based observing (atmospheric scintillation and extinction changes, weather, etc.) an Earth-orbiting observatory is necessary to achieve the precision and continuous monitoring required. A mission concept to use a 5-by-5 array of CCD detectors is being developed that would monitor the brightness of 6000 stars in a wide field of view telescope. With such a photometer, the detection of about sixty transits by Earth-size planets is expected during the first year of operation. During subsequent years repeat transits would be seen because of the high probability that the transits are from planets in short-period orbits. Continuation of the observation program would determine the orbital periods of the planets and confirm the regularity of the transits.

3.1. COMPARISON OF MULTIPLEX PHOTOMETRY TO OTHER DETECTION METHODS

Other approaches to planet detection include astrometry and radial velocity measurements. Astrometry requires the measurements of very small (sub-milliarc-second) displacements of the central star caused by the orbiting planet(s). Radial velocity measurements detect the small Doppler shifts (few m/sec) also caused by the orbiting planet(s). As shown in Figure 4, both of these techniques are limited to massive Jovian-like planets. On the other hand, photometry is better suited for detection of Earth-like planets in inner orbits, which if our solar system is archetypical, would be the planet type that could support life.

References

Barwig, H., Schoembs, R. and Buckenmayer, C.: 1987, *Astron. and Astrophys.* **175**, 327–344.

Baum, W.A., Hiltner, W.A., Johnson, H.L. and Sandage, A.R.: 1959, *Astrophys. J.* **130**, 749–763.

Borucki, W.J. and Summers, A.L.: 1984, *Icarus* **58**, 121–134.

Buffington, A., Hudson, H.S. and Booth, C.H.: 1990, *Publication Astronomical Society Pacific* **102**, 688–697.

Buffington, A., Booth, C.H. and Hudson, H.S.: 1991, *Publication Astronomical Society Pacific* **103**, 685–693.

Craig, W.W., Cook, K.H., Hailey, C.J. and Brodie, J.P.: (submitted to *Publication Astronomical Society Pacific*).

De Biase, G.A., Paterno, L., Pucillo, M. and Sedmak, G.: 1978, *Applied Optics* **17**, 435–441.

Geyer, E.H. and Hoffmann, M.: 1975, *Astron. and Astrophys.* **38**, 359–362.

Gilliland, R.L., Brown, T.M., Duncan, D.K., Suntzeff, N.B., Lockwood, G.W., Thompson, D.T., Schild, R.E., Jeffrey, W.A. and Penrase, B.E.: 1991, *Astron. J.* **101**, 541–561.

Grauer, A.D. and Bond, H.E.: 1981, *Publ. Astron. Soc. Pac.* **93**, 388–396.

Gray, P.M.: 1983, *Proceedings Society of Photo-Optical Instrumentation Engineers* **445**, 57–64.

Grønbech, B., Olsen, E.H. and Strömgren, B.: 1976, *Astron. and Astrophys. Suppl.* **26**, 155–176.

Hill, J.M., Angel, J.R.P., Scott, J.S., Lindley, D. and Hintzen, P.: 1980, *Astrophys. J.* **242**, L69–L72.

Hill, J.M., Angel, J.R.P., Scott, J.S., Lindley, D. and Hintzen, P.: 1982, *Proceedings Society of Photo-Optical Instrumentation Engineers* **331**, 279–288.

Lund, G.: 1984, *Very Large Telescopes, Their Instrumentation and Programs Proceedings IAU Colloquium 79*, 617–633.

Walker, E.N.: 1984, *Vistas in Astronomy* **27**, 421–432.

Young, A.T., Genet, R.M., Boyd, L.J., Borucki, W.J., Lockwood, G.W., Henry, G.W., Hall, D.S., Pyper-Smith, D., Baliunas, S.L., Donahue, R. and Epand, D.H.: 1991, *Publ. Astron. Soc. Pac.* **103**, 221–241.

SEARCHING FOR PLANETS BY DIFFERENTIAL ASTROMETRY
WITH LARGE TELESCOPES *

RICHARD DEKANY and ROGER ANGEL

Optical Sciences Center, University of Arizona, Tucson, Arizona, USA
Steward Observatory, University of Arizona, Tucson, Arizona, USA

and

KEITH HEGE and DAVE WITTMAN

Steward Observatory, University of Arizona, Tucson, Arizona, USA

Abstract. Traditional astrometric methods are limited in accuracy by the atmosphere in a way that does not show much improvement with increased telescope aperture. However, there is the potential for very high accuracy with large telescopes if advantage can be taken of these factors: First, the differential atmospheric distortion of images of closely adjacent stars is less with larger aperture; second, the diffraction limit is sharper, and third, photon statistics are improved. In this paper we analyze and give experimental tests of techniques that could be applied to the detection of planets with the mass of Jupiter or Uranus, if they are present in nearby binary star systems.

The atmospheric perturbation of the relative position of the energy centroids measured in short exposure images of binary stars depends on the effective height of the turbulent distortion. For a 4-meter telescope, the error in centroid determination of a 4-arcsec binary can be as small as 20 milliarcsec (mas) in a single 20-millisecond (msec) exposure. The relative position measured by cross-correlation of short exposure speckle images, as suggested by McAlister (1977b), may give even higher accuracy. In this case, Roddier (Roddier *et al.*, 1980) has shown that the atmospheric error depends on the thickness rather than the height of the layers that make the dominant contribution to the turbulence. Through Monte Carlo analysis we show that on occasions when the turbulence arises largely in a thin layer, a single 20-msec exposure of a 4-arcsec binary taken with a 4-m aperture can yield an astrometric accuracy of order 0.5 mas.

We report on experiments made at the Steward Observatory 2.3-m telescope which achieved accuracies corresponding to 1.7 mas in a 2.24-arcsec binary and 16.1 mas in a 6.0-arcsec binary with only 15 and 18 specklegram pairs respectively. We plan to use the 6.5-m converted MMT to obtain much higher performance, between 4.0 mas and 0.40 mas *per independent specklegram pair*, depending upon atmospheric conditions, for binaries of 4-arcsec separation. By cycling rapidly through perhaps 100 binaries, thus calibrating systematic errors through the average change in binary separation, Jupiter-mass planets may be detectable with small but regular access to the telescope.

1. Astrometry Through the Turbulent Atmosphere

Advances in detector technology and careful calibration techniques have made random image motion induced by atmospheric turbulence the dominant source of astrometric measurement error. For many years it has been known that relative, or narrow-field, astrometry enjoys an advantage over absolute, or wide-field, astrometry (Black, 1980). In absolute astrometry the time average motion of the centroid of a star image limits the astrometric accuracy. In order to obtain standard errors of less than 1 mas, integrations of an hour or more are typically required (Gatewood, 1987). For relative astrometry, the limitation to accuracy arises from

---

* Paper presented at the Conference on *Planetary Systems: Formation, Evolution, and Detection* held 7–10 December, 1992 at CalTech, Pasadena, California, U.S.A.

the time average difference of the centroid motion of two nearby stars. Since the centroid motion of two stars separated by less than the isoplanatic angle is well correlated, relative astrometric measurements suffer less atmospheric error.

In 1977, McAlister suggested an alternative to traditional astrometry performed through the measurement of star centroid locations (McAlister, 1977b). The principle underlying this technique is to measure the separation of two nearby stars by calculating the location of the peak of the cross-correlation of two short-exposure star speckle images. For two components of a binary star pair with separation of order the isoplanatic angle, the cross-correlation is a highly peaked function superimposed upon low, broad, background. The peak is due to superposition of similar speckle patterns at the true star separation. This peak has a width approximately the same as that of individual speckles, determined by the telescope diffraction limit. The low, broad, background results from the overlap of non-corresponding speckles in the two patterns that occurs at separations other than the true separation. The width of this background is approximately the width of the cross-correlation of the atmospheric long-exposure point spread function, or seeing disk.

Whereas, in long-exposure relative astrometry, the quantities to be calculated are exact centroids of seeing disks, the cross-correlation method involves the calculation of the centroid of the one-speckle-wide cross-correlation peak. The speckle cross-correlation method is superior to that of conventional relative astrometry in that utilization is made of high spatial frequencies not present in the long-exposure image. However, it breaks down at angular separations at which the speckle patterns become decorrelated. McAlister demonstrated separation measurement standard errors of 2 mas in each direction for a 1.6-arcsec binary averaged over 50 short exposure speckle observations with the 3.8-m stopped pupil of the KPNO Mayall reflector (McAlister, 1977a). When a set of such determinations was later compared to previously accurately known orbits, McAlister and DeGioia demonstrated residual errors of 0.3 ± 5.7 mas, indicating the absence of serious systematic errors in speckle astrometry. However, since then we have found no use of the method to search for astrometric motion caused by planets, and speckle techniques applied to binary stars have been used almost exclusively to resolve separations less than the seeing disk.

In the following analysis, we will examine the limits to accuracy of the differential astrometric methods set by atmospheric turbulence. We first review the analytical expressions for differential image motion, and then give numerical results for the accuracy of speckle cross-correlation based on two simple models of atmospheric turbulence.

2. Theoretical Astrometric Errors

The fundamental turbulence-induced limit to the error in an astrometric measurement using classical differential astrometry and the limit in a measurement using speckle cross-correlation technique differ. Physically, this difference arises because the two techniques depend in different manners upon the vertical distribution of turbulence in the atmosphere. Although the error in both techniques is critically related to the angular separation of the stars whose separations are to be measured, the angular separation at which the same level of error is achieved differs.

2.1. DIFFERENTIAL ASTROMETRY FROM ENERGY CENTROIDS

The turbulence-induced error in classical differential astrometry for a given binary separation depends upon the relative tilt between the two wavefronts incident from the two components of a binary star system. The change in apparent separation of two stars is determined by the differential wavefront tilt. The amount of differential wavefront tilt depends, in turn, upon the angle over which the two wavefronts suffer atmospheric phase aberrations that are correlated, point-for-point, in the collecting aperture. The relevant parameter characterizing the atmosphere is the angle at which the mean-squared phase difference between two wavefronts equals 1 rad^2; it is known as the isoplanatic angle, θ_0, and is defined in terms from atmospheric turbulence theory as (Fried, 1982),

$$\theta_0^{-5/3} = 2.91 \ k^2 \ \sec^{8/3} (\xi) \int C_n^2 (z) \ z^{5/3} \, dz \tag{1}$$

In Equation (1), k is the optical propagation wave number, $2\pi/\lambda$, ξ is the observing angle measured from zenith, $C_n^2(z)$ is the index of refraction structure function, and z is the height above the telescope. Physically, the wavefronts from two sources within the isoplanatic angle propagate through a mostly shared column of turbulent atmosphere. The isoplanatic angle, then, depends upon atmospheric conditions and is typically a few arcseconds at an observing wavelength $\lambda = 0.5 \ \mu$m.

Sandler et al. have calculated the magnitude of the centroid error resulting from differential tilt in the context of adaptive optics (Sandler et al., 1993). Defining the x-direction as the direction connecting the two stars under observation and the y-direction as normal to the x-direction, Sandler et al. found that the instantaneous differential centroid errors, σ_x and σ_y, are given by

$$\sigma_x = 0.217 \left(\frac{\lambda}{D}\right) \left(\frac{D}{r_0}\right)^{-1/6} \left(\frac{\theta}{\theta_0}\right) \tag{2}$$

$$\sigma_y = 0.125 \left(\frac{\lambda}{D}\right) \left(\frac{D}{r_0}\right)^{-1/6} \left(\frac{\theta}{\theta_0}\right) \tag{3}$$

where the domain of validity, $\theta \leq 0.5 \ (D/r_0)\theta_0$, depends linearly on aperture diameter, D. σ_x is in units of radians, unless otherwise noted. This is the uncertainty in the binary separation derived from energy centroids of individual snapshots. In these equations, the atmospheric coherence parameter (Fried's length), r_0, is defined by (Fried, 1965).

$$r_0^{-5/3} = 0.423 \ k^2 \ \sec (\xi) \int C_n^2 (z) \, dz \tag{4}$$

Intuitively, r_0 is the scale size over which the wavefront remains coherent. For $\lambda = 0.5\ \mu$m, r_0 is typically 5–20 cm. Long exposure images taken under conditions of $r_0 = 10$ cm have a FWHM of approximately 1 arcsec. In this paper, r_0 values shall always be given for wavelength 0.5 μm. Note that r_0 differs from θ_0 in that, whereas the latter depends upon the vertical distribution of turbulence, the former depends only upon the total aberration, independent of height.

We can relate the Sandler *et al.* result for instantaneous centroid errors to those derived by Lindegren for centroid integrations (Lindegren, 1980). Lindegren considers only motion in the direction joining the two stars (direction x in Sandler *et al.*). So, beginning with Equation (2), we proceed, assuming that the error in the centroid determination reduces as $(1/\sqrt{N})\ t^{-1/2}$, where N is the number of independent measurements per second. Further, let the atmospheric phase aberrations remain fixed and move across the pupil at the mean wind velocity of $\bar{v}$. If we assume that independent measurements can be obtained approximately every $D/\bar{v}$ seconds, we then find that the multi-exposure centroid error can be represented as

$$\sigma = 0.217\sqrt{\frac{D}{\bar{v}}}\ \left(\frac{\lambda}{D}\right)\left(\frac{D}{r_0}\right)^{-1/6}\left(\frac{\theta}{\theta_0}\right)t^{-1/2} \qquad (5)$$

Now, if we evaluate Equation (5) using the assumptions of Lindegren, namely the Hufnagel–Valley turbulence model (which yields $r_0 = 0.18$ m and $\theta_0 = 1.5$ arcsec at $\lambda = 0.5\ \mu$m) and an average wind speed $\bar{v} = 14$ m s^{-1}, we find

$$\sigma = 590\ D^{-2/3}\ \theta\ t^{-1/2} \qquad (6)$$

where, following Lindegren, we have normalized Equation (7) to yield σ in arcsec, given θ in radians. Lindegren's result (35a) for the same domain of validity is

$$\sigma = 540\ D^{-2/3}\ \theta\ t^{-1/2} \qquad (7)$$

To within a constant of order unity, we have recovered Lindegren's result with quite reasonable assumptions.

In general, we can relate the isoplanatic angle and the coherence parameter by defining an effective turbulence height, $\hat{h}$, where

$$\hat{h} = \left(\frac{\int z^{5/3}\ C_n^2(z)\,dz}{\int C_n^2(z)\,dz}\right)^{3/5} \qquad (8)$$

Then we have

$$\theta_0 = 0.31\ \left(\frac{r_0}{\hat{h}}\right) \qquad (9)$$

For the case of a single thin phase screen at altitude h, we have $C_n^2(z) = C_n^2\ \delta(z - h)$ and $\hat{h} = h$. In this case, the shear, $\Delta x = \theta h$, at the

TABLE I

Fundamental atmosphere-induced uncertainty in classical astrometric centroid measurement for a 4.0-arcsec binary star system.

Atmospheric Model	Telescope Diameter (m)	Total Observing Time	σ_x (mas)	σ_y (mas)
Single thin phase screen				
10 km above telescope	4.0	Instantaneous	15	9.0
$r_0 = 13.2$ cm	4.0	1.0 sec	4.2	2.5
$\theta_0 = 0.84$ arcsec	8.0	Instantaneous	6.6	3.8
$\lambda = 0.5$ μm	8.0	1.0 sec	2.6	1.5
Two thin phase screens				
(0- and 10-km heights)	4.0	Instantaneous	10	5.9
$r_0 = 13.2$ cm	4.0	1.0 sec	2.8	1.7
$\theta_0 = 0.84$ arcsec	8.0	Instantaneous	4.3	2.5
$\lambda = 0.5$ μm	8.0	1.0 sec	1.7	1.0

phase screen in the directions of the two stars must be less than order r_0 to remain with the isoplanatic angle. For the case of two thin phase screens, one at altitude h and the other at ground level, that have a combined effective turbulence characterized by r_0, we use $C_n^2(z) = \frac{1}{2} C_n^2 \, \delta(z - h) + \frac{1}{2} C_n^2 \, \delta(z)$.

From Equations (2) and (3), we can now calculate the fundamental atmosphere-induced uncertainty in the classical astrometric measurement of the separation of two components of a binary star and summarize these results in Table I.

In calculating the uncertainty in the astrometric measurement after one second of observing time, we have again assumed that $\bar{v}/D$ independent measurements can be obtained per second, but here use a value for wind speed taken from an extensive study on atmospheric conditions in Southern Arizona (Merrill and Forbes, 1987), $\bar{v}$ = 25 m s^{-1}. We find that the improvement in astrometric precision with increasing telescope diameter is rather modest, $D^{-2/3}$, as was noted by Shao and Colavita (1992). The speckle cross-correlation technique, however, enjoys greater rewards with increasing telescope aperture as we shall discuss next.

2.2. DIFFERENTIAL ASTROMETRY FROM SPECKLE CROSS-CORRELATION

The turbulence-induced error in the speckle cross-correlation technique depends not only upon the differential wavefront tilt, but upon the higher spatial frequency phase aberrations as well. At first this seems to indicate that the astrometric error must be larger for the speckle cross-correlation technique. This is not the case, however, because the speckle cross-correlation technique is not burdened with the restriction that the two stellar wavefronts remain correlated point-by-point. In fact, there will be some level of correlation between speckle patterns as long as any portion of the atmospheric column through which the two wavefronts pass is shared. This results in a larger isoplanatic angle for speckle interferometry, one we shall call the speckle isoplanatic angle.

Roddier *et al.* first described the appropriate speckle isoplanatic angle and contrasted it with the adaptive optics isoplanatic angle (Roddier *et al.*, 1980). Roddier showed that the isoplanatic angle is the angle at which the value of the wavefront cross-correlation evaluated at zero separation decays by 1/e, while the speckle isoplanatic angle is the angle at which the integral of the wavefront cross-correlation over all separations decays by 1/e. The value of the cross-correlation at zero separation represents a point-by-point correlation. The integral of the cross-correlation represents correlation contributed from all separations. Roddier defined the speckle isoplanatic angle as

$$\Delta\theta = 0.36 \left(\frac{r_0}{\Delta h} \right) \tag{10}$$

where

$$\Delta h = \left(\frac{\int dh\, h^2\, C_n^2(h)}{\int dh\, C_n^2(h)} - \left(\frac{\int dh\, h\, C_n^2(h)}{\int dh\, C_n^2(h)} \right)^2 \right)^{-1/2} \tag{11}$$

Note that Δh is a description, not of the effective height of the turbulence, but of the effective thickness of the turbulence. For a thin phase screen at height h, residual speckle correlation will be found for angular separation up to $\theta = D/h$. This is a fundamental advantage of the speckle cross-correlation method over conventional relative astrometry.

While analytical expressions, Equations (2) and (3), are available for relative motion of energy centroids, in order to study the astrometric errors arising in the speckle cross-correlation method, we turn to Monte Carlo simulations. Appropriate analytical expressions for the error in the cross-correlation method are currently being investigated.

3. Speckle Cross-Correlation Simulations

We conducted a series of simulations of imaging two components of a binary system through the atmosphere. These simulations assumed a simplified atmospheric model of single or multiple Kolmogorov phase screens of varying turbulence strength, with apertures up to 8 m.

The wavefront in the pupil was calculated as the sum of OPD contributions from the projection of the telescope pupil on each screen (when looking off-axis). The sampling in the image plane was approximately 3 pixels across a speckle FWHM. The cross-correlation was calculated in real-space over a 5×5 pixel subraster, with the location of the correlation peak, and hence the separation, found by fitting an asymmetric Gaussian to the cross-correlation function. These simulations ignored scintillation, which has a small effect on imagery.

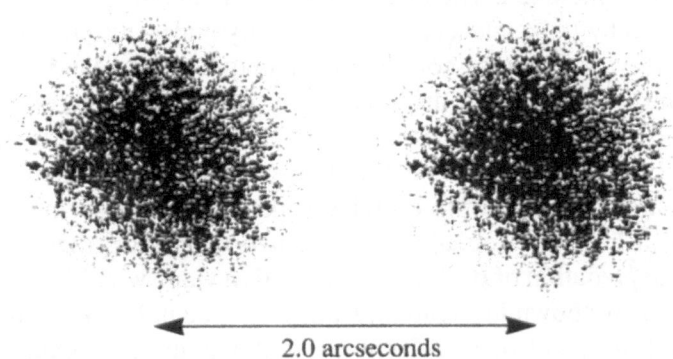

Fig. 1.   A typical 2-arcsec binary simulation, in this case for an 8-m telescope, $r_0 = 13.2$ cm, $\lambda = 0.5$ μm, half of the turbulence in a thin layer at 10 km, half at the surface.

The speckle cross-correlation accuracy depends strongly upon the thickness of the distorting atmospheric layers, largely independent of their height. To show this clearly, we give results for two simple atmospheric models that bracket the likely range of conditions. The first is for a single turbulent layer at high altitude h. Such conditions would be encountered occasionally at a well designed telescope, with no aberrations or dome seeing, at an excellent observing site with no ground layer turbulence. In the second model, we assume the turbulence to be divided between two thin layers, one at height h and one at the ground, each contributing equally to the total image degradation. The two layer turbulence model yields on-axis imagery similar to that of a single screen of effective r_0, given by

$$r_0^{-5/3} = r_{0_1}^{-5/3} + r_{0_2}^{-5/3} \tag{12}$$

where r_{0_i} for a single layer is the value of Equation (4), where the integral is taken over only that layer. Thus two layers of $r_0 = 20$ cm yield the same structure function as a single layer of $r_0 = 13.2$ cm. We find, however, that the cross-correlation effects are quite different. When looking off-axis the two turbulent layers are sheared with respect to one another.

A typical 2-arcsec binary simulation, in this case for an 8-m telescope, two phase screens, $r_0 = 13.2$ cm, $\lambda = 0.5$ μm, is shown in Figure 1. The phase screen generation was conducted with a sum-of-sines technique, forming a screen with Kolmogorov statistics. Details of this technique can be found in Colucci.

The quantities of interest extracted from these simulations are the rms uncertainties of the binary separation, as calculated by rms energy centroid motion (classical) and rms cross-correlation peak motion (speckle), found by fitting a narrow Gaussian to the sharp peak of the cross-correlation function. Here, we shall be concerned with the variation of these quantities with two parameters: the angular separation of the stars and the diameter of the telescope pupil. For these simulations no detector or photon noise was included. The effects of noise, including the more favorable conditions afforded by large telescopes, are currently under investigation.

The applicability of the cross-correlation technique to the search and study of planetary systems depends upon the availability of binary stars (or single stars with an assumed stationary reference, such as a distant galaxy). It is therefore important to analyze the degradation of the cross-correlation as a function of binary separation. The strength of the cross-correlation peak above the seeing background is an indication of how the isoplanicity of the two images degrades with binary separation. The degradation of peak cross-correlation value with binary separation for our two models is shown in Figure 2. Data presented in Figure 2 are for a 4-m telescope, $\lambda = 0.5$ μm, and $r_0 = 13.2$ cm. From this figure it is apparent that the peak correlation value is sensitive to the detailed distribution of turbulence with altitude. Thus, the thickness of the turbulent atmospheric layer becomes of primary importance for this technique, determining the peak correlation value and thus the ultimate limit to which this technique can be used in the presence of photon and detector noise. The case of two equal strength layers of turbulence (one high, one low) represents the worst case for the cross-correlation technique, because in this case, the projection of the telescope pupil on the two layers shears most rapidly with binary separation. Isoplanicity between the two images is lost approximately when the projection of the pupil onto the upper phase screen is sheared by about r_0 with respect to the projection of the pupil onto the lower phase screen. In Figure 2, the simulation results are plotted as points with error bars. The error bars indicate uncertainty in the simulation result due to the relatively few number of image pairs, 12, that were used for each point. Note that separations of less than 1 arcsec may not be usable for the cross-correlation technique, due to speckle overlap, but are useful to describe the nature of the Kolmogorov turbulence models. We are currently investigating non-linearities caused by speckle overlap.

Figure 3 summarizes the astrometric uncertainty results for the same simulations. We have plotted the rms uncertainty in the binary separation as a function of binary separation. Each of the four plots shown contains 3 curves. In each plot, the uppermost curve is the theoretical energy centroid motion, Equations (2) and (3). Very near to this in each plot is the rms energy centroid uncertainty resulting from the simulations. The lowermost curve in each plot is the rms speckle cross-correlation uncertainty.

Figure 4 summarizes the results of simulations using increasing telescope aperture, for a 4-arcsec binary and the model atmospheres considered above. The interpretation of these simulation results is twofold. First, when the effective thickness of the optical turbulence is small, it is possible to achieve unprecedented astrometric precision in the determination of binary star separation using the cross-correlation technique. For a 4-m telescope, a "thin" atmosphere can yield 0.30-mas accuracy from a single specklegram pair. For an 8-m telescope the single specklegram accuracy can be less than 0.10 mas. Second, the sensitivity of the cross-correlation technique precision to the thickness of the atmospheric turbulence is apparent.

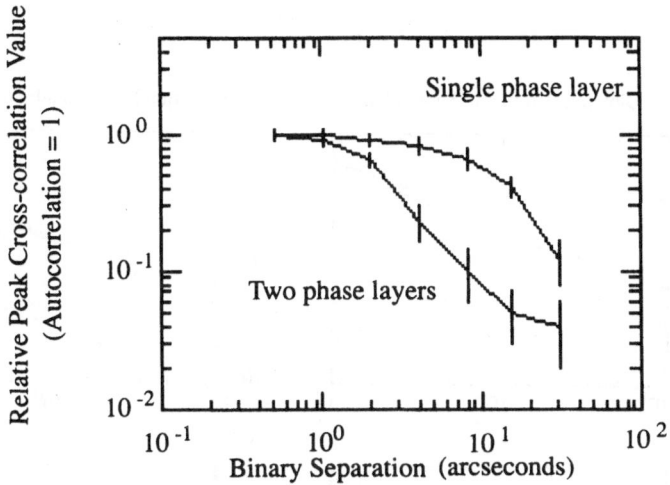

Fig. 2. Cross-correlation strength vs. binary separation for a 4-m telescope, $r_0 = 13.2$ cm, $\lambda = 0.5\,\mu$m, and all turbulence at 10 km or half at 10 km, half at surface.

TABLE II

Fundamental atmosphere-induced uncertainty in speckle cross-correlation measurement for a 4.0-arcsec binary star system.

Atmospheric Model	Telescope Diameter (m)	Total Exposure Time	σ_x (mas)	σ_y (mas)
Single thin phase screen				
10 km above telescope	4.0	Instantaneous	0.31	0.30
$r_0 = 13.2$ cm	4.0	1.0 sec	0.088	0.085
$\theta_0 = 0.85$ arcsec	8.0	Instantaneous	0.056	0.045
$\lambda = 0.5\,\mu$m	8.0	1.0 sec	0.022	0.018
Two thin phase screens				
(0 and 10 km above telescope)	4.0	Instantaneous	8.0	5.7
$r_0 = 13.2$ cm	4.0	1.0 sec	2.6	1.6
$\theta_0 = 1.7$ arcsec	8.0	Instantaneous	3.4	2.0
$\lambda = 0.5\,\mu$m	8.0	1.0 sec	1.4	0.80

Real atmospheric conditions should result in precisions between those of the two atmospheric models indicated. Data collected, to be described below, verify this situation for one observing run at Kitt Peak.

Some results of simulations as a function of pupil diameter are presented in Table II. We see that in the one phase screen simulation, the uncertainty in the speckle cross-correlation decreases more rapidly than D^{-2}. This is significantly stronger dependence than for classical differential astrometry. Heuristically, the

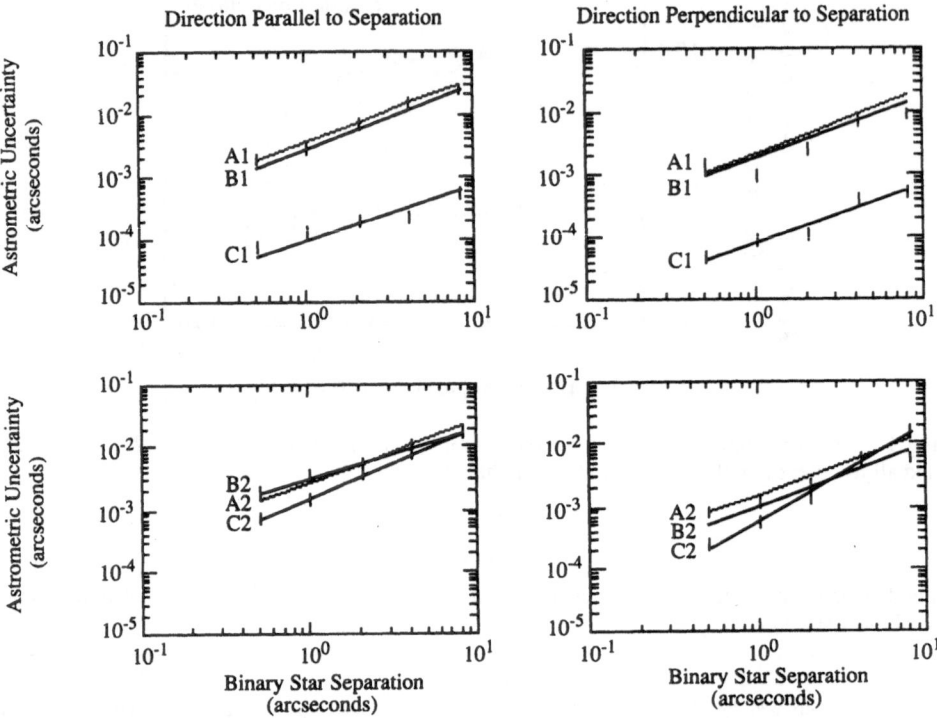

Fig. 3. Uncertainty in binary separation measurement for a 4-m telescope, $r_0 = 13.2$ cm, $\lambda = 0.5$ μm. Labels on curves refer to: A = theoretical centroid errors, B = simulated centroid errors, C = simulated speckle cross-correlation errors; 1 = one layer atmosphere model, 2 = two layer atmosphere model.

speckle cross-correlation scaling is determined by the accuracy with which one can determine the location of the cross-spectral peak (see Appendix A). This uncertainty is given approximately by

$$\sigma_{cc} \approx \frac{\text{width of cross-spectrum}}{\sqrt{\text{volume of cross-spectrum}}} \tag{13}$$

The width of the cross-spectrum is determined by the diffraction limit of the collecting aperture and decreases as D^{-1}. Lohmann and Weigelt (1979) have shown that the peak value of the cross-spectrum increases approximately as D^4 for small binary separations. This implies that the volume of the cross-spectral peak increases as D^2. The speckle cross-correlation uncertainty, therefore, is expected to improve approximately as D^{-2}. The development of a more rigorous analytical description of the speckle cross-correlation error is the basis for extensive future work which shall be discussed below.

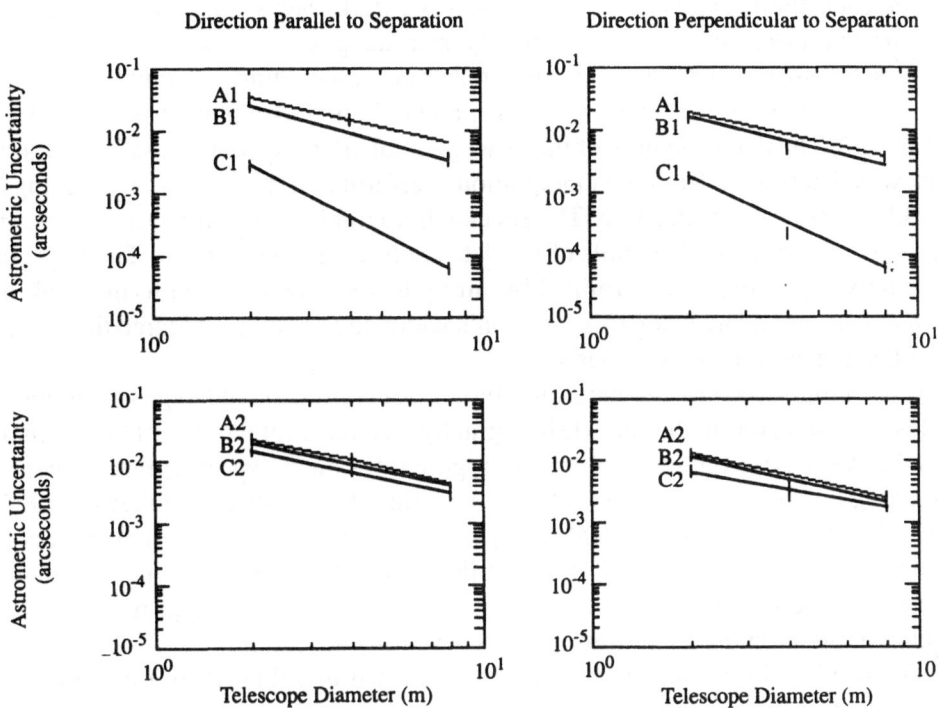

Fig. 4. Uncertainty in the measurement of the separation of a 4-arcsec binary vs. telescope diameter in two directions for $r_0 = 13.2$ cm, $\lambda = 0.5\ \mu$m, and two models of turbulence. Labels on curves refer to: A = theoretical centroid errors, B = simulated centroid errors, C = simulated speckle cross-correlation errors; 1 = one layer atmosphere model, 2 = two layer atmosphere model.

4. Observations and Data Reduction

Observations of several representative binary systems were made on November 11–12, 1992, with the Steward Observatory 2.3-m reflector at Kitt Peak. A Loral low-noise 800 lines × 1200 pixels/line CCD was used in an approximately f/50 focal plane, somewhat over-sampling the diffraction limit of the telescope aperture at 0.5 μm, giving a rectangular field of view of about 25-arcsec diagonal. A Uniblitz shutter provided accurate short exposure control, and on-chip cross-line image shifting with the shutter closed allowed multiple short exposure images (speckle-grams) in a single slow-scan read cycle. The detector read noise was about 9 e$^-$ per pixel rms. The observing bandpass was defined by optical filters and the spectral cut-off of the silicon detector. From the observations of the orbital positions predicted from five published orbits (Hirshfeld and Sinnott, 1985), the CCD image scale was computed to be 18.3 ± 0.6 mas/pixel.

For each multi-exposure frame of data the dark background was subtracted and the subrasters containing the binary star images were extracted. Figure 5 is extracted from a background corrected frame containing a sequence of four 30-ms specklegrams for μ Cyg (4.78 mag and 6.09 mag components separated by 1.9 arcsec). The determination of the exact location of the speckle cross-correlation peak was determined from data integrations performed in frequency space using a speckle holographic procedure. The speckle holography reconstruction results are presented, along with the details of the data analysis, in Appendix A. The value of the binary separation is determined by fitting a Gaussian function to the peak of the reconstruction and using this function's width and the mean-squared fit error to calculate the astrometric precision.

A tracking record of the measured binary separation versus exposure number yields similar results to the speckle holography technique for the case of high signal-to-noise ratio. An example is plotted in Figure 6 for the 24 specklegram sequence for μ Cyg obtained over an interval of about 9 minutes during November 12, 1992, 3:51:31 to 4:00:17 UT. Individual exposure times were 30 msec. The two curves in Figure 6 demonstrate the level of variability induced by the turbulent atmosphere in the measured binary separation. The two curves are for cross-correlation peak motion in the x- and y-directions (upper curve for y-direction).

The results of our data analysis are summarized in Table III. In this table, we present data concerning the star pairs observed, including their measured binary separation, θ, visual magnitude, the observing bands, and integration times. We then compare the measured standard error *per specklegram* with the centroid astrometry standard error predicted for these objects under similar atmospheric conditions by Equations (2) and (3) for classical differential astrometry. The standard error per specklegram is the standard error from the speckle reconstruction (see Appendix A) times the square-root of the number of observations that went into that reconstruction. To evaluate Equations (2) and (3), we used a value of $r_0 = 0.07$ m, $\theta_0 = 0.85$ arcsec.

The standard errors achieved for μ Cyg and γ Ori are comparable to or less than the standard errors originally reported by McAlister for a 1.6-arcsec binary taken at a 3.8-m telescope (McAlister, 1977b). McAlister reported 2 mas in each direction with 50 frames, or 14 mas in each direction per specklegram. We credit this improvement, at a smaller telescope, due to the superior quality of our detector.

We note that the speckle cross-correlation technique has, in an initial observing run, yielded experimental standard errors of similar magnitude to the theoretical centroid astrometry standard errors. The results are sensitive, however, to the instantaneous quality of the specklegram correlation. The most striking example of this can be seen in the results for γ And (A-B), where two data sets, taken only minutes apart, result in widely different standard errors. The better correlation between specklegrams present within the second data set is obvious from visual

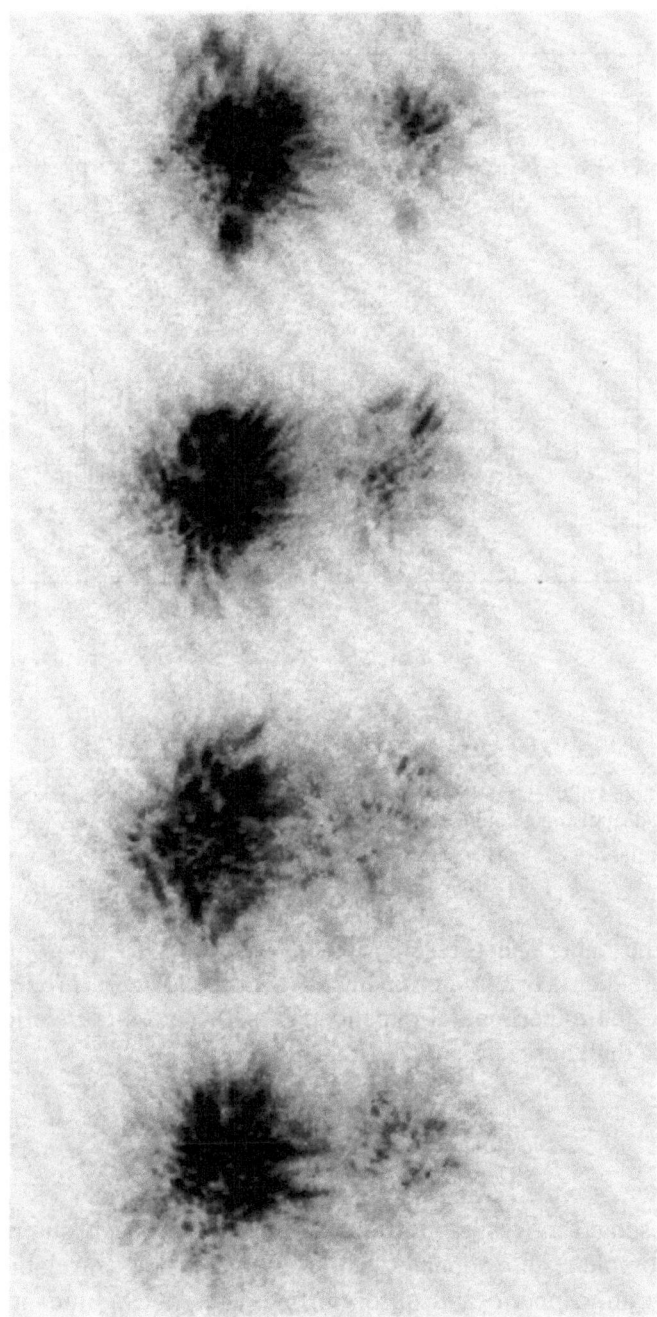

Fig. 5.   Sequence of four 30-msec exposures of μ Cyg (1.9-arcsec separation) with 900-nm (center)/100-nm (width) filter.

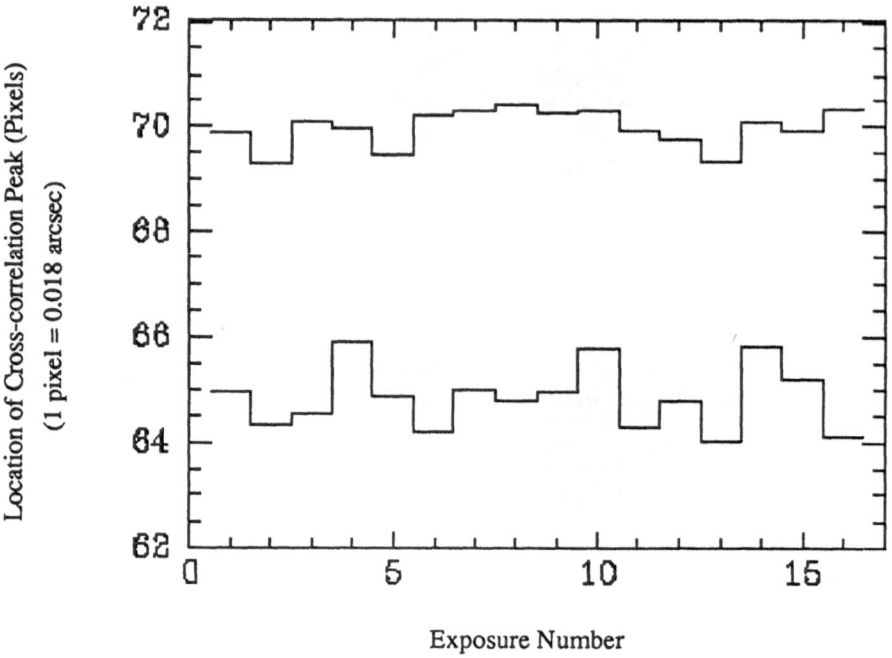

Fig. 6. Tracking record of the x- and y- centers (centroids) of a 5×5 pixel subraster centered on the peak of the cross-spectral image component $A = a^{**}l_{xs}$ of μ Cyg observed with 900-nm/100-nm filter.

inspection of the data. These results are consistent with a predominantly two-layer atmosphere that is briefly dominated by a single layer of turbulence. Further investigation of the experimental dependence of the cross-correlation technique upon turbulence thickness is planned.

5. Future Work

The results presented in this paper have stimulated significant interest in understanding the complete dependencies of the speckle cross-correlation technique upon the many atmospheric and observing parameters. Significant conclusions concerning the experimental behavior of the standard error as a function of integration time, filter bandwidth, signal-to-noise, and binary separation require better statistics. Therefore, we intend to follow our original observing run with another, scheduled for early June 1993, using equipment optimized for fast detector readout and storage.

TABLE III

Summary of observational speckle cross-correlation results and comparison to classical centroid differential astronomy theoretical predictions.

Object	θ (arcsec)	Visual Mag.	Filter Center/ Width (nm)	t (ms)	Experimental Speckle Cross-Correlation Errors		Theoretical Centroid Errors	
					σ_x (mas)	σ_y (mas)	σ_x (mas)	σ_y (mas)
μ Cyg	1.91	4.78/6.09	610/100	30	15	11	5.6	3.2
			900/100	30	4.8	6.0	5.2	3.0
ζ Ori	2.24	1.88/4.02	613/30	10	8.8	6.9	6.6	3.8
			600/60	10	7.2	13	6.6	3.8
			900/100	20	4.3	5.2	6.2	3.6
ζ Cnc (A-C)	6.02	5.6/6.2	600/60	10	60	32	18	10
ζ Cnc (B-C)	5.77	6.0/6.2	600/60	10	62	31	17	9.8
ξ Cep	8.13	4.4/6.5	900/100	50	210	170	23	13
				100	41	43	23	13
γ And (A-B)	9.19	2.3/5.5	600/60	30	140	68	27	16
				30	20	11	27	16
γ And (A-C)	9.7	2.3/6.3	600/60	30	66	28	29	17

We are planning future collaboration with Dr. James Beletic, who will provide such a Si CCD camera system developed at Georgia Tech (Beletic *et al.*, 1992). This camera consists of a 420 × 420 array of 17-μm square pixels (Lincoln Labs technology), which is read out with approximately 6–8 electrons of rms noise. The GTRI system can be programmed to read out simultaneous subarrays and log them in real time using 8-mm digital video tape (Exabyte). This should allow for the collection of 10^2 to 10^3 times the number of specklegrams obtained last November. We intend to utilize this ability to perform extensive investigation of the dependence of the speckle cross-correlation technique upon telescope diameter, the speckle isoplanatic angle, detector sampling, detector integration time, filter bandpass, and to verify the $1/\sqrt{N}$ improvement.

In the future, it will be desirable to avoid off-line reductions of tens of gigabytes of data per night of observation. Real-time calculation of the speckle cross-correlation is feasible in small dedicated signal processors using Intel i860 chips. For example, the CSPI SC-3XL dual chip i860 implementation (200 mFlop) can compute 256 × 256 discrete Fourier transforms at 50 Hz, easily accommodating the requirements for real-time reduction of speckle cross-correlations.

The extremely rapid rates at which the speckle cross-correlation technique can achieve submilliarcsecond accuracy allow us to relax the stability requirements for the instrumentation. We propose measuring the binary separations of up to 100 star pairs per night, a few nights a year. Due to the large number of independent motions, we attribute overall scale changes to systematic effects and subtract the average motion for each program object. Variations within the instrument occurring over time scales of seconds to minutes are apparent in derived separation tracking records (see Figure 6). Variations between observing runs are calibratable by considering the average of all binary separations to have not changed, or by observing distant, but bright, object pairs which are known to have very small proper motion.

6. Conclusions

We have reported on the investigation of a powerful technique of differential astrometry that depends uniquely upon the thickness of the atmospheric turbulence layer and not simply upon r_0 or the isoplanatic angle. We have shown that in favorable conditions, namely that of a single thin aberrating turbulence layer, the technique of speckle cross-correlation is capable of experimental astrometric standard errors of 10 mas per specklegram or 2 mas per second of observation for a 2-arcsec binary in moderate to poor seeing at a 2.3-m telescope, and has the potential to yield much smaller errors, less than 100 μas per specklegram, for favorable atmospheric conditions on new 6.5- to 8-m class telescopes. Originally proposed by McAlister, the technique of speckle cross-correlation is now fundamentally improved and technologically feasible due to the production of new very large telescopes and new small format, low noise, rapid readout CCDs.

Appendix A

SPECKLE HOLOGRAPHY AND THE CROSS-CORRELATION TECHNIQUE

In order to facilitate the calculation of the speckle cross-correlation, we chose to perform the cross-correlation in frequency space. By processing the data using speckle holography we obtain the tracking record of the cross-correlation peak as a by-product of the speckle reconstruction.

Speckle holography, first proposed by Bates, Gough, and Napier, requires simultaneous specklegrams for the object of interest and a nearby (within the isoplanatic angle) reference source. The technique, described by Hege (Hege et al., 1989), is facilitated by integration of the cross-spectral image computed from the simultaneous specklegram pairs. The output of a speckle holography reconstruction is known as a cross-spectral image, and is nothing more than the noise-filtered cross-correlation of an object and a reference, in our case the two specklegrams of the binary star. The cross-spectral image is defined as the inverse Fourier transform

$$i_{xs} = FT^{-1}\left(\frac{SR^*}{RR^*}\right) \tag{1}$$

where $S = FT(s)$ and $R = FT(r)$ are the complex two-dimensional Fourier spectra of the specklegram, s, and the corresponding reference point-spread function (PSF), r, respectively. FT^{-1} represents the inverse Fourier transformation and $*$ represents complex conjugation. Physically, the numerator, SR^*, is the Fourier transformation of the cross-correlation of the image and reference components, and the denominator, RR^*, is the power spectrum of the seeing. If the short exposure image is of a binary star, $s = a**PSF_a + b**PSF_b$, where $**$ represents convolution, and its Fourier spectrum is $S = A R_a + B R_b$. If both components are unresolved, then A and B are scalars representing the binary component intensities. R_a and R_b are the optical transfer functions (that is, Fourier transforms of the PSFs) for components a and b, respectively. If, for example, the reference OTF is $R = R_b$, then

$$i_{xs} = a**FT^{-1}\left(\frac{R_a R_b^*}{R_b R_b^*}\right) + b**FT^{-1}\left(\frac{R_b R_b^*}{R_b R_b^*}\right) \tag{2}$$

or

$$i_{xs} = a**l_{xs} + b\,\delta(r - r_b) \tag{3}$$

where l_{xs} is the cross-spectral response function

$$l_{xs} = FT^{-1}\left(\frac{R_a R_b^*}{R_b R_b^*}\right) \tag{4}$$

in the perfectly isoplanatic case $R_a = R_b$ in which case the binary is recovered exactly. The effect of partial anisoplanicity is to broaden the response function l_{xs} (low-pass filtering due to loss of coherence at high spatial frequencies) and to shift the location of the response (wave-front tilt differences).

Primont and Fontinella (Primot et al., 1988) and Gonglewski et al. (1990) have shown the benefits of averaging the numerator and denominator independently in accumulating the self-calibrated average cross-spectral image,

$$\langle i_{xs}\rangle = FT^{-1}\left(\frac{\langle SR^*\rangle}{\langle RR^*\rangle}\right) \tag{5}$$

The cross-spectral images (see below) were self-referenced, computed using one of the unresolvable image components of the binary as the reference point-spread function (PSF). A typical example, for μ Cyg, is shown in Figure 7. The sequence shows the results of 1, 2, 4, 8, and all 20 frames processed. The autocorrelation of

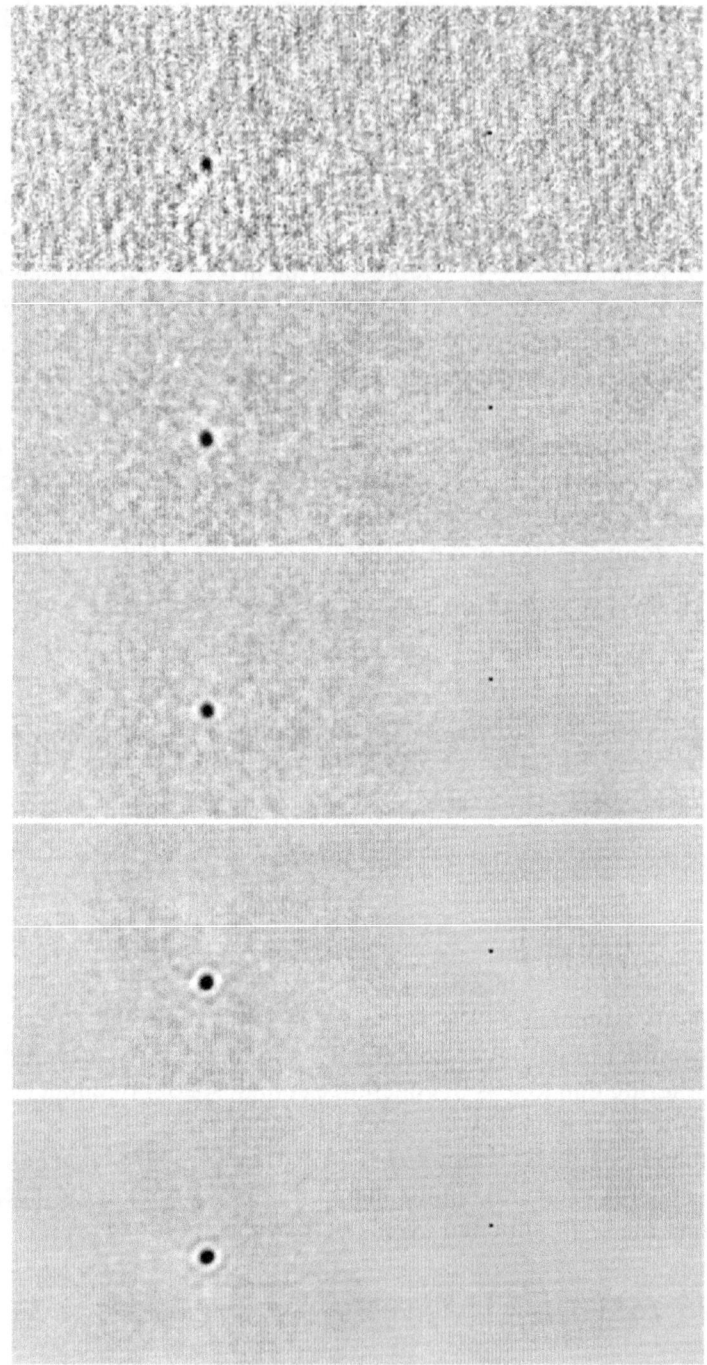

Fig. 7. Self-referenced cross-spectral image of μ Cyg observed with 900-nm/100-nm filter, computed using the fainter unresolved component as the reference point spread function. The sequence shows 1, 2, 4, 8, and all 20 frames processed.

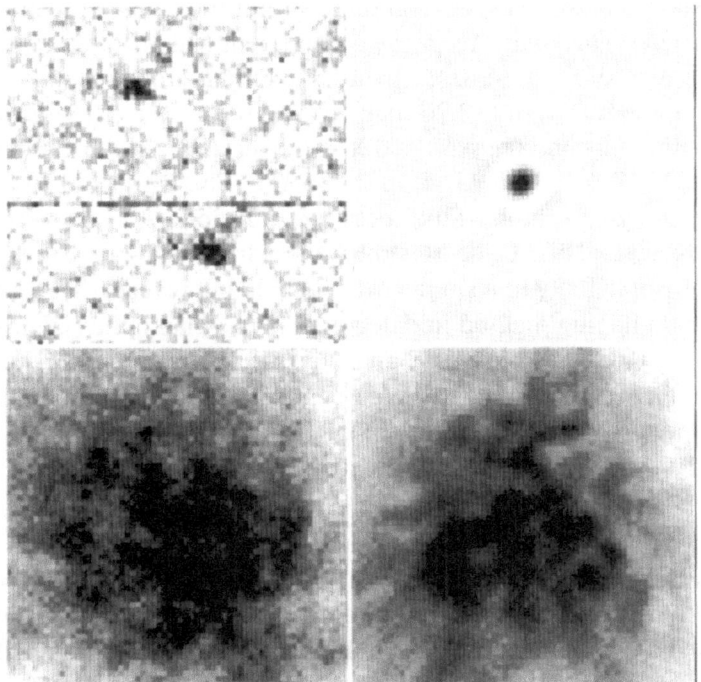

Fig. 8.   Cross-spectral response function for μ Cyg A (900-nm/100-nm filter), self-calibrated by μ Cyg B (lower left). Cross-spectral response function for resolved binary component ζ-Cnc BC (600-nm/60-nm filter), self-calibrated with ζ Cnc A (lower right). The corresponding long exposure images are at the top.

the fainter component, used as a PSF estimate, with itself (right) is always a single-pixel delta function. The cross-spectral response, l_{xs}, increases in signal-to-noise ratio as more frames are processed (left).

The standard errors presented in Table III are calculated by least-squares fitting a Gaussian function to a 5×5 subraster centered on the cross-spectral image reconstruction peak, and using the width of the Gaussian and the mean-squared fit error to form

$$\text{standard error in } x\text{-direction} = \sigma_x = \sigma_{x \text{ Gaussian}} \overline{\epsilon^2} \qquad (6)$$

where $\overline{\epsilon^2}$ is the mean-squared error of the fit. A similar definition gives σ_y. Thus, in Figure 7, the standard error decreases with increasing number of frames processed, due to higher signal-to-noise of the reconstruction and a correspondingly better fit.

From the results in Figure 7, we find that the standard deviations from the mean (x, y) separations are 1.19 mas in x and 2.21 mas in y. The mean cross-spectral response of the 24 specklegram result for the μ Cyg A response is shown in Figure 8 (left). This image response is consistent with the prediction based upon the tracking record. Figure 8 (right) shows a similar cross-spectral integration of 62 specklegrams for γ And (2.26 mag K0 and 4.84 A0 separated by 9.2 arcsec) observed with a 60-nm bandpass at 600 nm over an interval of about 7 minutes, November 11, 1992, 11:50:23 to 11:57:46 UT. This is, to our knowledge, one of the highest signal-to-noise ratio reconstructions of a 9.2-arcsec binary using a 10% filter to date. This result illustrates the performance of the algorithm near the limits of isoplanicity at a shorter wavelength, but reveals the sensitivity of the cross-spectral imaging method in detecting the $\sim$1 mag fainter component of the resolved ζ-Cnc BC image in this short integration. The characteristic tendency of partially compensated imaging to yield a sharp, nearly diffraction limited, response superimposed on a diffuse, nearly seeing extent, noisy background is illustrated.

Acknowledgements

This work is supported in part by NSF grant AST 92-03336. KH acknowledges support from E.M. Hege and the University of Arizona Foundation. The data reductions were assisted by Tucson Unified School District Professional Internship Program student S. Kobriger and by physics/astronomy undergraduate M. Cheselka.

References

Bates, R. H. T., Gough, P. T. and Napier, P. J.: 1973, *Astron. Astrophys.* **22**, 319–320.
Beletic, J. W., Zadnik, J. A., Tritsch, C. L. and DuVarney, R.C.: 1992, in : High-Resolution Imaging by Interferometry, Proc. ESO Conf. and Workshop #39.
Black, D.C.: 1980, In Search of Other Planetary Systems, *Space Science Reviews* **25**, 35–81.
Colucci, D.: 1992, *SPIE Proc.* **1688**, 527–535.
Fried, D.L.: 1965, *JOSA* **55, No. 11**, 1427–1435.
Fried, D.L.: 1982, *J. Opt. Soc. Am.* **72**, 52.
Gatewood, G.D.: 1987, *Astrophys. J.* **94**, 213.
Gonglewski, J.D., Voelz, D.G., Fender, J.S., Cayton, D.D., Spielbush, B.K. and Pierson, R.E.: 1990, *Appl. Optics* **29**, 4527–4529.
Hege, E.K.: 1989, in: D. M. Alloin and J. M. Mariotti, ed(s)., *Diffraction-Limited Imaging With Very Large Telescopes*, Kluwer Academic Publishers, pp. 113–124.
Hirshfeld, A. and Sinnott, R.W. (eds).: 1985, *Sky Catalog, 2000.0, Vol. 2, Epoch 1992.865*, Sky Publishing Corporation, Cambridge, Massachusetts.
Lindegren, L.: 1980, *Astron. Astrophys.* **89**, 41–47.
Lohmann, A.W. and Weigelt, G. P.: 1979, *Optik* **53**, 167.
McAlister, H. A.: 1977a, *Astrophys. J.* **215**, 159–165.
McAlister, H. A.: 1977b, *Icarus* **30**, 789–792.
McAlister, H. A. and DeGioia, K. A.: 1979, *Astrophys. J.* **228**, 493–496.
Merrill, K. M. and Forbes, F. F.: 1987, *Comparison of Astronomical Site Quality of Mount Graham and Mauna Kea*, NNTT Report #10, NOAO, Tucson, Arizona.
Primont, J., Rousset, G. and Fontanella, J.C.: 1988, in *Very Large Telescopes and Their Instrumentation I*, Proc. ESO Conf. and Workshop #30, 683–692.

Roddier, F., Gilli, J. M. and Vernin, J.: 1980, On the Isoplanatic Patch Size in Stellar Speckle Interferometry, *J. Optics* **13**, 63–70.
Sandler, D.G., Stahl, S., Angel, J.R.P, Lloyd-Hart, M. and McCarthy, D.: 1993, *JOSA A*, in press.
Shao, M. and Colavita, M. M.: 1992, *Astron. Astrophys.* **262**, 353–358.

ON THE SEARCH FOR O_2 IN EXTRASOLAR PLANETS *

JEAN SCHNEIDER

CNRS-UPR, L'Observatoire, Meudon, France

Abstract. The present work is a first attempt to investigate the feasibility of the optical detection of absorption lines in the atmosphere of extrasolar planets. Two cases are considered: (1) the "reflection" case, where the planet is observed by imaging via the light coming from the parent star and re-emitted by the planet; and (2) the "occultation" case, where the planet makes a partial occultation of the star. I find only in the latter case that there are planetary configurations where the absorption lines can effectively be detected with existing or forthcoming telescopes. The discussion is basically made for the O_2 A-band, which is of interest for exobiology.

1. Introduction

The search for extrasolar planets is difficult by any method. This difficulty is, for instance, illustrated in a recent paper by Walker *et al.* (1992) who point out that the spectral variations of the star γ Ceph are probably due to its intrinsic activity and not to its velocity variations. The search for small occultations of parent stars due to an orbiting planet, although difficult, is not much harder than other methods. This photometric method has been proposed by Struve (1952) and investigated more quantitatively by Rosenblatt (1971), Borucki and Summers (1984) and Schneider and Chevreton (1990). It suffers from the weakness that it requires much telescope time if no particular care is taken in the choice of the stars to be observed. In order to improve the geometrical probability, the temporal frequency and the photometric efficiency of the occultations, Schneider *et al.* (1991) have proposed to search for planets near dwarf stars.

If a planet were to be discovered by any method, the main question would be: "does it experience any complex organic chemistry?" The indirect discovery of a planet by the reflex motion of its parent star, like the two pulsar planets (Wolszczan and Frail, 1992), would not help to learn anything about the planet itself (apart from its mass and orbital characteristics). The direct imaging and occultations of the parent star by the planet would be the only ways to discern the composition of its atmosphere. Here I will focus on the search for a biochemical activity.

Lovelock (1975) suggests that the presence of oxygen in the atmosphere of a planet may be the sign of complex organic chemical processes. Owen (1980) has argued that ozone would also be a tracer for life and has suggested that both oxygen and ozone could be detectable spectroscopically. Angel (1988, 1990) and Burke (1988) have investigated the feasibility of detecting ozone in the IR for the case of direct imaging of a planet. The conclusion of Angel (1990) is that a 16-m telescope in orbit, with supersmooth mirrors and apodization, would be necessary for the

---

* Paper presented at the Conference on *Planetary Systems: Formation, Evolution, and Detection* held 7–10 December, 1992 at CalTech, Pasadena, California, U.S.A.

Astrophysics and Space Science **212**: 321–325, 1994.
© 1994 *Kluwer Academic Publishers.*

detection of ozone; this is certainly not an easy goal. Moreover Léger *et al.* (1993) have shown that ozone may be not a good tracer for life. In this paper I present a first attempt to investigate the feasibility of the spectroscopic detection of oxygen absorption lines in the optical spectrum of the parent star when the latter is partially occulted by an orbiting planet.

According to Owen (1980) the main feature useful for the present purpose in the absorption spectrum of O_2 is the A-band at 7600 Å. Its width of about 30 Å allows for moderate resolution spectroscopy, whereas the Schumann–Runge band at 2000 Å and the B-band at 6800 Å would need higher resolution. Let us then successively consider the 'reflection' case, where the observer takes a spectrum of the light reflected by the planet, and the 'occultation' case, where he takes a spectrum of the star slightly occulted by the planet and its atmosphere. Let N be the recorded photon counts per wavelength interval in the continuum part of the spectrum near the absorption band and $(1 - \eta)N$ be the recorded photon number per wavelength interval inside the absorption band. The pertinent signal is the difference ηN between these two quantities; η can be seen as an effective depth of the absorption band. We have to compare the signal ηN with the noise ΔN. In the framework of this first attempt, I consider only the Poisson noise of the signal. Since in the present case η is very small $(1 - \eta)N \approx N$ and thus the statistical noise inside the absorption band is roughly $\Delta(1 - \eta)N \approx \sqrt{N}$. Therefore the pertinent noise/signal ratio $\Delta S/S$ is

$$\frac{\Delta S}{S} = \frac{\sqrt{N}}{\eta N} = \frac{1}{\eta \sqrt{N}}. \tag{1}$$

2. Imaging or 'Reflection' Case

$N(= N_R)$ is the sum of two terms:

1. The number N_P of recorded photons reflected by the planet.
2. The number N_{Diffr} of photons present in the wings of the diffraction peak of the parent star.

The number $(1 - \eta_R)N_R$ is the difference between the contribution N_{Diffr} from the diffraction peak and the fraction $f N_P$ (f is the absorption coefficient of the planet atmosphere) of the photons absorbed in the band considered. Therefore $(1 - \eta_R)N_R = N_R - f N_P$ or $\eta_R = f N_P/N_R$. Since the planet is very faint compared to the wings of the diffraction peak then $N_R \approx N_{\text{Diffr}}$. For a planet with radius R_P, albedo A and in orbit with radius a we have

$$N_P = \frac{A}{4} N_\star \left(\frac{R_P}{a}\right)^2. \tag{2}$$

For a telescope diameter Δ and a star at distant D, an empirical approximation for N_{Diffr} is

$$N_{\text{Diffr}} = 0.2 \times N_{\star} \left(\frac{\lambda}{\Delta} \cdot \frac{D}{a} \right)^3 \tag{3}$$

where $N_{\star}$ is the incoming number of photons emitted by the star. Therefore

$$\eta_R = \frac{f N_P}{N_{\text{Diffr}}} \approx 1.2 \times f \left(\frac{R_P}{a} \right)^2 \left(\frac{\Delta}{\lambda} \cdot \frac{a}{D} \right)^3 \tag{4}$$

and

$$\left(\frac{\Delta S}{S} \right)_R = \frac{1}{\eta_R} \frac{1}{\sqrt{N_R}} = \frac{1.6}{f} \left(\frac{R_P}{a} \right)^{-2} \left(\frac{\lambda}{\Delta} \cdot \frac{D}{a} \right)^3 N_{\star}^{-1/2}. \tag{5}$$

3. Occultation Case

Here $N(= N_O)$ is just the recorded photon counts $N_{\star}$ from the star and the depth η_O of the absorption band is

$$\eta_O = f \frac{\Sigma_{\text{Atm}}}{\Sigma_{\star} - \Sigma_P} \approx \frac{\Sigma_{\text{Atm}}}{\Sigma_{\star}} \tag{6}$$

where $\Sigma_{\star}$, Σ_P and Σ_{Atm} are, respectively, the projected surface area of the star, the planet and the planet atmosphere. For a star with radius $R_{\star}$ and an atmosphere with height H the depth η_O is approximately given by

$$\eta_O = f \frac{2 H R_P}{R_{\star}^2}. \tag{7}$$

Thus the noise/signal ratio is

$$\left(\frac{\Delta S}{S} \right)_O = \frac{1}{\eta_O} \frac{1}{\sqrt{N_O}} = \frac{f R_{\star}^2}{2 H R_P} \cdot \frac{1}{\sqrt{N_{\star}}}. \tag{8}$$

4. Numerical Estimates

The photon number $N_{\star}$ recorded from a star with an apparent magnitude m in a bandwidth W in a time T_{int} is given by

$$N_{c,\star}^W = 0.8 \times 10^{13} \times \epsilon 10^{0.4 \, m} \left(\frac{\Delta}{1 \text{ m}} \right)^2 \left(\frac{W}{100 \text{ Å}} \right) \left(\frac{T_{\text{int}}}{10^4 \text{ sec}} \right) \tag{9}$$

for a given detection efficiency ϵ. To obtain a signal/noise ratio $S/\Delta S$ of 3, we find for the occultation case that

TABLE I
Configurations used for the evaluation of T_{int}.

Configuration	I	II	III	IV
R_*	$1\,R_\odot$	$0.3\,R_\odot$	$1\,R_\odot$	$0.3\,R_\odot$
M_*	$1\,M_\odot$	$0.3 M_\odot$	$1 M_\odot$	$0.3 M_\odot$
R_P	$1\,R_\oplus$	$1\,R_\oplus$	$1\,R_\oplus$	$1\,R_\oplus$
a	1 AU	0.03 AU	1 AU	0.03 AU
D	2 pc	10 pc	2 pc	10 pc
H	30 km	30 km	30 km	30 km
m	2	5	2	5
Δ	$2.4\,m$	$2.4\,m$	$16\,m$	$16\,m$

$$T_{int,O} = 8 \times 10^7 \times$$

$$10^{0.4\,m} \frac{f^2}{\epsilon} \left(\frac{H}{1\,\text{km}}\right)^{-2} \left(\frac{R_P}{R_\oplus}\right)^{-2} \left(\frac{R_*}{R_\odot}\right)^{4} \left(\frac{\Delta}{1\,\text{m}}\right)^{-2} \left(\frac{W}{100\,\text{Å}}\right)^{-2} \text{sec} \quad (10)$$

and for the reflection case

$$T_{int,R} = 2.5 \times 10^{10} \times 10^{0.4\,m} \frac{f^2}{A^2\epsilon}$$

$$\left(\frac{a}{1\,\text{AU}}\right) \left(\frac{\lambda}{8000\,\text{Å}}\right)^{3} \left(\frac{\Delta}{1\,\text{m}}\right)^{-5} \left(\frac{R_P}{R_\oplus}\right)^{-4} \left(\frac{D}{1\,\text{pc}}\right)^{3} \left(\frac{W}{100\,\text{Å}}\right)^{-2} \text{sec.} \quad (11)$$

Let us compare numerically these integration times for two configurations: an Earth-like planet near a Solar-type star and a planet orbiting close to an M dwarf. For both cases we consider, respectively, a 2.4-meter and a 16-meter telescope. The complete characteristics of these four combinations are listed by Table I. Taking $\epsilon = 0.3$, $A = 0.5$, $W = 30$ Å and $f = 1$ (saturated band) the resulting integration times are given in Table 2.

5. Discussion

One can see from Table II that the spectroscopic detection of O_2 in the image of an extrasolar planet in an orbit with radius of 1 AU or smaller would require unrealistic observation times with presently existing telescopes of the HST class. On the other hand, the spectroscopic detection of O_2 during the partial occultations of a star by an orbiting planet would be possible with a 2.4-meter telescope with standard optics. This should encourage observers to search for extrasolar planets by the photometric method.

TABLE II

Integration times for $S/\Delta S = 3$ for the different configurations.

Configuration	$T_{int,R}$ (sec)	$T_{int,O}$ (sec)
I	2.6×10^9	9.7×10^3
II	1.3×10^{10}	3.5×10^3
III	2×10^5	2.4×10^2
IV	3×10^8	10^2

References

Angel, R.: 1990, *Use of a 16m Telescope To Detect Earthlike Planets*, Steward Observatory, Preprint nr. 928.

Borucki, W. and Summers, A.L.: 1984, , *Icarus* **58**, 121.

Burke, B.F.: 1988, in G. Marx, ed(s)., *Bioastronomy–The Next Step*, 153.

Léger, A., Pirre, M. and Marceau, F.J.: 1993, , this volume.

Lovelock, J.E.: 1975, *Proc. Roy. Soc. Lond.* **B 189**, 167.

Owen, T.: 1980 , in Papagiannis, ed(s)., *Strategies for the Search of Life in the Universe*, 177.

Rosenblatt, F.: 1971, *Icarus* **14**, 71.

Schneider, J. and Chevreton, M.: 1990, *Astron. Astrophys.* **232**, 251.

Schneider, J., Chevreton, M. and Martin, E.L.: 1991, in: B. Battrick, ed(s)., *24th ESLAB Symposium*, ESA SP-315, 67.

Struve, O.: 1952, *The Observatory* **72**, 199.

Walker, G.A.H., Bohlender, D.A., Walker, A.R., Irwin, A.W., Yang, S.L.S. and Larsen, A.: 1992, *The Astrophys. J. Lett.* **396**, L91.

Wolszczan, A. and Frail, D.A.: 1992, *Nature* **355**, 145.

HOW TO EVIDENCE LIFE ON A DISTANT PLANET

A. LÉGER
Institut d'Astronomie Spatiale,
CNRS, Université Paris Sud, Orsay, France

and

M. PIRRE and F.J. MARCEAU
Laboratoire de Physique et Chimie de l'Environnement,
CNRS, Orléans, France

Abstract. Considering the future importance of the search for evidences of primitive life on a distant planet, we have revisited some points of the O_2 and O_3 detection criteria.

The budget of free oxygen and organic carbon on Earth is studied. If one includes the organic carbon in sediments, it confirms that O_2 is a very reactive gas whose massive presence in a telluric planet atmosphere implies a continuous production. Its detection would be a strong indication for photosynthetic activity, providing the planet is not in a runaway greenhouse phase.

In principle, the direct detection of O_2 could be possible in the visible flux of the planet at 760 nm (oxygen A-band) but it would be extremely difficult, considering the much larger flux from the star. The alternative search for the 9.7 μm absorption of O_3 may be easier as the contrast with the star is improved by 3 orders of magnitude. A simple atmospheric model confirms that the O_3 column density is not a linear tracer of the atmospheric O_2 content, as was found in the pioneer work by Paetzold (1962). However, the detection of a substantial O_3 absorption ($\tau > 30\%$) would probably indicate, within the validity of this model, an O_2 ground pressure larger than 10 mbar. The question is raised of whether this pressure is sufficient to indicate a photosynthetic origin of the oxygen. If the answer was positive, it would be an even *more sensitive test of photosynthetic activity than the detection of the oxygen A-band*. Further studies of these points are clearly needed before determining an observing strategy.

1. Introduction

The search for evidences of primitive life on extraterrestrial planets will probably be of increasing importance in the future. Specially if the development of the SETI program does not give indications of technological signals, an important issue will be to decide where is the bottleneck: from planetary habitable conditions to actual appearance of primitive life or from primitive life to evolved and technological life?

Although we should be prepared to search for life quite different from what we have on Earth, T. Owen (1980) has argued that, if based on chemistry, alien life is likely to rely on carbon chemistry and liquid water. Silicon chemistry and liquid ammoniac, for instance, seem less favorable. Looking for life implying photosynthesis may be not too restricted an approach.

A first idea is to search for the signature of chlorophyll-like molecules. However, we do not know where such a signature should be and, more important, this signature could be quite weak. In the case of the Earth, the absorption of the visible light reflected by the planet is only about 2% at the maximum of the chlorophyll absorption bands (420 and 660 nm).

Astrophysics and Space Science **212**: 327–333, 1994.
© 1994 *Kluwer Academic Publishers.*

Considering the situation in the Solar System, Lovelock (1975) concludes that the coexistence of gases in a mixture which is out of thermodynamical equilibrium (e.g., O_2 and CH_4) is a strong, although not certain, indication of biological activity.

T. Owen (1980) considers that the massive presence of O_2 in the atmosphere of a telluric planet with its reducing rocks (ferrous oxides) is indicative of photosynthesis activity but for the short period of a possible "runaway greenhouse" phase when the photolysis of H_2O can be a major, but time limited, source of oxygen. Otherwise, in a cool atmosphere, water is protected from getting in its upper parts, where it can be photolysed, by a cold trap that makes it fall down as snow or rain. Then, the massive presence of O_2 in an atmosphere (0.01–0.2 bar) and a moderate temperature (300 ± 50 K) would be a strong indication of photosynthesis-based life in a distant planet.

In this paper, we want to study the budget of the different oxygen and carbon reservoirs on Earth to check whether they are compatible with a biological origin of the atmospheric oxygen. We also discuss the relevance of detecting O_3 in the IR (9.7 μm) as a tracer of O_2 in the atmosphere.

2. Budget of Free Oxygen and Organic Carbon on Earth

The global reaction of photosynthesis can be written as:

$$CO_2 + H_2O + 2h\nu(\text{chlorophyll}) \rightarrow (CH_2O) + O_2 \tag{1}$$

(CH_2O) is symbolic for a typical organic molecule such as a glucide $(CH_2O)_n$. The organic carbon involved in such species is partially reduced and can be easily distinguished from oxidized carbon in CO_2 or in carbonates. The produced O_2 goes in the atmosphere and may oxidize rocks or organic carbon.

If atmospheric O_2 on Earth has a photosynthetic origin and organic C has not another sink than oxidation by oxygen, there must be more, or as much, C atoms in a reduced form than O_2 molecules in the atmosphere. This is not obvious if one considers the respective number of species in the atmosphere and in the biomass, and it warrants a more detailed study as has been done by Walker (1977) but using more recent estimates of the reservoirs.

The different reservoirs of O_2 and C can be measured by mass or equivalently by their thickness if spread in a uniform layer covering the whole Earth surface, with an arbitrary specific mass of 1 g cm^{-3}. One has the relation:

$$1 \text{ m} \leftrightarrow 5.15 \times 10^{20} \text{ g}$$

This way of measuring the reservoirs is rather pictorial. For instance, we can readily appreciate the amount of atmospheric O_2 or living biomass on lands. Table I shows the content of the different reservoirs of free oxygen and organic carbon as reported by Siegentharler (1986). As expected, the oxygen amount (2.3 m) is much larger than the carbon one contained in the biomass (6 mm) or in fossil

TABLE I

Major reservoirs of free O_2 and organic carbon (adapted from Siegenthaler, 1986).[1]

	Reservoir			h (m)
(1)	O_2 in atmosphere			2.3
	dissolved in oceanic water[2]			$< 8 \times 10^{-3}$
(2)	Organic carbon			
	– biomass	– land	(living)	1.1×10^{-3}
			(humus)	3×10^{-3}
		– ocean	(living)	6×10^{-6}
			(dissolved)	2×10^{-3}
	– fossil fuels			1.1×10^{-2}
	– sediments:	organic C		23
		(inorganic C)		(120)

[1] The amounts are expressed in thicknesses of a hypothetical layer of the species spread on the Earth's surface that would have a specific mass of 1 g cm^{-3}. The equivalence with the reservoir mass is 1 m $\leftrightarrow 5.15 \times 10^{20}$ g(-3) means 10^{-3}.
[2] Based on less than 5 ml of O_2 STP dissolved per liter of water when averaged over the ocean depth.

fuels (11 mm). But *the key reservoir of reduced carbon is organic in sediments* (23 m). Sediments contain a substantial fraction of organic carbon ($\sim 1\%$ for clays) besides the inorganic one (carbonates). As a result, the total reservoir of organic carbon is significantly larger than that of free oxygen and this is *compatible with a photosynthetic origin of atmospheric* O_2. This is in agreement with the conclusions by Walker (1977).

Reaction (1) implies that each time an organic carbon atom has been produced by photosynthesis in the past, an oxygen molecule has been released. The production of 23 m of reduced C has released 61 m of free O_2. This is much more than the present amount of atmospheric oxygen (2.3 m). Then, over 96% of the photosynthetic oxygen produced during the past has been removed from the atmosphere and fixed in rocks by oxidation of sulfurs and iron during weathering. A continuous production of oxygen appears to be necessary to maintain a substantial fraction of this gas in an atmosphere. This strengthens the argument by T. Owen (1980) that *the massive presence of* O_2 *in the atmosphere of a planet* which is not in a runaway greenhouse phase *is a strong indication of photosynthetic activity*.

The mere study of the IR emission by the planet could be insufficient to exclude that it is in a runaway greenhouse phase, because it may come from the cold upper parts of its atmosphere (e.g., Venus) but this phase is, anyway, of short duration and the detection of O_2 in several planets would be statistically meaningful.

3. Is O_3 a Good Tracer of Atmospheric O_2?

3.1. How To Detect O_2?

T. Owen (1980) has suggested to detect the presence of O_2 in a distant atmosphere by looking for its A-band absorption at 760 nm. However, this is a very difficult observation. Apart from the favorable case of a planet eclipsing a star (Schneider, 1992), one has to image the star and the planet, block the star light and make a spectrum of the planet light. The star-to-planet flux ratio is 5×10^9 in the case of Sun/Earth. This indicates how difficult would be such an observation.

3.2. An Interesting Alternative: The Detection of Ozone

Angel (1986) has pointed out that this ratio decreases by almost 3 orders of magnitude if fluxes are considered in the IR, around 10 μm. In this spectral range, he noticed that Earth is the only telluric planet that exhibits a strong O_3 absorption line ($\lambda = 9.7$ μm, $\tau \sim 0.6$). The formation of O_3 in the Earth atmosphere requires O_2 molecules and if O_3 can be considered as a good tracer of the presence of O_2 in an atmosphere, its detection would be easier. In fact, Angel proposed a telescope (16 m diameter in space, passively cooled to 80 K) which could be built in the middle range future and could make such an observation.

The point we want to examine here is the quality of O_3 as a O_2 tracer. It is clear that such a costly project could be considered only if we think that it relies on a safe basis.

Readily one can think of an objection. In the Earth atmosphere, O_3 is formed in the upper layers, down to an altitude such that the optical depth of the oxygen column density above it is a few units in the UV spectral range that dissociate O_2 into atomic O. As oxygen is abundant on Earth, this altitude is high (~ 20 km); but if oxygen was less abundant, one can imagine that to form a similar quantity of O_3 by lowering its location so that the oxygen column above it would be sufficient.

Ozone seems to be a nonlinear tracer of the amount of oxygen in an atmosphere so it requires a more quantitative study.

4. A Model for O_3 Column Density as a Function of O_2 Amount in an Atmosphere

To make the model simple, we assume an atmosphere which has several similarities with the Earth's. It has the same total pressure, same pressure and temperature distributions with altitude, and same turbulence coefficients for the dynamical mixing of gas layers. However, its fractional content in O_2 can vary from 10^{-7} to 1, the Earth case being described by 0.2. Gases are: O_2, N_2, and produced O, O_3. The incoming UV radiation flux is that of the Sun on the Earth. The O_3

formation/destruction is described by the Chapman cycle (1930) and the rate of fixation of O_3 in the ground is the same as on Earth (0.07 cm s^{-1}). In a second step, we have modified the temperature distribution to make it self-consistent with the heating due to the light absorption by ozone molecules at their actual altitude location.

The resulting O_3 column density as a function of the ground pressure of O_2 is shown in Figure 1. The model reproduces the case of Earth within a factor of 1.4 (catalytic destruction of O_3 has been neglected) and that of Mars within a factor of 2. The latter point gives some confidence in the model generality and its capability to describe the O_3 amount in an atmosphere which is not strictly Earth-like. Specifically, the O_2 amount seems to be a more crucial parameter to determine the O_3 column density than the total gas pressure or the star flux, as we have also checked.

The model shows clearly that O_3 *is a nonlinear tracer of the O_2 amount of an atmosphere*, in agreement with the pioneer work by Paetzold (1962). For instance, when the O_2 ground pressure varies from 10^{-2} to 1 bar, the O_3 amount varies only by a factor of 2.

However, it does not mean that the detection of O_3 in an extrasolar planet atmosphere would not be a good criterion for exobiology. Indeed, it would be very interesting to compute the optical depths of the O_3 band that result from the different column densities and T distributions found in our model in order to test the implications of an ozone detection in an exoplanet about the O_2 content of its atmosphere. This calculation is planned for the near future. Meanwhile, we can consider what would be these implications if the O_3 lines were optically thin and the 9.7-μm band depth proportional to the O_3 column density (this is true for column densities somewhat lower than the Earth's). To be detectable in difficult conditions, the 9.7-μm absorption would have to be rather strong, say larger than half that of the Earth's ((I/I_o)$_{\text{Earth}} \sim 60\%$). Then a 9.7-$\mu$m band with $\tau > 30\%$ would implies that *the oxygen pressure in such atmosphere is larger than 10 mbar.*

If the greenhouse runaway phase can be eliminated, would such a pressure be sufficient to indicate a continuous photosynthetic production of O_2? We think that this is an *open question* that should be worked out in the future. If its answer is positive, *the detection of the 9.7 μm band of O_3 would be an even more sensitive test for the presence of Life than the search for the oxygen A-band in the visible.* As a matter of fact, with $P_{O_2} = 10$ mbar, the latter would be quite weak, with an optical depth of 7×10^{-3} when measured with resolution R = 40 (Owen, 1980) and would not be detectable.

We also plan to calculate what would have been the optical depth of the 9.7-μm band if an observer had made spectra of the Earth during its geological past, searching for evidences of Life. We shall use the recent estimates of the O_2 content of the Earth atmosphere in the past (Holland, 1990), and our model for the resulting O_3 amount.

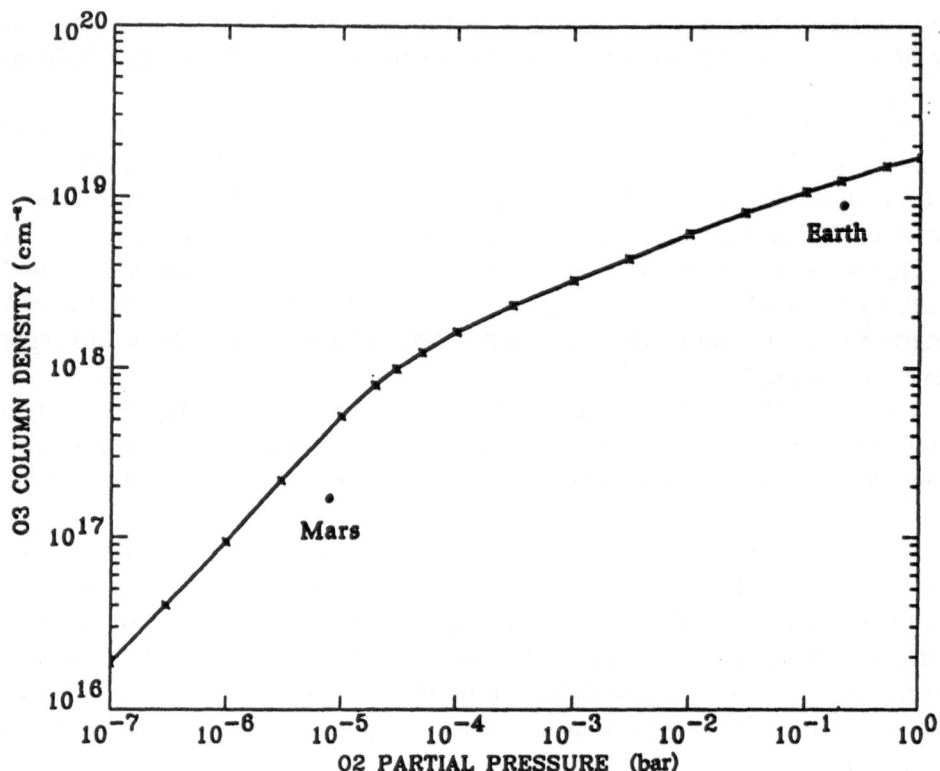

Fig. 1. Column density of ozone as a function of the ground pressure of oxygen as resulting from the atmospheric model. It reproduces the cases of Earth and Mars within a factor 1.4 and 2, respectively.

5. Conclusion

The budget of free oxygen and organic carbon on Earth confirms that O_2 is a very reactive gas whose massive presence in a telluric planet atmosphere implies a continuous production. Its detection would be a strong indication for photosynthetic activity, provided the planet is not in a runaway greenhouse phase.

In principle, the direct detection of O_2 could be possible in the visible flux of the planet at 760 nm (oxygen A-band) but it would be extremely difficult, considering the much larger flux from the star. The alternative search for the 9.7 μm absorption of O_3 may be easier as the contrast with the star is improved by 3 orders of magnitude. A simple atmospheric model indicates that the O_3 column density is not a linear tracer of the atmospheric O_2 content. However, the detection of a substantial O_3 absorption ($\tau > 30\%$) may indicate, within the validity of the model and provide a confirmation from a calculation of the band depth, a O_2 ground pressure larger than 10 mbar. The question is raised of whether this pressure is sufficient to indicate a photosynthetic origin of the oxygen. If the answer was positive, it would be an even more sensitive test of photosynthetic activity than the detection of the oxygen A-band. Further studies of these points are clearly needed before determining an observing strategy.

Acknowledgements

It is a pleasure to thank C. Camy-Peyret, G. Mégie, A. Poisson, C. Sagan and B. Schauer for providing us with precious pieces of information and an anonymous referee for constructive criticisisms.

References

Angel, J.R.P., Cheng A.Y.S. and Woolf, N.J.: 1986, *Nature* **322**, 341.
Chapman, S.: 1930, A Theory of Upper Atmospheric Ozone, *Mem. Roy. Meteorol. Soc.* **3**, 103–125.
Holland, H.D.: 1990, *Nature* **347**, 17.
Lovelock, J.E.: 1975, *Proc. Roy. Soc. London* **B189**, 167.
Owen, T.: 1980, in M.D. Papagiannis, ed(s)., *Strategies for the Search for Life in the Universe*, Reidel Publ. Co., 177–185.
Paetzold, H.K.: 1962, *Mem. Soc. Roy. Sci. Liège* vol. 7, La Physique des Planètes, 452.
Schneider, J.: 1992, in: J. Tran Thanh *et al.*, ed(s)., *Frontiers of Life*, éditions Frontière, Paris.
Siegenthaler, U.: 1986, in: P. Buat-Ménard, ed(s)., *The Role of Air Sea Exchange in Geochemical Cycling*, Reidel, NATO-ASI Series, 209–247.
Traub, W.A., Carleton, N.P., Connes, P. and Noxon, J.F.: 1979, *Ap. J.* **229**, 846.
Walker, J.C.G.: 1977, *Evolution of the Atmosphere*, MacMillan, New York.

THE PHOTOMETRIC METHOD OF EXTRASOLAR PLANET

DETECTION REVISITED *

ALAN HALE
Southwest Institute for Space Research,
Alamogordo, New Mexico, USA

and

LAURANCE R. DOYLE
SETI Institute, NASA Ames Research Center,
Moffett Field, California, USA

Abstract. We investigate the geometry concerning the photometric method of extrasolar planet detection, i.e., the detection of dimunition of a parent star's brightness during a planetary transit. Under the assumption that planetary orbital inclinations can be defined by a Gaussian with a σ of 10° centered on the parent star's equatorial plane, Monte Carlo simulations suggest that for a given star observed at an inclination of exactly 90°, the probability of at least one Earth-sized or larger planet being suitably placed for transits is approximately 4%. This probability drops to 3% for a star observed at an inclination of 80°, and is still $\sim 0.5\%$ for a star observed at an inclination of 60°. If one can select 100 stars with a pre-determined inclination $\geq 80°$, the probability of at least one planet being suitably configured for transits is 95%. The majority of transit events are due to planets in small-a orbits similar to the Earth and Venus; thus, the photometric method in principle is the method best suited for the detection of Earthlike planets.

The photometric method also allows for testing whether or not planets can exist within binary systems. This can be done by selecting binary systems observed at high orbital inclinations, both eclipsing binaries and wider visual binaries. For a "real-world" example, we look at the α Centauri system ($i = 79°.2$). If we assume that the equatorial planes of both components coincide with the system's orbital plane, Monte Carlo simulations suggest that the probability of at least one planet (of either component) being suitably configured for transits is approximately 8%.

In conclusion, we present a non-exhaustive list of solar-type stars, both single and within binary systems, which exhibit a high equatorial inclination. These objects may be considered as preliminary candidates for planetary searches via the photometric method.

1. Introduction

The unequivocal detection of planets around other stars would rank as one of the most important and exciting discoveries in the history of astronomy, indeed in all science. At the present time technology is beginning to reach the point such that potentially successful searches can be undertaken. Search methods can be grouped into two broad categories: *direct* detection, wherein a signal from a planet is itself detected; and *indirect* detection, wherein a planet's presence is inferred from observations of its effects upon its parent star. Direct detection methods include all means, interferometric and otherwise, of imaging the planet directly in one of various spectral regions. Indirect detection methods include astrometric, spectroscopic (i.e., differential radial velocity) and photometric search programs.

---

* Paper presented at the Conference on *Planetary Systems: Formation, Evolution, and Detection* held 7–10 December, 1992 at CalTech, Pasadena, California, U.S.A.

Astrophysics and Space Science **212**: 335–348, 1994.
© 1994 *Kluwer Academic Publishers.*

Details of the various methods and discussions of their various strengths and weaknesses are given in the review by Black (1980) as well as in recent scientific committee reports (NRC 1990; SSED 1992).

The photometric method primarily involves the detection of the drop of a star's brightness as it is transited by an orbiting planet. The detected signal may either be a straight drop in the photon count received from the star, or in the form of a color signature as described by Rosenblatt (1971).

There are, of course, some difficulties with the photometric method which might serve to limit its usability. One of these would seem to be the strength of the signal required to unambiguously detect the effect of the transiting planet. The drop in the parent star's brightness is directly proportional to the surface area of the planet in question; for a star of 1 $R_\odot$ this is approximately 10^{-2} magnitude for a Jupiter-sized planet and 10^{-4} magnitude for an Earth-sized planet. The former value is obtainable without significant difficulty using present technology, and in principle the latter value can be detected if sufficiently long integration times are employed.

A more serious difficulty is encountered when the requisite geometry is investigated. In order for a transit to occur, the planet's orbit must lie almost exactly in the line of sight from the earth. The minimum value of the inclination i depends upon the size of the planet's orbit, and is given by

$$\tan i_{min} = \frac{a}{R_*} \tag{1}$$

where a is the planet's semimajor axis and R_* is the parent star's radius. The relevant geometry is shown in Figure 1. (In practice, i_{min} would be somewhat larger, since this discussion is ignoring any effects on the transit signal due to limb darkening and other similar phenomena.)

For a given star, the probability that any accompanying planets will have their orbital planes lying within the critical inclination is of course quite low. When one considers the additional difficulties created by the relatively short durations of the transit events (on the order of 1 day for a planet in a Jupiter-like orbit) and the long time intervals between them ($\sim$10–20 years for Jupiter-like planets, if one assumes our planetary system is typical), the effort involved in creating an effective search program which could lead to the unambiguous detection of a planetary signal would appear to be prohibitive. Several investigations (e.g., Black 1980; NRC 1990) have concluded that as many as 10^4 to 10^5 stars would have to be continuously monitored in order to ensure a reasonable chance of success.

Some headway against these rather large numbers could be made if one can preselect those stars which are observed at high inclination, thus increasing the likelihood that accompanying planets' orbits might lie within the critical range of inclinations. Several authors (e.g., Doyle et al., 1984; Borucki and Summers,

$$\tan (90° - i) = \frac{z}{a} \quad \longrightarrow \quad \tan i = \frac{a}{z}$$

$$z = \frac{R_*}{d}(d - a)$$

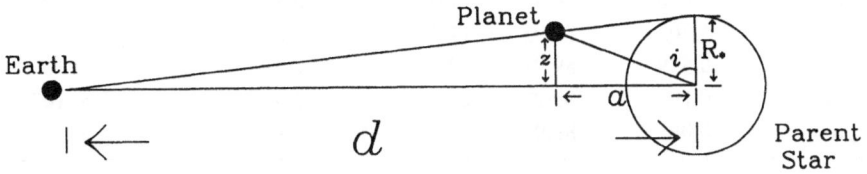

$$\tan i_{min} = \frac{a}{\dfrac{R_*(d-a)}{d}} = \frac{ad}{R_*(d-a)} \approx \frac{ad}{R_* d} = \frac{a}{R_*}$$

Fig. 1. Geometry of the photometric method, and derivation of i_{min}.

1984; Campbell and Garrison, 1985) have pointed out that, in principle, a star's inclination can be found from:

$$\sin i = \frac{v \sin i}{(2\pi R_*/P_{rot})} \tag{2}$$

where $v \sin i$ is the star's projected rotational velocity and P_{rot} is its rotational period. The use of this equation was specifically applied to the photometric method by Borucki and Summers (1984), who hoped to identify stars with $i \geq 82°$ as potential candidates for planetary searches.

A serious difficulty with this process was pointed out by Soderblom (1985a, 1985b) who noted that the observational uncertainties associated with the quantities in equation (2) — in particular, the determination of $v \sin i$ for solar-type stars — coupled with the nature of the sine function itself, are such as to preclude an accurate determination of the inclination, particularly when i is close to 90°. As an example, Soderblom points out that isolating those stars with $i \geq 82°$, as Borucki and Summers hope to do, requires an accuracy in the measurement of $\sin i$ better than 1%, a value that is not presently achievable and which will probably remain so for the foreseeable future. On the other hand, equation (2) does allow one to at least discriminate between high-inclination and low-inclination objects, although Soderblom also points out that this does not reduce the necessary sample by much more than half.

TABLE I
Planetary transit parameters

Planet	Orbital Distance (AU)	i_{min} (°)	Interval (years)	Maximum Duration (days)	Magnitude Drop
Venus	0.723	89.63	0.62	0.46	8.5×10^{-5}
Earth	1.000	89.73	1.00	0.54	9.4×10^{-5}
Jupiter	5.203	89.95	11.9	1.23	1.1×10^{-2}
Saturn	9.539	89.97	29.5	1.67	7.7×10^{-3}
Uranus	19.19	89.99	84.0	2.37	1.5×10^{-3}
Neptune	30.06	89.99	165.0	2.96	1.4×10^{-3}

A second question which might be raised concerns the fact that equation (2) produces the parent star's equatorial inclination, when what in fact is desired is the orbital inclination of the accompanying planets. It is perhaps logical to assume that planets will orbit in or near a star's equatorial plane, and such is the case within our planetary system: the average plane of the planets' orbits differs by only 7°.25 from the sun's equator. Hale (1992, 1994) examined solar-type stars within binary systems and concluded that, over distance scales similar to those of planetary orbits within the solar system, the equatorial planes of the respective components are aligned with the orbital planes of the given systems to within ±10 degrees. If one then assumes that these results apply to planetary systems as well, we may expect the bulk of planets' orbits to lie within 10° of their parent star's equator.

2. Investigation

In an effort to examine the geometry associated with the photometric method of extrasolar planet detection and to determine whether or not equation (2) might in fact be useful in selecting search candidates, various Monte Carlo simulations were performed on candidate parent stars with varying equatorial inclinations. The six planets of our solar system which are potentially detectable via the photometric method were placed at their respective distances around the parent stars. Relevant information on these planets is given in Table I; the duration between transit events is, of course, equivalent to the planet's orbital period.

The inclinations of the planets' orbits with respect to the parent star's equatorial plane were randomly selected from a Gaussian probability distribution centered on the parent star's equator; this was accomplished via interpolation of the Gaussian integral values given in Table C-2 of Bevington (1969). Following Hale's results, a σ of 10° was chosen for this Gaussian. A stellar radius of 1 $R_\odot$ was used for all

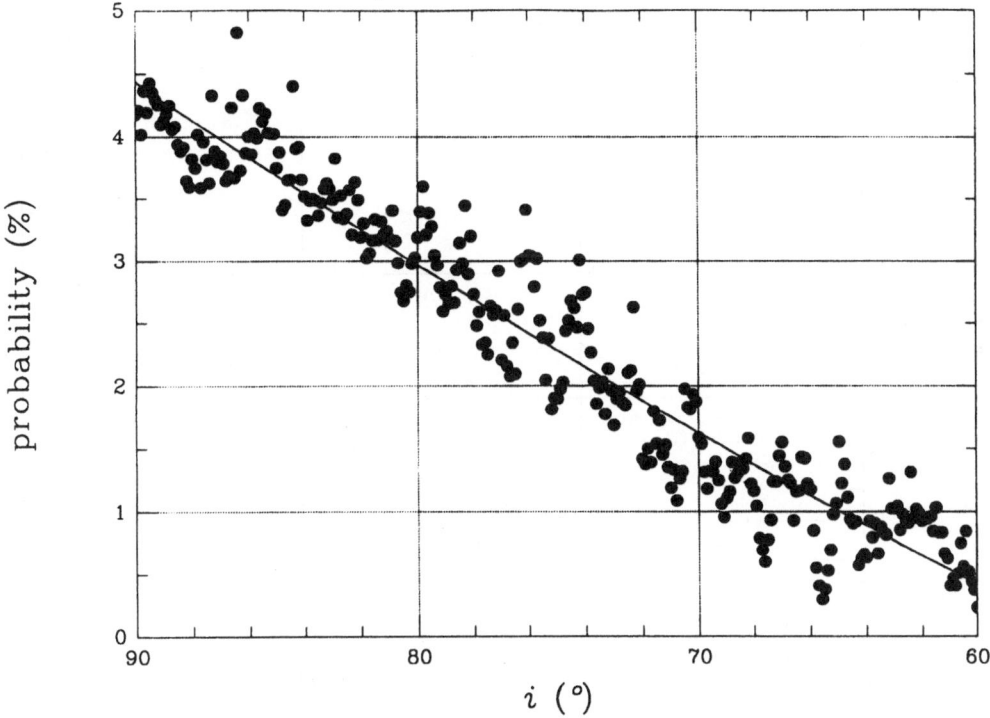

Fig. 2. Results of Monte Carlo simulations; probability of a "hit" as a function of stellar equatorial inclination. The line is the quadratic fit given by equation (3).

simulations. A total of 20,000 such systems were "created" at each stellar equatorial inclination between 90° and 60° (equivalent to 3 σ) at intervals of 0°.1.

Results of these simulations are shown in Figure 2. For a star viewed at an inclination of exactly 90°, the probability that at least one planet is suitably configured for transit events is approximately 4.5%. This probability drops to ~3% for stars viewed at an inclination of 80°, and even for stars viewed at an inclination as low as 60° the probability of a planet being suitably configured is still ~0.5%. The line in Figure 2 is a quadratic fit to the Monte Carlo results; according to this fit, the probability of at least one planet producing a "hit" is given by:

$$P(\text{"hit"}) = 6.53 \times 10^{-6} i^2 + 3.61 \times 10^{-4} i - 4.10 \times 10^{-2} \tag{3}$$

where i is given in degrees.

Figure 3 shows the number of stars viewed at a given inclination that would have to be examined in order to produce a 95% probability of at least one planet being suitably configured for transit events, based upon the probability function given in equation (3). In practice, one might select a particular inclination as a lower limit to be applied to the range of candidate stars, thus the values shown in

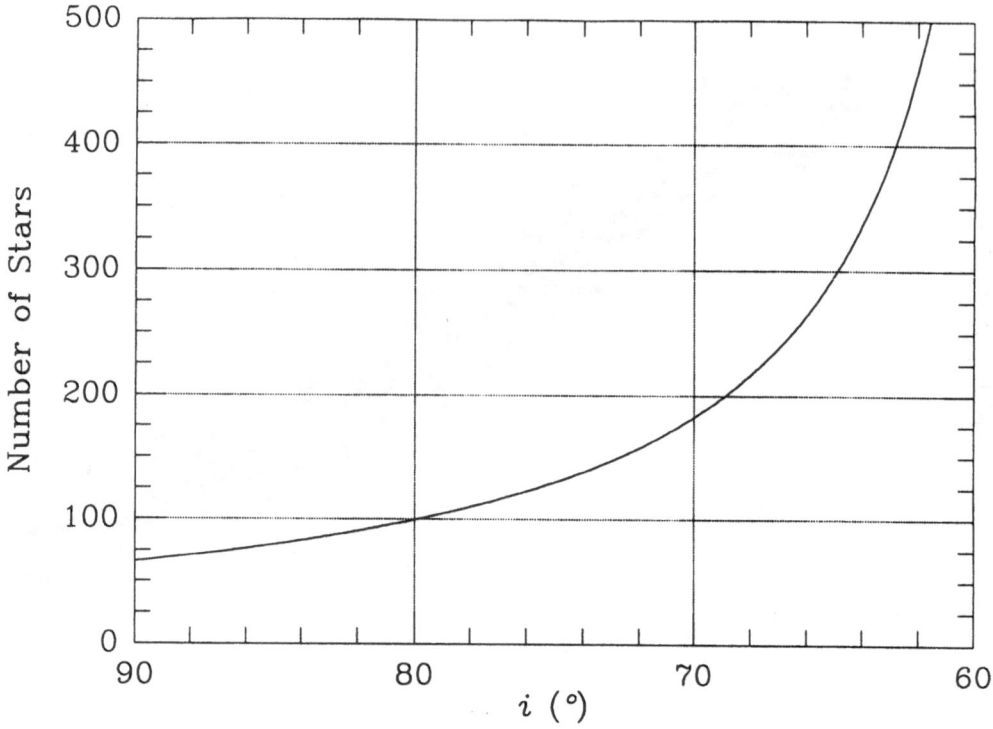

Fig. 3. Number of stars at a given equatorial inclination required to be examined for a 95% probability of a "hit", based upon the probability function given in equation (3).

Figure 3 represent an upper limit to the number of stars which must be examined in order to reach the 95% probability level. If one selects this lower inclination limit to be 80°, then only 100 stars need to be examined to produce a 95% chance of a "hit"; however this requires an unrealistic accuracy of 1.5% in the determination of $\sin i$. On the other hand, only 300 stars are required to be examined for a limiting inclination of 65°; this requires an accuracy of $\sim$10% in the $\sin i$ determination which, based upon the experience of both authors, is not unreasonable.

It thus appears that the photometric method may indeed be used in the search for extrasolar planets without the prohibitively large number of candidate stars which previous studies had concluded were necessary. An additional benefit of the photometric method is shown in Figure 4, wherein the relative number of "hits" are given for each planet (as determined from a simulated sample of 2,000,000 planetary systems where the parent star's inclination is $\geq$ 80°). Because of their relatively small semimajor axes, planets in Venus-like and Earthlike orbits are far more likely to be suitably configured for transits than are the outer planets in Jupiter-like orbits and larger. (The values shown in Figure 4 are in fact given rather closely by the inverse of the planets' respective semimajor axes.) From this consideration,

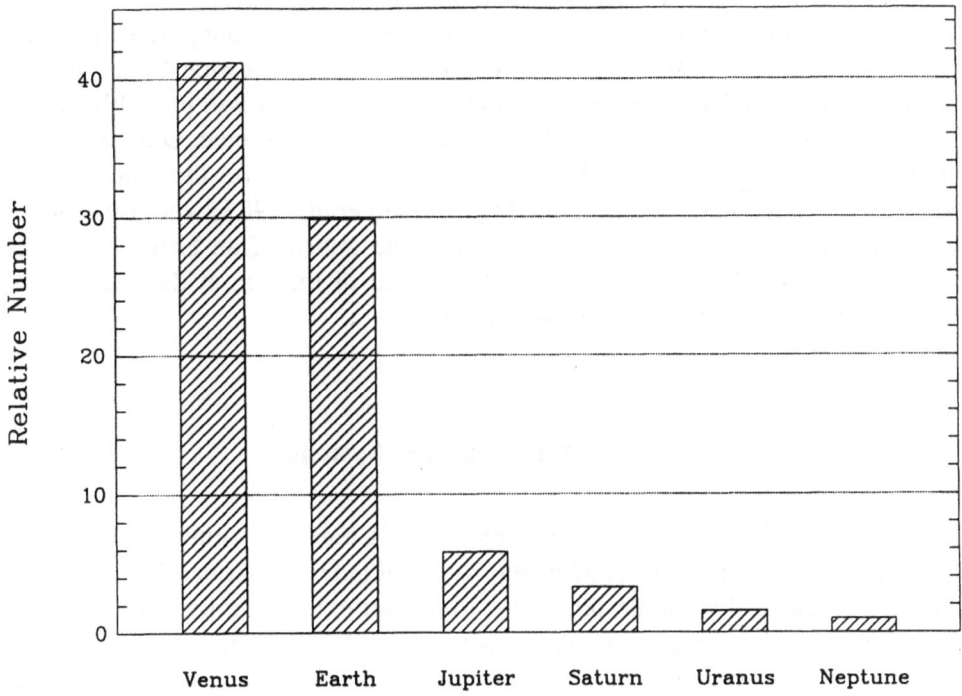

Fig. 4. Relative number of "hits" per planet resulting from 2,000,000 simulated stars with equatorial inclinations $\geq 80°$. The results of the simulations have been normalized to those of Neptune (the relative number of "hits" = 1).

then, the photometric method is well-suited for the detection of Earthlike planets, and in fact is the only one of the methods currently discussed which offers such a possibility within the foreseeable future. Because of the relatively short intervals between transit events — on the order of 1 year — confirmation and cataloguing of such planets would not require prohibitively long timescales, as opposed to the timescales involved in confirming the transit of a Jovian planet. If a Jovian planet were to be detected around a high-inclination star via one of the other methods, the photometric method is ideally suited to search for Earthlike planets within that system.

The significant drawback to the use of the photometric method in the search for Earthlike planets is the small signal due to the transit; as discussed above, this is approximately 10^{-4} magnitude. In principle, this can be compensated for by the use of suitably long integration times. The formulae in Appendix A of NRC (1990) suggest that, for a G2V star located at a distance of 10 parsecs, an integration time of ~4 minutes is required; a similar star located at 50 parsecs would require a probably unmanageable integration time of 1.6 hours. Another serious problem is the masking of the signal due to other phenomena producing variability at the 10^{-4} magnitude level. Atmospheric effects are a significant potential source of such "noise"; it may be possible to overcome these somewhat by the use of slaved telescopes as discussed by Rosenblatt (1971), but they will probably not be eliminated until a space-based search platform can be constructed. One additional source of "noise" is intrinsic stellar variability at this level over the timescales in question. This issue has been examined by Borucki *et al.* (1985) who, from *Solar Maximum Mission* (*SMM*) data, conclude that the transit of an Earthlike planet across a star with an activity level similar to that of the sun at the peak of its activity cycle can in fact be marginally detected.

3. Planets Within Binary Systems

The issue of whether or not planets can exist within binary systems is pertinent to the study of extrasolar planets, and is unresolved at the present time. Many studies have demonstrated that stable planetary orbits can exist within binary systems, depending upon the separation and masses of the stellar components, but these studies are unable to address the question of whether or not such planets could actually *form* within these systems.

The photometric method is well-suited to test binary systems for the existence of planets. Based upon Hale's (1992, 1994) results, wherein it was found that binary systems with component separations $\leq \sim 20$ AU can be expected to have the stellar equatorial planes roughly aligned with the system's orbital plane, one can therefore select binary systems with high orbital inclination within this separation range as potential search candidates. A similar approach was proposed by Schneider and Chevreton (1990), who advocated the use of eclipsing binaries as search candidates. It should be noted that the Hale study determined that the presence of a short-period system within a longer-period binary was often associated with noncoplanarity between the inner and outer orbits; whether or not such a situation would also occur within planetary systems cannot be stated.

For a "real-world" example, we may examine the α Centauri system ($a = 24$ AU, $i = 79°.2$; USNO catalogue, Worley and Heintz, 1983). A stability analysis of this system by Benest (1988) suggests that planets out to a distance of ~ 5 AU (i.e., Jupiter) may exist in stable orbits around either component. Monte Carlo simulations similar to those described in section 2, except for being limited to the planets Venus, Earth, and Jupiter, and using radii of 1.23 and 0.80 $R_\odot$ for components A and B, respectively (Soderblom 1986), suggest that component A has a 5% chance of having at least one detectable planet, and component B has a 3% chance, for a combined probability of $\sim 8\%$. If one assumes that similar results apply to other binary systems similar to α Centauri, only 35 such systems need to be searched in order to raise the probability level to 95%.

4. Tables of Candidate Stars

We here present some candidate high-inclination solar-type stars which have been identified by us in previous studies. (We define "solar-type" as including the spectral type range F5V-K5V.) Table II is a list of individual objects for which we have measured equatorial inclinations $\geq \sim 65°$. The parameters used to determine these inclinations may be found in their respective parent catalogues, either the Hale study or in Doyle *et al.* (1993). Table III is a listing of nearby solar-type binary systems with orbital inclinations $\geq \sim 75°$; this listing is restricted to those systems with an orbital period between 10 and 150 years. Systems with periods less than 10 years would perhaps not be likely to have an Earthlike planet in a stable orbit, whereas those with periods longer than 150 years may not exhibit coplanarity between the orbital and stellar equatorial planes, as per the Hale study. (It will be noticed that some objects are common to both Tables II and III.) Table IV lists some solar-type eclipsing binaries (and other short-period binary systems with a high inclination) selected from the catalogue by Batten *et al.* (1989), for use in examining the approach given by Schneider and Chevreton (1990). (The inclination values given in Table IV are those cited by the Batten *et al.* catalogue or by the references given therein.) This list may be taken as complementary to the example binaries given by Schneider and Chevreton, all of which are of spectral types B and A.

The listings given in Tables II, III, and IV are not intended to be exhaustive; rather they reflect a sample of the inclination determination work done to date. A more complete census of solar-type stars within the solar neighborhood, including an attempt to pre-determine equatorial inclinations for as many as is possible, is being planned at this time.

5. Conclusions

Under the assumptions that our planetary system is representative of planetary systems in general, and that the distribution of planetary orbital inclinations about a parent star can be represented by a Gaussian with a σ of 10° centered on the parent star's equator, Monte Carlo simulations suggest that a star viewed at an inclination of exactly 90° has approximately a 4% probability of having at least one planet suitably configured for transit events. This probability drops to 3% for stars seen at an inclination of 80°, and to 0.5% for stars seen at 60°. If one can select a list of 300 candidate stars with pre-determined inclinations $\geq 65°$ — a process which requires an accuracy of $\sim 10\%$ in the determination of $\sin i$ — the probability of having at least one suitably configured planet is 95%. Since a large majority of "hits" are due to planets in orbits similar to those of Venus and the Earth, and since these objects would exhibit relatively short duration times between transit events, the photometric method is a viable means to engage in a search for Earthlike planets. Because of the small signal inherent in these events (10^{-4} magnitude), "noise" due

TABLE II

Solar-type stars with high equatorial inclination

HD	Name	Spectral Type	$\sin i$	i (°)
1835	9 Cet	G2V	1.28 ± 0.14	$\sim$90
3443		G7V	1.89 ± 0.45	$\sim$90
6920	44 And	F8V	2.33 ± 0.37	$\sim$90
16895	θ Per	F7V	1.02 ± 0.17	90^{+0}_{-32}
19373	ι Per	G0V	0.96 ± 0.30	75^{+15}_{-33}
19994	94 Cet	F8V	3.39 ± 0.64	$\sim$90
20010	α For	F8IV	1.23 ± 0.57	$\sim$90$^{+0}_{-48}$
20630	κ^1 Cet	G5V	1.10 ± 0.18	90^{+0}_{-23}
22484	10 Tau	F9V	1.53 ± 0.38	$\sim$90
25893	V491 Per	K1V	1.04 ± 0.20	90^{+0}_{-33}
27406	vB 31	G0V	1.26 ± 0.18	$\sim$90
27859	vB 52	G1V	1.23 ± 0.19	$\sim$90
28099	vB 64	G2V	0.90 ± 0.18	64^{+26}_{-18}
28344	vB 73	G1V	1.01 ± 0.15	90^{+0}_{-31}
29310	vB 102	G1V	1.11 ± 0.18	90^{+0}_{-22}
38392	γ Lep B	K2V	1.26 ± 0.68	90^{+0}_{-54}
38393	γ Lep A	F6V	1.35 ± 0.26	$\sim$90
39587	χ^1 Ori	G0V	0.95 ± 0.08	71^{+19}_{-11}
64096	9 Pup	F9V	1.21 ± 0.24	$\sim$90$^{+0}_{-15}$
81809		G2V	1.28 ± 0.45	$\sim$90$^{+0}_{-34}$
86728	20 LMi	G2V	1.55 ± 0.43	$\sim$90
95128	47 UMa	G0V	1.06 ± 0.29	90^{+0}_{-40}
102870	β Vir	F9V	1.05 ± 0.24	90^{+0}_{-36}
114378	α Com	F5V	1.08 ± 0.13	90^{+0}_{-17}
114710	β Com	G0V	1.01 ± 0.20	90^{+0}_{-36}
115404		K1V	1.29 ± 0.34	$\sim$90$^{+0}_{-19}$
115617	61 Vir	G6V	1.18 ± 0.61	90^{+0}_{-55}
120136	τ Boo	F6IV	1.20 ± 0.28	$\sim$90$^{+0}_{-23}$
122106		F8V	1.73 ± 0.20	$\sim$90
124115		F7V	2.46 ± 0.28	$\sim$90
126660	θ Boo	F7V	2.00 ± 0.23	$\sim$90
128620	α Cen A	G2V	1.21 ± 0.44	$\sim$90$^{+0}_{-40}$
128621	α Cen B	K1V	1.24 ± 0.65	$\sim$90$^{+0}_{-54}$

TABLE II

(continued)

HD	Name	Spectral Type	$\sin i$	$i\,(°)$
131156B	ξ Boo B	K4V	1.20 ± 0.54	$\sim 90^{+0}_{-49}$
141004	λ Ser	G0V	1.21 ± 0.43	$\sim 90^{+0}_{-39}$
142860	γ Ser	F6V	1.12 ± 0.15	90^{+0}_{-14}
143761	ρ CrB	G2V	2.12 ± 0.76	~ 90
154905	μ Dra A	F7V	1.07 ± 0.19	$\sim 90^{+0}_{-28}$
162004	ψ Dra B	G0V	1.44 ± 0.21	~ 90
165341	70 Oph A	K0V	1.43 ± 0.49	$\sim 90^{+0}_{-19}$
177474	γ CrA	F8V	0.90 ± 0.42	65^{+25}_{-36}
176051		G0V	1.34 ± 0.31	~ 90
179957		G4V	1.13 ± 0.38	90^{+0}_{-41}
179958		G4V	1.42 ± 0.27	~ 90
185144	σ Dra	K0V	0.99 ± 0.55	81^{+9}_{-48}
186427	16 Cyg B	G2.5V	1.35 ± 0.52	$\sim 90^{+0}_{-34}$
187013	17 Cyg	F7V	0.96 ± 0.25	74^{+16}_{-29}
197076		G5V	1.91 ± 0.40	~ 90
204121		F5V	0.97 ± 0.12	76^{+14}_{-18}
206860	HN Peg	G0V	0.99 ± 0.14	82^{+8}_{-24}
212698	53 Aqr A	G1V	1.42 ± 0.32	~ 90
212754	34 Peg	F7V	1.56 ± 0.22	~ 90
213429		F8V	0.94 ± 0.43	70^{+20}_{-39}
217014	51 Peg	G5V	1.04 ± 0.50	90^{+0}_{-57}
222368	ι Peg	F7V	0.97 ± 0.16	76^{+14}_{-22}

to atmospheric effects and to intrinsic stellar variability at this level is the primary issue which must be considered prior to the initiation of a comprehensive search program.

The photometric method may also be used to test for the existence of planets within binary systems. If one selects high-inclination systems with orbital periods long enough to allow for stable planetary orbits and short enough such that the equatorial and orbital planes can be expected to be relatively aligned, one can examine these objects for planetary transit events in a manner similar to that which can be used for single stars. Monte Carlo simulations suggest an 8% probability

TABLE III

Visual binary systems with high orbital inclination

HD (BD)	ADS (or other)	Name	Spectral Types	Period (yrs)	i (°)
3443	520		G7V + G8V	25.0	78.0
6582		μ Cas	G5V + M6V	22.1	71.8
10009	Kpr7		F7V	27.7	70.5
10307			G1.5V + M2V	19.6	76
13872	Cou 79	21 Ari	F6V	24.5	75.8
18025	Hu 1562		G5	65.0	77.0
22262	B 52		F5V	19.4	84.4
27989	3210	vB 58	G4V + G6V	27.9	81.8
28363	3248	vB 75	F7V + G0V	39.9	85.5
32127	3614		G7V + G8V	82.0	72.0
39587		χ^1 Ori	G0V + M6V	14.2	86.6
47230	φ 19		G0V	29.0	80.3
64096	6420	9 Pup	F9V + G4V	23.3	79.7
65123	6483		F7V	57.0	79.0
73752	6914		G3V	123	82.9
79969	7284		K3V + K3V	34.2	77.0
81809	B 2530		G2V + K1V	34.4	84.0
114378	8804	α Com	F5V + F5V	25.8	89.9
128620		α Cen	G2V + K1V	79.9	79.2
144892	9932		F7V	55.0	84.9
156023	10421		F5V	98.2	72.5
158614	10598		G8IV	46.4	81.3
1602769	10660	26 Dra	G0V + K3V	76.0	74.3
163077	10871		G8V	37.4	72
(+14°3507)	11300		G0V + G5V	60.0	84.0
214810	Kpr 114		F8V	54.2	87.4
217166	16417		G2V	26.5	90

that at least one planet within the α Centauri system is suitably configured for transits (if one assumes that such planets exist); the examination of 35 systems similar to α Centauri allows the probability level to reach 95%.

Acknowledgements

AH thanks the Astronomy and Computer Science departments at New Mexico State University for providing the computer facilities necessary for the completion of this research.

TABLE IV

High-inclination short-period systems

HD (BD)	Name	Spectral Types	Period (days)	i (°)
6980	AI Phe	F7V + K0IV	24.59	88.5
(+47°781)	LX Per	G5IV + G5IV	8.04	~90
22403		G2V + K	1.93	≥65
27130	vB 22	G6V + K6V	5.61	≥85
34335	CD Tau	F7V + F5IV	3.44	87
37513	UX Men	F8V + F8V	4.18	90
77137	TY Pyx	G2 + G2	3.20	88
92109	UV Leo	G0V	0.60	83
93486	RZ Cha	F5IV–V + F5IV–V	2.83	83
107760	AS Dra	G0V + K0V	5.41	—
114519	RS CVn	F4IV–V + K0IV	4.80	84
120734	V757 Cen	G0V + G0V	0.34	65
123423	DM Vir	F6–7V	4.67	90
143313		K2V	9.01	~90
150708	WW Dra	G5V	4.63	81.4
185912	V1143 Cyg	F5V	7.64	87
188088		K3V + K3V	46.82	—
195987		G9V + (K5V)	57.32	> 77
200391	ER Vul	G0V + G5V	0.70	70
214686		F7V + F7V	21.70	~90
(+52°3383A)	RT And	F8V + G7V	0.63	87
219113	SZ Psc	K1IV + F8V	3.97	~77

References

Batten, A.H., Fletcher, J.M. and MacCarthy, D.G.: 1989, *Publ. D. A. O.* **24**, Eighth Catalogue of the Orbital Elements of Spectroscopic Binary Systems.

Benest, D.: 1988, *Astron. Astrophys.* **206**, 143.

Bevington, P.R.: 1969, *Data Reduction and Error Analysis for the Physical Sciences*, McGraw-Hill, New York.

Black, D.C.: 1980, *Space Sci. Rev.* **25**, 35.

Borucki, W.J. and Summers, A.L.: 1984, *Icarus* **58**, 121.

Borucki, W.J., Scargle, J.D. and Hudson, H.S.: 1985, *Ap. J.* **291**, 852.

Campbell, B. and Garrison, R.F.: 1985, *Publ. A. S. P.* **97**, 180.

Doyle, L.R., Wilcox, T.J. and Lorre, J.J.: 1984, *Ap. J.* **287**, 307.

Doyle, L.R., Baliunas, S.L. and Backman, D.E.: 1993, in preparation.

Hale, A.: 1992, *Ph.D. dissertation: Orbital Coplanarity in Solar-Type Binary Systems: Implications for Planetary System Formation and Detection*, , New Mexico State University.

Hale, A.: 1994 *Astron. J.* **107**, (in press).

National Research Council (NCR): 1990, *Strategy for the Detection and Study of Other Planetary Systems and Extrasolar Planetary Materials: 1990–2000*, National Academy Press, Washington, D.C.

Rosenblatt, F.: 1971, *Icarus* **14**, 71.

Schneider, J. and Chevreton, M.: 1990, *Astron. Astrophys.* **232**, 251.

Soderblom, D.R.: 1985a, *Publ. A. S. P.* **97**, 57.

Soderblom, D.R.: 1985b, *Icarus* **61**, 343.

Soderblom, D.R.: 1986, *Astron. Astrophys.* **158**, 273.

Solar System Exploration Division: 1992, *TOPS: Toward Other Planetary Systems*.

Worley, C.E. and Heintz, W.D.: 1983, *Publ. U.S.N.O. (2nd series)* **24** (**Part VII**), Fourth Catalog of Orbits of Visual Binary Stars.

TODCOR: A TWO-DIMENSIONAL CORRELATION TECHNIQUE TO ANALYZE STELLAR SPECTRA IN SEARCH OF FAINT COMPANIONS *

T. MAZEH and S. ZUCKER

School of Physics and Astronomy, Tel Aviv University,
Tel Aviv, Israel

Abstract. TODCOR is a new TwO-Dimensional CORrelation technique to measure radial velocities of two components of a spectroscopic binary. Assuming the spectra of the two components are known, the technique correlates an observed binary spectrum against a combination of the two spectra with different shifts. TODCOR measures *simultaneously* the radial velocities of the two stars by finding the maximum correlation.

One of the advantages of TODCOR is its ability to detect a very faint companion in a combined spectrum, and to measure its radial velocity. We performed numerous tests in which we applied TODCOR to simulated spectra which were prepared as combinations of two spectra with various luminosity ratios, together with random noise. These tests show that TODCOR can detect a very faint secondary spectrum and measure correctly its velocity, even with a luminosity ratio of 1000, provided the combined spectrum has enough spectral coverage and high S/N. Measuring the radial velocity of the faint secondary will enable us to estimate the companion mass, a very useful tool in the search for brown dwarfs and giant planets around nearby stars.

1. Introduction

In a recent paper (Mazeh and Zucker, 1992) we introduced TODCOR — a new algorithm to measure simultaneously the radial velocities of the two components of a spectroscopic binary. The new algorithm (Zucker and Mazeh, 1993) is a generalization of the cross-correlation method developed by Tonry and Davis (1979), which has been frequently used to measure Doppler shifts of digitized celestial spectra (e.g., Latham, 1992). In the present paper we show that TODCOR is able to detect and measure the radial velocity of a very faint companion included in the observed spectrum, given a high enough ratio of signal to noise (S/N). To demonstrate this ability of TODCOR we analyzed simulated spectra, out of which the algorithm detected a faint secondary with intensity as small as 0.001 of the primary. Using the one-dimensional cross-correlation technique, such a detection is unimaginable. We suggest using TODCOR to directly detect the spectra of brown dwarfs and giant planets in the IR, where the intensity ratio between the faint companion and its parent star is the highest.

---

* Paper presented at the Conference on *Planetary Systems: Formation, Evolution, and Detection* held 7–10 December, 1992 at CalTech, Pasadena, California, U.S.A.

2. TODCOR: The TwO Dimension CORrelation Algorithm

The original one-dimensional cross-correlation algorithm calculates the correlation between the observed spectrum and an assumed template as a function of the shift between the two spectra (Tonry and Davis, 1979). The relative radial-velocity of the observed spectrum is found as the location of the correlation maximum. Since this technique uses all the information embedded in the spectra *simultaneously*, it finds the correct radial velocity even for extremely low S/N spectra (Latham, 1985), and was therefore applied extensively to study spectroscopic binaries (e.g., Latham, 1992).

When the stellar spectrum is composed of two components, the cross-correlation function displays two peaks, which correspond to the radial velocities of the two components. However, if the two velocities are too close, the two peaks might blend and become unresolvable.

To demonstrate the difficulties to resolve the two peaks with the one-dimensional technique we simulated a double-line spectrum, composed of two calculated spectra of an A and a G star (Kurucz, 1991). The spectra covered the range from 5164 Å to 5213 Å and consisted of 2048 pixels. The rotational velocities of the two stars were chosen to be 40 km s^{-1}, with relative intensity of 0.25. The A and G spectra were shifted with radial velocities of 0 and 50 km s^{-1}, respectively. To mimic real observations we added normally distributed noise with S/N of 20 (see Mazeh and Zucker, 1992 for details). In Figure 1 we show the cross-correlation of the simulated spectrum against the A and the G spectra, respectively. The primary peak is obviously seen, but the secondary appears only as a small shoulder, to the right of the main peak. Obviously, the secondary's shift is very difficult to evaluate.

To overcome this difficulty we developed TODCOR—a TwO-Dimensional CORrelation algorithm (Zucker and Mazeh, 1993). The new algorithm assumes that the observed spectrum is composed of two known spectra with unknown shifts. It therefore calculates the correlation of the observed spectrum against a combination of the two templates with all possible radial-velocity shifts. The correlation is thus a two-dimensional function of the velocity shifts of the two templates. The location of the correlation maximum corresponds to the radial velocities of the two components of the observed system.

A straightforward application of the algorithm would require an enormous amount of computation time. We therefore developed an efficient way to perform the algorithm, by using the Fourier transforms of the spectra involved (Zucker and Mazeh, 1993). The use of the transforms rendered the application of TODCOR possible on present workstations.

To demonstrate the ability of TODCOR, we applied it to the same simulated spectrum of Figure 1. We show in Figure 2 a contour map of the two-dimensional parameter space around the maximum, which was found by TODCOR at velocities of 1 ± 1 and 51 ± 1 km s^{-1} for the primary and secondary, respectively. The contour map shows a clear maximum at the correct velocities.

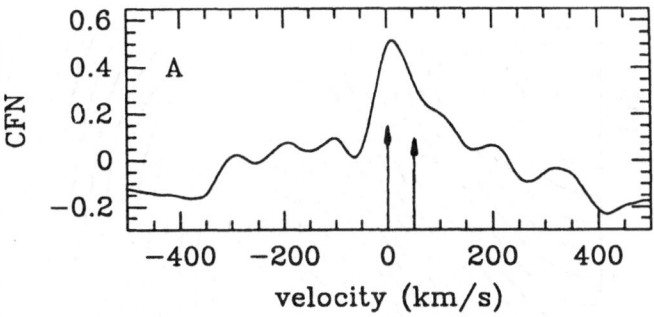

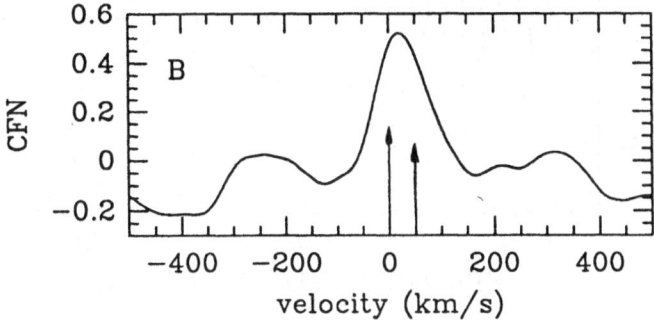

Fig. 1. One-dimensional cross-correlations of a simulated spectrum composed of an A-star and a G-star, at luminosity ratio of 0.25. The simulated velocities are 0 and 50 km s^{-1}. A is the cross-correlation against the primary template and B is the cross-correlation against the secondary template. The arrows indicate the velocities used in the simulation.

Figure 3 displays two cross sections of the correlation function, taken parallel to the two axes at maximum. These cross sections are actually one-dimensional cross-correlations of the composite spectrum against each of the templates, freezing the other template at a given velocity. For the correct velocity of the primary, the peak of the secondary is dramatically intensified relative to the one-dimensional correlation of Figure 1, illustrating the ability of TODCOR to separate the information relevant to each spectrum.

3. Application to the Detection of Faint Companions

In an effort to explore the limits of TODCOR's capabilities, we have run numerous tests with simulated spectra, each composed of two spectra with different intensity ratios. In each case we checked whether TODCOR could retrieve the correct radial

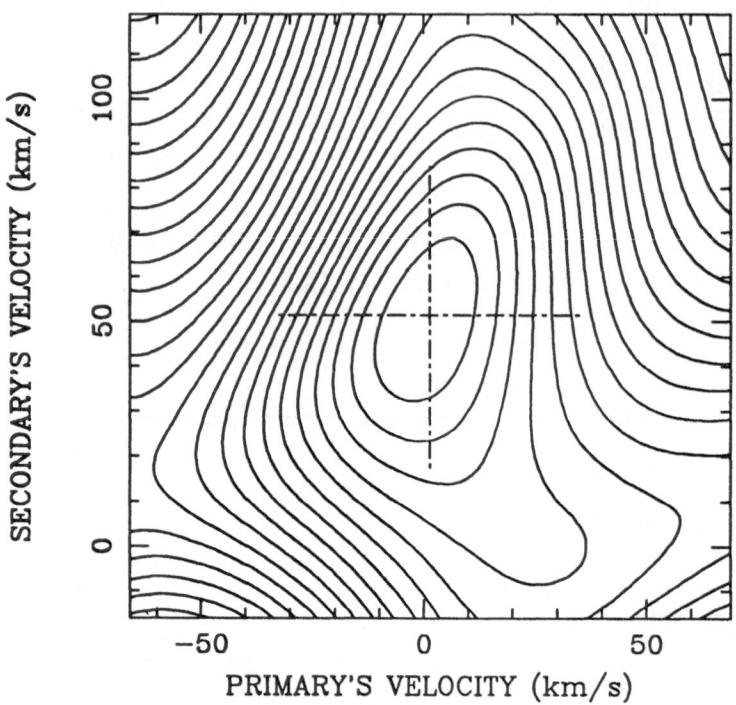

Fig. 2. A contour plot of the two-dimensional correlation function around the maximum. The dashed lines are parallel to the axes and go through the maximum. The observed spectrum and the templates are the same as in Figure 1.

velocities and intensity ratio of the two spectra. The main result—relevant to the detection of planetary systems—is the low limit of the ratio between the secondary and the primary luminosity. We found this limit to be less than 0.001! Obviously, this luminosity ratio requires a very high S/N, because the effective noise for the secondary luminosity is the noise of the simulated spectrum multiplied by the luminosity ratio.

We show in Figures 4 and 5 the result of one such simulation. In this simulation we used observed IR spectra of a G-type star and a K-type star in the 2.2 μ window (Kleinmann and Hall, 1986). The spectral window was centered at 2.2108 μ, its width was 0.3974 μ, and the spectra consisted of 4096 pixels. The two spectra were shifted by 0 and 300 km s^{-1}, respectively, and were then superposed with relative weight of 0.001. Random normal noise, at S/N of 3000, was added. Figure 4 displays the one-dimensional cross-correlation of the simulated spectrum against

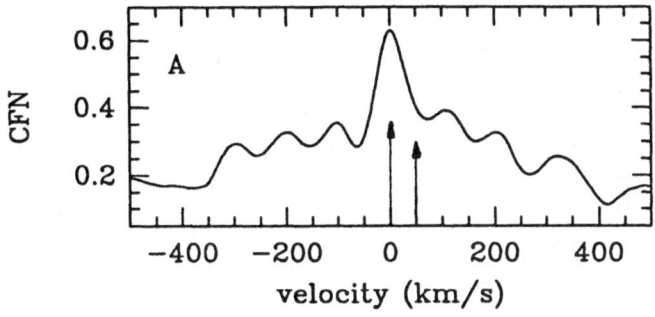

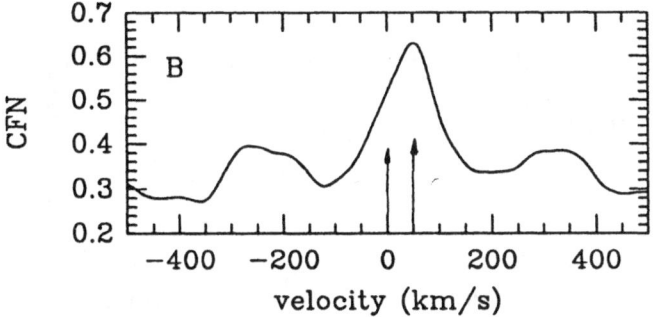

Fig. 3. Cross sections of the two-dimensional correlation function, taken along the dashed lines of Figure 2. A is the cross section taken parallel to the primary's velocity axis and B is the cross section taken parallel to the secondary's velocity axis. The arrows indicate the velocities used in the simulation.

each of the two templates. The two cross-correlations do not show even a hint of the presence of a secondary spectrum, let alone indicating its velocity. On the other hand, TODCOR located the maximum of the two-dimensional correlation at 0 ± 3 and $302 \pm 16 \, \mathrm{km \, s^{-1}}$. Figure 5 shows the cross sections of the correlation function, the second of which displays a prominent peak at $302 \, \mathrm{km \, s^{-1}}$, corresponding to the correct secondary velocity.

4. Conclusions

We have demonstrated here how TODCOR may be used to detect and study very faint companions to nearby stars. This capability of TODCOR can be used to turn known single-line spectroscopic binaries into double-line binaries, rendering the mass ratio of the system observable directly. This will allow us to derive the

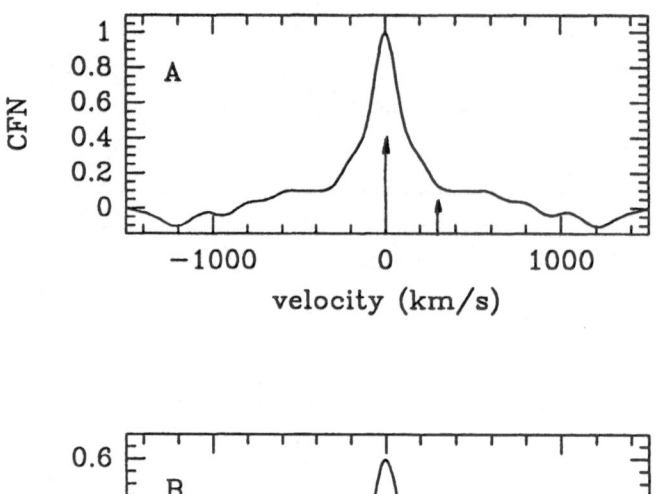

Fig. 4.   One-dimensional cross-correlations of a simulated spectrum composed of a G-star and a K-star, at luminosity ratio of 0.001. The simulated velocities are 0 and 300 km s^{-1}. A is the cross-correlation against the primary template and B is the cross-correlation against the secondary template. The arrows indicate the velocities used in the simulation.

secondary mass, provided the primary mass can be estimated. This method can be very important in exploring the distribution of the secondary mass (Mazeh and Goldberg, 1992; Mazeh et al., 1992), its low-mass region in particular.

TODCOR can be also used to detect secondary brown dwarfs and giant planets in the spectra of their primaries. Here again, measuring the radial velocity of the faint companion will offer a unique opportunity to estimate its mass, a most important parameter in the study of low-mass companions. In the case of an M-dwarf primary and a brown dwarf secondary, the luminosity ratio can be as high as 0.01 in the K window (Henry, 1991). In such a case, the secondary velocity can be measured with the present facilities operating in the infrared. In fact, the CSHELL of the IRTF can secure IR spectra with S/N substantially higher than 100. Such a high S/N will allow, in principle, to detect even relatively faint brown dwarfs

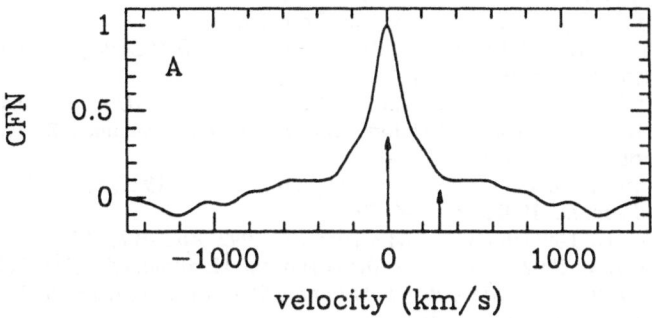

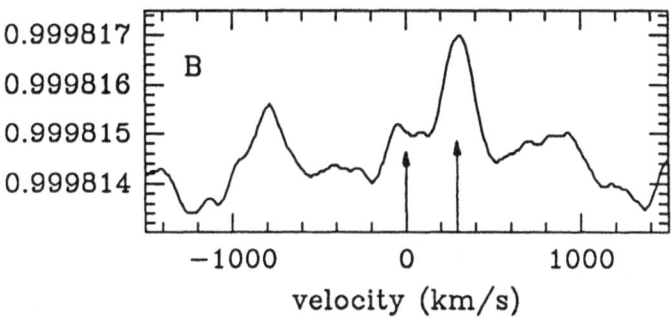

Fig. 5.   Cross sections of the two-dimensional correlation function, taken at the maximum. The observed spectrum and the templates are the same as in Figure 4. A is the cross section taken parallel to the primary's velocity axis and B is the cross section taken parallel to the secondary's velocity axis. The arrows indicate the velocities used in the simulation.

as secondaries of M-star primaries, or bright brown dwarfs as faint companions of G-dwarf primaries.

As a first step, we plan to apply TODCOR to a few nearby M stars which are suspected to have a low-mass companion. We also will try and measure the radial velocity of the secondary of HD114762—a G-type star known to have a low-mass companion (Latham *et al.*, 1989). These measurements will enable us to find the nature of the unseen secondaries—a late M-dwarf, a brown dwarf, or even a giant planet.

References

Henry, T.J.: 1991, Ph.D. Thesis, University of Arizona.
Kleinmann, S.G. and Hall, D.N.B.: 1986, *Ap JS* **62**, 501.

Kurucz, R.L.: 1991, in: A.G.D. Philip, A.R. Upgren and K.A. Janes, ed(s)., *Precision Photometry: Astrophysics of the Galaxy*, Schenectady, New York, 27.

Latham, D.W.: 1985, in: A.G.D. Philip and D.W. Latham, ed(s)., *IAU Colloquium 88, Stellar Radial Velocities*, Schenectady, New York, 21.

Latham, D.W.: 1992, in: H.A. McAlister and W.I. Hartkopf, ed(s)., *IAU Colloquium 135, Complementary Approaches to Double and Multiple Star Research*, Astronomical Society of the Pacific Conference Series, Vol. 32, 110.

Latham, D.W., Mazeh, T., Stefanik, R.P., Mayor, M. and Burki, G.: 1989, *Nature* **339**, 38.

Mazeh, T. and Goldberg, D.: 1992, *Ap J* **394**, 592.

Mazeh, T., Goldberg, D., Duquennoy, A. and Mayor, M.: 1992, *Ap J* **401**, 265.

Mazeh, T. and Zucker, S.: 1992, in: H.A. McAlister and W.I. Hartkopf, ed(s)., *IAU Colloquium 135, Complementary Approaches to Double and Multiple Star Research*, Astronomical Society of the Pacific Conference Series, Vol. 32, 164.

Tonry, J. and Davis, M.: 1979, *A J* **84**, 1511.

Zucker, S. and Mazeh, T.: 1993, *Astrophys. J.*, (in press).

DEVELOPMENT OF ABSOLUTE ACCELEROMETRY *

PIERRE CONNES

CNRS, Service d'Aéronomie, Verrières, France

Abstract. Absolute accelerometry is a technique for detecting small radial velocity changes involving lasers. The final output is a beat frequency similar to that from a Doppler radar. A progress report is presented on the development which began three years ago. While a suitable stellar échelle spectrograph is being built at Observatoire de Haute Provence, a demonstration of the main features on laboratory sources and the Sun is attempted at Verrières. Partial results are presented.

1. Introduction

Planetary detection through radial velocity (RV) changes of the parent star so far makes use of traditional techniques. Basically, a grating spectrograph analyzes star light which has been passed through either a gas absorption cell or a Fabry-Perot (FP) étalon, and relative line shifts are deduced from spectrograph and FP parameters. The order-of-magnitude error for a single observation is 10 m s^{-1}; this is but slightly reduced when many observations are combined. The quality of the results varies little with stellar magnitude; even on the Moon or Sun it is not markedly better. Altogether, we have a textbook example of systematic errors dominating over photon/detector noises. A full calculation of these noises has been given by Connes (1985a); for instance, on a $V = 10$ solar-type star, a 1 m telescope should be able to give a 1 m s^{-1} RMS error in 1 hour with a dedicated CCD spectrograph.

All of which points to the present weakness but also to the ultimate strength of the RV search: this is one clear case where only moderate-size telescopes are needed. Hence plenty of observing time should be available, provided we understand and eliminate systematic errors. Unfortunately, the best of all techniques on that count, heterodyne spectroscopy, is ruled out on that of photon-noise. The second best, Fourier spectroscopy (see Guelachvili (1981) for 15 cm s^{-1} accuracy laboratory results), falls far behind the CCD spectrograph in efficiency for the violet-to-green range where most of the RV information is present for the more interesting spectral types.

Much can be learned by scrutinizing asteroseismology, a field in which great instrumental advances have been made recently. Solar seismic lines with amplitudes of a few cm s^{-1} are now routinely detected, by integrating over many days. However, this is true only within the 5 min oscillation band; when the analysis was extended to much lower frequencies, spurious oscillations were found with periods of 160 min and 13 days (see discussion below).

---

* Paper presented at the Conference on *Planetary Systems: Formation, Evolution, and Detection* held 7–10 December, 1992 at CalTech, Pasadena, California, U.S.A.

Absolute Astronomical Accelerometry (AAA) was first proposed as a way to combine 1) optimal efficiency in treating the photon-limited RV information, and 2) maximal reduction of systematic errors. It was intended both for planetary searching and stellar seismology; a full description (Connes, 1985a) and two brief accounts (Connes, 1984, 1985b) were given. No support was obtained at the time, no laboratory facilities were available for even partial testing, and no interest in a collaboration was found either in France or abroad.

However, about three years ago, the same team that produced the well-known CORAVEL* announced the construction at Observatoire de Haute Provence (OHP) of a SUPERCORAVEL spectrograph (variously called ELODIE), which appeared suitable (but not optimal) for our project. Hence, we were enabled to present a proposal which did not include the CCD spectrograph, i.e., the most expensive but least original element of AAA. Also, a collaboration became feasible with the Bureau International des Poids et Mesures (BIPM) and the Laboratoire de l'Horloge Atomique (LHA). At the Service d'Aéronomie (SA) in Verrières, a complete system is now being built; it is intended to demonstrate the AAA principle on laboratory sources and on the Sun. If this first stage is successful, the equipment will be taken to OHP, and tried on stars with ELODIE. A progress report is presented here, together with results from the so-far completed fraction of the system.

2. Short Summary of Principle

Figures 1 and 2 describe AAA. A CCD spectrograph observes either the stellar spectrum or the channelled spectrum of a FP interferometer from a white-light laboratory source. This operation is first performed at some epoch 1 (also called "reference"), when the star radial velocity is V_1 and the FP spacing SP_1. At a later epoch 2, the velocity has changed to V_2 and the spacing to SP_2. The consequences on both spectra are similar: they are stretched along the frequency axis. There is one new spacing SP_2 which makes both stretching factors equal, and the game will be adjusting the FP to achieve this result.

Hence the AAA operation: First, both reference spectra are stored in a computer. Next, the measurements of RV change may begin. An optical commutator alternately sends the two spectra to the CCD. The computer compares each new spectrum with the corresponding reference one. Two separate error signals are immediately computed. The first (proportional to the stellar shift) is sent to the spectrograph, and stabilizes the lines on the pixels. The second (the difference between the stellar and FP shifts) tunes the FP, and the channelled spectrum is constrained to follow the stellar one.**

---

* M. Mayor (Obs. de Genève), A. Baranne (Obs. de Marseille) and an OHP team collaborate on the project. The spectrograph has a 40 × 10 cm échelle, a 1k × 1k CCD and will be fiber-fed from the 193-cm telescope.

** This CCD-measured correction is applied stepwise, after the end of each cycle. If the stellar exposure time exceeds a few minutes, both the FP and the spectrograph will also track continuously the precomputed Earth-velocity changing Doppler shift.

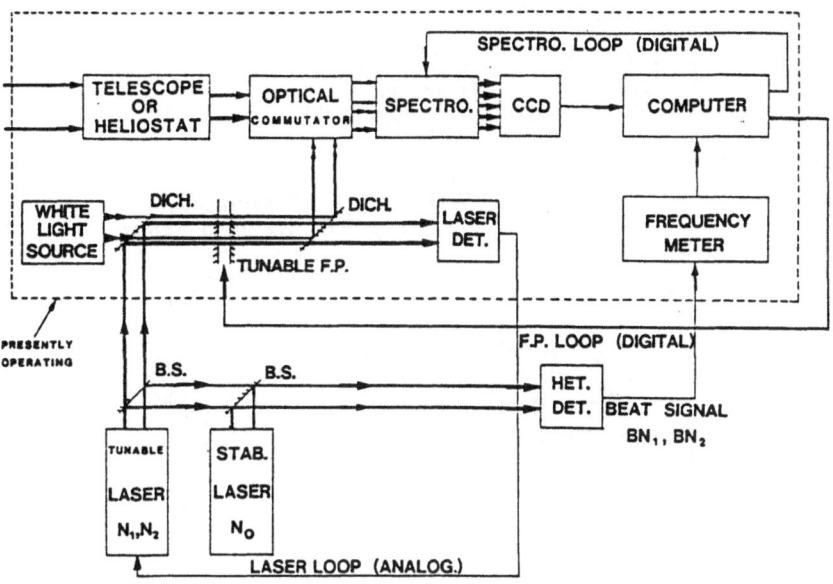

Fig. 1. AAA block diagram. The completed and tested portion is that within the dashed rectangle.

A tunable laser beam passes through the FP. An analog servo-loop locks the laser line to the FP channelled spectrum; hence the frequency shifts from N_1 to N_2. A fraction of the beam is mixed with that of a stabilized laser (of frequency N_0), and beats of frequency BN generated by a heterodyne detector. The final result is

$$\frac{(V_1 - V_2)}{c} = \frac{(BN_1 - BN_2)}{N_0}.$$

The beat frequency change is precisely what one would get if one laser had been on the star, and the other on Earth, or again half that given by a Doppler radar. However, both techniques would measure velocities; here, we can only get velocity changes.

No calibration is needed; spectrograph parameters, FP spacings, and line center-wavelengths or profiles vanish. The spectrograph-CCD-computer subsystem operates as a null-checking device: it is never asked to measure a line shift, only to estimate sign and order-of-magnitude of an infinitesimal residual shift, in order to derive the next similarly-infinitesimal FP-spacing correction. That is the essence of absolute techniques.

We believe such features to be essential for planetary detection, because expected velocity changes are extremely small compared with Earth-motion induced ones. Of course, these last are accurately known, but such knowledge is of little help if the measuring technique itself is not calibrated in RV with equal accuracy. A key

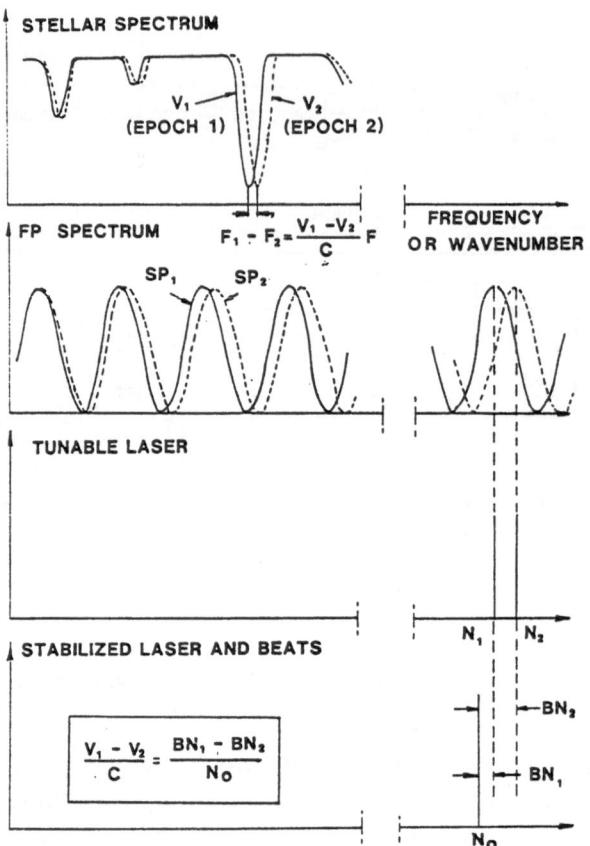

Fig. 2. Operation of AAA. The laser and astronomical frequency ranges may be quite different.

point is generally missed: having a stable zero-RV check is not enough. This failure is responsible for many of the unaccounted-for systematic errors. A clear-cut
illustration of the difficulty has been given by the "discovery" of a 13-day solar
oscillation from RV measurements by the potassium-resonance technique. A 5-m/s
amplitude signal with a 13.035-day periodicity was found, which remained coherent over 7 years; this result was taken as evidence for a fast-rotating solid solar
core. We showed the effect to arise from the 28th harmonic of the year (Connes,
1986).

The close analogy of solar-core detection with planetary searching is obvious.
All present RV techniques share the same difficulty, because the calibration curve
is laboriously extracted from the data themselves. Similar artifacts are bound to
surface whenever one attempts to look for smaller and smaller terms from presently
available data. If the RV calibration curve is imperfectly known but perfectly
stable, harmonics or subharmonics of the terrestrial day/year will be generated;

if it is unstable, these will fuse into a continuous spectrum, and mimic random errors. Merely increasing photon-limited SNR, either by improving efficiency or by accumulating more observations, does not solve the problem. With AAA, no calibration is involved, nor is any difficulty in subtracting out the huge Earth-RV terms: the stellar lines do not move relative to pixels nor to any laboratory fiducial marks.

The above-described procedure, in which the same CCD pixels are used alternately by the two beams (which pass through the same fiber), gives the best feasible cancellation of all long-term effects arising from slow changes in optical adjustments and/or CCD parameters; in principle, it should even be allowed to put in a new CCD at some stage of the program. However, this switched-beam technique is unavoidably slow (because of CCD readout-time), and leaves partly uncorrected the very-short-term spectrograph fluctuations, e.g. those produced by internal turbulence. Hence, a second beam from the same tunable FP must be fed to the spectrograph (through a second fiber), and a second FP spectrum interleaved with the stellar one must be projected on the CCD. This second FP spectrum is identical with the first, but the stellar and FP integrations are now simultaneous instead of successive; the additional error signals are intended solely for checking short-term residuals.

The second (but better-known) difficulty of RV planetary detection is that the hoped-for shifts are very small relative to the line widths; this implies limited SNR from photon and detector noises. In our set-up, stellar light directly reaches the entrance slit of the spectrograph (and time lost on the FP is small). Moreover, we make use of a demonstrably optimal algorithm which integrates all the RV information present in a given spectral profile and given number of photons; in principle, the performance quoted in the Introduction may be achieved. This is far from being the case when the light is passed through an absorption cell or FP étalon. Even compared with CORAVEL (the most efficient device so far), a large SNR gain is to be expected.

This gain is actually so large that it becomes feasible to subdivide the spectrum into arbitrary sets of lines, preselected solely from astrophysical criteria; the signal from one given set may be used for servoing, and the others recorded for a posteriori analysis*. This is the best possible answer to the standard RV-technique objection, i.e., that one does not know in advance how intrinsically stable the stellar lines will prove to be**. Here, because the entire spectrum is observed in a standard manner, we have built-in control for at least all the differential shifts. Of course, telluric lines are eliminated by the same procedure.

---

* It is also feasible to choose at will not only lines but also portions of line profiles, i.e., to get separate error signals for line tips, bases, etc. However, this procedure requires a higher resolution spectrograph, and hence implies a reduced spectral range and solid angle on the sky. The technique (mostly of interest for stellar oscillations) is limited to the brighter stars, and will not be accessible with ELODIE.

** R.S. McMillan et al., Ap. J. (Feb. 1993, in press), give the best and most comforting result so far (no significant shift exceeding ± 4 m s^{-1} in the solar spectrum over 5 years).

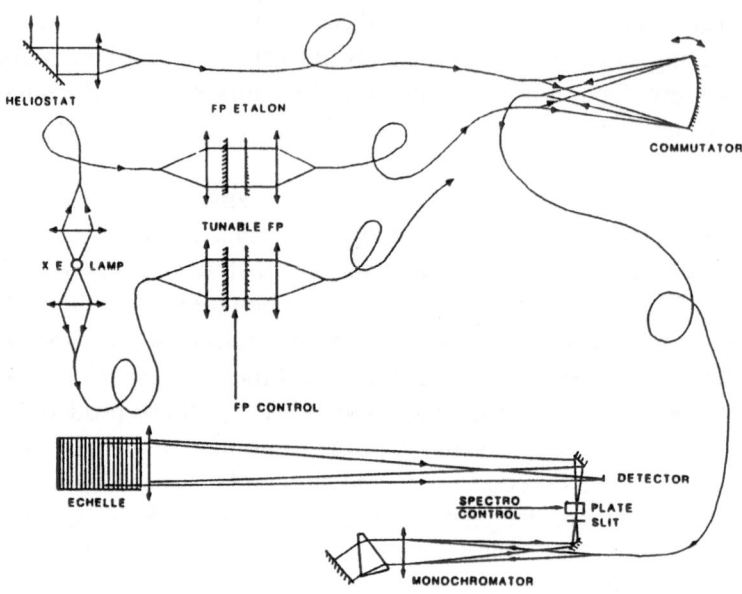

Fig. 3. Presently operating optical system. The two commutator inputs are used to compare the two FP spectra to each other, or either of them to the Sun.

3. Description of Present System

A simplified optical diagram is found in Figure 3. The heliostat 3-cm beam feeds a 1-mm silica fiber with integrated sunlight. A Xe high-pressure lamp illuminates two FP interferometers of the same 6-cm diameter and 1-cm spacing. The first is an étalon with ZERODUR spacers; it is used for preliminary testing only. The second is a homemade tunable FP; it includes three PZT spacers, three displacement transducers, and parallelism control from three laser beams. Special multidielectric layers were computed and fabricated by E. Pelletier (Lab. des Couches Minces), providing 0.9 reflectivity at 633 nm, and about 0.4 in the astronomical range (400–550 nm). Electro-mechanical and optical performance of the FP is adequate, but the electronics for the servo-loop are not complete.

The optical commutator, a flipping concave mirror, feeds a fiber going to the entrance of a prism monochromator and Littrow-type "solar spectrograph." This uses a single (selectable) order from a 63° blaze angle échelle with 73 grooves/mm on a 200×100 mm blank. The lens is a 100 mm-diameter 1 m focal length achromat, and the detector a HAMAMATSU MOS with 1024 25 μm pixels and 3000 e⁻ RMS readout noise. The spectral range is 3.2 nm at 500 nm, and the velocity scale 1800 m s⁻¹ pixel⁻¹. Unlike a matrix CCD, the present detector does not have spare pixels for recording the second FP beam; however, readout is fast enough to make these unnecessary. A 1 mm-thick tilting glass plate close to the input slit provides fine velocity tuning.

The specific AAA programs have been developed from the already described algorithms (Connes, 1985) by M. Kabyl. A COMPAQ 386/25 PC controls the whole system; it stores the 12-bit 1k-sample spectra, computes the two error signals which are sent to DA converters, and triggers the commutator and frequency meter. It also provides 6 analog inputs for recording temperatures, etc. Lastly, it computes the astronomical Doppler shift (DOS) with programs written by F. Chollet. There are four main separate operations (and corresponding programs): 1) REGLAGES checks spectrograph and CCD performance. 2) REFERENCE stores one pair of reference spectra and the corresponding beat frequency. 3) With MESURE, which lasts a full day, up to 2048 pairs of spectra are taken and immediately used for deriving the error signals. The DOS is subtracted from the results. All computing operations take about 3 s within each cycle. 4) CORRECT is used a posteriori; cloud-induced gaps are filled in, and Fourier transforms and various statistical operations performed.

A first He-Ne laser pair has been tested at BIPM by P. Jeseck under the direction of J.M. Chartier. The tunable laser is homemade, and derived from that described by Chartier (1990). With an avalanche diode and 1 GHz BW amplifier, a heterodyne signal adequate for the 1.2 GHz frequency counter is obtained. So far, a borrowed two-mode stabilized laser (Newport NL1) is used, and, as a first step, has been thoroughly checked against the BIPM lasers. Frequency fluctuations were recorded during runs of up to 60 hours over a two-month period. Fourier transforms of the recordings were computed, and showed that the NL1 error-contribution during observations devoted to stellar oscillations would be negligible. However, some tests of the same laser performed 4 years ago were available and showed that a 10^{-8} drift (equivalent to 3 m s^{-1}) has taken place. Hence (and as should be expected from two-mode stabilization), the NL1 is by itself unsuitable for planetary detection, but may become suitable through periodic checks with iodine-stabilized lasers; these have a proven stability of 10^{-11} (3 mm s^{-1}) over years.

The least-advanced part of the AAA setup is the generation of the error signal from the FP for the laser servo-loop. One appealing method is that described, e.g., by Hough (1984). An electro-optical modulator (EOM) phase-modulates the tunable-laser beam before passing it through the FP. After detection by an avalanche diode and synchronous demodulation, a DC error signal is produced. As a first step, an EOM has been built and tested by E. Samain at LHA under the direction of P. Cerez, who provided the design. It is a lithium tantalate crystal fitted within a 160 MHz resonant cavity, with a Q of about 50, in order to reduce drive power. However, the overall system is difficult and expensive for whoever must start with no VHF experience nor measuring equipment. Hence, we are also trying the simpler procedure of oscillating the tunable FP spacing in the LF range.

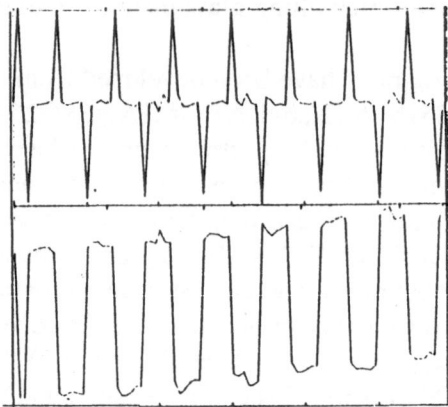

Fig. 4. Digital servo response to a square wave perturbation. Upper curve: residual velocity error; lower curve: correction. The vertical scale is arbitrary: the error is proportional to the amplitude of the perturbation. The result is independent of the spectrum used.

4. Preliminary Results

The lasers not yet being usable, we were not able to control the tunable FP and no overall check of AAA has so far been made. However, we have (1) operated all programs and processed spectra through the specific AAA procedure, (2) checked spectrograph performance against the FP étalon, and (3) recorded the shift of solar spectra relative to the étalon. The commutator has been used in 2 modes: 1) when stopped, any spectrum may be compared to itself; noise limitations and spectral "quality factors" (Connes, 1985a) are measured; 2) when flipping, the spectrum from the Sun, or an absorption cell, is compared to that of the FP, or the two FP's to each other.

The spectrograph digital servo-loop has been well checked. Figure 4 illustrates the residual velocity-error when a square wave is fed directly to the tilting plate. Any spectrum can be used. Both curves are almost identical to those given by numerical simulation of the loop: the first sample after the perturbation is off (because the correction is not yet known), and the next one falls back to the rest level. The FP digital loop, while not yet checked, is expected to behave similarly.

Figures 5 and 6 illustrate results when the commutator alternates between the solar and FP étalon beams with 1 s integrations and a 5 s cycle. The servo locks the solar lines to the pixels, irrespective of solar RV changes and spectrograph drift. In Figure 5 the FP is not involved: the plot merely shows the residual servo error. Even close to saturation, photon noise barely exceeds readout noise with the HAMAMATSU, and the $1.5\,\mathrm{m\,s^{-1}}$ RMS residual is about what can be expected from their sum. The typical mean error for such records is $5\,\mathrm{cm\,s^{-1}}$. Note that the spectral range is small (515–518 nm), and contains only about 20 "strong" lines.

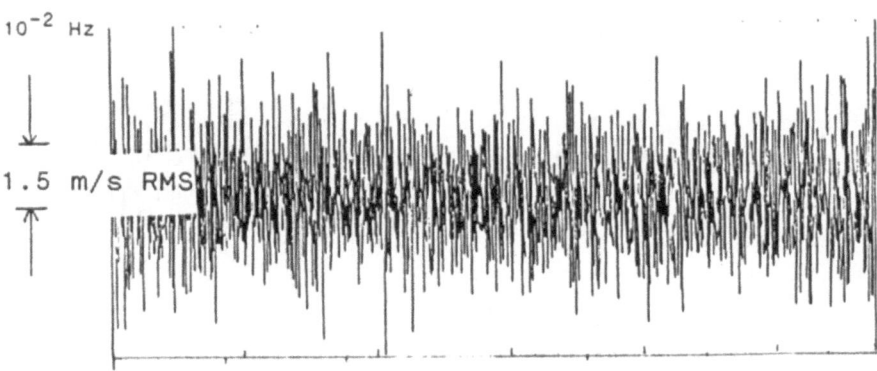

Fig. 5. Residual velocity-error signal with digital spectro loop closed (on solar spectrum). Individual 5 s measuring cycles (each containing two 1 s integrations) are seen. The same test in the same range with the FP spectrum (which is filled with lines) gives a 0.2 m s^{-1} RMS error.

Better results would be recorded around 400 nm, but the lens chromatic aberration is too large.

From the same test, Figure 6 gives the FP error-signal, which simply follows the solar surface RV: if solar lines do not move relative to pixels, then those from a fixed-spacing FP have to. The upper curve is the recording. The velocity scale is calibrated in standard fashion from the DOS (which has been subtracted out); hence we stress again that the present result is not absolute. The lower curve is the square root of the power FT. While the 5-min oscillation band is visible, it is contaminated by atmospheric transparency-gradient noise, which dominates at still lower frequencies. Integrated-sunlight velocities are very sensitive to such gradients; we have shown the effect to be almost negligible at a good site (Appourchaux, 1984), but Verrières is sadly different. This 3-hour recording (which is much too short for getting the oscillation peaks to stand out clearly) is the least-bad one we obtained from May to September; cumulus clouds invariably come over at midday.

Above 10^{-2} Hz, the Figure 6 FT shows pure instrumental noise, with an RMS dropping to 9 mm s^{-1} (i.e., 5×10^{-6} pixel). It is comforting to see a commonplace spectrograph, handled in the specific AAA manner, performing at that level. However, we have not yet tackled a key point: behavior with a stellar beam. In Connes (1985a), we proposed a double-fiber scrambler for feeding star- and FP-light into the spectrograph. A first version has since been demonstrated by Brown (1989) with highly encouraging results; still, more work is required.

A few words about near-term plans. The red Ne line provides only a 1-GHz maximum tuning range, which corresponds to a ± 600 m s^{-1} velocity shift. This is adequate for solar work, and also for tests on stars provided they are either not too far from the pole of the ecliptic, or followed for a limited number of nights

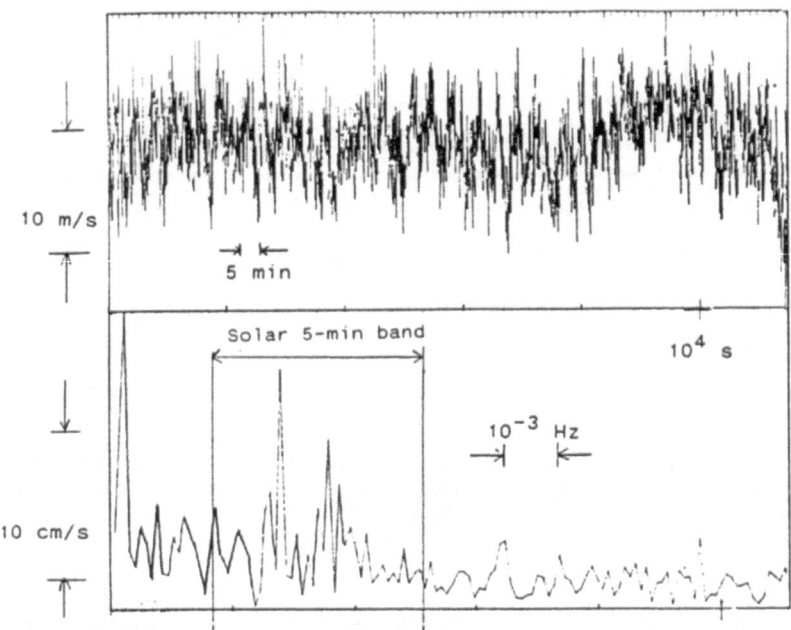

Fig. 6.   Solar surface velocity relative to FP étalon, after subtraction of Earth motion. Upper curve: direct measurements (as in Figure 5). Lower curve: square root of power Fourier transform, i.e., with linear velocity scale.

only. We originally proposed dye-lasers, but the far simpler diode laser has since much improved. The ECL (extended-cavity laser) is clearly suitable, and the DFB (distributed-feedback laser) possibly so; a recent review is that of Labachelerie (1992). Hence, we plan to build a pair of 780 nm AlGaAs ECL lasers*, one of which will be stabilized on the Rb resonance line (a standard technique). The full ± 30 km s^{-1} Earth-induced velocity change (requiring 64 GHz tuning) will be easily accommodated. The maximum beat frequency will be 32 GHz; commercial detectors, frequency meters and wave analyzers cover that range.

5. Conclusion

RV planetary detection has not yet attracted up-to-date technology (in particular, lasers) in the same way the astrometric method did. In this last case, long-base interferometry clearly promised radical improvements, from an approach very similar to ours: measurement of angles is reduced to counting laser fringes, and usual factors such as plate scale are eliminated; early expectations have been well confirmed in practice, and more is to follow (Shao and Colavita, 1992). Our

---

* Suitable commercial versions of the key AAA building-blocks exist. These include tunable FP's, iodine-stabilized lasers, and ECL diode lasers. Each is far too expensive for the present funding.

AAA attempts to fill that gap. However, the present effort is clearly understaffed and underfunded, and progresses far too slowly. The most worrying issues are insufficient technical help, the lack of Ph.D. students so far, and the clouded future of OHP itself. Even so, the first tests have checked two necessary points: specific treatment of the spectra, and spectrograph performance. Far more will be required for a complete demonstration.

Acknowledgements

The present research is supported by an INSU grant, but would not be feasible without much outside aid. We thank T.J. Quinn, P. Cerez and P. Véron for the valuable help from BIPM, LHA and OHP, respectively, and also the whole ELODIE-SUPERCORAVEL team for its collaboration. The contributions of J.M. Chartier, F. Chollet, P. Jeseck, M. Kabyl, E. Pelletier, and E. Samain have proved essential, as discussed in the text. P. Bouchareine, M. Dreyfus, A. Golman, and P. Luc helped with equipment and/or advice. At SA, J.L. Bertaux loaned the Hamamatsu detector, while J. Millard assembled most of the electronics.

References

Appourchaux, T.: 1984, in: F. Praderie (ed.), *Space Research Prospects in Stellar Activity and Variability*, Meudon Observatory, p. 117.
Brown, T.: 1989, *Astr. Soc. Pac. Conf. Ser.* **8**, p. 335.
Connes, P.: 1984, *IAU Coll.* **88**, p. 131.
Connes, P.: 1985a, *Astrophys. Space Sci.* **110**, 211.
Connes, P.: 1985b, *IAU Symp.* **112**, 9.
Connes, P.: 1986, in: : D.O. Gough, ed(s)., *Seismology of the Sun and Distant Stars*, D. Reidel Pub. Co., p. 229.
Guelachvili, G.: 1981, *App. Opt.* **20**, 2121.
Hough, G. *et al.*: 1984, *App. Phys. B* **33**, 179.
Labachelerie, M. *et al.*: 1992, *Jour. Phys. III France* **2**, 1557.
Shao, M. and Colavita, M.: 1992, *Astron. Astroph.* **262**, 353.

EXTRASOLAR PLANET DETECTION *

R.P. KORECHOFF, D.J. DINER, E.F. TUBBS and S.L. GAISER

Jet Propulsion Laboratory, California Institute of Technology,
Pasadena, California, USA

Abstract. This paper discusses the concept of extrasolar planet detection using a large-aperture infrared imaging telescope. Coronagraphic stellar apodization techniques are less efficient at infrared wavelengths compared to the visible, as a result of practical limitations on aperture dimensions, thus necessitating additional starlight suppression to make planet detection feasible in this spectral domain. We have been investigating the use of rotational shearing interferometry to provide up to three orders of magnitude of starlight suppression over broad spectral bandwidths. We present a theoretical analysis of the system performance requirements needed to make this a viable instrument for planet detection, including specifications on the interferometer design and telescope aperture characteristics. The concept of using rotational shearing interferometry as a wavefront error detector, thus providing a signal that can be used to adaptively correct the wavefront, will be discussed. We also present the status of laboratory studies of on-axis source suppression using a recently constructed rotational shearing interferometer that currently operates in the visible.

1. Introduction

Although there is mounting circumstantial evidence for the existence of extrasolar planetary systems, the unambiguous detection of such a system would be an achievement of profound significance. The detection of extrasolar planets around many stars would provide data on the frequency of planetary formation, the stellar environments which favor planetary formation, and possibly the conditions which result in terrestrial and gas-giant type planets. Direct detection systems, that is, systems that observe radiation reflected or emitted by the planet, have rigorous requirements on sensitivity, spatial resolution, and dynamic range. A number of direct detection schemes have been suggested, including visible imaging (Davies, 1980; KenKnight, 1977; Terrile and Ftaclas, 1989), infrared imaging (Angel *et al.*, 1986; Diner, 1989; Watson *et al.*, 1991), visible interferometry (Burke, 1986), infrared interferometry (Bracewell, 1978; Shao, 1989), and submillimeter aperture synthesis (Jones and Diner, 1989).

Direct detection of extrasolar planets requires contending with extreme brightness contrast (10^{10} in the visible; 10^7 in the infrared; and 10^5 in the submillimeter), and small angular separations (typically < 1 arcsec). The planet image, which resides in the wings of the central star point-spread function (PSF), must be detected over a background of diffracted and scattered starlight, zodiacal light and, in the infrared, thermal radiation from telescope mirrors. Of these background sources, the suppression techniques discussed in this paper only affect diffracted starlight. Trade-off studies performed assuming a range of values believed to be achievable for the suppression of diffracted starlight, telescope primary mirror

---

* Paper presented at the Conference on *Planetary Systems: Formation, Evolution, and Detection* held 7–10 December, 1992 at CalTech, Pasadena, California, U.S.A.

Astrophysics and Space Science **212**: 369–383, 1994.

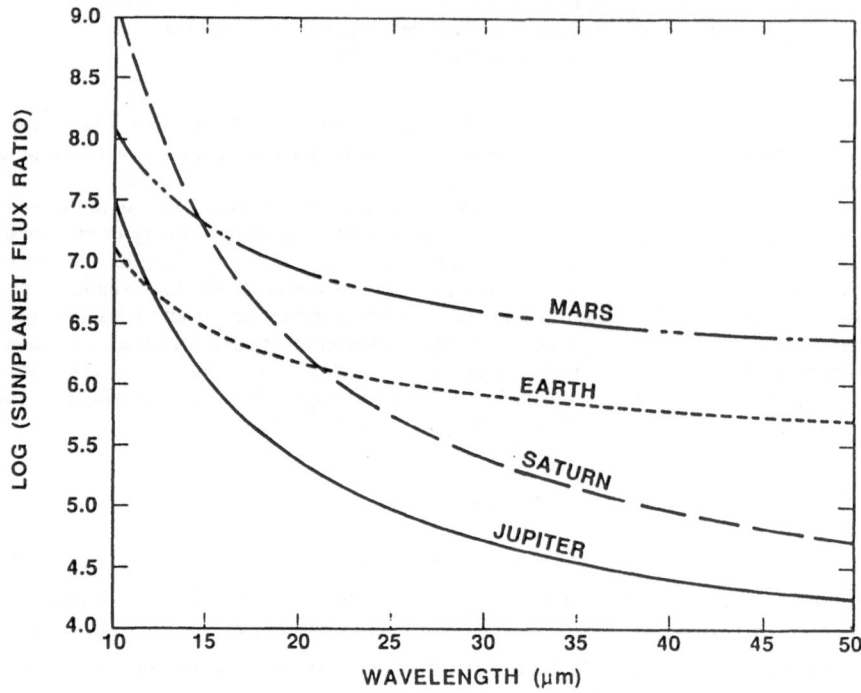

Fig. 1.   Solar-type star-to-planet flux ratios for various planets in our solar system, as a function of wavelength.

smoothness (scattering), and telescope temperature (thermal radiation) have led us to conclude that the optimal spectral range for detecting both terrestrial and gas-giant type planets is 10-15 μm. Although the star/planet ratio decreases at longer wavelengths (Figure 1), for terrestrial-type planets the zodiacal light/planet ratio increases by more than an order of magnitude between 10 and 50 μm (Figure 2). In addition, the telescope cooling requirements become considerably more difficult as the wavelength increases (Figure 3). Finally, the telescope aperture size required to achieve the desired angular resolution would exceed 20 m if wavelengths longer than 15 μm are considered.

The telescope requirements are similar to those described by Angel *et al.* (1986) and Watson *et al.* (1991), who proposed using cooled, large aperture infrared telescopes with special apodization masks to suppress diffracted starlight. For planet detection, diffracted starlight must be suppressed by four to five orders of magnitude. Angel *et al.* (1986) suggested masking a 16-m diameter primary mirror with an opaque annular ring. The effect of this apodization pattern is to produce a deep annular null in the stellar PSF. At a wavelength of 10 μm, the chosen dimensions of the annular ring result in a null extending from 0.2 to 0.3 arcsec.

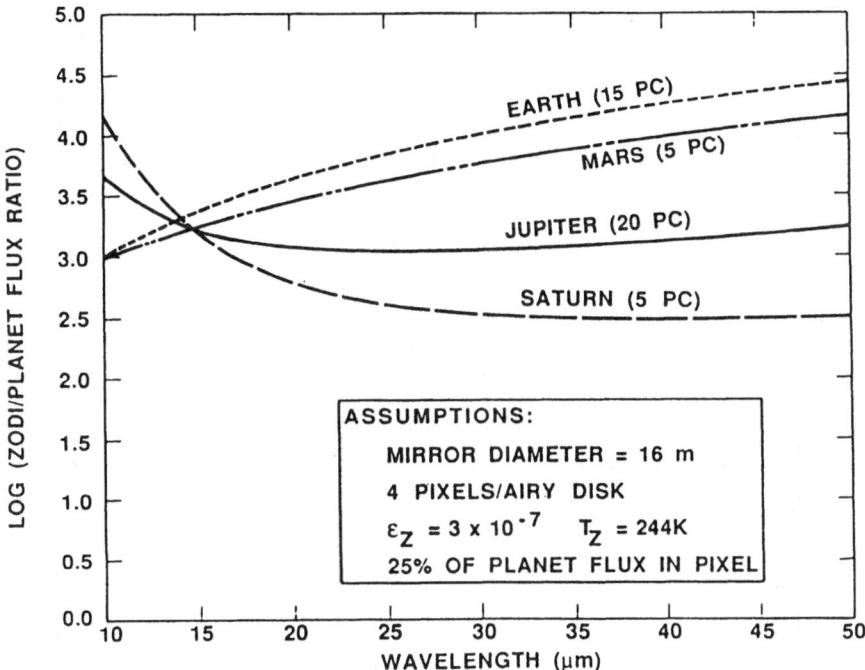

Fig. 2. Ratio of zodiacal light to planet flux for various planets in our solar system as a function of wavelength. The temperature and emissivity of the zodiacal cloud correspond to the values at the ecliptic. The distances shown in parentheses correspond to the farthest distance for which a planet of the specified type is calculated to be detectable. ϵ_z and T_z refer to the assumed emissivity and temperature of the zodiacal cloud used.

This is optimal for a planet 1 A.U. from the central star and at maximum elongation when observed at a distance of 4 parsecs. If one considers a search volume out to 25 parsecs, and angular separations that result in planetary equilibrium temperatures comparable to our solar system (a few tenths to several arcsecs), then an annular apodization pattern is inadequate. Watson *et al.* (1991) proposed using a square aperture telescope and coronagraphic apodization utilizing specially shaped masks. Their results show that the desired suppression can be achieved with considerably less sacrifice of search space around the star.

This paper describes an optical signal-processing scheme that increases the search space by supplementing a modest coronagraphic apodizer with a modified rotational shearing interferometer (Breckinridge, 1974; Roddier, Roddier, and DeMarcq, 1978). Adaptive optics, in the form of a tip-tilt mirror to control pointing jitter and a deformable mirror for wavefront correction, are also envisioned as part of the optical train. We refer to this set of optical components as the Interferometer Based Imaging System (IBIS). The use of a folding shear interferometer was

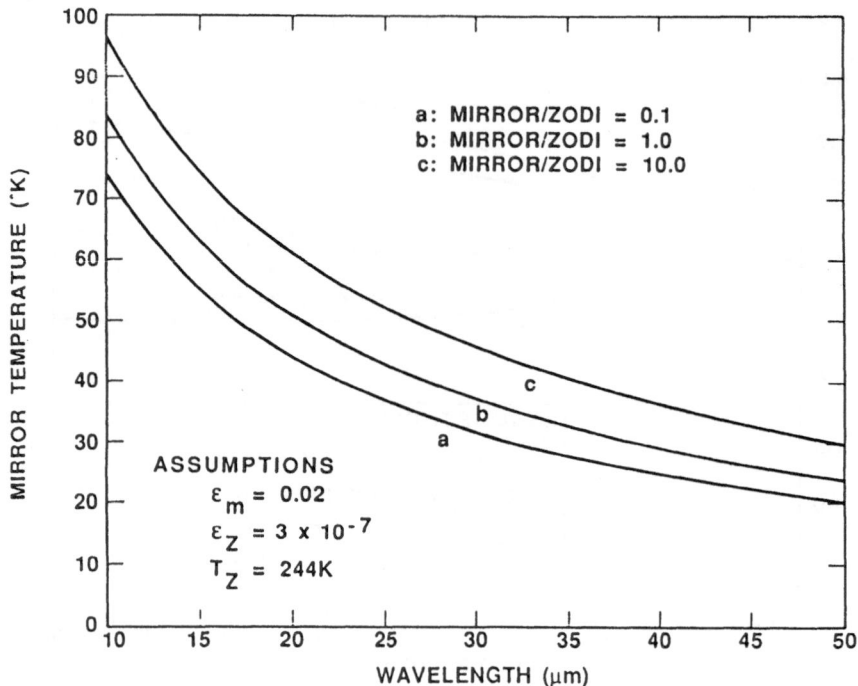

Fig. 3. Required mirror temperature as a function of wavelength such that the mirror thermal background is equal to 0.1, 1.0, 10.0 times the zodiacal background. ϵ_m is the assumed mirror emissivity.

suggested by KenKnight (1977) for extrasolar planet detection in the visible. The rotational shearing interferometer (RSI) offers the advantage of starlight suppression that is independent of the polarization state of the incident radiation. As with the approaches of Angel *et al.* (1986) and Watson *et al.* (1991), IBIS must work in conjunction with a large-aperture ($\approx$16-m diameter) space-based telescope. Such a telescope would be shared by other instruments designed for general astrophysical and solar system research in order to justify the extensive resources.

2. The IBIS Concept

As described above, the IBIS instrument is attached to a large aperture telescope (see Figure 4). To mitigate tight requirements on primary mirror figure errors, we envision a two stage system similar to the design approach taken with the Large Deployable Reflector (Swanson *et al.*, 1986). The first stage consists of a Cassegrain-type design that forms an intermediate image at a central aperture in the quaternary mirror. This image is relayed by the second stage (tertiary and

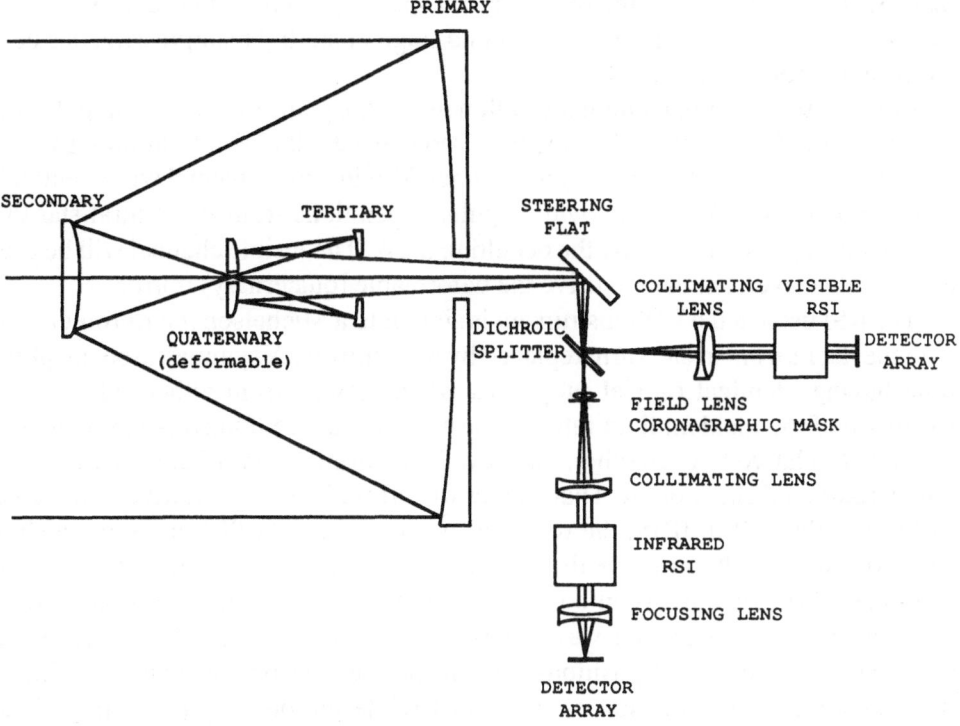

PRIMARY

SECONDARY

TERTIARY

STEERING
FLAT

COLLIMATING VISIBLE
LENS RSI

DETECTOR
ARRAY

QUATERNARY
(deformable)

DICHROIC
SPLITTER

FIELD LENS
CORONAGRAPHIC MASK

COLLIMATING LENS

INFRARED
RSI

FOCUSING LENS

DETECTOR
ARRAY

Fig. 4. Schematic of the telescope/IBIS Optical System, showing the two-stage telescope with the deformable mirror, the small steering flat, dichroic splitter, visible RSI for wavefront sensing, Lyot coronagraph, and IR RSI for planet imaging.

quaternary) to a second focal plane behind the telescope primary. The telescope is designed such that the secondary and tertiary mirrors image the primary (entrance pupil) on the quaternary. The idea of the two-stage telescope is to correct primary mirror figure errors at the relatively small aperture, deformable quaternary mirror. Since the quaternary is a pupil, the correction will be independent of field angle.

The first stage of starlight suppression in IBIS is a Lyot coronagraph. The coronagraph contains an occulting mask and field lens at the second focal plane of the telescope. The occulting mask, which is centered on the optical axis of the instrument, blocks the central portion of the stellar PSF. The field lens re-images the telescope pupil (quaternary) to a plane containing a Lyot stop and a collimating lens. The largest contribution to diffracted light comes from the edge of the pupil. The field lens images the pupil edge on a second mask (Lyot stop), thus preventing it from propagating further. For a 16-m diameter telescope operating at 12 μm, an occulting mask of radius 0.2 arcsec will block the Airy disk. Coupled with a Lyot stop in the pupil plane, the Lyot coronagraph suppresses the starlight by one to two orders of magnitude. Greater suppression would require a larger occulting mask which would reduce the planet search space. Therefore, in order to preserve as much search space as possible, a second stage of starlight suppression has been added in the form of an RSI.

Both the star and planet light are collimated at the plane of the Lyot stop. Before entering the RSI, a dichroic beamsplitter separates the incident light into a visible channel and an infrared (10–14 μm) channel. The infrared channel will eventually form an image of any planets with angular separations from the central star that exceed the angle subtended by the occulting mask. The visible channel will be used to generate wavefront data to drive the deformable (quaternary) mirror.

The RSI used in the IBIS instrument is similar to a Michelson interferometer but with the flat mirrors in each arm replaced by roof mirrors. A symmetric beamsplitter (one having identical optical properties when viewed from either side) directs roughly half the incident light into each arm. When the roof mirrors are rotated by an angle θ relative to each other, the wavefronts from each arm are superimposed with a rotational shear of 2θ. That is, one wavefront is rotated relative to the other by 2θ. For the RSI in IBIS, the roof mirrors are rotated by 90° providing a shear angle of 180°. If the arms of the interferometer have equal optical paths, there is a phase difference between corresponding points of π. For an off-axis source, the two rotationally sheared wavefronts will be imaged in the final focal plane at diametrically opposed positions with respect to the optical axis. One image represents the actual location of the planet while the other (ghost) image is an artifact of the RSI. As the source moves toward the optical axis, the real and ghost images superimpose and null each other. In this manner the central star, assumed to be located on the optical axis, is suppressed while any planets present are imaged without interference.

In order to detect faint stellar companions in reasonable observing times it is necessary for IBIS to operate over a broad spectral band without regard to the polarization of the incident light. Analysis has shown that if the surfaces of the roof mirrors are perfect conductors, both the S- and P-polarization states interfere destructively at the output of the interferometer. Since real metal coatings do not behave in an ideal manner, the degree of destructive interference is limited. This difficulty can be overcome by introducing an additional fold mirror in each arm of the interferometer (see Figure 5). The effect of the additional fold mirrors is to create, for each eigenpolarization state of the system, two S-type and two P-type 45° reflections in each arm of the fold and roof mirrors. Since the arms have equal path length and the beamsplitter is optically symmetric, the symmetry in the reflections removes all wavelength and polarization dependence between the two arms. Hence, the performance of the interferometer is limited only by the finite size of the source, the deviation of the input wavefront from 180° degree symmetry, and optical component fabrication and alignment errors. Because of the symmetry inherent in the RSI, it is not necessary to use mirrors coated with a near-perfect conductor. In fact, the fold and roof mirrors could be fabricated from any reflective material, including total-internal-reflection (TIR) prisms for the roof mirrors. The

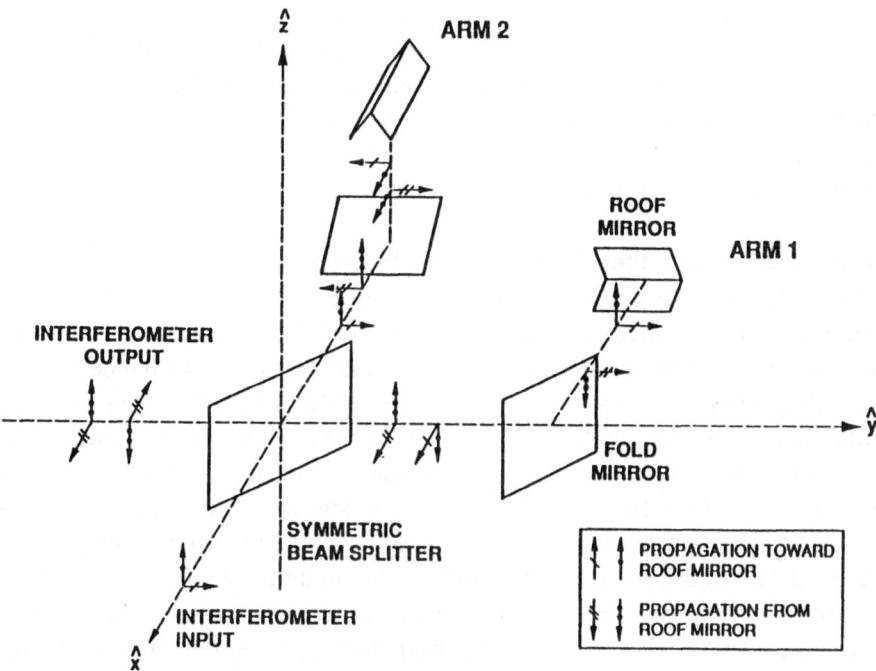

Fig. 5. Schematic of wavelength and polarization independent, rotational shearing interferometer. The roof mirrors are oriented to produce 180° shears. The arrows trace the phases of eigenpolarization states through the interferometer.

polarization symmetry discussed above only exists for rotational shear angles of 180°. Breaking the symmetry by operating the interferometer at shear angles other than 180° would degrade the suppression of the on-axis source.

Total destructive interference requires the source to be located on the RSI optical axis. Thus, the finite size of the central star limits the expected suppression. An analysis shows that the theoretical suppression achievable is given by $(\pi D\theta_*/\lambda)^2/8$, where D is the telescope diameter, λ is the wavelength, and θ_* is the angular diameter of the star. For a 16-m diameter telescope and a typical stellar angular subtense of 0.002 arcsec, three orders of magnitude of suppression are theoretically possible. However, approaching the theoretical suppression limit implies keeping the star centered on the RSI optical axis. Controlling a large aperture telescope to milli-arcsec accuracy would be a formidable task. A more practical solution is to introduce a small steering mirror into the optical train. A defocused image of the star in the visible would provide the centroiding information to control the steering mirror.

A sensitivity analysis was carried out to determine the effect of fabrication and alignment errors on on-axis suppression in the RSI. To simplify the analysis a monochromatic, point source was assumed. The RSI achieved a suppression of three orders of magnitude at 12 μm when, simultaneously, the difference in interferometer arms was <50 nm, the roof mirror dihedral angle differed from 90° by <0.3 arcsec, and the tip/tilt error in the mirror orientations was <0.5 arcsec. These fabrication and alignment requirements are within the current state of the art.

As described above, the RSI interferes light from diametrically opposed points on the incident wavefront. Thus, the only requirement on the wavefront of on-axis sources, for satisfactory cancellation in the RSI focal plane, is symmetry under a rotation of 180° about the optical axis. Hence, the performance of the RSI is completely unaffected by azimuthally symmetric aberrations such as defocus and spherical aberration, or aberrations with the appropriate rotational symmetry such as astigmatism. The fact that only the phase difference, and not the absolute phase, between points diametrically opposed is significant means that the figure tolerances on the primary mirror can be relaxed by about an order of magnitude. To relax the surface figure requirements further, the quaternary mirror is deformable. The visible starlight redirected by the dichroic beamsplitter (see Figure 4) into an RSI similar to the one used to form planetary images, detects surface figure errors on the primary mirror by measuring the starlight extinction over the pupil. Primary mirror surface errors will show up in the visible RSI as bright spots. The intensity and locations of these bright spots will be used to drive the quaternary deformable mirror. The requirements on the deformable mirror are benign, with respect to temporal response and stroke, compared to deformable mirrors that perform atmospheric corrections in real time. The primary mirror figure required to achieve the necessary RSI suppression is approximately $\lambda/800$ ($\lambda_{vis}/40$) without a deformable mirror. When a deformable mirror is implemented to correct wavefront asymmetries, the primary figure requirement is relaxed to about $\lambda/80$ ($\lambda_{vis}/4$).

3. Aperture Configurations

The optimum telescope configuration with regards to collecting area, dynamic range, and angular resolution is a filled, circular primary mirror aperture. However, due to the large aperture diameter required for planet detection in the infrared (16 m), such a system would have to be segmented and deployed in orbit. By compromising the optical performance to some extent, savings in weight and cost can be incurred by implementing a dilute, or unfilled, aperture. For this reason we have compared the optical performance of some unfilled geometries to a filled, circular aperture of the same collecting area.

One drawback of unfilled apertures is that they produce PSFs that have azimuthal dependence. A solution to this problem is to rotate the telescope about its optical axis continuously during the observation period while simultaneously decoupling the final focal plane so that the real and ghost images of the planet continue to fall on the same pixels. The effect of the telescope rotation is to produce a stellar PSF that is azimuthally symmetric. In addition, any irregularities in the PSF (resulting, for example, from mirror figure errors) will be azimuthally averaged (smeared), making it less likely that they would be confused with signals from real planets. Since the RSI rotationally shears the telescope wavefront by 180°, only aperture geometries which are invariant under 180° rotations need be considered. We have considered three such aperture geometries:

1. A filled circular aperture, 16 m in diameter with a central obscuration 3.4 m in diameter;
2. A 4-arm cross-shaped aperture, with arm dimensions of 12×4 m; and
3. A 6-arm configuration with arm dimensions of 8×4 m.

Each aperture geometry has a collecting area of 192 m^2. The telescopes are assumed to operate with a spectral bandwidth of $\Delta\lambda/\lambda = 0.4$ centered at $\lambda = 12$ μm. Figure 6 shows the azimuthally averaged PSFs for each aperture geometry listed above. In general, the unfilled apertures take energy from the Airy disk and redistribute that energy in the outer diffraction rings. This effect obviously decreases the planet-to-star flux ratio in the wings of the stellar diffraction pattern. The redistribution of energy can be seen more clearly in Figure 7 which is a plot of encircled energy for a point source. Again, the reduced encircled energy at a given radius for the unfilled apertures compared to the filled aperture implies degraded planet detection performance.

We have calculated the theoretical starlight suppression achievable with IBIS for each telescope aperture geometry. In each case, the coronagraphic mask is assumed to be a circular stop 0.2 arcsec in radius. However, the Lyot stop must be tailored to the individual aperture and we assume it to be the same shape as the primary mirror and covering the outer 30 percent of the pupil. This stop reduces the effective collecting area of the telescope from 192 to 134 m^2 for the circular aperture and to 94 m^2 for the unfilled apertures. We assume that the RSI, which follows the Lyot coronagraph, suppresses the starlight by three orders of magnitude. The resulting azimuthally averaged PSFs are shown in Figure 8. Although the unfilled apertures result in higher starlight backgrounds in the likely planet search space compared to the filled aperture, the degraded performance is still acceptable for planet detection as will be shown in the next section.

4. Planet Detection Calculations

A critical assumption in determining the performance of a particular extrasolar planet detection scheme is the model of star-planet system incorporated in the calculation. We define an extrasolar planetary system as one in which the planets have the same equilibrium temperatures and sizes as the planets in our solar system. The star-planet angular separations are then determined by the stellar type and the distance to the star. Using the Woolley catalog of stars within 25 parsecs, we

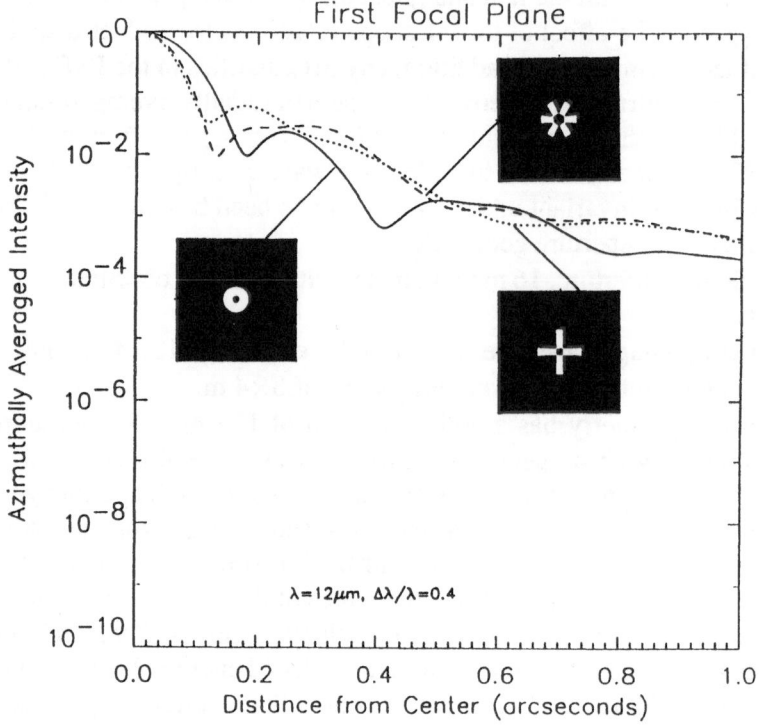

Fig. 6. Azimuthally averaged PSFs for the three aperture geometries given in the text. The broadband PSFs are calculated by integrating monochromatic calculations using Gaussian quadrature.

computed the star-planets separations using this model. These angular separations were then multiplied by $\pi/4$ to give the angular separation for a circular orbit averaged over all possible viewing angles.

The temperature of the optics is assumed to be 70 K, the system emissivity 0.1, the system quantum efficiency 20%, and total observing time of 50 hours per star. The zodiacal background used is an average of the values in the plane of the ecliptic and perpendicular to this plane. The criteria for planet detection are (1) the ratio of planet flux to shot noise from all background sources (central star, zodiacal light, thermal emission) exceeds 5; (2) the average angular separation between the star and the planet exceeds 0.3 arcsec; and (3) the ratio of residual starlight to planet signal, within the pixel where the planet is detected, does not exceed 100. As an example, consider a 275 K "Earth" orbiting Fomalhaut, an A-type star at a distance of 7 parsecs. Using the collecting area of a multi-arm aperture with the appropriate Lyot stop, combining the fluxes from the two planet images (real and ghost) produced by the RSI, taking into account losses in the interferometer and coronagraph and assuming that 40% of the remaining planet flux is in the detection pixel (a value calculated from the encircled energy performance predicted for the

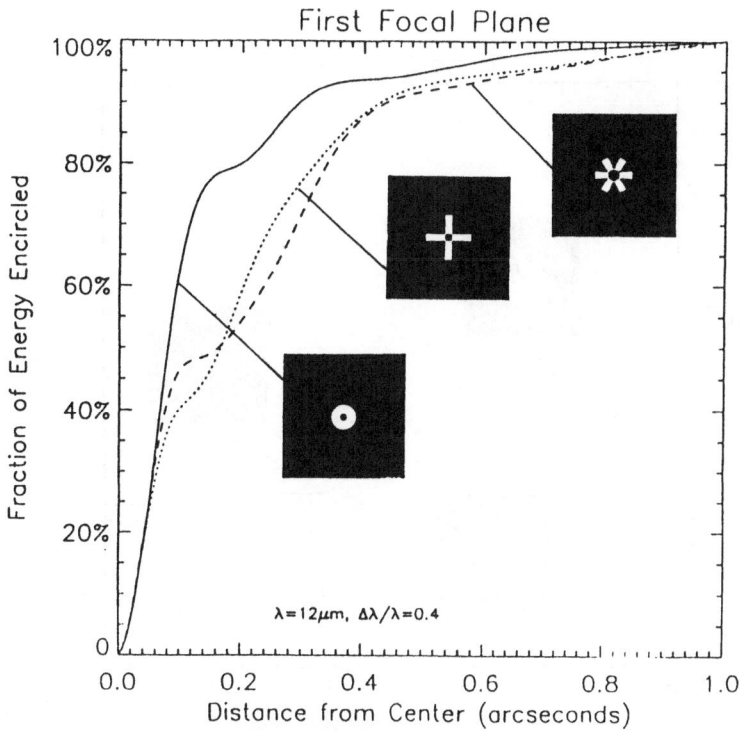

Fig. 7. Encircled energy as a function of field angle for the three aperture geometries given in the text. In this paper we have assumed square pixels approximately 0.2 arcsec on a side. Because the pixels are square, the incident energy per pixel will differ slightly from the encircled energy represented by these curves. In calculating these curves we have made the approximation that 100% of the encircled energy is contained within a radius of 1.0 arcsec.

multi-arm apertures), we calculate a planet signal of 37 detected photons/sec. The zodiacal and mirror thermal backgrounds are 3430 and 7980 detected photons/sec respectively, assuming square pixels measuring 0.189 arcsec on a side and also taking into account the combination of fluxes from diametrically opposed pixels. The central star flux is about 500 photons/sec at a star-planet angular separation of 0.6 arcsec based on the suppression shown in Figure 8. Assuming shot noise limited detection, the signal-to-noise ratio reaches 5 after 4 minutes of integration and around 140 in 50 hours of integration. Similar calculations have been repeated for stars in the Woolley catalog out to 25 parsecs. Because the catalog is believed to undersample the K and M stars in the solar neighborhood, the number of such stars were scaled, according to their distance from the Sun, by the population densities given in Allen (1973). The predicted number of planets detected is shown in Table I.

Table I also includes the number of planet detections for a filled circular aperture. The improved performance of the filled aperture, compared to the multi-arm aperture, is the result of three factors: (1) greater encircled planet energy in the

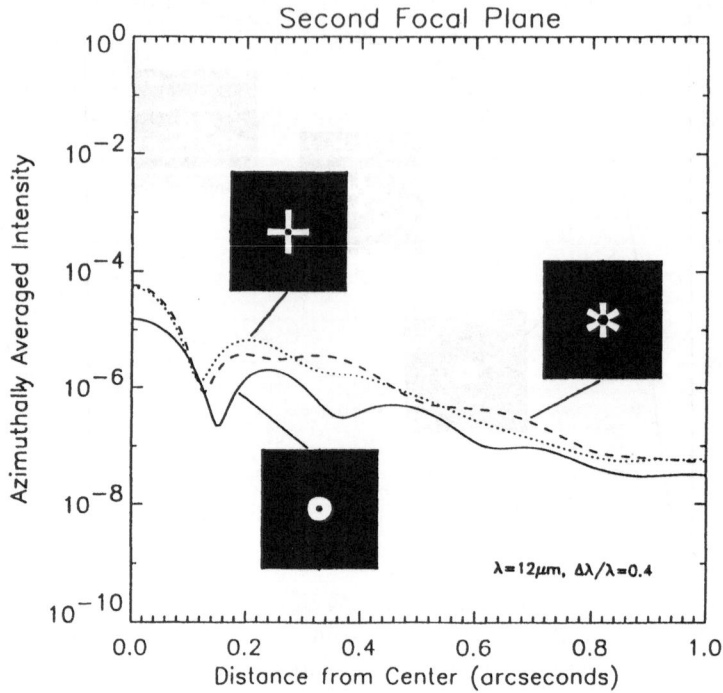

Fig. 8. Azimuthally averaged PSFs for the three aperture geometries given in the text, after application of the IBIS diffraction suppression scheme.

TABLE I

Predicted planet detection performance by planet type.

Planet Type	Mercury	Venus	Earth	Mars	Jupiter	Saturn
Radius (km)	2500	6100	6400	3400	70,000	60,000
Temperature (K)	450	335	275	225	125	95
No. detectable (multi-arm)	2	10	14	5	295	10
No. detectable (circular)	2	12	21	13	313	46

TABLE II

Predicted "Jupiter" detection performance by stellar type.

Stellar Type	O	B	A	F	G	K	M
No. detectable (multi-arm)	0	0	76	132	54	33	0
No. detectable (circular)	0	5	85	136	54	33	0

detection pixel; (2) better starlight suppression; and (3) greater collecting area after coronagraphic apodization. Nevertheless, the results in Table I support the feasibility of planet detection with unfilled apertures.

Table II provides a breakdown of Jupiter-type planet detections by stellar spectral type for both multi-arm and filled circular apertures. The capability of IBIS to detect such planets around both early- and late-type stars is significant for understanding the planetary system formation process.

5. Laboratory Rotational Shearing Interferometer

A laboratory version of the RSI described in section 2 is currently being tested. The objectives of the tests are to measure the suppression of an on-axis point source in both broad and narrow spectral bandwidths, and to understand the sensitivity of the suppression to fabrication and alignment errors in preparation for the design of a second-generation interferometer. A schematic of the RSI optical layout is shown in Figure 9.

The design approach taken for the laboratory RSI was to work in the visible and near infrared (silicon detection region) and use commercially available components wherever possible. The source used is a 100 W, short-arc mercury lamp, which produces a continuum of radiation from about 600 to 900 nm. The light from the source is passed through a Kodak Wratten 29 filter which cuts on at 600 nm and has a peak transmission beyond 700 nm. The silicon detector is sensitive in the infrared to about 1000 nm. The required "symmetric" property of the beamsplitter is approximated by using an appropriate index-matching oil between an uncoated glass plate and a plate coated with a thin metallic layer. To minimize polarization-dependent phase shifts, angles of incidence at the beamsplitter are made as near normal as possible. The path length of one arm of the interferometer is controlled by a piezoelectric translator. Protected aluminum-coated optics have been used for the initial alignment and preliminary measurements. These elements will eventually

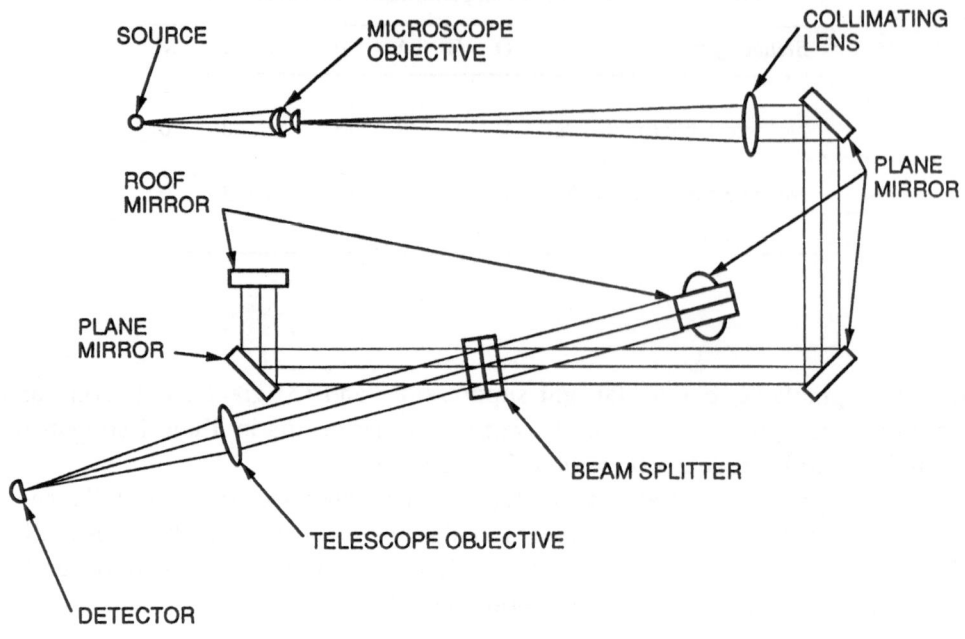

Fig. 9. Optical layout of the laboratory rotational shearing interferometer.

be replaced by optics coated with unprotected gold which have better reflectivity in the infrared.

Quantitative data regarding suppression levels are in the process of being collected. However, we can report at this time that a white-light, broad (fluffed out) null has been observed for an on-axis point source.

6. Conclusions

In this paper we have shown that the IBIS instrument (Lyot coronagraph, RSI), operating in conjunction with a large (16-m characteristic dimension), deployable primary mirror in a two-stage telescope, has the capability of detecting both terrestrial and gas-giant type planets around stars of varying spectral types in reasonable observing times.

Acknowledgements

We thank J.F. Appleby for assistance with the stellar statistics, Y-Y. Cheng for interferometer tolerance analysis, E. Ribak for suggesting the use of the rotational shearing interferometer in this application, and M. Shao for proposing the polarization-compensating fold mirrors in the interferometer design and for providing the stellar population density scaling factors. We also thank the Journal of the British Interplanetary Society for permission to use figures and tables which appeared in the October 1991 issue.

This research was carried out by the Jet Propulsion Laboratory, California Institute of Technology, under contract with the National Aeronautics and Space Administration.

References

Allen, C.W.: 1973, *Astrophysical Quantities*, Athlone Press, University of London, London 249.

Angel, J.R.P., Cheng, A.Y.S. and Woolf, N.J.: 1986, *Nature* **322**, 341.

Bracewell, R.N.: 1986, *Nature* **322**, 780.

Breckinridge, J.B.: 1974, *Applied Optics* **13**, 2760.

Burke, B.F.: 1986, *Nature* **322**, 340.

Davies, D.W.: 1980, *Icarus* **42**, 145.

Diner, D.J.: 1989, *Proceedings of the Workshop on the Next Generation Space Telescope*, Space Telescope Science Institute, Baltimore, Maryland, 133.

Jones, D.L. and Diner, D.J.: 1989, *Nature* **337**, 32.

KenKnight, C.E.: 1977, *Icarus* **30**, 422.

Roddier, F., Roddier, C. and DeMarcq, J.: 1978, *J. Optics (Paris)* **9**, 145.

Shao, M.: 1989, *Proceedings of the Workshop on the Next Generation Space Telescope*, Space Telescope Science Institute, Baltimore, Maryland, 160.

Swanson, P.N., Breckinridge, J.B., Diner, A., Freeland, R.E., Irace, W.R., McElroy, P.M., Meinel, A.B. and Tolivar, A.F.: 1986, *Optical Engineering* **25**, 1045.

Terrile, R.J. and Ftaclas, C.: 1989, *Proceedings of the Workshop on the Next Generation Space Telescope*, Space Telescope Science Institute, Baltimore, Maryland, 225.

Watson, S.M., Mills, J.P., Gaiser, S.L. and Diner, D.J.: 1991, *Applied Optics* **30**, 3253.

INDIRECT PLANET DETECTION WITH GROUND-BASED
LONG-BASELINE INTERFEROMETRY *

M.M. COLAVITA and M. SHAO

Jet Propulsion Laboratory, California Institute of Technology,
Pasadena, California, USA

Abstract. Narrow-angle astrometry with long-baseline infrared interferometers can provide extreme-ly high accuracies as required for indirect planet detection. Narrow-angle astrometric interferometry exploits the properties of atmospheric turbulence over fields smaller than the interferometer baseline divided by the atmospheric scale height. For such fields, accuracy is linear with star separation, and nearly inversely proportional to baseline length. To exploit these properties, the interferometer observes a relatively bright (< 13 mag_K) target in the near infrared at 2.2 μm, and uses phase referencing to find a reference star within the 2.2-μm isoplanatic patch. With this technique faint references can be found for most targets. With baselines > 100 m, which also minimize photon-noise errors, and with careful control of systematic errors by using laser metrology, accuracies of tens of microarcseconds/$\sqrt{}$hour should be possible.

1. Introduction

Indirect detection of exoplanets senses the presence of a planet around another star by observing the motion of the star about the system barycenter. These planetary signatures are generally quite small. A Jupiter–Sun system observed from 10 pc would exhibit a peak-to-peak signature of 1 mas with a 12-yr period. The signature is clearly smaller for a less massive planet, a more massive star, or a faster orbital period. Narrow-angle astrometry on the ground using astrometric telescopes with modern detectors exhibits accuracies of the order of 2–3 milliarcseconds in one night (Monet and Dahn, 1983; Gatewood, 1987), limited by the atmosphere, which is inadequate for any comprehensive planet search. However, the accuracy of these current techniques is far from the ultimate accuracy that can be achieved when observing through atmospheric turbulence.

The key to obtaining high astrometric accuracies is a detailed understanding of the error sources in the measurement — fundamentally, for ground-based measurements, atmospheric turbulence — and the design of a measurement technique to control the error or to exploit the particular characteristics of the error. One technique that can reach the ultimate accuracies allowed by the atmosphere is long-baseline interferometry. The key features of long-baseline interferometry in this application are dual-beam operation to perform a simultaneous differential measurement between a target and reference star, long baselines ($\geq$ 100 m) to reduce atmospheric and photon-noise errors, infrared observations with phase referencing to increase sensitivity in order to locate nearby reference stars, and laser metrology to control systematic errors. The atmospheric limit for an optimized

---

* Paper presented at the Conference on *Planetary Systems: Formation, Evolution, and Detection* held 7–10 December, 1992 at CalTech, Pasadena, California, U.S.A.

Astrophysics and Space Science **212**: 385–390, 1994.
© 1994 *Kluwer Academic Publishers.*

system can reach 10 $\mu as/\sqrt{hr}$, with photon noise and systematic errors controlled to a similar level. A system with performance close to these limits could conduct a comprehensive search for Jupiter-mass planets around stars of various spectral types, and for short-period Uranus-mass planets around nearby M and K stars.

2. Atmospheric Limitations

The detailed expression for the atmospheric error in a differential astrometric measurement follows from a straightforward derivation using conventional models of atmospheric turbulence (Lindegren, 1980; Shao and Colavita, 1992). The key parameters in the model are the baseline length (or telescope diameter) B, and the separation vector θh, where θ is the angle between the two stars and h is the atmospheric height, $\sim$ 5–10 km. In an ordinary differential astrometric measurement, the separation θh is much larger than the telescope diameter. In this case the astrometric error is only weakly dependent on the star separation (as $\theta^{1/3}$), and not dependent on the baseline length; this dependence has been confirmed by measurements with the Multichannel Astrometric Photometer (Han, 1989).

However, when θh is smaller than the baseline length, the error behavior changes significantly.[*] In this case the error variance, σ_δ^2, of a t-second differential measurement is given by

$$\sigma_\delta^2 \simeq 5.25 B^{-4/3}\theta^2 \int dh\, C_n^2(h)h^2 V^{-1}(h)t^{-1}, \qquad \theta \ll B/h, \quad t \gg B/V, (1)$$

where $C_n^2(h)$ and $V(h)$ are the turbulence and wind-speed profiles. With the substitution of atmospheric parameters appropriate to an excellent site like Mauna Kea, this expression yields (Shao and Colavita, 1992)

$$\sigma_\delta \simeq 300 B^{-2/3}\theta t^{-1/2} \text{ arcsec}, \qquad \theta \ll B/3000, \quad t \gg B/10, \tag{2}$$

with B in meters and θ in radians. The key features in this very-narrow-angle regime are the linear dependence of the error on star separation, and the nearly linear decrease in error with increasing baseline length. Figure ?? plots the complete error behavior for several different baseline lengths to show more clearly the dependencies of the error. From the graph or from the expressions above, it is clear that with long baselines and small star separations, very small astrometric errors can be achieved; e.g., with a 200-m baseline and a 15″ star separation, the atmospheric error is 10 $\mu as/\sqrt{hr}$.

There are fewer confirming measurements in this very-narrow-angle regime than in the more conventional narrow-angle regime. However, recent star-trail measurements at Mauna Kea (Gatewood, private communication, 1991) show a marked break from $\theta^{1/3}$ behavior at small separations. In addition, recent measurements of wide ($\sim$3″) binaries with the Mark III stellar interferometer on Mt.

---

[*] The breakpoint between regimes, $\theta = B/h$, is sometimes referred to as the isokinetic angle.

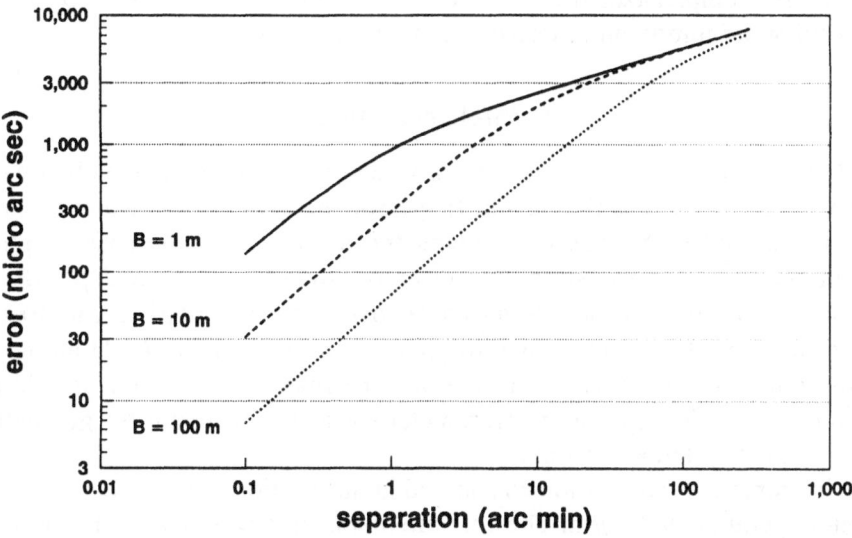

Fig. 1.   Narrow-angle and very-narrow-angle astrometric errors for several baseline lengths using measured Mauna Kea turbulence profiles and an integration time of 1 hr.

Wilson, modified to simultaneously detect the fringe packets from the primary and secondary, have yielded preliminary results which are consistent with the expressions above.

There are some additional points about the error behavior which should be emphasized: (1) As is clear from the $t^{-1/2}$ dependence in the expression above, the error is white for long integration times; this is unlike the behavior of an absolute astrometric measurement. (2) The derivation above does not depend upon perfect frozen flow between apertures, and is based on the most conservative orientation of the baseline and star-separation vectors. In addition, this derivation uses a conservative (infinite outer scale) Kolmogorov model. (3) The result above is essentially identical for telescopes and interferometers. While it is clear that, unlike an interferometer, a telescope acts as a spatial filter for spatial scales smaller than the telescope diameter, a detailed examination of the integral leading to Equation (1) shows that most of the residual error for long integration times $(t \gg B/V)$ is attributable to large spatial scales, rather than to those filtered by the telescope. Thus, there is no significant difference in the astrometric performance between interferometers and telescopes of the same size; although, in practice, the long baselines necessary to exploit the characteristics of the atmospheric error are most easily achieved with an interferometer. It is also worth recalling that the position of the centroid of a star's position as measured with a telescope depends only on a contour integral around the periphery of the telescope (Hogge and Butts, 1976); i.e., the interior phase distribution does not matter: an annular telescope has precisely the same statistics of centroid motion as a filled aperture of the same diameter.

Thus, this fact emphasizes that it is the long baselines of the telescope's (u, v) coverage that are important to its astrometric performance.

3. Implementation

To exploit the atmospheric limits given above requires the reduction of photon noise and systematic errors to similar levels. Interferometers offer significant advantages over conventional telescopes in reducing these errors. For the case of photon noise, the contribution to astrometric error is proportional to (λ/B)/SNR, i.e., to the diffraction limit of the interferometer over the photometric signal-to-noise ratio, so that long baselines reduce the photon-noise error (as B^{-1}) along with the atmospheric error (as $B^{-2/3}$ in the very-narrow-angle regime). In addition, compared with a telescope, an interferometer has an easily monitored geometry in order to control systematic errors.

The instrument configuration considered in this section is a long-baseline interferometer operating at 2.2 μm. The aperture diameter is chosen as ~ 1.5 m, which is readily phased for operation at that wavelength with simple tip-tilt correction. There are several reasons for choosing the 2.2-μm band. One is that for typical cool stars, say K5, the fringe-tracking limiting magnitude of a 2.2-μm interferometer with large apertures is several magnitudes greater than that of a visible-wavelength interferometer. A second reason is that the isoplanatic patch at 2.2 μm is much larger than at visible wavelengths: ~ 15–$20''$ rather than ~ 3–$4''$. While the isoplanatic angle is quite different from the isokinetic angle discussed above, it does play a role in achieving high astrometric accuracies. This is because the magnitude of a reference star which yields a photon-noise error consistent with the desired atmospheric error is significantly fainter than the fringe-tracking limiting magnitude of the interferometer. Thus, to use faint reference stars near the target star in order to minimize the atmospheric error, a means of increasing the coherent integration time is needed. The technique used is phase referencing, whereby the fringe position of the target star is used to stabilize the fringe position of the reference star, allowing coherent integration times well in excess of those allowed by atmospheric turbulence. However, like adaptive optics, this technique is only applicable within the isoplanatic patch, and thus the large 2.2-μm isoplanatic patch offers significant advantages in finding reference stars.

A detailed numerical example is worked in Shao and Colavita (1992); in summary, a 200-m baseline interferometer using 1.5 m telescopes was assumed. For a $15''$ separation between the target and reference stars, the atmospheric limit is 10 μas/$\sqrt{\text{hr}}$. The photon noise from the target star, assumed brighter than 13 mag$_K$, is negligible compared to the atmospheric noise. Using the target star to phase-reference within the isoplanatic patch, the reference star can be as faint as 17.5 mag$_K$ and still achieve a photon-noise error no larger than 10 μas/$\sqrt{\text{hr}}$. Assuming average star distributions, scaled according to spectral type, it can be shown that reference stars of this magnitude should be available within the isoplanatic patch over most of the sky.

4. Implementation Details and Future Instruments

To achieve the performance described above requires care in implementation, but does not require instrumental components beyond the current state of the art. The key features of an instrument which can demonstrate high-accuracy narrow-angle astrometry are phase-coherent fringe tracking at 2.2 μm with high sensitivity and a dual-star feed system. The latter device sends two beams from each interferometer aperture through separate delay lines to separate two-way beam combiners. A corner-cube retroreflector is mounted directly in front of (or on the surface of) the siderostat mirror, and its position is monitored using laser metrology from each beam combiner to control systematic errors to the required precision. The relative accuracy that must be achieved is $\sim$ 10 nm, which is well within the capabilities of laser metrology systems. However, because of the narrow fields involved, many errors are common mode. For example, the absolute baseline vector need only be known to $\sim$ 100 μm—a factor of 100 less accurate than what has been achieved in wide-angle astrometry with the Mark III stellar interferometer (Shao *et al.*, 1988; Shao *et al.*, 1990).

In order to develop the technology for narrow-angle interferometric astrometry, as well as to validate the underlying atmospheric theory, NASA has recently funded, as part of the TOPS (Toward Other Planetary Systems) program (TOPS Report, 1992), the TOPS-0 Testbed Interferometer. This interferometer is being specifically optimized for narrow-angle astrometry, using a dual-star feed, laser metrology, and high-sensitivity phase-coherent fringe tracking using array detectors at 2.2 μm. The instrument will use modest collecting apertures, $\sim$ 40-cm clear aperture, and will nominally be installed at Palomar Mountain. Compared to a more optimized instrument, the major compromise in this low-cost prototype is the aperture diameter, which will limit sensitivity and hence the fraction of the sky over which nearby reference stars can be found. However, the instrument will demonstrate the necessary technology and validate the atmospheric models to the limit set by the site. It will also be able to begin a modest observational program.

As a testbed, technology development is the primary goal of the Palomar interferometer: the lessons learned and the technology developed will transfer directly to the proposed Keck Interferometric Array (KIA) (TOPS Report, 1992). The KIA would combine the two Keck 10-m telescopes with four movable 1.5-m outrigger telescopes to form a powerful array for synthesis imaging as well as for astrometry. In its astrometric mode, the array would use the four outrigger telescopes in an orthogonal array with baselines of $\sim$ 120 m. With the excellent seeing at the site, plus the sensitivity allowed by the large apertures in order to select nearby reference stars, the outrigger array should be able to achieve astrometric accuracies of less than 30 μas/$\sqrt{}$hr; better performance is possible by incorporating one or both 10-m telescopes.

5. Conclusion

Over small fields the atmospheric limits for narrow-angle astrometry are surprisingly small. However, to achieve the ultimate limits set by the atmosphere requires the long baselines that can be achieved only with an interferometer. An interferometer with a 200-m baseline at a good site can measure the angle between two stars 15″ apart to 10 μas/$\sqrt{}$hr. To achieve the sensitivity to find suitable references near an arbitrary target requires the use of modest ($\sim$ 1.5 m) apertures and phase referencing at 2.2 μm to increase the coherence time within the isoplanatic patch. To control errors to the appropriate level requires a dual-star feed with appropriate laser metrology. The recently funded TOPS-0 Testbed Interferometer will demonstrate the technology for narrow-angle interferometric astrometry, and begin a modest detection program for selected targets. The technology and techniques would then be transferred to the proposed Keck Interferometric Array, which should be able to achieve astrometric errors of less than 30 μas/$\sqrt{}$hr. This instrument could conduct a comprehensive planetary-search program for Jupiter- and Saturn-mass planets around nearby stars of spectral types M–F, and for Uranus-mass planets in 10-yr orbits around nearby M and K stars.

Acknowledgements

This work was performed at the Jet Propulsion Laboratory, California Institute of Technology, under a contract with the National Aeronautics and Space Administration.

References

Gatewood, G.D.: 1987, *Astron. J.* **94**, 213.
Han, I.: 1989, *Astron. J.* **97**, 607.
Hogge, C.B. and Butts, R.R.: 1976, *IEEE Trans. Antenna Propagat.* **AP-24**, 144.
Lindegren, L.: 1980, *Astron. Astrophys.* **89**, 41.
Monet, D.G., Dahn, C.C.: 1983, *Astron. J.* **88**, 1489.
Shao, M., Colavita, M.M. and Hines, B.E. *et al.*: 1990, *Astron. J.* **100**, 1701.
Shao, M., Colavita, M.M. and Hines, B.E. *et al.*: 1988, *Astron. Astrophys.* **193**, 357.
Shao, M. and Colavita, M.M.: 1992, *Astron. Astrophys.* **262**, 353.
NASA: 1992, TOPS: Toward Other Planetary Systems: A Report by the Solar System Exploration Division.

DIRECT IMAGING OF PLANETARY SYSTEMS WITH A

GROUND-BASED RADIO TELESCOPE ARRAY *

DAYTON L. JONES
*Jet Propulsion Laboratory, California Institute of Technology,
Pasadena, California, USA*

Abstract. The National Radio Astronomy Observatory's proposed Millimeter Array (MMA) will bring unprecedented sensitivity, angular resolution, and image dynamic range to the millimeter wavelength region of the spectrum. An obvious question is whether such an instrument could be used to detect planets orbiting nearby stars. The techniques of aperture synthesis imaging developed for centimeter wavelength radio arrays are capable of producing images whose dynamic ranges greatly exceed the brightness ratio of a solar-type star and a Jupiter-like planet at sub-millimeter or millimeter wavelengths. The angular resolution required to separate a star and planet at a few pc distance can be obtained with baselines of several km. The greatest challenge is sensitivity. At the highest possible observing frequencies ($\sim$ 300 GHz for typical high, dry sites, and $\sim$ 900 GHz from the Antarctic plateau), the proposed MMA will be unable to detect the thermal emission from a Jupiter-like planet a few pc away. An upgraded MMA operating near 300 GHz with twice the currently proposed number of antennas, a 20% fractional bandwidth, and improved receivers could detect Jupiter at 4 pc in a few months. Building such an array on the Antarctic plateau and operating at $\approx$ 900 GHz would allow Jupiter at 4 pc to be detected in approximately one day of observing time.

1. Introduction

Wavelengths longer than about 20 microns offer two main advantages over shorter (optical/near-IR) wavelengths for the detection of planets orbiting nearby stars. The first advantage is that the star/planet brightness ratio is reduced by orders of magnitude, greatly reducing the dynamic range required to detect planets through direct imaging. Second, the ability to amplify signals without loss of phase information allows powerful self-calibration algorithms developed for radio interferometer arrays at centimeter wavelengths to be used. These algorithms permit aperture synthesis images made by interferometers to have far greater dynamic range than images made by an individual telescope (Pearson and Readhead, 1984).

The goal of this paper is to explore the possibility of detecting planets orbiting stars within several pc of the Sun with a ground-based interferometer array operating at millimeter or sub-millimeter wavelengths. If this proves practical, a significant reduction in cost over space-based planet detection instruments may be possible. Direct imaging has the advantage of showing all (sufficiently bright) planets in a system at once, and the orbit of each planet can be determined from multiple images over a period of months to years. The ability of imaging to detect planets is not degraded if the system contains several large planets of comparable mass. This situation would be unfavorable for indirect detection techniques such

---

* Paper presented at the Conference on *Planetary Systems: Formation, Evolution, and Detection* held 7–10 December, 1992 at CalTech, Pasadena, California, U.S.A.

as precise astrometry or radial velocity measurements, as the signal being searched for would not have a clear periodicity.

This study was limited to gas giant planets (Jupiters), which are considerably easier to detect than terrestrial type planets. We consider first the capabilities of the Millimeter Array (MMA) proposed by the National Radio Astronomy Observatory (Hughes, 1990).

2. Prospects for Planet Detection with the MMA

The Millimeter Array is planned to have 40 antennas, each 8 meters in diameter. The shortest observing wavelength is $\approx$ 1 millimeter ($\approx$ 300 GHz) and the longest baseline is 3 km, so the maximum angular resolution is about 0.06 arcsecond. This resolution corresponds to 1 AU at a distance of 17 pc.

The flux density of Jupiter at 4 pc is only 0.3 μJy at a wavelength of 1 mm, so very high sensitivity is needed. The MMA bandwidth will be 2 GHz and the total system temperature is predicted to be $\sim$ 200 K. With these values, the standard sensitivity calculation for an interferometer array (e.g., Thompson, Moran, and Swenson, 1986) shows that it would take centuries of integration time to detect Jupiter at 4 pc.

The MMA will have adequate angular resolution and probably dynamic range to image planets orbiting nearby stars, but it will not have sufficient sensitivity to accomplish this in a reasonable amount of observing time. Are there realistic ways to increase the continuum sensitivity of the MMA?

3. Planet Detection with an Upgraded MMA

The sensitivity of the MMA could be improved by increasing the number (or size) of the antennas, increasing the bandwidth, or reducing the total system temperature. Since we need a large increase, all three of these strategies will be used. The antennas represent the largest single expense in the estimated cost of the MMA, so it is not practical to increase the number (or size) of antennas by a large factor. Similarly, the quantum limit on system temperature ($T_{sys} > h\nu/k$) prevents us from reducing this parameter by a large factor. That leaves bandwidth as the parameter to concentrate on.

Consider an upgraded MMA with twice the original number of antennas (80), a 20% fractional bandwidth, and total system temperatures of 50 K. If these improvements could be made, a Jupiter-like planet would be just detectable (SNR > 5) at 4 pc in 50 days. This is still far too long to be appealing. Much wider bandwidths or lower system temperatures are unlikely, so we need still more antennas or a site which allows observations at wavelengths < 1 mm. Using the atmospheric window near 0.34 mm instead of 1 mm would increase the flux density of the planet by nearly an order of magnitude, reducing the required integration time by nearly two orders of magnitude.

4. A Sub-Millimeter Array on the Antarctic Plateau

Probably the best site for ground-based observations at sub-millimeter wavelengths is the high, very dry plateau in central Antarctica (Bally, 1989; Townes and Melnick, 1990). An MMA-size array on this plateau should be able to operate at wavelengths near 345 microns ($\approx$ 870 GHz) much of the time. Precipitable water vapor values below 0.2 mm have been measured at Vostok station during the winter (Burova *et al.*, 1986), and it is possible that values as low as 0.05 mm occur at higher elevations on the plateau (Bally, 1989). Although the temperatures at this location are very low and elevations exceed 4000 meters, the ground wind velocities are also low (Parish, 1988) which eliminates the need for protective antenna radomes.

An upgraded MMA as described above, located on the Antarctic plateau and operating near 345 microns, should be able to detect (SNR > 5) a Jupiter-like planet at 4 pc in less than one day. This is actually rather optimistic because the assumed system temperature (50 K) is only slightly above the quantum limit at this higher frequency and the atmosphere will contribute more to the total system temperature, but the required integration times are still measured in days rather than months. (The bandwidth assumed here remains at 60 GHz instead of 20% of the observing frequency due to the width of the atmospheric window at 345 microns.)

Note that the relatively strong, unresolved star will serve as an excellent real-time phase calibration source. Thus, long coherent integrations need not be degraded by variations in atmospheric conditions. Multiple images—separated by months—would allow the unambiguous determination of planetary orbits.

5. Cost Estimates

Using the cost equations in the NRAO proposal for the MMA, the cost of the sensitivity upgrade described here is dominated by the additional antennas ($\approx$\$50M) and the increase in correlator bandwidth ($\approx$\$80M). Together these double the estimated cost of the MMA.

This is less than the cost of even a very modest space mission, and the resulting array would have a useful lifetime of decades. In addition, it would be accessible for periodic repairs and improvements. Finally, an increase in MMA sensitivity would obviously benefit many areas of astrophysics in addition to the detection of other planetary systems.

The additional site-related costs involved in building and operating an array on the Antarctic plateau instead of a continental U.S. site are difficult to estimate. These extra costs would be minimized if an international scientific station of the sort proposed by Lynch (1989) were already established on the Antarctic plateau above 4000 meters. Such a station could serve as a technology testbed for many types of astronomical instruments destined for eventual deployment on the Moon (e.g., Jones and Diner, 1989).

6. Summary

Only the last option considered here, an upgraded MMA located on the high central Antarctic plateau, would be able to detect Jupiter-like planets at distances of a few pcs in a reasonable amount of observing time ($\sim$ a day). The cost of such an instrument would be increased by the logistical difficulties of the site, but could still be substantially less than a dedicated space-based instrument.

Increasing the sensitivity of the MMA at its proposed southwestern U.S. site, although not as attractive an option for planet detection, would be a good initial step. Nearby planets larger or hotter than Jupiter could be detected with an MMA-type array operating at a wavelength of $\approx$ 1 mm.

Acknowledgements

This work was done at the Jet Propulsion Laboratory, California Institute of Technology, under contract with the National Aeronautics and Space Administration.

References

Bally, J.: 1989, in: D.J. Mullan, M.A. Pomerantz and T. Stanev, ed(s)., *Astrophysics in Antarctica*, AIP Conf. Proc. 198, 100.,

Burova, L.P. *et al.*: 1986, *Soviet Astron. Lett.* **12**, 339.

Hughes, R.E.: 1990, *The Millimeter Array, A Proposal to the National Science Foundation Submitted by Associated Universities, Inc.*, NRAO, 153.

Jones, D.L. and Diner, D.J.: 1989, *Nature* **337**, 51.

Lynch, J.T.: 1989, in: D.J. Mullan, M.A. Pomerantz and T. Stanev, ed(s)., *Astrophysics in Antarctica*, AIP Conf. Proc. 198, 249.

Parish, T.R.: 1988, *Rev. Geophys.* **26**, 169.

Pearson, T.J. and Readhead, A.C.S.: 1984, in: G. Burbidge, D. Layzer, and J. Phillips, ed(s)., *Ann. Rev. Astron. Astrophys.* **22**, Annual Reviews: Palo Alto, 97.,

Thompson, A.R., Moran, J.M. and Swenson, G.W.: 1986, *Interferometry and Synthesis in Radio Astronomy*, John Wiley & Sons: New York, 164.

Townes, C.H. and Melnick, G.: 1990, *P.A.S.P.* **102**, 357.

INFRARED AND SUB-MILLIMETER SEARCHES FOR EXTRA-SOLAR PLANETARY SYSTEMS FROM ANTARCTICA *

JOHN BALLY

Center for Astrophysical Research in Antarctica,
Center for Astrophysics and Space Astronomy
University of Colorado, Boulder, Colorado

March 22, 1994

Abstract. Large single-dish telescopes and long baseline interferometers operating at mid-infrared to sub-millimeter wavelengths may provide the best opportunity to search for planets and proto-planets during the next several decades. The interior of Antarctica may be the best place to deploy such instrumentation, providing a viable compromise between the costs associated with space-based deployment and site quality.

1. The Search for Extra-Solar Planetary Systems

The search for extra-solar planets is easiest at wavelengths on the Rayleigh-Jeans portion of the planetary blackbody spectrum where the ratio between the planetary flux and the stellar flux is at a maximum. Thus, the search wavelength needs to be longer than $\lambda \approx 0.3/T_{planet}$ K, which implies $\lambda > 10$ μm for terrestrial planets or $\lambda > 30$ μm for gas giants.

Direct detection requires an angular resolution smaller than the orbital separation between the parent star and the object. Thus

$$D > 206 \frac{\lambda_{100} d_{10\ pc}}{r_{AU}} \quad \text{(meters)}$$

where D is the minimum aperture or baseline, λ_{100} is the observing wavelength in units of 100 μm, $d_{10\ pc}$ is the distance to the star in units of 10 pc, and r_{AU} is the planetary orbit radius in astronomical units.

The total collecting area must be large enough to detect photons from the planet above the background. In space, the background flux is determined by the emission from interstellar dust, cool stars, and zodiacal dust emission toward the line-of-sight. Near 100 μm, this background is of order 1 MJy sr^{-1} and varies as a complex function of position and wavelength. On the ground, the thermal emission from the atmosphere is given by $B_a = \int_{\Delta\lambda} \epsilon(\lambda) B_\lambda(T_{atm})\, d\lambda$ where $\Delta\lambda$ is the bandwidth of the observation, $\epsilon(\lambda)$ is the atmospheric emissivity function, and $B_\lambda(T_{atm})$ is the Planck function for the mean temperature of the atmosphere, T_{atm}. The minimum collecting area needed to get a signal-to-noise ratio of 1 on a planet located at

---

* Paper presented at the Conference on *Planetary Systems: Formation, Evolution, and Detection* held 7–10 December, 1992 at CalTech, Pasadena, California, U.S.A.

a distance $d_{10} = 10$ pc, which has a radius $r_9 = 10^9$ cm, with a temperature $T_p = 300$ K can be shown to be

$$A = 8B_{\text{MJy}}^{.5}\lambda_{100}^{1.5}T_p^{-1}d_{10}^2r_9^{-2}t_4^{-.5}\Delta\nu_{11}^{-.5} \quad (\text{meters}^2)$$

where B_{MJy} is the background flux in MJy sr^{-1}, λ_{100} is the observing wavelength in units of 100 μm, t_4 is the integration time in units of 10^4 sec, and $\Delta\nu_{11}$ is the observing bandwidth in units of 100 GHz. For a background flux of 1 MJy sr^{-1}, expected from high Earth orbit or the surface of the Moon, 41 m^2 of collecting area is needed at 350 μm or 1.3 m^2 at 30 μm. From the ground, assuming an emissivity of 0.1 from a 200 K atmosphere, more than 10^5 m^2 of collecting area are needed at 350 μm and 4×10^3 m^2 at 30 μm.

Extra-solar planet searches at mid-infrared or longer wavelengths require baselines of order 100 m or more and large collecting apertures. The expense of deploying the required instrumentation in space or on the Moon almost certainly precludes such a project during the next several decades. Although ground-based searches require much larger apertures to overcome atmospheric emission, deployment of a large instrument on the ground may be cheaper than deployment of a smaller required aperture in space. As discussed below (Section 4), there are several indirect methods for searching for extra-solar planets that can be performed by smaller infrared and sub-millimeter wavelength instruments. Such telescopes need to be built prior to a large facility capable of direct searches. Therefore, a good ground-based site for sub-millimeter and infrared observations, such as the Antarctic interior, may provide the best near-term opportunity for progress in the search for extra-solar planets. Recent measurements (Chamberlin and Bally, 1993) demonstrate that the interior of the Antarctica Plateau may be the best site on Earth for infrared and sub-millimeter observations.

2. Antarctica as a Sub-Millimeter and Infrared Site

The Center for Astrophysical Research in Antarctica (CARA) is developing the first permanent observatory in Antarctica. Three groups of telescopes are nearing completion and will be deployed at the South Pole: AST/RO (Antarctic Sub-Millimeter Telescope and Remote Observatory), a 1.7-m-diameter telescope whose primary mission is to perform the first survey of the 492-GHz fine-structure line of neutral atomic carbon (C I) in the interstellar medium; SPIREX (South Pole Infra-Red EXplorer), a 24" telescope for ultra-deep imaging in the 2-μm region; COBRA (COsmic Background Radiation Anisotropy), designed to measure small angular scale fluctuations in the cosmic microwave background.

Preliminary results from these experiments and site testing performed by the ATP (Advanced Telescope Project) experiments indicate that the South Pole is the best site on Earth tested so far for astrophysical measurements in the infrared to sub-millimeter (1-μm to 1-mm) spectral region.

Figure 1 shows a set of measurements of the 230 GHz atmospheric transmission during the first 180 days of 1992 as measured by a heterodyne atmospheric radiometer on loan to CARA from NRAO (National Radio Astronomy Observatory). Figure 2 (top) shows the month-to-month averages of cloud cover over the South Pole. Although cloud cover statistics indicate that about 40% of the nighttime hours are overcast, most of the clouds consist of ice crystals with little liquid water content. Frozen clouds are relatively transparent at wavelengths longer than the particle size, resulting in high sub-millimeter transmission even during cloudy periods. Figure 2 (bottom) shows the month-to-month averages of precipitable water vapor over the South Pole and Vostok (located 800 miles from the Pole) along with the extreme high and low values observed during each month. Figure 3 shows the average wind speed and surface temperature at the South Pole on a month-by-month basis.

At infrared wavelengths, the low temperature of the Antarctic environment implies that when clear, the sky will be darker than at temperate latitude sites such as Mauna Kea (Harper, 1989). On the Wien side of the black-body spectrum the thermal emission from telescope optics, the surrounding environment, and the sky decreases rapidly with decreasing temperature. In the 2-μm window, the emission from the Antarctic sky is expected to be more than a hundred times lower than at a site where the temperature is near 0°C (at wavelengths away from the OH airglow lines). Thus, near the long wavelength part of the 2-μm K-band atmospheric window (where there are no OH airglow lines), a given telescope, camera, and exposure time combination is expected to reach stars 5 magnitudes deeper in Antarctica than from a temperate latitude site. At mid-infrared wavelengths, the combination of a very dry atmosphere and a low physical temperature will result in the lowest thermal background attainable from any ground-based site.

3. Dome-A: A 4,200-m Plateau at Latitude −82°

Lynch (1989) has pointed out that the high elevation plateau at latitude −82° near the middle of the Antarctic continent is potentially the best site for infrared to sub-millimeter wavelength observations. A 200-mile-diameter flat plateau is located at an altitude in excess of 4,000 m directly beneath the centroid of the permanent high pressure zone which develops over the Antarctic Plateau. Due to the low atmospheric temperature, the barometric pressure at the surface is equivalent to what is found near the altitude of 5,000 m in temperate latitudes (pressure altitude of dome-A is thus 5,000 m).

Dome-A may be the best IR to sub-millimeter site for four important reasons:

1. Low water vapor. Atmospheric model calculations indicate that water vapor overburdens as low as 50 μm are to be expected. Figure 4 shows a model atmospheric calculation of the expected transmission over Dome-A (Bally, 1989).

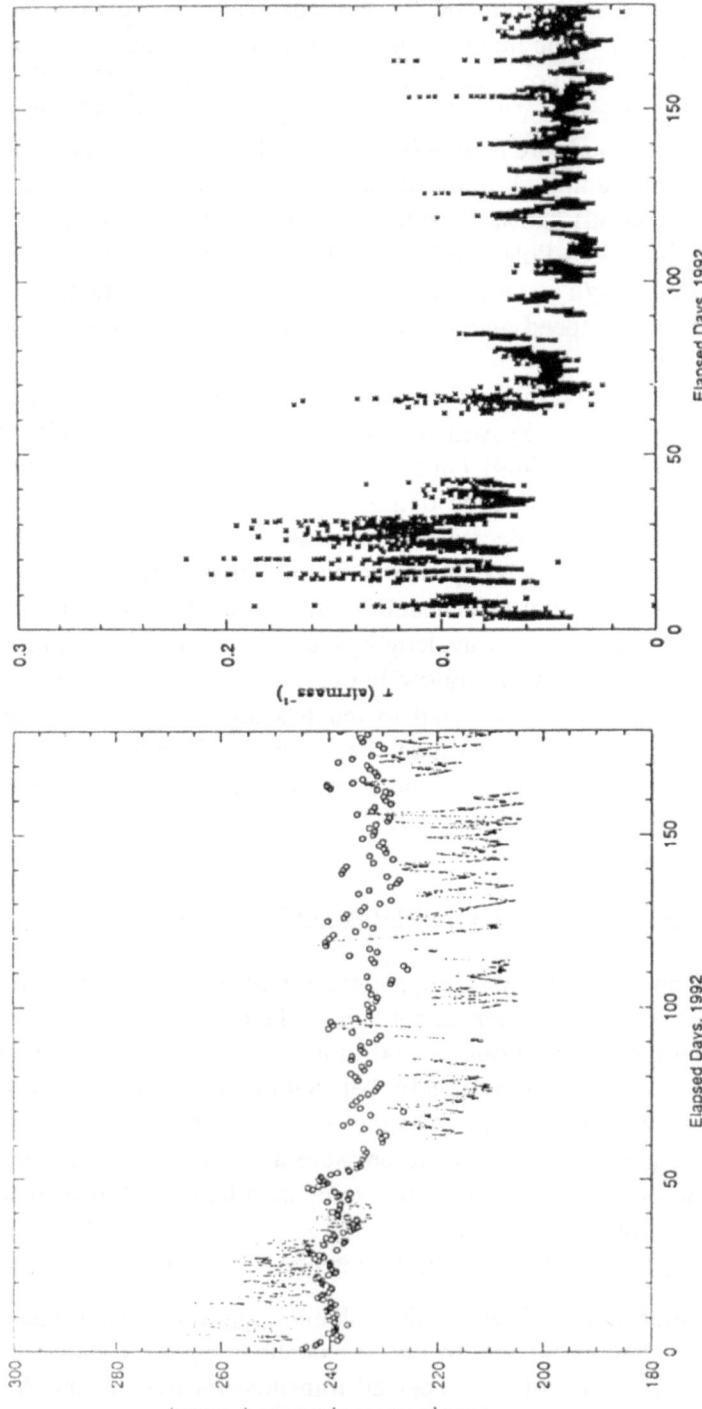

Fig. 1a. *Left*: Atmospheric temperature determined from a 230 GHz radiometric measurement (dotted curve). Surface temperature at the radiometer site at the South Pole (open circles). Taken from Chamberlin and Bally (1993). *Right*: Atmospheric optical depth at 230 GHz determined by fitting a model atmosphere to the measurements of the sky brightness as a function of zenith angle at 230 GHz. Taken from Chamberlin and Bally (1993).

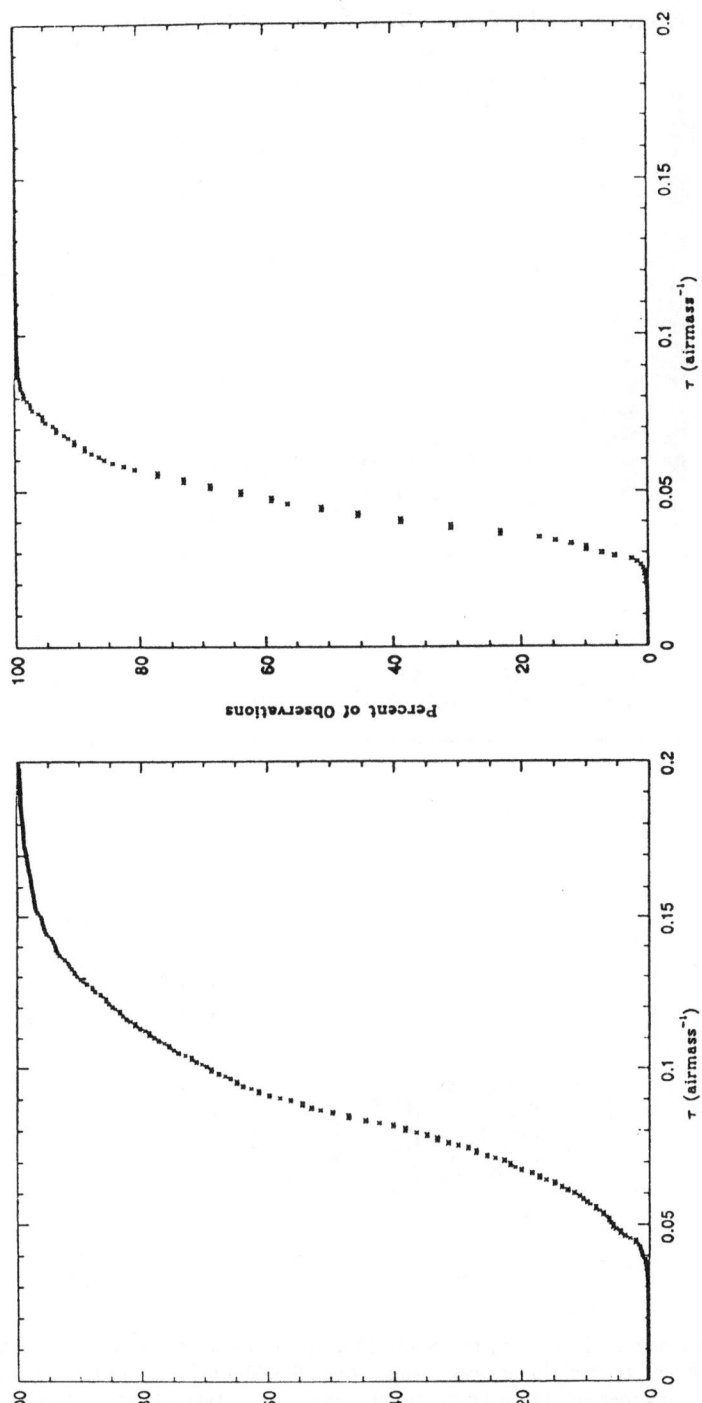

Fig. 1b. *Left*: The percentage of measurements for which $\tau(225\ GHz)$ was less than a given value between day 1 and day 70 in 1992. Quartiles are 0.068, 0.080, 0.103, 0.186. Taken from Chamberlin and Bally (1993). *Right*: The percentage of measurements for which $\tau(225\ GHz)$ was less than a given value between day 71 and day 180 in 1992. Quartiles are 0.036, 0.042, 0.051, 0.083. Taken from Chamberlin and Bally (1993).

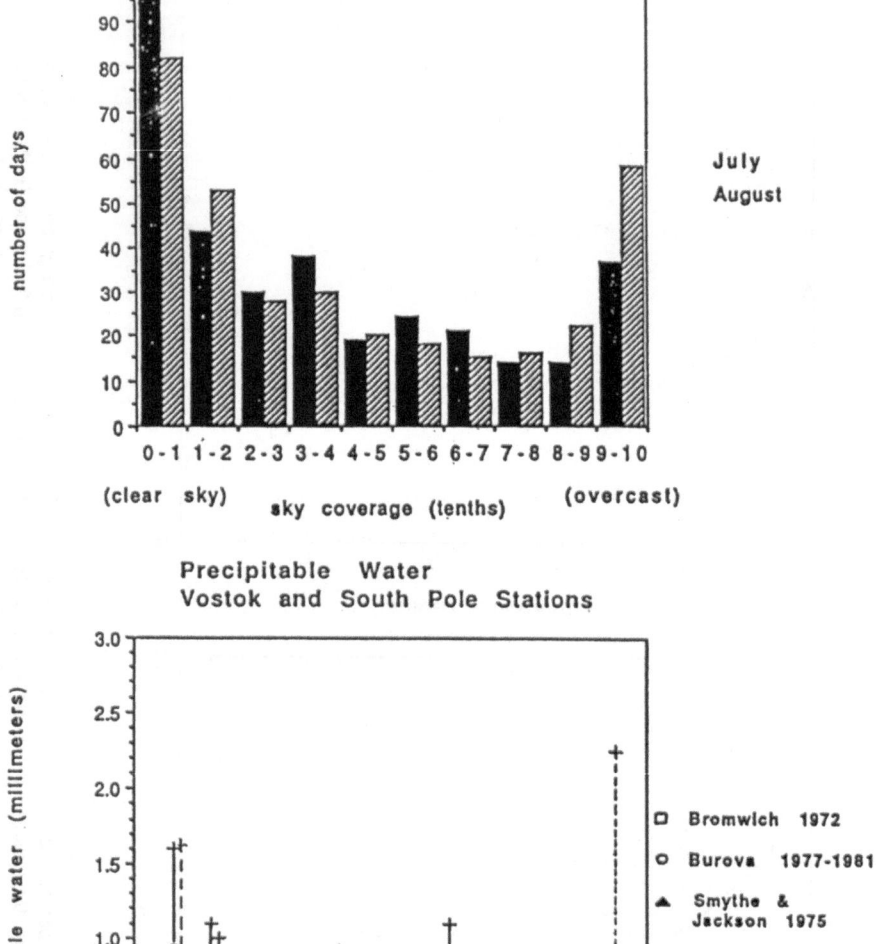

Fig. 2. *Top*: Cloud cover statistics over the South Pole from visual observations by South Pole personnel. *Bottom*: Average precipitable water vapor over the South Pole and over Vostok. The highest and lowest values measured in each month are also shown. Taken from Bromwich (1988), Burova *et al.*, (1986), and Smythe and Jackson (1977) for the years indicated in the figure.

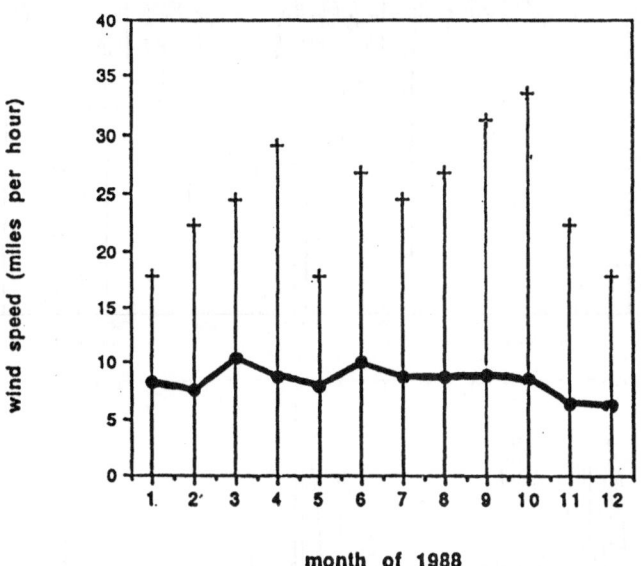

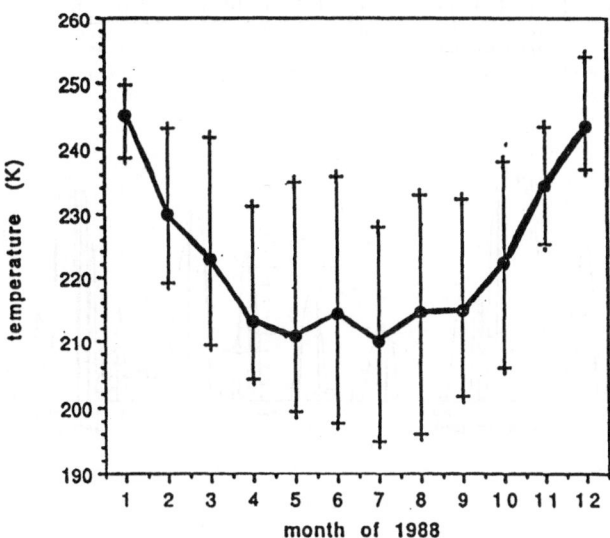

Fig. 3.   Wind speed (*top*) and surface air temperature (*bottom*) statistics over South Pole.

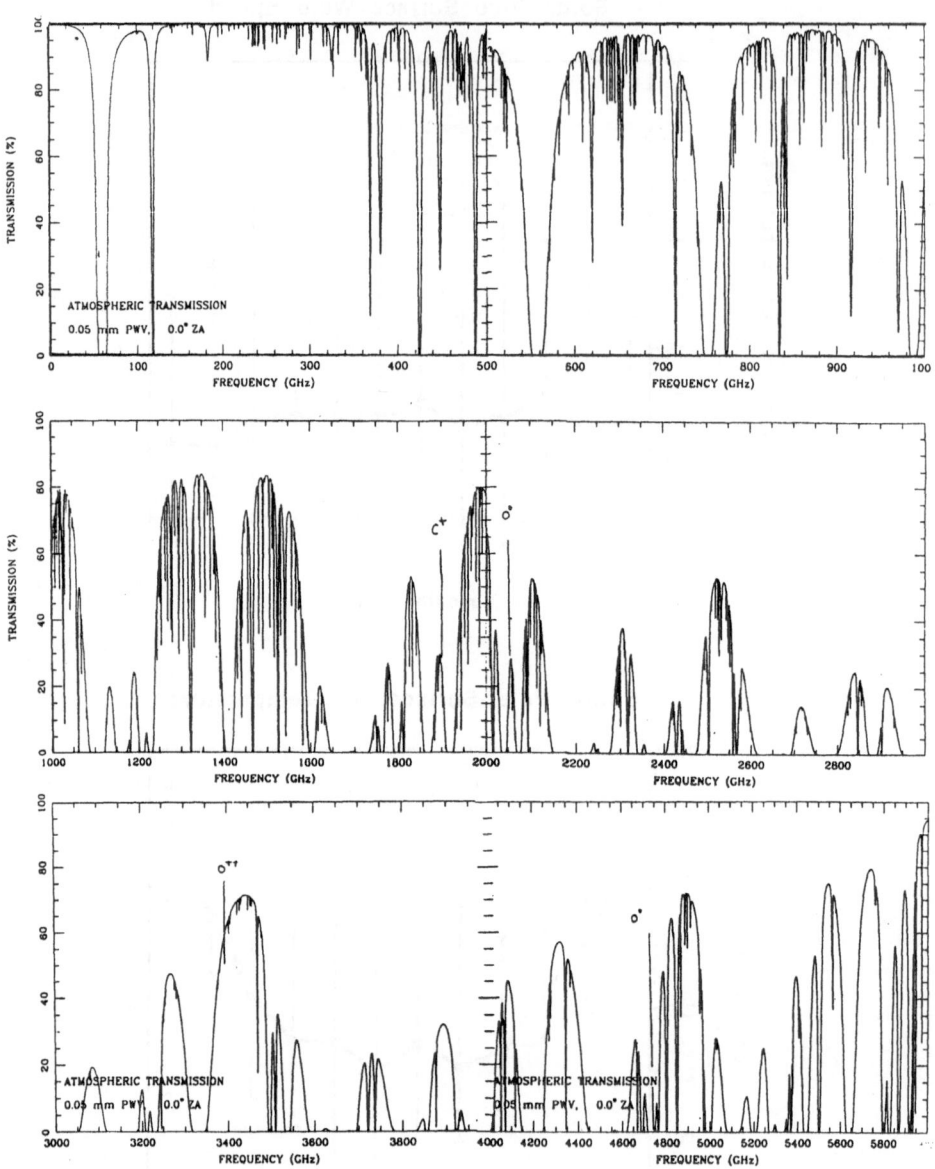

Fig. 4.   Model atmospheric transmission computed for Dome-A for 50 μm of precipitable water vapor.

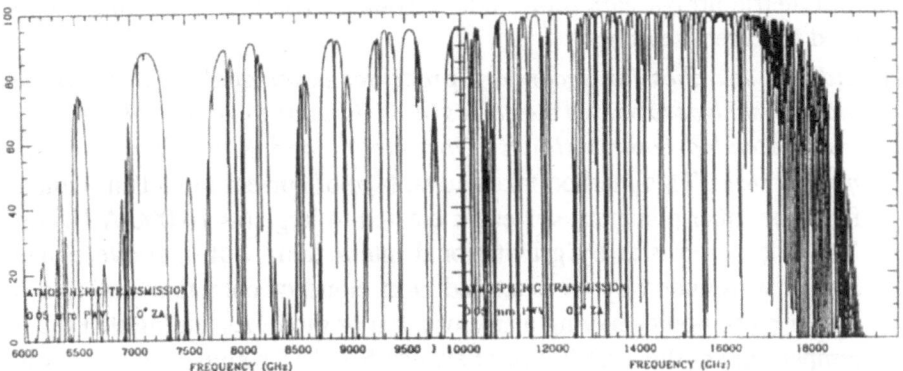

Fig. 4. Continued.

2. The high pressure altitude implies less broadening of atmospheric lines. This implies that atmospheric windows are broader and more transparent than at lower altitudes.
3. Average and peak wind velocities are expected to be very low because Dome-A is located near mean position of the permanent Antarctic high pressure zone where dry stratospheric air descends from above.
4. Surface temperatures are very low (under 200 K at times), giving very low thermal backgrounds, especially in the near infrared. The total amount of precipitation is less than 10 cm equivalent of liquid water.

Lynch (1989) has proposed that an international base be established at this site to take advantage of the exceptional conditions to be found there.

Dome-A may be a unique place for the development and testing of advanced infrared and sub-millimeter instrumentation. The flat ice surface is ideal for the deployment of long baseline interferometric arrays such as is required for the search for extra-solar planets. The remoteness, high altitude, and low temperature of Dome-A make it the best analog to space on Earth. Many of the techniques required for operation of a Lunar base are required for operations at dome-A, including pressurized and heated working environments and long logistical supply lines. However, experience indicates that the cost of developing an Antarctic base is more than 3 orders of magnitude cheaper than accomplishing the same task in space.

4. Ground-Based Search Strategies for Extra-Solar Planets

Extra-solar planet searches can be grouped into the following categories in order of increasing difficulty. The next generation of infrared and sub-millimeter ground-based telescopes may be able to perform the first few experiments. Much larger interferometric arrays with thousands of square meters of collecting area will be required for direct searches.

1. *Indirect searches for secondary processes associated with the existence of planets, such as extended dust discs (e.g., β Pictoris), or the collective emission from extended comets clouds (Oort clouds and Kuiper belts) surrounding nearby stars.* This method takes advantage of current models in which debris from the inner planetary system is ejected to large ($\approx$ 10,000 A.U.) distances from the parent star by gravitational interactions with massive planets. The Sun is surrounded by an extended Oort cloud of comets containing perhaps 10^9 or more bodies having a size of a km or more. Assuming that the average temperature of these bodies lies between 5 and 15 Kelvins, and that comets are the upper scale of a fragmentation spectrum which can be extrapolated down to millimeter-size bodies, the collective sub-millimeter emission from an entire Oort cloud may be detectable with the current generation of cosmic background anisotropy experiments, such as the COBRA experiment now proceeding at the South Pole.

2. *Searches for evidence of proto-planetary discs and planet formation surrounding proto-stars and young stellar objects.* Circumstellar discs of dust and gas with properties similar to what is expected of planetary systems in formation have already been detected from the vicinity of young stellar objects (YSOs). A large fraction of YSOs exhibit circumstellar dust emission at sub-millimeter wavelengths with total masses ranging from 0.001 to 0.5 $M_\odot$ (Beckwith et al., 1990; Weintraub *et al.*, 1989, 1991; Adams *et al.*, 1990). This gives us confidence that the conditions for planetary system formation are commonplace. The detection of planetary mass objects requires several orders-of-magnitude improvement over present-day angular resolution and sensitivity, implying larger apertures and physical dimensions than that of available telescopes. Several sub-millimeter and infrared telescopes are being built or planned (the Smithsonian sub-millimeter wave array—SMA, NRAO millimeter-wave array—MMA, Keck, and VLT) that can be used to study protoplanetary discs. High resolution continuum images may reveal condensations or other signs of planet formation. Velocity resolved spectroscopic observations of emission and absorption lines in proto-planetary discs may provide clues to the composition, velocity field, and physical conditions. Lines of OH, CO, CI, OI, and many other species have been detected in star forming regions. Future high resolution (R > 100,000) spectroscopic observations (best performed with heterodyne receivers in the sub-millimeter and in the mid-IR) could be obtained from Antarctic observatories. Such instruments are relatively insensitive to the residual thermal emission from the atmosphere.

3. *Astrometric searches for periodic oscillations in the position of a star induced by the gravitational attraction of a planet.* This requires high precision determinations of stellar positions in a given field of view. Astrometric precision

is limited by both the telescope beam size (resolution) and atmospheric turbulence, which decreases with increasing wavelength. Long baseline interferometry at the longest wavelength where stars are detectable promises to provide the most accurate astrometric determination of stellar positions. A mid-infrared array may be the best tool to use for astrometric searches for planets.

4. *Periodic radial velocity fluctuations induced by planetary gravity.* The expected velocity perturbations are very small, of order 10 ms^{-1}. At optical wavelengths these subtle velocity fluctuations may be masked by the effects of stellar surface activity.

5. *Direct imaging searches for planets surrounding main sequence stars.* This method requires the greatest sensitivity and resolution and is therefore the most difficult. At optical and near-infrared wavelengths, planets shine by reflected light only, resulting in a flux ratio (flux from parent star divided by flux from planet) of order 10^9 or more. At mid-infrared and sub-millimeter wavelengths, planets emit thermal emission, and the brightness ratio is of order 10^5 making long wavelength searches easier than short wavelength searches. However, direct searches from the ground are limited by the emission from the atmosphere which requires that very large collecting areas be used. Searches from space require smaller collecting areas.

Most methods of searching for extra-solar planets work best at long wavelengths in the mid-infrared to sub-millimeter portions of the spectrum. The large apertures and long baselines that are needed for such searches make deployment of such instruments in space very expensive and unlikely in the near future. Therefore, construction of mid-IR and sub-millimeter instruments at the best ground-based sites provides the best strategy in the search for extra-solar planets.

5. Conclusions

It is shown that direct searches for extra-solar planets require large single-mirror or long-baseline interferometric telescopes. Several methods can be used to search for indirect evidence of extra-solar planets in orbit around nearby stars. Some of these techniques, such as the search for extra-solar comet clouds, can be used on the present generation of sub-millimeter telescopes. Others, such as the investigation of the physical properties of protoplanetary discs associated with young stellar objects, will require the next generation of infrared and sub-millimeter instruments. Space provides a low background environment where direct searches require apertures dictated mostly by the need for angular resolution sufficiently large to resolve a planet from its parent star. Although ground-based searches require much larger collecting apertures to overcome the greater flux of radiation emitted by the atmosphere, the cost of deployment on the ground is orders of magnitude less expensive. Therefore, ground-based efforts in relatively transparent mid-IR and sub-millimeter atmospheric windows should be exploited to develop the technology needed for the search for extra-solar planets.

Preliminary measurements conducted under the CARA South-Pole site testing program indicate that the high, cold, and dry conditions which prevail in the interior of Antarctica provide the best place on Earth for the future deployment of large instruments in the infrared to millimeter-wavelength spectral range. High angular resolution, high sensitivity, and high spectral resolution observations can be performed throughout the mid-IR and sub-millimeter bands from high sites in the interior of Antarctica. Developing large telescopes that can operate in Antarctica is a logical step towards the development and eventual deployment of this technology in large space or Lunar observatories in the far future and may provide the first evidence for the existence of extra-solar planets.

References

Adams, F.C., Emerson, J.P. and Fuller, G.A.: 1990, *Astroph. J.* **312**, 788.

Bally, J.: 1989, in: D.J. Mullan, M.A. Pomerantz and T. Stanev, ed(s)., *Astrophysics in Antarctica*, AIP Conference Proceedings 198, AIP Press, New York, 100.

Beckwith, S., Sargent, A.I., Chini, R. and Gusten, R.: 1990, *A. J.* **99**, 924.

Bromwich, D.H.: 1988, *Reviews of Geophysics* **26**, 149.

Burova, L.P., Gromov, V.D., Lukyanchikova, N.I. and Sholomitskii, G.B.: 1986, *Soviet Astron. Lett.* **12**, 339.

Chamberlin, R. and Bally, J.: 1993, *Applied Optics* in press.

Lynch, J.T.: 1989, in: D.J. Mullan, M.A. Pomerantz and T. Stanev, ed(s)., *Astrophysics in Antarctica*, AIP Conference Proceedings 198, AIP Press, New York, 249.

Harper, D.A.: 1989, in: D.J. Mullan, M.A. Pomerantz and T. Stanev, ed(s)., *Astrophysics in Antarctica*, AIP Conference Proceedings 198, AIP Press, New York, 123.

Smythe, W.W. and Jackson, B.V.: 1977, *Applied Optics* **16**, 2041.

Weintraub, D.A., Sandell, G. and Duncan, W.: 1989, *Astroph. J.* **340**, L69.

Weintraub, D.A., Sandell, G. and Duncan, W.: 1991, *Astroph. J.* **382**, 270.

SIRTF: CAPABILITIES FOR THE STUDY OF PLANETARY SYSTEMS *

D.P. CRUIKSHANK

Astrophysics Branch, NASA Ames Research Center,
Moffett Field, California, USA

M.W. WERNER

Jet Propulsion Laboratory, Pasadena, California, USA

and

D.E. BACKMAN

Physics and Astronomy Department, Franklin and Marshall College,
Lancaster, Pennsylvania, USA

Abstract. The Space Infrared Telescope Facility, to be launched into a near-Earth heliocentric orbit in the year 2001, will open broad new vistas for the study, at infrared wavelengths, of the objects in the Solar System and planetary systems around other stars. This paper focuses on the study of Kuiper-belt comets and circumstellar planetary debris disks.

1. SIRTF: The Space Infrared Telescope Facility

SIRTF is a one-meter-class cryogenically cooled observatory for infrared astronomy from space which will be the infrared component of NASA's family of Great Observatories. SIRTF, currently planned for a launch in the year 2001, will be operated as a facility for the entire scientific community. SIRTF is to operate in an Earth-trailing heliocentric orbit for a minimum of three years. The new SIRTF mission concept which has been developed for this orbit, and more general descriptions of SIRTF's scientific programs, are presented by Werner (1991, 1993) and by Rieke *et al.* (1986). Dramatic advances in infrared detector technology mean that SIRTF will gain factors of 100 to 10,000 in sensitivity over its predecessors. An additional enormous increase in capability results from the availability of these highly sensitive detectors in large-format arrays containing tens of thousands of pixels.

The scientific potential of a cryogenic space observatory equipped with state-of-the-art infrared arrays is so compelling that SIRTF was designated in 1991 both as the highest priority astronomy mission for the 1990s by the National Academy of Sciences, and as NASA's highest priority "flagship" scientific mission by the interdisciplinary Space Sciences Advisory Committee.

Only a cryogenically cooled telescope operating from above the atmosphere is fully capable of overcoming the obstacles to infrared observations imposed by the partial and variable opacity of the Earth's atmosphere and the high level of thermal background radiation from the sky and a ground-based telescope. IRAS, which surveyed the sky to unprecedented sensitivity levels from 8 to 120 μm

---

* Paper presented at the Conference on *Planetary Systems: Formation, Evolution, and Detection* held 7–10 December, 1992 at CalTech, Pasadena, California, U.S.A.

during its ten-month flight in 1983, was the first such space telescope and gave us our first look at the full splendor of the infrared sky. Important contributions to planetary science by IRAS include the detection of numerous zodiacal dust bands (Sykes, 1990), the most complete and least biased asteroid survey ever made (Matson *et al.*, 1989), and the discovery of the Vega and Beta Pictoris debris disks (Gillett, 1986; Backman and Paresce, 1993). SIRTF will follow up on and go beyond IRAS' contributions to these planetary science questions, and also will build upon the planetary investigations of the European Space Agency's Infrared Space Observatory (ISO) to be launched in 1995 (Encrenaz and Kessler, 1992).

The performance characteristics expected for SIRTF are given in Table I. SIRTF will cover the entire spectral region from 2.5 to 200 μm, and the optical system will provide diffraction-limited images at wavelengths longward of 5 μm over a 5 arcmin field of view.

TABLE I

SIRTF system parameters.

Parameter	Value
Mirror diameter	85 cm
Wavelength coverage	2.5–200 μm
Image quality	50% encircled energy in 2 arcsec at 3.5 μm
Pointing stability/accuracy	0.25 arcsec/0.25 arcsec
Sensitivity (one sigma in 500-sec integration)	
10 μm	6 μJy
60 μm	100 μJy
Number of detectors	100,000
Spectral resolving power	2000
Lifetime	Minimum of 3 years

Three focal plane instruments have been selected for SIRTF and are under definition study. The basic capabilities of the instruments are given in Table II. The Infrared Array Camera (IRAC) will use large area, two-dimensional infrared array detectors in formats up to 256 × 256 pixels to provide wide-field (5 arcmin) diffraction-limited imaging over the spectral region 2.5–28 μm. The Multiband Imaging Photometer for SIRTF (MIPS) will provide background-limited wide-field (5-by-5 arcmin) imaging and photometry over the spectral range 15–120 μm, and 0.6-by-5 arcmin imaging and photometry in the range 120–200 μm, also using detector arrays. The very high sensitivity and complete spatial sampling of IRAC and MIPS, plus the high quality, stable images to be provided by features of the spacecraft and telescope design, will permit the use of super-resolution techniques to improve the angular resolution of the images by a factor of two or more at

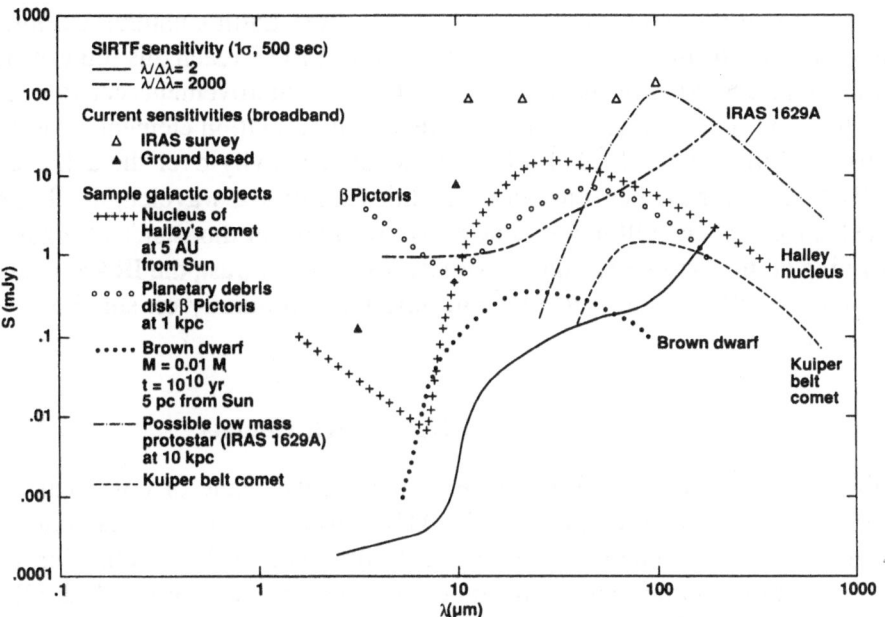

Fig. 1. The sensitivity of SIRTF (1 sigma in 500-sec integration time) in photometric and high resolution spectroscopic modes is compared with that of IRAS (survey mode) and large ground-based telescopes. The limiting sensitivity in the far infrared of current airborne telescopes is about the same as that shown for IRAS. The SIRTF predictions are based on demonstrated detector performance or current expectations and also include both natural background and confusion limits. Also shown are the fluxes of known and predicted extra-solar and solar system objects of the types to be studied by SIRTF. The Kuiper belt comet shown in this diagram has a diameter of 200 km.

wavelengths longward of 20 μm over that implied by the Rayleigh criterion. IRAC and MIPS will also have the capability to measure polarization. The Infrared Spectrograph (IRS) is a grating spectrograph covering the wavelength interval from 4–200 μm using two-dimensional array detectors, with resolving power varying from 1000 to 2500. Spectroscopic capabilities shortward of 4 μm will be available by use of a grism in the InSb camera, with resolving power 100–250.

The expected performance of the SIRTF instruments is compared in Figure 1 with the brightness of potential targets, the flux levels reached in the IRAS all-sky survey, and the capabilities of current and planned ground-based facilities for infrared astronomy. Note that SIRTF will easily be able to obtain complete spectra at resolution 1000 + of even the faintest IRAS survey sources — sources that are at or below the limit of detectability even in broad spectral bands with current infrared facilities.

Of utmost importance is the fact that SIRTF's instruments will make extensive use of the newest infrared detector array technology available to the astronomical community. SIRTF's goal is to have natural-background limited sensitivity; that is,

SIRTF's detectors should be limited in sensitivity only by the fundamental statistical fluctuations in the faint infrared background of the Earth's natural astrophysical environment. The detectors currently in hand for SIRTF reach this limit over most of the entire 2.5–200 μm spectral band, and further improvements can be expected over the next several years. On a per-pixel or per-resolution element basis, SIRTF realizes a full 1000 to 10,000-fold increase in sensitivity over the achievements of IRAS and over the best current capabilities at infrared wavelengths (Figure 1). SIRTF's focal plane will incorporate arrays with tens of thousands of pixels, each operating at this extremely high sensitivity level. In contrast, the IRAS focal plane incorporated 62 discrete detectors, and ISO relies primarily on small arrays and discrete detectors.

2. Detection of Kuiper Belt Comets

Indirect evidence for a disk of planetesimals in the plane of our Solar System beyond the orbit of Neptune (r > 35 AU) comes from the orbital statistics of the Jupiter family comets, which appear to originate on low-inclination planet-crossing orbits with perihelia in the outer planetary system (Kuiper, 1951; Duncan et al., 1988). This hypothetical disk is sometimes referred to as the Kuiper belt (Duncan *et al.*, 1988; Tremaine, 1990). Observation of objects in the Kuiper belt would verify this conjecture and provide an important connection between what is studied in our Solar System and the planetary debris disks to be studied by SIRTF around nearby stars.

In 1992, Jewitt and Luu (1992, 1993) discovered an object designated 1992QB1, a very red solid body with an orbital semi-major axis of about 44 AU, and with an orbit of low inclination (2.2 deg) and moderate eccentricity (0.11). The perihelion distance is 40 AU and the period is 296 years. A second object of similar brightness found in 1993 in the same search was reported by Luu (1993). Designated 1993FW, the semi-major axis is $\approx$ 42.5 AU and the inclination $\approx$ 8 deg. If the geometric albedo of these two bodies is 0.04 (at 0.55 μm wavelength), their diameters are about 250 km. These objects appear to confirm the predicted existence of a Kuiper belt that is at least several degrees wide. The search statistics suggest that objects of this size may occur at the rate of about one per square degree on the sky in a belt of some undetermined width centered on the ecliptic. There may be several thousands of these bodies of diameter on the order of 100–200 km, amounting in total mass to a few Pluto masses.

At the expected equilibrium temperature of 50 K, the 100-μm flux of 1992QB1 at 41 AU from the Sun is 2 mJy, which will be detectable by SIRTF in a standard 500 sec integration (Figure 1). Thus SIRTF has the ability to detect the larger Kuiper belt objects. SIRTF's capabilities can be compared with the limits set by current optical surveys for slow-moving objects in the ecliptic plane, which are summarized in Figure 2, adapted from Tremaine (1990). The visual magnitude of 1992QB1 is about 23.5. Figure 2 shows that a SIRTF search for this and related

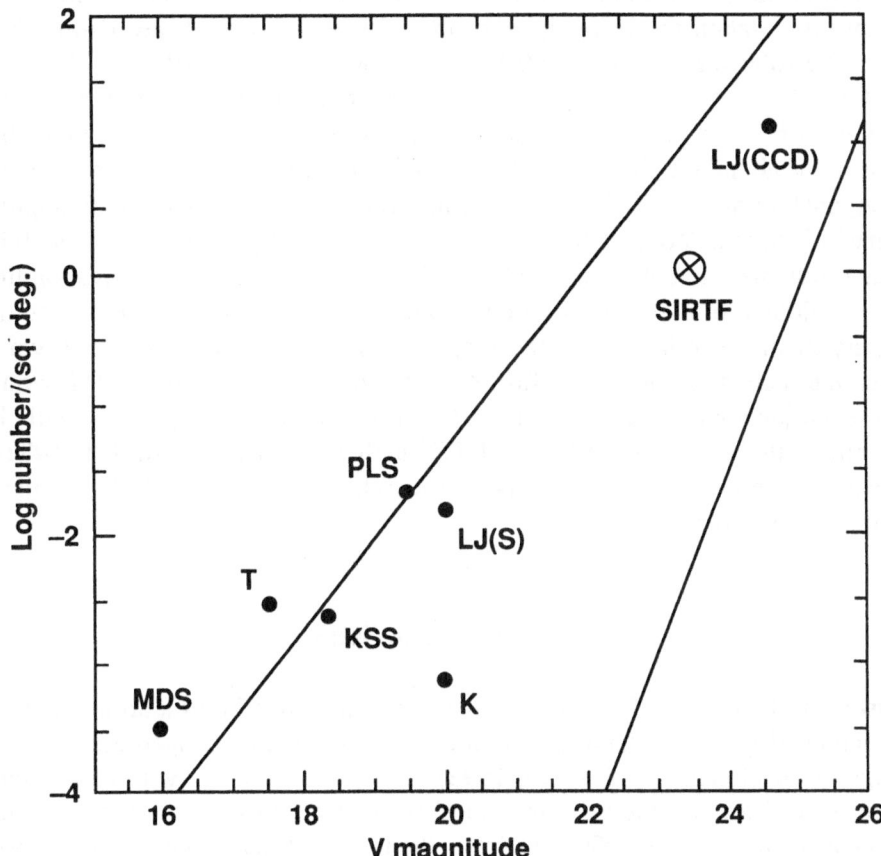

Fig. 2. Limits on the number density per square degree of Kuiper belt objects brighter than visual mag V. The data points are 99% confidence upper limits. Key to symbols: SIRTF = Space Infrared Telescope Facility described in this paper; MDS = Kuiper *et al.* (1958); T = Tombaugh (1961); PLS = van Houten *et al.* (1970); KSS = Ishida *et al.* (1984); LJ = Luu and Jewitt (1988); K = Kowal (1989). The upper and lower diagonal lines represent estimates of the expected number of objects per unit solid angle brighter than a given V-mag for two extreme distributions of the sizes of collisional fragments and accreted planetesimals, all assumed to have geometric albedos of 0.05. Adapted from Tremaine (1990).

objects should be capable of detecting one object/square degree in order to be scientifically useful.

SIRTF can, in fact, fairly quickly search 1 square degree for Kuiper belt objects as faint as 1.5 mJy. The SIRTF 100-μm camera will have a field of view of 5×5 arcmin, so that 1 square degree requires 144 exposures, or 20 hours of observation at 500 seconds/frame. At the end of 20 hours, any objects in the first exposure which are in fact 40 AU from the Earth and Sun will have moved 1.4 arcmin, due to the orbital motion of the spacecraft. This is large compared to the SIRTF angular resolution but small compared to the field of view, so a second 20-hr set of 144 exposures, taken immediately after the first, would reveal any Kuiper belt objects. The total time required, 40 hours, is consistent with SIRTF's ability to point at a single position on the sky for days on end. An alternate strategy would break the 40 hours into longer blocks of time in order to search for fainter but perhaps more abundant objects. In the case of two 20-hour exposures at a single position, each corrected for image smearing produced by the predictable angular motion of a Kuiper belt object, the limiting flux decreases by a factor of 12, or about 2.7 mag (equivalent to detecting an object of diameter 60 km). The corresponding visual wavelength limit would be slightly fainter than 26th magnitude, where the sensitivity of the search [now in the range of 100 objects/square degree] would again be interesting. To achieve a full 99% confidence upper limit, SIRTF would have to examine about 5 square degrees. Improved strategies will undoubtedly be used for the actual observations from SIRTF, but these examples show that SIRTF's high sensitivity and large arrays will permit searches for previously hidden objects in the outer Solar System.

3. Matter Around Stars

The discovery by the IRAS mission of disks of solid debris in orbit around nearby stars such as Vega, Fomalhaut and Beta Pictoris, radically changed the study of cosmogony (Backman and Paresce, 1993). The number of known or suspected cases exceeds 100, and they occur around stars of all spectral classes. Ranging in distance from 10 to 10,000 AU from their central stars, these debris disks have characteristic temperatures of 40–150 K and masses ≥ 0.1 M_{Earth}. Their luminosities are from $< 10^{-5}$ to 10^{-3} those of the central stars. The debris in these disks has particle sizes in the range $1–100^+$ μm, which is much larger than the grains in the interstellar medium.

SIRTF will answer in detail the fundamental question which has arisen from the discovery of this phenomenon: How are the "Vega-type" disks related to planetary systems?

For example, SIRTF will give a much more complete census of material around nearby stars by improving the sensitivity limit of the diagnostic infrared flux excesses to > 100 times better than IRAS. At the IRAS flux limit some 80% of nearby main-sequence stars have no detected debris, but the corresponding limit of $L_{dust}/L_{star} = 10^{-5}$ is 100 times the luminosity of the Sun's zodiacal cloud. The number of cases of dust disks rises sharply toward the IRAS detection limit, suggesting that many more should occur toward the SIRTF limit of $L_{dust}/L_{star} \leq 10^{-7}$. SIRTF extends the limit at which substantial disks like Vega's (10 times the photosphere flux density at 60 μm) could be detected from stars of visual magnitude + 3.5 for IRAS to > 10th mag, thus bringing 10^5 stars within range.

For the most prominent systems, such as Vega and Fomalhaut, SIRTF's images will show the orientation, structural features, and detailed morphology of the disks. The inner dust-depleted regions (Gillett, 1986; Backman and Paresce, 1993) and possible asymmetries deduced from IRAS data will be revealed in great detail. The effects of companions, both stars and massive planets, on the dust disks could be discernible using SIRTF's high-precision images and super-resolution techniques. SIRTF will show if the dust depletion zones in the nearby disks, the radii of which are presently uncertain, represent a phase transition (e.g., evaporation of H_2O or other ice), or if it is a (dynamically determined) structural feature of the disks. Furthermore, SIRTF will tell us if young stars ($t < 0.5$ Gy) have more material at planetary temperatures which may disappear as planets form, as well as cold Vega-disk material which persists for the system lifetime.

Among the 12 stars in Campbell *et al.*'s (1988) radial velocity survey there are three cases of overlap with stars known to have circumstellar dust. Of these, at least one (ϵ Eri) has both dust and suspected perturbations from an object with planetary mass. As the radial velocity and high-precision astrometric surveys expand and improve in the 1990s, more stars with companions of planetary mass should be found; SIRTF will search for dust disks to high sensitivity levels in all the radial velocity and astrometric stars.

SIRTF will not only image the circumstellar debris disks in several infrared wavelengths, but will obtain spectra for determining their compositions. The comparison of the volatile and refractory components of the disks with comets will be of extraordinary interest.

SIRTF will be the first mission to combine the intrinsic sensitivity of a cryogenically cooled telescope for infrared astronomy with the tremendous imaging and spectroscopic power of large-format detector arrays; as such, it represents an enormous increase in our capability to explore the Universe at infrared wavelengths. Through its detailed photometric and spectroscopic study of a large sample of stars with circumstellar debris disks, SIRTF will yield a greatly improved understanding of the processes by which planetary systems form.

Note Added in Proof

Since this article was written, SIRTF has undergone a major design revision to bring its cost within stringent guidelines consistent with astrophysics space missions of the 1990s. The new design retains the sensitivity characteristics outlined in this paper, and SIRTF will be able to make the scientific contributions to the study of the outer Solar System and matter around stars discussed here. Its reduced capabilities will make it less suitable for certain other Solar System applications, however.

Also since this article was written, four additional objects which may be Kuiper Belt planetesimals (called comets in this paper) have been discovered, as shown in the following table, which lists all six known objects:

Kuiper Belt Planetesimals

Designation	Aphelion Distance (AU)	Eccentricity	Inclination (degrees)	Reference
1992 QB1	39.998	0.0876	2.213	IAUC 5855*
1993 FW	42.145	0.0407	7.745	IAUC 5856
1993 RO	32	0	2.53	IAUC 5865
1993 RP	35	0	2.79	IAUC 5867
1993 SB	33.15	0	2.28	IAUC 5869
1993 SC	34.45	0	5.58	IAUC 5869

* IAUC = International Astronomical Union Circular

Acknowledgements

This work was carried out in part at the Jet Propulsion Laboratory, California Institute of Technology, under contract with the National Aeronautics and Space Administration.

References

Backman, D.E. and F. Paresce: 1993, in: E.H. Levy and J.I. Lunine, (eds.), *Protostars and Planets III*, Univ. of Arizona Press, (in press).
Campbell, B., G.A.H. Walker and S. Yang: 1988, *Astrophys. J.* **331**, 902.
Cruikshank, D. P., M. W. Werner and D. E. Backman: 1992, *Adv. Space Res.* **12** *11*, 187–193.
Duncan, M., T. Quinn and S. Tremaine: 1988, *Astroph. J.* **328**, L69.
Encrenaz, T. and M. Kessler: 1992, *Infrared Astronomy With ISO*, Nova, Commack.
Gillett, F.C.: 1986, in: F.P. Israel, (ed.), *Light on Dark Matter*, D. Reidel, Dordrecht, 61.
Ishida, K., T. Mikami and H. Kosai: 1984, *Publ. Astr. Soc. Japan* **36**, 357.
Jewitt, D. C. and J.X. Luu: 14 Sept. 1992, 1992QB1, *International Astronomical Union Circular (IAUC)* **5611**, see also IAUC 5633 and IAUC 5684.
Jewitt, D. C. and J.X. Luu: 1993, *Nature* **362**, 730.
Kowal, C.T.: 1989, *Icarus* **77**, 118.
Kuiper, G.P.: 1951, in: J.A. Hynek, (ed.), *Astrophysics*, McGraw-Hill, New York, p. 357.
Kuiper, G.P., Y. Fujita, T. Gehrels, I. Groeneveld, J. Kent, G. van Biesbroeck and C.J. van Houten: 1958, *Astrophys. J. Suppl.* **3**, 289.
Luu, J.X.: 1993, *1993 FW International Astronomical Union Circular (IAUC)* , 5730.
Luu, J.X. and D. Jewitt: 1988, *Astron. J.* **95**, 1256.
Matson, D.L., G.J. Veeder, E.F. Tedesco and L.A. Lebofsky: 1989, in: R. Binzel, T. Gehrels and M.S. Matthews, (eds.), *Asteroids II*, Univ. of Arizona Press, Tucson, 269.
Rieke, G.H. *et al.*: 1986, *Science* **231**, 807.
Sykes, M.V.: 1990, *Icarus* **84**, 267.
Tombaugh, C.W.: 1961, in: G. Kuiper and B. Middlehurst, (eds.), *Planets and Satellites*, Univ. of Chicago Press, Chicago, 12.
Tremaine, S.: 1990, in: D. Lynden-Bell and G. Gilmore, (eds.), *Baryonic Dark Matter*, Kluwer, Dordrecht, 37.

van Houten, C.J., I. van Houten-Groeneveld, P. Herget and T. Gehrels: 1970, *Astron. Astrophys. Suppl.* **2**, 339.

Werner, M.W.: 1991, *Adv. Space Res.* **11** 2, 279.

Werner, M.W.: 1993, in: B.T. Soifer, (ed.), *Sky Surveys: Protostars to Protogalaxies*, Astron. Soc. Pacific Conference Series.

PLANETARY SYSTEM EVOLUTION AND THE VEGA STARS: THE POTENTIAL FOR ESA'S INFRARED SPACE OBSERVATORY *

ROBERT E. STENCEL

Department of Physics and Astronomy, University of Denver,
Denver, Colorado, USA

and

DANA E. BACKMAN

Department of Physics and Astronomy, Franklin and Marshall College,
Lancaster, Pennsylvania, USA

Abstract. ESA's Infrared Space Observatory [ISO], scheduled for launch within the next 2–3 years, will place a complement of powerful infrared imagers and spectrometers into high orbit, with an operational life anticipated to be about 18 months. During this time, numerous scientific investigations of every conceivable astrophysical target will be made. The purpose of this paper is to consider the instrumental complement in terms of specific observations of Vega-like systems with cold, infrared excesses, in order to investigate problems relating to the evolution of planetary systems, and to optimize the scientific results possible with ISO on such topics.

1. Introduction

The Infrared Space Observatory satellite is an ESA project which will place a supercooled 64 cm aperture, f/15 Richey–Cretien telescope, diffraction limited at 5 microns, into an elliptical orbit (Kessler *et al.*, 1992). Launch is presently scheduled for no sooner than late 1995, and mission life with the roughly 2300 liters of superfluid helium is anticipated to last 18 months. While the most significant discoveries to be made with ISO probably cannot be predicted now, the instrument PIs and their teams have organized a long list of the galactic and extra-galactic investigations which they are eager to pursue. In addition to this ESA "Central Programme" and selected U.S. and Japanese "Key Projects", it is anticipated that there will also be Guest Observer opportunities. The long-wave infrared (45 to 180 microns), which is almost entirely blocked by the earth's atmosphere, will be opened up for discovery and utilization of new gas and dust diagnostic spectral features, as well as for the systematic study of dust at temperatures of 20 to 100 K. Because this thermal regime includes sublimation points for ices associated with planet formation, ISO data can be used to directly inventory and assay the nature of the putative planetary debris disks, such as that observed in β Pictoris and surmised in related systems.

Analysis of the β Pic disk by Backman *et al.* (1992) using ground-based photometry, IRAS photometry and scan profiles, and the optical scattered light distribution

---

* Paper presented at the Conference on *Planetary Systems: Formation, Evolution, and Detection* held 7–10 December, 1992 at CalTech, Pasadena, California, U.S.A.

Astrophysics and Space Science **212**: 417–422. 1994.

showed that: (1) the color temperature and angular scale of the disk are consistent with emissions from small ($\sim 1\mu$m) grains; (2) models with distinct inner and outer structural components joined at about $r = 80$ AU provide the best match to the observations; (3) the inner component represents a gross deficit of material relative to the outer component; (4) the temperature of typically sized grains at the boundary between the components is about 90–140 K, roughly corresponding to the temperature of the transition from slow to rapid sublimation of water ice: sublimation would proceed more rapidly than other destruction mechanisms for small ice grains inside the ~ 80 AU boundary, and could be responsible for the relative depletion of grains in the inner disk; (5) most of the material in the inner component is at temperatures where sublimation of small water-ice grains would occur so rapidly that predominance of refractory grains is indicated; (6) the 10 micron observations appear to require that the inner component extends inward only to a poorly discerned limit somewhere between about 1 and 30 AU from the star, leaving an innermost void in terms of small particles.

Among the questions raised by this analysis, that we believe ISO observations can address, are these:

1. Does β Pic represent success or failure in forming planets? The unambiguous detection of disk inner edge structure could allow us to distinguish between sublimation and/or dynamical processes at work in this and related systems. This investigation calls for spatial resolution slightly better than that of ISO, but can be addressed with high S/N spectral energy distribution of the warmest dust.

2. How frequent are less dense, less extensive versions of the β Pic disks among normal main-sequence stars? Broad-band ISO photometry should help establish the presence or absence of longwave infrared excesses among volume- and brightness-limited samples to a level one hundred times fainter than IRAS.

3. Does disk density anti-correlate with stellar age? IR excesses among stars in clusters of various ages have already begun to imply this, but a census of far-IR excesses can help delineate whether the disk dissipation rates are precipitous or gradual, perhaps depending on planetesimal coalescence factors. Alternative views of formation and stripping of protostellar disks (Lissauer and Griffiths, 1989) argue that other processes may also affect the observable quantities.

4. Do Kuiper belts and Oort clouds play a role in post-main sequence evolution, as stars begin their ascent of the red giant branch, in terms of brief but significant augmentation of observed infrared excesses? Again, longwave IR photometry will demonstrate the frequency of such excesses among yellow giant stars to a level much fainter than IRAS could detect.

2. Potential Observing Projects for ISO

Weidenschilling (1984) argues that planetesimals can form only under quiescent conditions, and that such conditions are rare during the lifetime of an accretion disk, stating that "the large scale structure of the solar system may have been determined by the behavior of microscopic particles during its formation". This work suggests that systems with the smallest total energy may be those which are most efficient at forming planets, i.e., the lower mass stars. A brightness-limited sampling of the IRAS data (SAO stars vs. Point Source Catalog, e.g., Stencel and Backman, 1991) found that upper main sequence stars (A and F types) more frequently reveal far infrared excesses, compared with lower main sequence stars (G and K types). However, a volume-limited sampling (Backman and Gillett, 1986) shows prevalence may not be a strong function of main sequence spectral type. Is disk detectability, instead, a function of stellar age and luminosity?

At the time of this writing, the estimated observing efficiency for ISO is being revised, but the ESA Proposal Generation Aid software suggests that the overheads may be considerable. However, this remains in a state of flux and we used published sensitivities and made no assumptions about overheads in estimates for our meeting poster. Our purpose was to outline some of the areas under consideration, to engender further discussions about making best use of the precious time ultimately available. We of course will welcome comments and ideas about any of this as we plan our Key Project observing along the following lines:

2.1. SPATIALLY-RESOLVED STUDIES

Perhaps the most significant work of ISO in this area will involve improving on the IRAS resolution of disk systems at infrared wavelengths not accessible from the ground or KAO. ISO's diffraction limit and aperture selections will not permit subarcsecond resolution, but a combination of photometric and spectroscopic techniques can improve on the IRAS information. One experiment might involve using ISOCAM, with 15 micron broadband imaging and 6 arcsec p.f.o.v., to map the inner, dense edge of the "main cloud" (1 pix, diam. away from β Pic, α PsA and Vega). In terms of exposure times, given recent sensitivity figures, Fomalhaut: $T(\text{disk}) = 60$ K, $\tau = 10^{-4}$ would yield $S/N = 10$ in 1 hr, with ISOCAM pixels corresponding to 40 AU. Some of the questions addressed in this way include: are source shapes really disks in projection? If yes, how is disk orientation connected to the stellar rotation axis (determined from v sin i and photometric rotation period)? Is there any azimuthal asymmetry? Is position of edge determined by ice sublimation transition or by planetary perturbations?

2.2. ROTATION CURVES FOR β PIC

One of the questions regarding the β Pic disk is the radial profile of the gas to dust ratio. Ultraviolet spectroscopy argues that considerable amounts of infalling gas, interpreted as cometary kinematics, exist within a central region near the star (Beust et al., 1991; Boggess et al., 1991). In principle, the gaseous component in the outer disk should exhibit Doppler-shifted spectral features in observations on opposite sides of the star. Ground-based attempts have failed, but far IR lines and the $>10^4$ resolution mode with ISO's SWS/LWS Fabry-Perots may succeed.

2.3. Spectral Energy Distribution [SED] Statistics

It is not absolutely clear whether the sources of far-infrared excess in the IRAS-defined, 100-plus candidate systems are the same as in the three resolved prototypes despite similar excess color temperatures and fractional luminosities, in terms of $L_{IR\ excess}/L_{star}$. When many of the 100-plus candidates have been resolved, statistical and evolutionary studies will be possible and the story behind the origin, maintenance and future of such circumstellar disks will become clearer. ISOPHOT could be used to conduct photometric scans of nearby stars with IR excesses detected by IRAS, but with disk scale too small or flux too low to have been resolved by IRAS. ISOPHOT, 50–100 micron broadband, repeated slow scans across sources at peak wavelengths and various azimuths would achieve this. These data will be deconvolved versus comparison with point source scans. The main questions to be addressed: are these sources centered on the stars (i.e., not background); can these sources be judged to be disks in projection; can we estimate the grain size from angular scale vs. temperature?

2.4. Search for New Examples

The optical depth of the faintest disks detected by IRAS is of order 10^{-5}. This is two orders of magnitude thicker than the optical depth of the zodiacal dust cloud in our solar system. Nevertheless, at the IRAS sensitivity limit, 20–50% of nearby stars, and up to 10% of a magnitude-limited (SAO) survey of stars, depending on spectral type, are found to have cool far-IR excesses (Backman and Gillett, 1986; Stencel and Backman, 1991). This suggests that a more sensitive instrument might find even higher percentages. ISO's sensitivity for detection of these systems in the 50–100 μm range is up to 100 times better than that of IRAS. One could use ISOPHOT, 50–100 micron broadband for photometry in the far-IR of the 80% of nearby stars for which IRAS failed to find excesses. An example exposure for 0.02 Jy at 70 microns yields S/N of 10 in 100 sec, corresponding to an A3V star at 30 pc, G2V at 15 pc. In this way we can search for 30% excesses above photospheric flux levels, corresponding to disk optical depths of $> 10^{-6}$. The key questions: are there many cases of cool excess with disk optical depths below IRAS limits (we suspect so!); do M dwarfs (aside from Ross 128) also show this phenomenon?

2.5. Assessing Age and Composition

As demonstrated by Telesco and Knacke (1991), β Pic reveals a 10 micron silicate emission feature similar to that observed in comets. Near-IR broadband colors of the outer disk are also comparable to some observations of comet comae (Backman and Witteborn, this conference). If these studies in fact relate to the surface conditions of the material in the disk of such systems, one could use ISOCAM-CVF over the 1–6 micron range, with 1.5 arcsec p.f.o.v. to perform reflectance "mineralogy" of the β Pic disk in the near IR. We estimate 5e-5 Jy at 2–3 microns will yield S/N of 10 in 3600 sec, which reaches $K = 18$. Assuming the disk has K-band surface brightness of about 15 mag per sq. arcsec at 6 arcsec (corresponding to 100 AU), we can can work in the region $100 < r < 300$ AU imaged originally by Smith and Terrile. However, some care with the scattered light from the much brighter star ($K = 3.5$ at 6 arcsec distance) needs to be taken. A particular feature of interest studied by Nuth, Moore and Tanabe (1992) involves changes to the 4.5 micron silicate complex as a function of oxidization of attached -SiH functional groups. Just as solar wind ions have been shown to chemically reduce exposed surfaces on the moon, it seems reasonable to expect similar processes to act on the refractory orbiting particles, which will affect the strength and spectral position of the -SiH fundamental stretch, and thus be proxy for the relative age of surfaces. A survey of the β Pic disk in detail and other disks in integrated light, at 3 microns, will yield S/N of 10 in 3600 sec, which could prove quite revealing in terms of planetary processing and disk evolution. It will be very important towards understanding these systems if grain exposure ages can be compared with predicted grain lifetimes, which will be determined by a balance between grain destruction by various processes and grain production from planetesimal collisions and cometary activity.

2.6. RELATED DISK SYSTEMS

There are several additional classes of stars that merit investigation in the context of the overall formation, evolution and destruction of planetary systems, but space does not permit more than brief mention. These include the Herbig Ae stars (Grady, Thé and collaborators, this conference); λ Boo systems (Charbonneau, 1991); G and K type ("yellow") giant stars (Judge, Jordan, and Rowan-Robinson, 1987; Stencel and Backman, 1991), and white dwarfs.

Carol Grady and colleagues have identified a possible evolutionary sequence in terms of disk density among stars such as R CrA (densest), HD 176386 and HR 5999 (intermediate) and 51 Oph (least). β Pic would be even less dense in this ordering. Furthermore, there is evidence for variable/clumpy accretion among some of these stars (Graham, 1992; Welty, Barden, and Huenemoerder, 1992) Far infrared spectroscopic observations of selected objects are warranted to explore this proposed disk density sequence.

A few of the λ Boo stars listed by Gray (1988) appear in the IRAS catalog, and of these, λ Boo and HD31295 show a 60 micron excess qualitatively similar to the Vega stars. Non-LTE model atmosphere analysis by Lemke (1990) and others has revealed the non-solar composition of Vega and other 'normal' A stars. Holweger (1991, preprint) reports an anticorrelation between C and Si, and suggests this [C/Si] variation among A stars may indicate a variable gas/dust ratio due to gas-dust fractionation. Charbonneau (1991) modeled the gas-dust segregation in accretion for such stars to explain this observation.

Presuming that the β Pic disk is more substantial than our own planetary ensemble, what would we expect to see, in terms of disk destruction, when that star leaves the main sequence, evolves into a giant star and begins to engulf its domain? A few yellow giants have far IR excesses like the β Pic stars, but **do not** feature strong 12 and 25 micron excesses of red giant stars actively losing mass. Some of the best examples include δ And, δ Phe and ϵ Col (cf. Jura, 1990).

Finally, is there any prospect for white dwarfs to retain shards of their prior Oort clouds? Deep, far IR images might just turn up a surprise or two, especially considering the recent ROSAT and far UV discoveries of relatively high metal opacities in selected white dwarf photospheres. Could the metals have been added to the surface from incoming solid phase materials?

Acknowledgements

We are pleased to acknowledge partial support for these efforts under NASA grant NAGW-3680 to the University of Denver, and express appreciation for hospitality while the first author was on a sabbatical visit to the University of Michigan, Department of Physics. Useful discussions with Fred Adams, Carol Grady, Joseph Nuth and Francesco Paresce are gratefully acknowledged, as are the heroic efforts of NASA personnel in creating the possibility for U.S. observers to use ISO.

References

Backman, D. and Gillett, F.: 1986, in : J. Linsky and R. Stencel, ed(s).,*Fifth Cambridge Workshop on Cool Stars*, Springer-Verlag, Heidelberg, 340.
Backman, D., Gillett, F. and Witteborn, F.: 1992, *Astrophys. J.* **385**, 670.
Beust, H., Gry, C., Lagrange-Henri, A. and Ferlet, R.: 1991, *Astron. Ap.* **247**, 505.
Boggess, A., Bruhwieler, F., Grady, C., Ebbetts, D., Kondo, Y., Trafton, L., Brandt, J. and Heap, S.: 1991, *Astrophys. J.* **377**, L48.
Charbonneau, P.: 1991, *Astrophys. J.* **372**, L33.
Gray, R.: 1988, *Astron. J.* **95**, 220.
Judge, P., Jordan, C. and Rowan–Robinson, M.: 1987, *Monthly Notices Royal Astron. Soc.* **224**, 93.
Jura, M.: 1990, *Astrophys. J.* **365**, 317.
Kessler, M., Metcalfe, L. and Salama, A.: 1992, *Space Sci. Rev.* **61**, 45.
Lemke, A.: 1990, *Astron. Ap* **240**, 331.
Lissauer, J. and Griffiths, C.: 1989, *Astrophys. J.* **340**, 468.
Nuth, J., Moore, M. and Tanabe, T.: 1992, *Icarus* **98**, 207.
Stencel, R. and Backman, D.: 1991, *Astrophys. J. Suppl.* **75**, 905.
Telesco, C. and Knacke, R.: 1991, *Astrophys. J.* **372**, L29.
Weidenschilling, S.: 1984, *Icarus* **60**, 553.

THE EDISON INFRARED SPACE OBSERVATORY

AND THE STUDY OF EXTRA-SOLAR PLANETARY MATERIAL *

H.A. THRONSON, JR.
Wyoming Infrared Observatory and the Royal Observatory, Edinburgh

T.G. HAWARDEN
Joint Astronomy Centre and the Royal Observatory, Edinburgh

J. BALLY
University of Colorado, Boulder, Colorado, USA

D. RAPP
Jet Propulsion Laboratory, Pasadena, California, USA

and

S.A. STERN
Southwest Research Institute, USA

Abstract. *Edison* is a proposed large-aperture, radiatively-cooled space observatory planned to operate at wavelengths between 2 and 130 μm or longer. Current estimates for the telescope allow an aperture of 1.7 m which will achieve a final equilibrium temperature of about 30 K, although use of cryocoolers may permit temperatures below 20 K. *Edison* will be a powerful tool to investigate our Solar System, as well as planetary material around distant stars. At near- and mid-infrared wavelengths, where planetary material emits most of its radiation, *Edison* will be the most sensitive photometric and spectroscopic observatory under current consideration by the space agencies. With its large aperture, *Edison* will be able both to resolve the structure in nearby circumstellar "Vega disks" and to discriminate faint IR emission in the crowded environment of the galactic plane. With its long lifetime, *Edison* will allow extensive follow-up observations and increase the likelihood of catching transient events. We propose *Edison* as a precursor to elements of a future space-based IR interferometer.

1. Background: Infrared Observations of Planetary Material

The search for and study of extra-solar planetary material is one of the most exciting prospects for future infrared space missions. Solids within a few hundred AU of more or less normal stars equilibrate at temperatures in the range of 20–500 K, which means that the bulk of the radiation emitted by planets, comets, asteroids, moons, dust, and other circumstellar material will emerge at wavelengths between about 5 and 200 μm. A number of proposed programs of extra-solar planetary detection and study require distinguishing between the feeble emission from a star and that of planets. The contrast between the emission from a star and that of planets such as the Earth or Jupiter, although extremely small, is greatest at mid-infrared wavelengths. Finally, although many presentations at this conference emphasize *detection* of planetary material, a great deal more work will be required to analyze and to understand whatever will be found orbiting distant stars. This

---

* Paper presented at the Conference on *Planetary Systems: Formation, Evolution, and Detection* held 7–10 December, 1992 at CalTech, Pasadena, California, U.S.A.

Astrophysics and Space Science **212**: 423–431, 1994.

will require spectroscopic observations of key gaseous and solid state diagnostic features in the infrared.

Edison, with its large aperture and operating at the celestial background limit, will have as part of its key scientific justification, the detailed study of planetary material around other stars.

2. Current Design for the *Edison* Observatory

2.1. RADIATIVE COOLING OF SPACE OBSERVATORIES

Radiative cooling for infrared space observatories has several advantages over the alternative—cooling via large tanks of liquid cryogens: (1) current designs indicate that for the same overall spacecraft size, a telescope approximately twice the diameter of that of a cryogenic observatory is possible; (2) a radiatively-cooled observatory has no built-in lifetime; (3) without the massive cryogen tanks, the spacecraft is substantially lighter and may be less complex, translating into savings in time and money; and (4) a radiatively cooled observatory is likely to be more robust on orbit against potential catastrophic failures, such as inadvertent pointing toward one of the many hot objects in our Solar System. These characteristics have been recognized for some years, leading to the design and development of a number of IR/sub-millimeter observatories which will be cooled at least in part via radiation: the *Cosmic Background Explorer* (COBE), the *Sub-Millimeter Wave Astronomical Satellite* (SWAS), and the *Far-Infrared and Sub-Millimetre Space Telescope* (FIRST).

At the same time, radiative cooling has clear technological hurdles, including the necessity for additional cooling of some elements of the instruments, notably long-wavelength detectors, and the possibility that without the massive cryogen tanks, there may be spatial and/or temporal temperature fluctuations in the optical system.

2.2. BASIC OUTLINE OF *Edison* DESIGN

Large solid bodies deep in space equilibrate at temperatures in the range of 5–15 K, depending upon their composition and their proximity to sources of heat. Such temperatures indicate that, if the radiation environment of deep space can be duplicated for spacecraft, radiative cooling is the future technological direction for infrared and sub-millimeter observatories. The first goal, then, of proper passive cooling requires that the radiation field incident upon the optical system approach as closely as possible that of deep interstellar space.

Hawarden *et al.* (1992) outline the basic principles of an effective radiatively cooled space telescope. These have been incorporated into our current design for *Edison*, which is shown in cross section in Figure 1. The concentric radiation shields/gap radiators are shown surrounding a telescope which we estimate can be as large as 1.7 m for the resources available for future medium-size space missions. The telescope/radiator assembly is shown mated to the SOHO service module, which is the baseline bus designated for the mission.

Critical to the design of cold observatories in space is location: the further away from sources of heat, the longer the lifetime (for cryogenic observatories such as the European Space Agency's Infrared Space Observatory (ISO) or the National Aeronautics and Space Administration's (NASA's) Space Infrared Telescope Facility (SIRTF)) or the lower the equilibrium temperature (for radiatively-cooled observatories such as COBE or *Edison*). Our baseline location is a halo orbit surrounding the L2 point, about 1.5 million km anti-sunward from the Earth, which is energetically relatively easy to enter. At this location, Earth contributes almost nothing to the spacecraft heat load. The equilibrium optical system temperature calculated for the design shown in Figure 1 at L2 is about 30 K, which means that *Edison* will be limited in sensitivity by the celestial thermal noise at all wavelengths shortward of about 30 μm. At far-infrared wavelengths, broadband observations will be limited ultimately by source confusion, which requires a large aperture for increased sensitivity. Consequently, for broadband observations of point sources, *Edison* is the most sensitive infrared observatory under consideration by the space agencies. Operation at L2 also allows alternative spacecraft designs, in particular taking advantage of having the dominant heat source (the Sun) constantly on one side of the satellite. In the first place, the sun shield/solar array assembly, which could take advantage of recent developments in inflatable structure technology, would be on one side of the spacecraft, while the other will be exposed only to cold space. Taking advantage of this, Hawarden *et al.* (1992) described a schematic design in which large anti-sunward sections of the concentric radiation shields are removed, significantly increasing the radiating area. Recent estimates of the achievable temperature for such designs are in the range of 15–20 K. Furthermore, such estimates may be conservative, as near-future performance of mechanical cooling systems under development at Rutherford Appleton Laboratory (RAL), the Jet Propulsion Laboratory (JPL), and elsewhere allow significant additional cooling power beyond that possible via radiation alone.

2.3. CURRENT STATUS OF THE *Edison* PROJECT

The *Edison* infrared space observatory has been proposed to the European Space Agency (ESA) for Assessment/Phase A study in response to that agency's M3 opportunity. A parallel and supporting proposal to investigate the enabling technology has been submitted to NASA. The results of the ESA Assessment/Phase A are expected to be released in Spring 1996, with an expected M3 launch date of 2003.

Near-future work on the project includes (1) an advanced, detailed thermal model of our baseline design, as well as models for innovative geometries; (2) an advanced optical systems design and analysis; (3) continuing assessment and preliminary design on the cryocoolers; and (4) an increasingly detailed and complete model for the limits to sensitivities for IR/sub-millimeter observations.

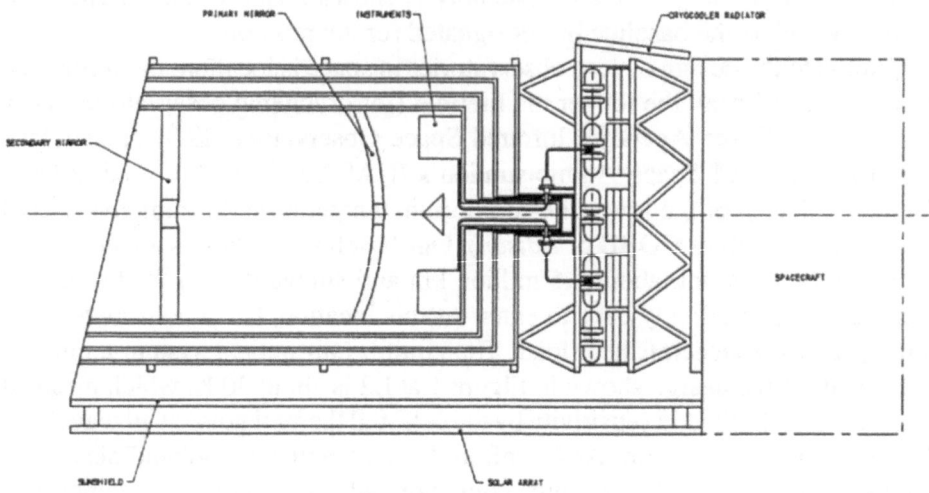

Fig. 1.   Cross-sectional diagram of our current design for the *Edison* infrared space observatory, drawn approximately to scale, where the telescope primary is 1.7 m in diameter. The solar array panel and exterior sunshield are shown stored along the bottom of the figure. In this design, a set of 5 closed-cycle refrigerators are mounted along the top of the service module. In practice, the coolers will be enclosed. This present design has adopted the SOHO bus as the baseline service module.

2.4. *Edison*: THE NEXT STEP TO FUTURE IR OBSERVATORIES

Although *Edison* as proposed will be the most sensitive space telescope operating throughout most of the infrared, far-future telescopes in space will be expected to perform even better. Two technological directions seem attractive based upon presentations at this workshop: (1) increasing aperture, thus permitting major gains in raw sensitivity, and (2) spatial interferometry from orbit or on the moon. In the former case, radiative cooling of large volumes of the future observatory will be required to achieve low temperatures while allowing large light-collecting optics. Presumably, just as with *Edison*, segments of future IR space observatories will use additional onboard cryocooling systems. For the latter case, that of spatial interferometry, *Edison* may be considered as a prototype for one element of an array of space-based IR telescopes. Just as on Earth, interferometry is unsurpassed for resolving small-scale structure, which presumably will be required for study of material surrounding distant stars and to probe deeply into crowded fields. To achieve reasonable sensitivities, individual elements of interferometers will require as large light-collecting apertures as is feasible. This, in turn, requires radiatively-cooled designs.

In sum, *Edison* may properly be thought of as the first step in the direction of future space observatories working at infrared and sub-millimeter wavelengths.

3. Investigation of Extra-Solar Planetary Material

3.1. *Edison* AND THE STUDY OF OUR SOLAR SYSTEM

One major program for *Edison* will, of course, be the study of our own planetary system (Encrenaz 1992). Within the next few years, ESA's *Infrared Space Observatory* (ISO) satellite, and within a decade, NASA's *Space Infrared Telescope Facility* (SIRTF), will be able to produce a complete spectral atlas of the outer planets, which should be a major step forward in the studies of, for example, the chemical evolution of planetary atmospheres and surface compositions. However, the angular resolution and lifetime of both facilities will be a serious limitation to such programs as study of the spatial variations in atmospheric chemistry or the opportunity to study more than the few comets likely to enter the inner Solar System during the lifetime of a cryogenic mission.

Edison, in contrast, will not be so limited. As one example, the pointing constraints of cryogenic observatories are very severe: a glancing view of the sun is likely to significantly reduce the mission lifetime as liquid helium would rapidly boil away. A radiatively-cooled observatory should also be pointed away from heat sources, but a brief view of the sun may do no more than heat large areas of the optical system. This almost certainly means a significant reduction in sensitivity due to the consequent high background and significant thermal gradients within the system that may create serious mechanical stress. However, both may be manageable and, in any case, it is unlikely that mission duration would be affected. If this is the case, it may be feasible to deliberately observe with *Edison* close to the sun, obtaining detailed infrared spectra of Venus, for example, or of comets during their most chemically active phases. The possibility of such observations will have to be considered and modeled carefully during the development phase of this space observatory.

3.2. DETECTION OF EXTRA-SOLAR PLANETS

The *Edison* observatory would appear in the title of these conference proceedings, rather than being discussed in one of many papers, were we able to demonstrate unambiguously that this space observatory would be able to detect reasonably normal planets around neighboring stars. With an aperture of about 1.7 m, *Edison* certainly has the sensitivity to detect planets such as Jupiter or the Earth out to distances of a few parsecs, given integration times of many hours. However, detection of such planets also requires the ability to discriminate between the feeble emission of the planet from that of the central star. This appears to require an optical system with a baseline or an aperture at least a few times larger than that which we are proposing for *Edison*. Some estimate of the required aperture for detection may be found, for example, in the work of Rapp (1992, 1993).

The agglomeration of circumstellar heavy elements into planets makes life close to stars possible, but at the same time significantly reduces the infrared signatures that indicate the presence of solid state material. Dusty disks and shells surrounding

stars are easier to detect and study than are planets. Consequently, it will be more
fruitful in the near future to search for dusty material around more or less normal
stars, than it will be to search for the evidence of planets. In any case, so far as
we can surmise, planets form out of dusty circumstellar material, so that studies
of dusty material around young stars will be critical to a complete understanding
of the evolutionary process for planets. One intriguing result is the interpretation
by Marsh and Mahoney (1992, 1993) of depressions in the mid-infrared spectra
of T Tauri stars as being due to "gaps" in the circumstellar material. These gaps,
if real, are suggested to be the result of missing material in a dusty circumstellar
disk. Such missing matter could be the result of gravitational perturbation by one
or more sub-stellar bodies. If this interpretation is correct, the technique of Marsh
and Mahoney may be profitably applied to extremely large numbers of T Tauri-like
stars, surveying the creation of planets. If, that is, an observatory has sufficient
sensitivity and powers of discrimination. *Edison* will have sufficient sensitivity to
detect easily T Tauri objects with high signal-to-noise throughout the Milky Way.
At a distance of about 140 pc, T Tauri has a 10 μm flux density of about 130 Jy.
Edison's 10 μm sensitivity as a function of integration time is approximately F_ν
(mJy) $\sim 0.1\,[\lambda/\Delta\lambda]^{1/2}\,[S/N]\,t^{-1/2}$ in the plane of the Milky Way (Thronson *et al.*,
1993), where the right side of the expression includes an adopted spectral resolution
and desired signal-to-noise ratio. A spectral resolution of 20 and a signal-to-noise
ratio of 10 should be sufficient to determine whether or not dusty T Tauri objects
possess depressions in their spectrum. A little algebra allows us to estimate that
Edison could obtain observations sufficient to determine the presence of gaps in a
dusty disk for objects at a distance of 30 kpc within a few seconds. In practice, the
most likely limitation to this type of study will be source confusion in the plane
of our galaxy and/or in crowded fields of young stellar clusters. As emphasized
in the opening sections of this presentation, *Edison* has been specifically proposed
to possess the largest aperture of any future IR space observatory, which will be
of significant advantage for observations in confused fields. However, the plane
of the Milky Way will still be a difficult region in which to work at the extreme
sensitivities of which *Edison* is capable. Instead, it may be more profitable to study
the formation of planets outside the Milky Way.

3.3. PLANET FORMATION IN OTHER GALAXIES

It will be most interesting to identify the formation of planets around the nearest
stars, thus demonstrating that our neighborhood in the cosmos is not devoid of
life and, consequently, hopelessly boring. Moreover, planets around nearby stars
are presumably being born under somewhat similar conditions to those which
prevailed when our own Earth formed. Such observations would supply a key data
point in our understanding of the Solar System's creation. However, observations
of more distant objects will be necessary to achieve a complete picture of the
process of planetary system creation. As noted in the previous section, detection of
distant circumstellar material may be very difficult in the plane of our galaxy, due

to the serious issue of source crowding. As a consequence, we suggest that study of T Tauri stars and related objects in Local Group galaxies may be easier and will certainly provide important information to a broad general picture of planet formation. Here we assume that the interpretation is correct of the depressions in mid-IR spectra from T Tauri stars as being due to perturbations by sub-stellar (planetary?) objects (Marsh and Mahoney 1992, 1993). Observations of planet-forming circumstellar material around T Tauri objects in other galaxies will be interesting for at least two reasons: (1) we hypothesize that planetary formation—frequency, rate, planetary mass—is dependent in some fundamental way upon the metallicity of the interstellar medium (ISM) out of which the stellar/planetary system formed; and (2) we hypothesize that planetary formation also depends upon the dynamics of the ISM: large-scale flows, compression waves, and/or shocks. These gross characteristics vary significantly from object to object in the Local Group. In addition, as noted already, due to source crowding, it may be easier to observe young stars in other galaxies than in our own.

In Figure 2, we present the spectrum of T Tauri at the distance of the Large Magellanic Cloud (LMC) (55 kpc), along with the expected sensitivity of *Edison* for a three-hour, low-resolution observation. This calculation includes the effects of confusion by distant background galaxies, but *not* the effects of confusion within the LMC itself, which could be severe: at 10 μm, *Edison* will have a diffraction-limited angular resolution of about $1''7$ or a spatial resolution of about 0.5 pc at the distance of the LMC. This may be larger than the separation between stars in young clusters.

In any case, *Edison* may be sensitive enough to survey the low-mass star-forming regions in the LMC for evidence of creation of planetary material. Comparison of such a census with a similar survey in the Milky Way will be a major step in understanding the gross galaxian parameters which govern the formation of planets. Furthermore, *Edison* has the theoretical sensitivity to detect an object such as T Tauri throughout a large fraction of the Local Group. After an integration time of 1 day (about 10^5 s), *Edison* will just about be able to obtain a low-resolution spectrum of an object such as T Tauri in the Andromeda Galaxy. This volume of space also includes a number of metal-poor, star-forming dwarf galaxies. Not only is stellar creation in these objects intrinsically interesting, but they also offer the opportunity to investigate formation of stars and planets under conditions which might roughly approximate that of the very early universe.

4. Summary

Edison has been proposed to the European Space Agency as a radiatively-cooled, large-aperture, long-lived infrared space observatory to follow the current generation of cryogenic missions. Current design is for a 1.7 m telescope to be launched by an Ariane 5, Proton, or Atlas. Preliminary thermal models indicate that it is straightforward to cool the optical system via radiation to temperatures of 30 K.

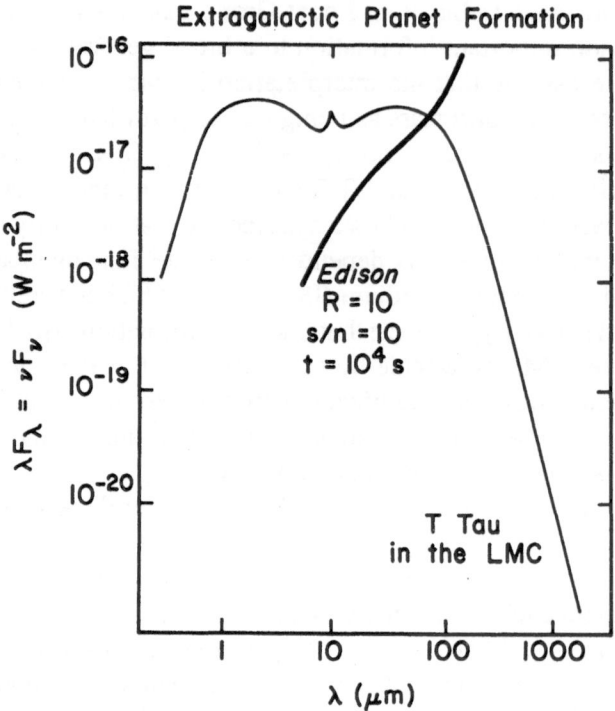

Fig. 2. It may be possible to detect planetary system formation in other galaxies with *Edison*. Here we show the spectrum of T Tauri if observed at the distance of the Large Magellanic Cloud (55 kpc), along with the expected sensitivity of *Edison* for a $S/N = 10$ detection with a spectral resolution of 10 after an integration time of about 3 hours. The calculated sensitivity does not include the effects of source confusion within the LMC.

Improved thermal design and/or the use of near-future cryocoolers may allow temperatures as low as 20 K.

A large-aperture infrared telescope operating at the celestial background in space will allow sensitive investigations into a wide range of scientific programs related to extra-solar planetary material, including studying the composition of circumstellar disks throughout much of the Milky Way, high-resolution spectroscopy of brown dwarf atmospheres, and low-resolution spectroscopy of T Tauri stars as part of a search for extragalactic planet formation. NASA has also been requested to participate with ESA at least in the early stages of the assessment of the *Edison* project, with the possibility of formal partnership with ESA at a future date.

Acknowledgements

This work was supported in part by NASA Grant NAG 8-899, Rutherford Appleton Laboratory (RAL), and the Royal Observatory, Edinburgh. We appreciate the support of the *Edison* Study Office at RAL, where current design work is taking place.

References

Encrenaz, T.: 1992, *Space Sci. Rev.* **61**, 13.
Hawarden, T. G., Cummings, R. O., Telesco, C. M. and Thronson, H. A.: 1992, *Space Sci. Rev.* **61**, 113.
Marsh, K. A. and Mahoney, M. J.: 1992, *Astroph. J.* **395**, L115.
Marsh, K. A. and Mahoney, M. J.: 1993, *Astroph. J. Lett.*, in press.
Rapp, D.: 1992, Potential for Active Structures Technology to Enable Lightweight Passively Cooled IR Telescopes, JPL Internal Report D-9449.
Rapp, D.: 1993, The *Edison* International Infrared Space Observatory: European Space Agency M3 Proposal, Rutherford Appleton Laboratory Report.
Thronson, H. A. *et al.*: 1992, *Space Sci. Rev.* **61**, 145.
Thronson, H. A., Rapp, D., Bailey, B. and Hawarden, T. G.: 1993, *PASP*, submitted.
Werner, M. W.: 1993, in: : B. T. Soifer, ed(s)., *Sky Surveys: Protostars to Protogalaxies*, SIRTF Surveys, Astronomical Society of the Pacific: San Francisco, in press.

THE ASTROMETRIC IMAGING TELESCOPE: DETECTION OF

PLANETARY SYSTEMS WITH IMAGING AND ASTROMETRY *

STEVEN H. PRAVDO and RICHARD J. TERRILE
Jet Propulsion Laboratory, California Institute of Technology,
Pasadena, California, USA

CHRIST FTACLAS
Hughes Danbury Optical Systems, Inc., Danbury, Connecticut, USA

GEORGE D. GATEWOOD
Allegheny Observatory, University of Pittsburgh,
Pittsburgh, Pennsylvania, USA

and

EUGENE H. LEVY
Lunar and Planetary Laboratory, University of Arizona, Tucson, Arizona, USA

Abstract. The Astrometric Imaging Telescope (AIT) is a proposed spaceborne observatory whose primary goal is the detection and study of extra-solar planetary systems. It contains two instruments that use complementary techniques to address the goal. The first instrument, the Coronagraphic Imager, takes direct images of nearby stars and Jupiter-size planets. It uses a telescope with scattering-compensated optics and a high-efficiency coronagraph to separate reflected planet light from the central star light. Planet detections take hours; confirmations occur in months. With a program duration of about 2 years, about 50 stars are observed. The second instrument, the Astrometric Photometer, shares the same telescope and focal plane. It uses a Ronchi ruling that is translated across the focal plane to simultaneously measure the positions of each target star and about 25 reference stars with sufficient accuracy to detect Uranus-mass planets around hundreds of stars. Enough stars of several spectral types are observed to obtain a statistically significant measurement of the prevalence of planetary systems. This observing program takes about 10 years to complete. The combination of both instruments in a single telescope system results from a number of innovative solutions that are described in this paper.

1. Introduction

Planetary systems are thought to commonly occur around stars like the Sun. If this hypothesis is correct then the fact that no other planetary systems are known is due only to our inability to detect them with current astronomical observatories and instruments. A major goal of the new NASA program, Toward Other Planetary Systems (TOPS), is to develop the instruments and make the observations to detect, characterize, and study other planetary systems (Burke *et al.*, 1992).

The new instruments must feature both an increased sensitivity to the presence of planets around nearby stars and the capability to perform a broad survey of stars so that even an unlikely null result is statistically significant. The Astrometric Imaging Telescope (AIT) is a space-based observatory designed to fulfill these requirements. It consists of two instruments that use different techniques, imaging and astrometry,

---

* Paper presented at the Conference on *Planetary Systems: Formation, Evolution, and Detection* held 7–10 December, 1992 at CalTech, Pasadena, California, U.S.A.

to obtain complementary data on planetary systems. In the following sections we will describe the AIT instruments, observations, and a mission and flight system designed to accommodate the instruments and fulfill the TOPS goals.

2. Instruments and Observations

2.1. IMAGING INSTRUMENT

The AIT imaging instrument is the Coronagraphic Imager (CI). It consists of a hybrid coronagraph and a Charged Coupled Device (CCD) camera operating in the visible range. The coronagraph uses graded occulting masks to reduce the diffracted light in the wings of target stars by at least a factor of 1000. A laboratory version of the coronagraph has demonstrated this performance. To take advantage of this coronagraph, AIT features a "scattering-compensated" optical system that reduces scattered light by the same factor in the small-angle regime. Sub-scale mirrors have been fabricated that demonstrate the performance required by AIT. These capabilities of the AIT/CI are unprecedented in an orbiting telescope but are technologically mature as shown by the laboratory verifications.

Other important AIT/CI features have been proven in past space programs. For example, the stability required by the AIT telescope pointing control system to keep the target star on the occulting mask during observations has been demonstrated in orbit. Similarly CCD imaging cameras are known to work well in space.

The CI can detect planets in the imperfect world of the space environment. Models of CI results in the presence of effects such as pointing jitter and pixelization, have shown it can perform successfully. Models have also shown that the coronagraph controls telescope aberrations just as it controls diffraction (Ftaclas et al., 1992) allowing performance goals to be met. Another paper in this volume describes the CI in further detail (Terrile et al., 1993).

2.2. IMAGING OBSERVATIONS

The CI detects Jupiter-size planets around nearby stars by imaging the region within a couple of arcseconds of the stars. The hybrid coronagraph and the scattering-compensated optics reduce the light from the star so that the reflected light from the planet is no smaller than 1% of the residual starlight at the planet's position. Planets are detected from the CCD images in less than 10 hours of observations each.

With the AIT 1.5-m-diameter primary mirror the CI detects Jupiter-size planets around approximately 50 target stars. The number of potential targets for planetary imaging increases sharply with primary diameter. With a larger primary, not only does the number of planetary photons increase as the diameter squared, but also the stellar background decreases due to the smaller diffraction spot size.

Each target is observed several times per year. If a detection is made, a confirmation will occur in several months when the star field is reobserved to eliminate spurious background objects. Color photometry of detected planets is performed with filters over periods of days. The entire observing program is completed in about 2 years.

Imaging yields the following information: the planet brightness (albedo/size), star-planet separation, multiplicity of planets, and orbital phase. Prominent spectral bands may be detected via filter photometry. Orbital data is also available, but in this case, over orbital period time scales. The detection and study of circumstellar material including proto-planetary disks are other TOPS 1 goals that are accomplished with the CI.

2.3. ASTROMETRIC INSTRUMENT

The AIT astrometric instrument is the Astrometric Photometer (AP). It consists of a high-precision Ronchi ruling and several photometric detectors operating in the visible range. The ruling is moved across the field of view at the telescope focal plane, and its thousands of transparent and opaque line pairs modulate the light from a target star and 25 reference stars. Movable optical fibers are positioned to capture the light from each star and direct it to photon-counting detectors from which the stellar signals are extracted. The relative phases of the target star signal and the reference star signals give an accurate determination of the target position in one dimension. Later an orthogonal measurement is made by rotating the ruling 90 degrees.

The fundamental metric in the AP is the Ronchi ruling. Errors in the ruling line edge positions limit the accuracy of the phase measurements. If the line edge errors are randomly distributed then measurement accuracy improves with the square root of the number of edges. For the AP 2500 lines with random errors of amplitude 50 nm result in an overall accuracy of 1 nm. This corresponds to an accuracy of 10 μarcsec in the target positions. Systematic errors in the line edges do not average down and are thus only tolerable at the final required accuracy. Laboratory tests of currently available rulings have demonstrated that the random ruling line edge errors are within the required accuracy. Further tests using Moiré techniques and laser metrology are under way to measure the systematic errors.

Other sources of error include spacecraft pointing jitter, contamination of optical surfaces, and telescope aberrations. The relative nature of the measurements and the modulation of the stellar signals at the ruling frequency reduce much of the effect of pointing jitter. The telescope optics are designed to be tolerant of contamination and aberrations so that astrometric performance is not compromised by errors such as secondary mirror tilt, decenter, and defocus. Another paper in this volume considers many of these error sources and shows how the required astrometric accuracy is achieved in a realistic end-to-end model (Shaklan et al., 1993).

2.4. ASTROMETRIC OBSERVATIONS

The AP is designed to detect Uranus-mass planets around hundreds of nearby stars. The RGO (Woolley et al., 1970) and Gliese (1969) catalogs contain more than 2000 candidate stars within 25 pc. About 800 of these stars are suitable AP targets since

Uranus-mass planets would cause stellar wobbles larger than 10 μarcsec with orbital periods of less than 10 years, the nominal AIT mission lifetime. The target stars range in spectral type from A to M, with significant numbers of G (89), K (276), and M (350) stars. With hundreds of stars observed, the statistics of planetary systems are determined. A statistically significant null result (no planetary systems) presents a major problem for the current theory of planetary system formation.

An AP observation achieves 10 μarcsec accuracy in 30–60 minutes. Targets are observed several times each year and must be followed over at least an orbital period, in most cases. The astrometric program is completed after about 10 years.

Astrometry yields the following information: the planet-to-star mass ratio, orbital elements, planet multiplicity, and star distance (from parallax).

3. Spacecraft and Mission

The size of the AIT spacecraft is driven by two considerations. First, the number of imaging targets suitable for planet detection is a strong function of the diameter of the primary mirror (see Section 2.2). Thus the largest diameter primary within the constraints is chosen. Second, the "distortion-free" design for the astrometric investigation (see below) pushes the optical design toward larger primary–secondary mirror separation. For best performance the AIT optical system is both wide and long.

The AIT launch vehicle, the Atlas II rocket, provides a fairing into which the spacecraft fits. Figure 1 shows AIT in the stowed position in the Atlas fairing. The primary mirror is located near the main ring assembly. A metering truss connects to the secondary mirror support structure and holds it in place with respect to the primary. The secondary mirror is located at the end of the spacecraft toward the tapered part of the fairing. Instruments are packaged behind the primary and in front of the launch adapter.

In this design solar arrays and a Sun shield are deployed in orbit after the fairing has been jettisoned. The solar arrays provide the power required for spacecraft and instrument operations. The Sun shield prevents sunlight and earthlight from shining on the secondary mirror, both to eliminate stray light and to better control the thermal environment. Thermal control of the optical bench and focal plane is accomplished passively with multi-layer insulation and actively with heaters. The electronics bus contains all or parts of other spacecraft subsystems such as command and data handling, telecom, and attitude control. Figure 2 illustrates AIT in the flight configuration.

The orbit is chosen to be a 700-km altitude, Sun-synchronous, polar orbit, again to better control the thermal environment. At 700 km the orbit will not decay significantly within the 10-year nominal mission. The South Atlantic Anomaly radiation zone is at a minimum at this altitude. At higher altitudes the size and intensity of the radiation zone increases resulting in both decreased observing times and increased radiation damage to the spacecraft and instruments. The Atlas

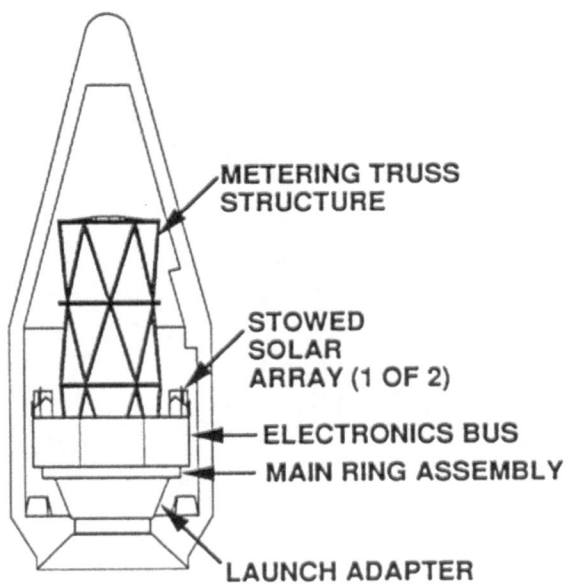

AIT LAUNCH CONFIGURATION
ATLAS IIAS FAIRING

Fig. 1. AIT launch configuration: Atlas IIAS fairing.

launch vehicle is incapable of placing a payload such as AIT above the radiation zones and into high Earth orbit.

Imaging and astrometric observations are performed in series. Figure 3 illustrates the AIT focal plane. Both the Ronchi ruling and the occulting masks are on a common translation stage. For AP observations the ruling is moved back and forth over the field of view with a frequency of 10–100 line pairs/s. The optical probes are placed behind the ruling at the star positions and direct the starlight to the detectors. Spacecraft pointing stability during an observation of 0.3 arcsec reduces this source of measurement error to an acceptable level (see also Shaklan et al., 1993).

For CI observations an occulting mask is placed on the optical axis. The AP optical probes are removed from the center of the field of view allowing the light of the target star and planet to reach the coronagraph and CCD camera. These observations require spacecraft pointing stability of 0.01 arcsec. The translation

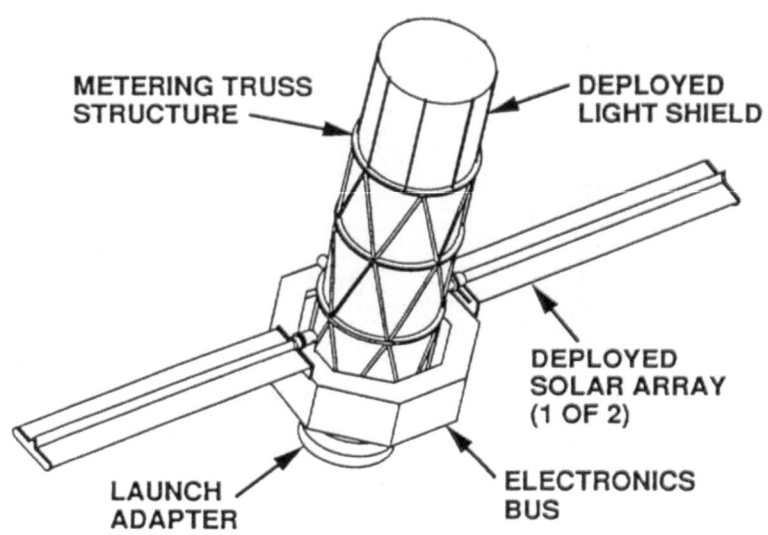

AIT FLIGHT CONFIGURATION

Fig. 2. AIT flight configuration.

mechanism is designed to return to the imaging position in case of failure so that imaging observations could continue.

The AIT optical design accommodates both instruments. The AP requires a "distortion-free" optical system (Korsch, 1989). This insures that the astrometric measurements are insensitive to changes in the optics due to in-orbit effects such as the contamination of optical surfaces or mirror misalignments. These designs generally perform better with longer focal lengths or larger mirror separations. In addition, a family of such designs exists which can be parametrized by the ratio of secondary to primary mirror diameters. The larger this ratio, the more tolerant the system is for the AP.

The CI makes less demands on the optical design but is photon-limited. The imaging investigation prefers a smaller secondary which results in a smaller central obscuration and larger effective collecting area. The AIT telescope with a mirror diameter ratio between 0.25–0.5 is thus a compromise between the instrument requirements.

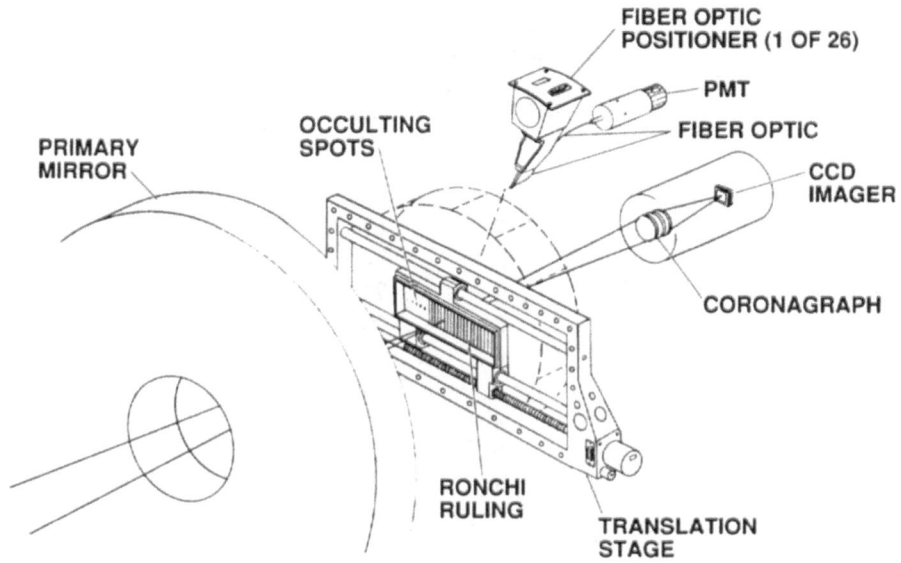

FIBER OPTIC
POSITIONER (1 OF 26)

PMT

OCCULTING
SPOTS

FIBER OPTIC

PRIMARY
MIRROR

CCD
IMAGER

CORONAGRAPH

RONCHI
RULING

TRANSLATION
STAGE

AIT FOCAL PLANE

Fig. 3. AIT focal plane.

For both instruments the telescope is rolled after an observation and data is then taken in the rolled orientation to eliminate instrumental effects such as a fixed speckle pattern. For the AP, observations are made in orthogonal directions. Observing constraints such as Sun- and Earth-avoidance and the South Atlantic Anomaly are expected to reduce the observational duty cycle to no more than 50%. The instrument data rates are less than 1 Mbps. Onboard data storage capacity is about 16 gigabits of solid state memory. In the current design data is telemetered to the 26-m Deep Space Network antennas on the ground with a maximum rate of 5 Mbps.

The science teams then analyze the data. Imaging data may reveal the presence of Jupiter-size planets within hours. Astrometric discoveries take longer with shorter period planets detected first. Since the planet astrometric detectability increases with orbital period (all other things being equal) it is interesting to note that the AP has sufficient sensitivity to detect even low period planets. Table I shows the number of target stars from the catalogs for which a Uranus-mass planet would be detected as a function of the orbital period. At the end of the nominal 10-year mission an inventory of circumstellar material, planets, and planetary systems is compiled.

TABLE I

Number of stars observed
vs. orbital period.

Period (years)	Stars
1	62
2	67
3	72
4	74
5	90
6	89
7	89
8	89
9	86
10	80

4. Summary

The Astrometric Imaging Telescope is designed to definitively address the question of the existence and prevalence of other planetary systems. It provides complementary data from two instruments which offer both images of other planets and a statistically significant survey of nearby stars. It is a technologically feasible and operationally robust mission.

Acknowledgements

This work was carried out in part at the Jet Propulsion Laboratory, California Institute of Technology, under a contract with the National Aeronautics and Space Administration.

References

Burke, B.F. *et al.*: 1992, *TOPS: Toward Other Planetary Systems*, report by the Solar System Exploration Division, NASA.
Ftaclas, C. *et al.*: 1992, *Astrometric Imaging Telescope Project Report*, Hughes Danbury Optical Systems report to the Jet Propulsion Laboratory.
Gliese, W.: 1969, *Veroffentilichungen des Astronomischen Rechen-Instituts*, **22**, Heidelberg.
Korsch, D.: 1989, in S. Pravdo, ed(s).*,Astrometric Telescope Facility FY'89 Final Report*, JPL D-7113 (internal report, Jet Propulsion Laboratory, Pasadena, CA).
Shaklan, S. *et al.*: 1993, this volume.
Terrile, R. *et al.*: 1993, this volume.
Woolley, R.V.D.R. *et al.*: 1970, *Roy. Obs. Ann.*, **5**.

THE CIRCUMSTELLAR IMAGER: DIRECT DETECTION OF
EXTRA-SOLAR PLANETARY SYSTEMS *

CHRIST FTACLAS and ANDREAS L. NONNENMACHER
Hughes Danbury Optical Systems, Inc., Danbury, Connecticut, USA

RICHARD J. TERRILE and STEVEN H. PRAVDO
*Jet Propulsion Laboratory, California Institute of Technology,
Pasadena, California, USA*

GEORGE D. GATEWOOD
Allegheny Observatory, University of Pittsburgh, Pittsburgh, Pennsylvania, USA

and

EUGENE H. LEVY
Lunar and Planetary Laboratory, University of Arizona, Tucson, Arizona, USA

Abstract. The Astrometric Imaging Telescope (AIT) is designed to probe the circumstellar environment by both direct imaging and indirect astrometric measurements. The Circumstellar Imager (CI) is a coronagraphic camera and is the direct imaging component of the AIT. The CI is designed to obtain high-sensitivity images of the circumstellar region. It provides crucial non-inferential information relating to the frequency, origin, and evolution of planetary systems and all forms of circumstellar matter. Such imaging is usually limited by the scattered and diffracted light halos of the star itself, which are greatly suppressed in the CI by mating a novel high-efficiency coronagraph with a phase-compensated optical system. For faint point sources in the circumstellar region, the CI will have a sensitivity in excess of 5 magnitudes fainter than the as-designed Hubble Space Telescope (HST). Laboratory data are shown for the coronagraph, which, in a diffraction-limited environment, is capable of suppressing the stellar diffraction sidelobes by several orders of magnitude without significant sacrifice of field of view. In order to realize the high rejection levels inherent in the coronagraph design, it is necessary to limit scatter in the optical systems, imposing a mid-spatial frequency figure error requirement an order of magnitude smaller than that of the HST. Experimental data directed toward meeting this requirement are also shown.

1. Introduction

The Astrometric Imaging Telescope (AIT) (Pravdo *et al.*, 1993) is an Earth orbiting experiment comprised of a 1.5-meter-diameter telescope with a focal plane shared by two instruments: the Astrometric Photometer (AP) and the Circumstellar Imager (CI). The primary science objective of the AIT is to survey and characterize planets and protoplanetary material orbiting nearby stars. The Circumstellar Imager employs a coronagraph to suppress diffracted light, allowing direct detection of Jovian planets. In the visible, as seen by an external observer, Jupiter is about 10^9 times fainter than the Sun. This ratio becomes more favorable at longer infrared wavelengths (Bracewell and MacPhie, 1979), but the penalty incurred by the resulting larger diffraction pattern can only be compensated for by going to larger apertures. For a fixed aperture in going from 0.5 μm to 10 μm, the Airy disk

---

* Paper presented at the Conference on *Planetary Systems: Formation, Evolution, and Detection* held 7–10 December, 1992 at CalTech, Pasadena, California, U.S.A.

has 400 times the area over which stellar background is collected. Alternatively, to obtain the same resolution as the AIT working at 0.5 μm, a 30-meter telescope is required when operating at 10 μm. The CI images at visible wavelengths and operates on the principle that both diffraction and scatter can be controlled to levels that allow the direct detection of planetary photons.

2. The Proposed Experiment

In the region immediately adjacent to any bright point source lies a previously unexplored part of the Universe. It is in this region that knowledge of the existence, origin, and formation of planetary systems lies hidden—masked by the scattered and diffracted light halo of the central source. The problem is not unlike trying to see fine detail while looking into a spotlight. The AIT, with unique capabilities to overcome this blinding light, is a combined direct and indirect assault on this region to discover and characterize planetary and protoplanetary systems.

The indirect approach is implemented by an astrometric instrument discussed elsewhere in this volume (Shaklan et al., 1993). The imaging component, the Circumstellar Imager (CI) described here, teams a newly developed, high-efficiency coronagraph (working against diffracted light) with an optical system capable of compensating for scattered light.

Imaging is a powerful adjunct to all indirect methods. Indirect techniques are most sensitive to a single massive planet whose orbital period is shorter than the program lifetime and insensitive to symmetrically distributed material that produces no stellar wobble. By design, and unlike indirect methods, imaging is extremely sensitive to circumstellar material such as zodiacal dust distributions or proto-planetary disks. Moreover, in the case of multiple planets, the image information recorded from each planet is spatially distinct, and detectability is not a function of the number of planets. In other words, all planets in a given system that are above the detectability threshold for that system are detected independent of one another within the same images. Even when no detections occur, upper limits can be placed on the absence of planets larger than a given diameter at a given orbital radius, which would help to resolve ambiguous, indirectly measured signals. Thus, the two approaches nicely complement one another.

Along similar lines, the imaging is sensitive to long period planets that are difficult to detect by indirect means simply because one has to wait too long. The period of Saturn at twice Jupiter's orbital distance is already 29.4 years. Yet we could detect a Jovian planet at Saturn's orbit and beyond for the closest stars in hours of observation, and as always provide upper limits on nondetections. This will provide insight into the processes that formed the biggest planets. Confirmation as a member of a planetary system can come in days from proper motion studies or in months from parallax studies without the need to observe for a significant fraction of the orbit.

Drawing from our explorations of binary systems in ground-based astronomy we know that spectroscopy, photometry and imaging all provide different and necessary inputs to our understanding of the dynamics of binary stars. Similarly, direct observation of planetary positions provides important input on stellar mass when combined with astrometric positional data or spectroscopic velocity data. Thus, direct imaging and indirect detection schemes nicely complement one another as tools for the investigation of the circumstellar environment.

Acting together in the CI, the high-efficiency coronagraph and the phase-compensated optical system combine to produce a reduction of the light halo around a bright star by three orders of magnitude. Taking into account the relative aperture sizes and optical figures, this amounts to a hundred-fold improvement in background level over comparable levels in the Hubble Space Telescope (HST). This dramatic reduction in background is necessary to image extra-solar Jovian planets, but it also has implications for the study of the circumstellar environment. We will be able to image fine detail in that region and directly observe subtle features that would otherwise totally exhaust the dynamic range capabilities of typical imaging systems.

For example, detecting the existence of a circumstellar disk whose intensity is a hundred times fainter than the star-induced background is a challenging but possible imaging task for a visible light telescope. But to observe fluctuations in that disk of an order of a few percent corresponding to structure is virtually impossible. Such an observation requires discerning fluctuations in the total observed intensity of one part in ten thousand. The CI's ability to suppress the very high background level reduces the problem of imaging fine-scale structure in a circumstellar disk to one comparable to seeing structure in a galaxy with a normal telescope.

The background light in the CI is suppressed rather than subtracted off. This is a crucial distinction, since the corresponding noise in the background, which ultimately limits system performance, is also suppressed. Thus, for example, spectroscopy of the circumstellar material now becomes possible, leading to dynamical and compositional characterization.

3. Science Overview

One important lesson of the Voyager program has been the central importance of imaging to the planetary program. Clearly, no other form of data collection allows a greater potential for the unexpected. In the study of the formation and distribution of planetary systems around other stars, there is no reason to believe that imaging will play a less valuable role. However, since the bulk of the expected phenomena will occur close to some bright central parent star, traditional approaches to imaging will be of limited use, since their dynamic range will be driven by the scattered and diffracted light halo around the star. The CI will circumvent this issue by suppressing the background, permitting imaging of faint features very close to the parent star. This will be a valuable tool in both the detection of planetary systems and in the study of protoplanetary systems.

The idea of planetary formation from a gaseous nebula suggests that the largest gas giant planets in the system should form at the closest point in the proto-planetary–planetary nebula where water ice condenses. Indirect approaches to the planetary detection problem may not test this hypothesis fully. The rapid increase in orbital period with orbital radius means that planets much beyond the orbit of Jupiter have periods longer than the time scales of most planned experiments, making their detection uncertain. For direct detection by imaging, however, the scattered light halo is expected to fall off with the cube of the field angle, but the planet's brightness only falls off quadratically with field angle. Thus a given planet will appear fainter but relatively more visible against the background as its orbital radius increases. The absolute number of photons is diminished, leading to longer integration times, but large planets remain detectable at the orbit of Saturn and beyond. Thus, the sensitivity of the CI at large orbital radii allows a direct test of an important aspect of the theory.

In its early stages the protoplanetary nebula is expected to be opaque to visible radiation but becomes increasingly transparent as it evolves. Visible/near IR imaging would discover and investigate nebulae at the later stages, when infall has largely terminated. The question of the minimum mass solar nebula could be investigated, and motions and structures within the nebula could be observed. If, as expected, the velocities, within the nebula are comparable to planetary orbital velocities, then the dynamical time scale of the system is comparable to the duration of the experiment, and one can imagine virtually making movies of dynamics within the nebula based on repeated observations. Spectroscopy of the nebular material would also greatly assist in understanding its dynamics.

The dust disk around β-Pictoris probably represents a later stage in the evolution of planetary systems. This disk was rendered visible from the ground, but this observation was due largely to the fortuitous edge-on orientation. More typical disk orientations become challenging detections, even for a restored HST. In particular, the question of evacuation of the central region, which is most easily determined in fainter, near pole-on orientations will be most difficult to see with traditional imaging tools.

The CI has been designed to image a region of space never before imaged. As with all new instruments, it is difficult to define in advance what it will discover. But even if there are no significant changes made to the theory of planetary formation, the CI will make an important contribution to the establishment and verification of that theory and will be a powerful tool in the search for extra-solar planets.

4. The CI Coronagraph

The classical Lyot coronagraph has been used in astronomy since 1934 and, of course, most recently played a crucial role in the discovery of the circumstellar disk around the star β-Pictoris (Smith and Terrile, 1984). Coronagraphs require occultation of the central star as a first step. Consideration of the scale of potential planetary systems under study suggests that a radius of order 0.5 arcsecond is about the maximum field of view that can be lost. For a meter-class telescope operating in the visible, the resulting potential diffraction reduction obtained from a classical Lyot coronagraph is limited to approximately a factor of 100.

The CI coronagraph currently under development has been optimized for the circumstellar problem. It produces significant reduction in scattered light without significant loss of either field of view or throughput. This performance gain is made possible by the use of an apodized (graded transmission) occulting mask that very compactly concentrates the stellar diffracted light in an annulus at the edge of the reimaged telescope pupil, allowing for very efficient rejection. Since the occulting mask is apodized and partly transparent all the way to its center, imaging inside of the mask diameter is possible. Even when transmission losses from the occulting mask are considered, the net reduction in the background is nearly uniform right to the central maximum of the diffraction pattern.

To date, a coronagraphic breadboard has been built and functions in complete agreement with theory. Diffraction sidelobe reductions in excess of one thousand have been observed with no deviation from model predictions (Figure 1). In a flight system the occulting mask could be reflective, permitting the use of the central star image for fine guidance. For nominal AIT system parameters, the occulting mask must have its transmission controlled over scales of tens of microns. These dimensions are well within the domain of microlithographic fabrication techniques, and apodized metal masks of the required dimension are presently being developed using existing technology.

5. The CI Foreoptics

The coronagraph developed for the CI can be tuned over a wide performance range. The diffraction reduction of one thousand predicted by our baseline analytical models and demonstrated in the laboratory, increases to one million when the system wavelength is decreased by 40%. But in any optical system, in addition to the ideal diffraction pattern, there is a second component of light, usually termed scatter, that has been redirected by figure and alignment errors. There is no point in reducing the diffracted light level far below that set by the scattered light against which the coronagraph has limited effect. Once the system makes the transition from diffraction-limited to scatter-limited there is no significant gain from further diffraction reductions. The objective is then to produce a system with the lowest possible scattered light level and then to tune the coronagraph to reduce diffraction to just below the scatter level. From a system point of view, the coronagraph represents a significant performance potential that can only be realized to the degree that scatter is suppressed from the optical system. This assures a minimum impact on throughput and field of view.

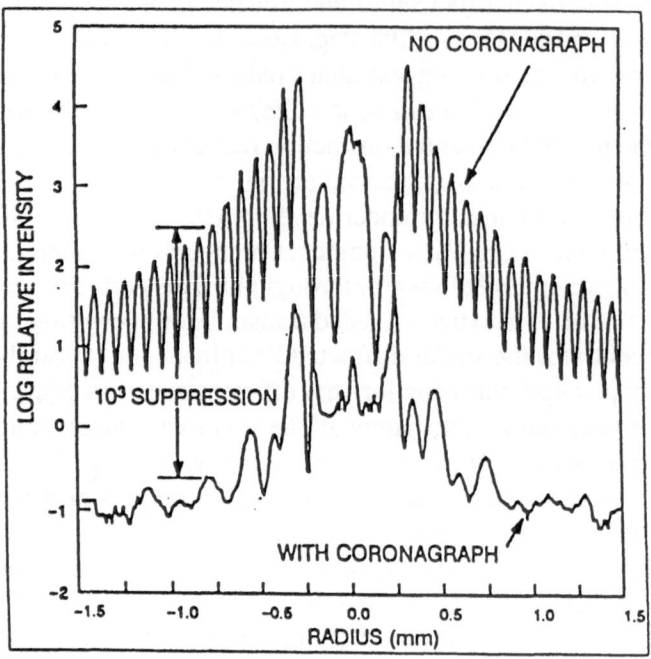

Fig. 1. Laboratory demonstration of high-efficiency coronagraph diffraction sidelobe reduction of 1000.

Scatter can arise from either figure errors or misalignments of the optical system. In analyzing the performance of the coronagraph we have found that it is also effective in reducing the long scatter tails associated with low-order aberrations so that the alignment aberrations (focus, coma and astigmatism) are reduced by factors comparable to that reducing diffraction (Ftaclas *et al.*, 1992). The net result of this beneficial by-product of the coronagraph is that, with the exception of mid-frequency figure error, the telescope has in all other regards, most notably alignment and pointing, tolerances that are typical of a diffraction-limited telescope of its size.

For the CI the principal challenge is scattered light in the range 0.5–5 arcseconds, which arises from mid-spatial frequency figure errors of 5–50 cycles per aperture in the pupil plane. Our goal of total background suppression by a factor of one thousand comes from looking at the quality of the best mirrors currently being fabricated, by considering the factors currently limiting their performance and estimating the potential performance gains that can be made. The optical system we require would have a total rms wavefront error of $\lambda/400$, about an order of magnitude better than the HST. The Advance X-Ray Astrophysics Facility (AXAF) optics fabrication effort currently represents state of the art in polishing and metrology technology. Its largest cylindrical mirrors are 1.2 meters in diameter and 1 meter tall with 25-millimeter wall thicknesses. Its mid-spatial frequency performance requirement is more than twice as stringent as that achieved on HST (Cerino, 1993).

The mid-spatial frequency figure problem is one that is relatively general and will be increasingly important as the science community becomes more experienced in diffraction-limited imaging and attempts to build larger telescopes. Even reaching the diffraction limit, in the sense of having the scatter sidelobes appreciably below diffraction sidelobes, severely constrains the mid-spatial frequency error content of optics. By significantly reducing scatter sidelobes, through a controlled fabrication process we can achieve the performance levels of a larger optic and even compensate for the smaller collecting area with a more favorable signal-to-noise ratio.

To date, we have carried out a mirror development program to identify and eliminate the causes of mid-spatial frequency figure errors and to determine what rms figure levels might be achievable (Ftaclas *et al.*, 1989). Working with a 10-inch-diameter spherical mirror from a microlithography instrument, we reduced lap vibrations and modified the manner in which the mirror was held. Taking away the turned edge at the outer diameter of the mirror, the central 8 inches were essentially at the CI requirement of having the scatter 1000 times less than diffraction over the metrology bandwidth (1–30 cycles per aperture). Thus, the goal we have set is not without precedent.

There are several strategies that can reduce the cost and difficulty of fabricating the required optics. The most viable approach, called phase-compensated optical fabrication, takes maximal advantage of the fact that the extremely high quality wavefront is required over a very limited field of view. This approach consists of first separately fabricating both the primary and secondary mirrors to high quality, i.e., $\lambda/75$–$\lambda/100$ rms (surface). The mirrors are then brought together in a system test, and the resulting telescope is tested in double-pass. One of the mirrors is then figured to the negative of the measured telescope wavefront such that the phase errors in the two mirrors cancel at the 10% level, leaving a thousandth wave system. There are several immediate gains from such an approach. First, the need for two metrology setups operating well below the $\lambda/1000$ rms absolute accuracy level is eliminated. This implies that only a single entry in the telescope wavefront error budget need be made for the telescope optics. Moreover, a double-pass test in autocollimation requires no null corrector assembly and calibration of only a reference sphere and autocollimating flat. Since it is a double-pass test, the metrology error signal is doubled whereas the reference optic signal is not, thereby increasing the tolerance on the reference optics.

One problem with any fabrication scheme based on system-level metrology is that the resulting telescope has a high-quality wavefront over a very small field of view. That is, as one moves off-axis the secondary mirror map is sheared with respect to the primary mirror map, degrading the cancellation that led to the thousandth wave system wavefront. We have analyzed the way in which the scattered light degrades with field angle and have found that the change in the power spectrum is proportional to the square of the field angle with the constant of proportionality depending quadratically on the degree of cancellation between the two optics and the shear between the primary and secondary mirror maps. Thus, combining two hundredth wave mirrors to yield a thousandth wave system is one hundred times more forgiving than asking two tenth wave mirrors to produce a thousandth wave system.

Based on the system parameters we have assumed and starting with two one hundredth wave mirrors, we estimate that approximately 200 μm of secondary decenter would produce only a factor of two increase in the scattered light background over a 5-arcsecond field of view. Since the scatter degradation increases quadratically with distance from the bright central source, most of it appears at the edge of the field, where additional scattered light is considerably more tolerable. The corresponding tilt tolerance is of order tens of arcminutes. An alignment requirement of this magnitude is quite achievable, based on alignment experience with the HST. Thus, there appear to be significant system and cost advantages to this approach to optical fabrication.

6. Detecting Extra-Solar Planets

Without question, the ability to do high-dynamic-range imaging within a few Airy rings of bright astronomical sources will contribute significantly to solving many problems in astronomy and planetary science in particular. Among the areas made feasible by this capability, the most technically demanding is the direct detection of extra-solar planets. In such a measurement, two types of noise corrupt the data and need to be taken into account: random and systematic. The random element is typified by shot noise, which represents a fundamental quantum limit on any detection. The systematic aspect of greatest concern is the subtraction of the background.

If we have a background count of N in time t, then when we say that the noise is given by $\sqrt{N}$, we have assumed a perfect background subtraction, and the noise in the measurement will grow with the $\sqrt{t}$. If we fail to subtract the background perfectly, leaving a residual of ΔN, then the noise is given by

$$\text{total noise} = \sqrt{\Delta N^2 + N} \tag{1}$$

By integrating longer we can always improve the balance of signal-to-shot noise photons, since the signal photons are presumed to grow linearly with time; but if ΔN is a fixed fraction of N, the systematic noise grows as rapidly as the signal, and there is no further improvement in signal-to-noise with integration time. Ultimately, most detections are limited either by the level of systematic error in the experiment or noise sources that grow with time ($1/f$ noise).

In the case of AIT, the systematic and random errors of detection are expressed by integration time T, and flux ratio Φ (Brown and Burrows, 1990). The flux ratio is the ratio of the number of photons per second arriving from the background compared to the number of photons per second arriving from the planet and is a measure of the visibility of the planet against the background. Φ and T are not entirely independent parameters. The integration time is calculated assuming we are in the shot noise limit, so, as the flux ratio improves, the integration time decreases. The integration time can be written as:

$$\text{Integration Time} = \left(\frac{\text{signal}}{\text{noise}}\right)^2 \cdot \left(\frac{\text{background}}{\text{planet}^2}\right) = \left(\frac{\text{signal}}{\text{noise}}\right)^2 \cdot \left(\frac{\Phi}{\text{planet}}\right) \quad (2)$$

where '(signal/noise)' is the desired signal-to-noise ratio and 'background' and 'planet' are the photon count rates from the background and planet, respectively. It is clear from this equation that if we hold the planet signal constant and improve, i.e., reduce, the flux ratio by reducing the background, we will reduce the integration time. This is the goal of the combined coronagraph and phase-compensated optical telescope assembly in which we expect to reduce the background and hence the integration time by a factor of 1000.

Using the above equation, and the assumption that each observation is shot-noise limited, an integration time can be obtained for any given signal-to-noise ratio. Bounding integration time provides a reality constraint on the system in that we reject those observations whose required integration time are judged to be too long. Bounding the flux ratio, on the other hand, provides a different type of constraint on detection. It is meant to characterize our ability to detect a faint source against a bright background and essentially quantifies the difficulty of doing a background subtraction. Results from detailed models show that, in general, integration time limits detection at large field angles and flux ratio limits detection at small field angles. To date the project has been using bounds of order 100 for flux ratio based on the general feeling for the photometric precision of area detectors and past experience in ground-based astrometry. However, whether it is one, two or three hundred makes a significant impact on the number of stars we can see, since it is the flux ratio that ultimately sets our ability to detect planets around more distant stars.

The background level is a continuous distribution of diffracted and scattered light produced by the parent star. In order to calculate the flux from the background, it is necessary to assume a collecting area equivalent to a pixel in which the detection is to be made. Therefore, the planet flux is just the encircled energy within the collection area, and the background flux is just proportional to the collection area. A detailed examination of the way in which encircled energy grows with aperture shows that one can choose a pixel size to optimize either flux ratio or integration time. In particular, it is possible to make significant improvements to planet visibility without compromising integration time. This point is illustrated for a circular aperture in Figure 2. The relative integration time to a fixed signal-to-noise ratio and flux ratio are plotted as a function of collecting radius measured in Airy radii. Both plots have been normalized to their values at 1.22 Airy rings, the

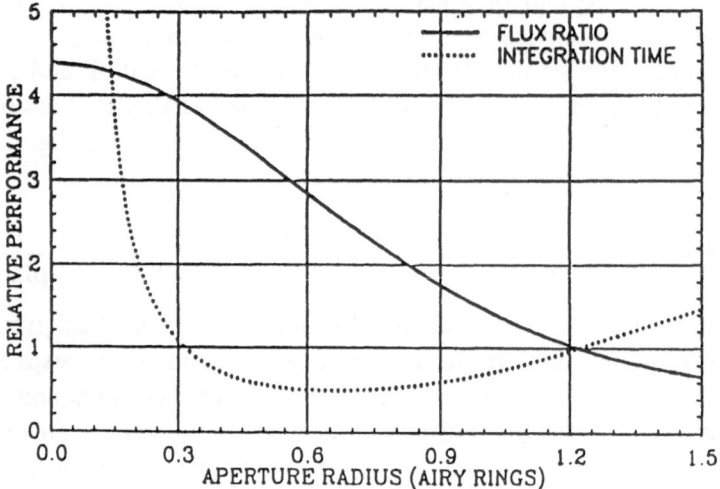

Fig. 2. Relative flux ratio and integration time as a function of aperture size. All parameters have been normalized to unity at the Airy disk edge.

edge of the central diffraction lobe and the canonical aperture traditionally used for signal-to-noise models.

Inspection of Figure 2 shows that, since the intensity of the point-spread function peaks at the origin, the encircled energy never grows as quickly as the collecting area, so the flux ratio peaks at the origin (zero pixel size). The integration time behavior is more complicated. As the collecting area is closed down from the edge of the Airy disk, fewer planet photons are being collected. However, the planet-to-star photon mix is growing more favorable, so the integration time initially decreases. As the aperture is closed down further, the improvement in the flux ratio is not as significant, and the overall loss of photons begins to dominate, driving the integration time to infinity. Most significantly, however, is that by using smaller pixels we can greatly improve planet visibility with little penalty (or even a gain) in integration time.

A typical performance plane illustrating the relative flux ratio and integration time contour for a Jovian planet shining by reflected light and in orbit about a solar-type star is shown in Figure 3. For some region of this discovery space, the planet would be within comfortable integration time and visibility bounds. By applying fixed bounds on visibility and integration time, and by specifying the planet to be Jovian, we can generate plots like this for each of the nearest known stars and can compile detection statistics for the catalog. Typical results are shown in Table I for various assumptions about the system. While numbers will vary in detail with different assumptions, there are certainly tens of stars that are candidates for direct planet detection.

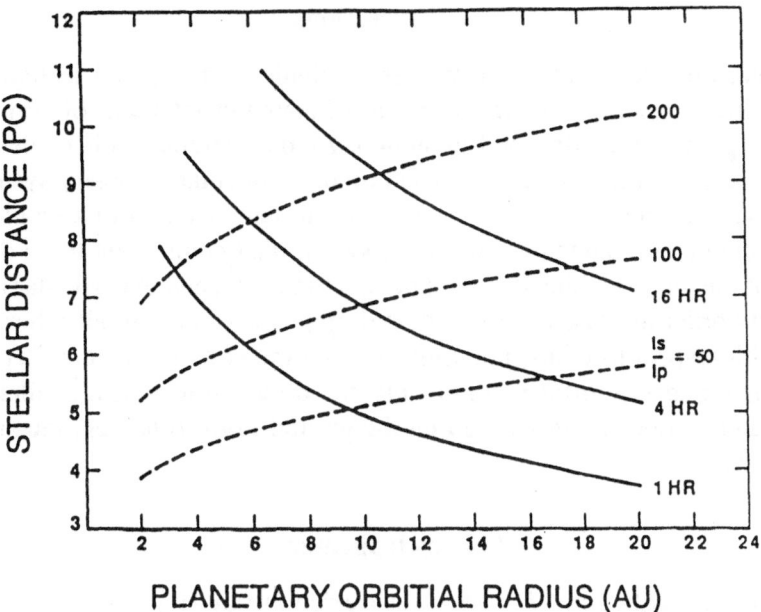

Fig. 3. Discovery space for Jupiter detection around a solar-type star.

TABLE I
Number of stars with detectable planets.

Central Wavelength:		$\lambda = 600$ nm		$\lambda = 550$ nm		$\lambda = 500$ nm	
Flux Ratio:		$F = 100$	$F = 200$	$F = 100$	$F = 200$	$F = 100$	$F = 200$
Jupiter in	$S/N = 5$	22	30	25	36	36	47
10 hours	$S/N = 3$	34	50	39	56	50	71
Jupiter in	$S/N = 5$	28	41	35	50	49	65
20 hours	$S/N = 3$	34	54	41	65	57	82

Loosely speaking, detections around nearby stars are relatively easy and grow rapidly more difficult with distance. The flux ratio increases as D^a where a is the power law slope of the scattered light background of the optics, where usually $2 < a < 3$, and D is the distance. The integration time scales are $D^{(a+2)}$. Based on realistic parameters and a real-sky catalog, we estimate that completion of the observation program will take about 1 to 2 years of shared observing time.

7. Conclusions

The circumstellar imaging problem poses a challenge to optical technology but offers significant scientific return, particularly in our understanding of the processes that led to planet formation and evolution. At the extreme limit of demand on circumstellar imaging is the direct detection of extra-solar planetary systems. We believe that the critical technologies have been demonstrated that would permit the construction and launch of an optical system that could meet these demands. Because of the way in which the coronagraph functions, and in particular its action against low-order aberrations, the required optical system is in all other respects normal with respect to focus, pointing, and alignment requirements. Because the system has modest on-orbit requirements, the principal technical challenges can be made and demonstrably verified on the ground prior to launch, thus reducing mission risk.

Acknowledgements

The research described in this paper was carried out by Hughes Danbury Optical Systems, Inc. and the Jet Propulsion Laboratory, California Institute of Technology, under contract with the National Aeronautics and Space Administration.

References

Bracewell, R.N.: 1978, *Nature* **274**, 780–781.
Bracewell, R.N. and MacPhie, R.H.: 1979 , *Icarus* **38**, 136–147.
Brown, R.A. and Burrows, C.J.: 1990, *Icarus* **87**, 484–497.
Cerino, J.: 1993, AO-93-7813, , Danbury, Connecticut.
Ftaclas, C., Krim, M.H. and Terrile, R.J.: 1989, *SPIE Proceedings* **1113**, 56–64.
Ftaclas, C. *et al.*: February 1992, Final Report for the AIT Metrology Definition Study, HDOS Project
       Report, *PR B11-0348*, Danbury, Connecticut.
Pravdo, S.H. *et al.*: 1993, this volume.
Shaklan, S. *et al.*: 1993, this volume.
Smith, B.A. and Terrile, R.J.: 1984 , *Science* **226**, 1421.
Terrile, R.J. and Ftaclas, C.: 1989, *SPIE Proceedings* **1113**, .

MODELLING OF A SPACE-BASED RONCHI RULING EXPERIMENT

FOR HIGH-PRECISION ASTROMETRY *

S. SHAKLAN and S. PRAVDO

Jet Propulsion Laboratory
California Institute of Technology,
Pasadena, California, USA

G. GATEWOOD

Allegheny Observatory, Observatory Station,
Pittsburgh, Pennsylvania, USA

and

C. FTACLAS

Hughes Danbury Optical Systems, Inc., Danbury, Connecticut, USA

Abstract. An end-to-end modelling program for an astrometric telescope employing a Ronchi ruling has been developed. The program models the aberrated images formed anywhere in the field-of-view. It then determines apparent centroids by simulating the motion of a Ronchi ruling across the field. Photo-electron statistics are included. A 6-term plate-constant model is used to determine the apparent motion of a target star within a reference frame as the object is re-observed through both ideal and perturbed optics. The modelling code is accurate at the submicroarcsecond level.

1. Introduction

Ronchi rulings have been successfully employed in ground-based astrometric programs by several groups (Gatewood, 1987; Buffington, 1990). Stellar positions are measured by sliding the ruling across the image plane, repeatedly eclipsing the images of a target star and several reference stars. The number of lines between stars gives the approximate separation (in one dimension), while the phase difference between the periodic signals provides high precision.

On the ground, random differential motions caused by atmospheric turbulence have limited astrometric precision to about 1 milli-arcsecond. Improvements will come with larger telescopes located at superior sites. Shao and Colavita (1992) predict that with a 10-meter telescope located at Mauna Kea, one can achieve sub-100 microarcsecond precision.

Current technology for the manufacture of Ronchi rulings should allow at least an order of magnitude improvement beyond 100 microarcseconds. To realize micro-arcsecond precision, it will be necessary to launch an orbiting 1.5-m class telescope. A mission to do that called the Astrometric Imaging Telescope (AIT) has been proposed, and is described elsewhere in these proceedings (Pravdo *et al.*, 1993).

There are several important differences between ground- and space-based systems. On the ground, atmospheric turbulence can be counted on to smooth images

---

* Paper presented at the Conference on *Planetary Systems: Formation, Evolution, and Detection* held 7–10 December, 1992 at CalTech, Pasadena, California, U.S.A.

Astrophysics and Space Science **212**: 453–463, 1994.
© 1994 *Kluwer Academic Publishers.*

to a resolution well below the diffraction limit. Ground-based algorithms for determining the position of the star depend upon both image smoothness (no sidelobes) and random image motion (Gatewood, 1987; Buffington, 1990). In space, images are diffraction limited and stable. The signal passed by the Ronchi ruling may have local minima due to diffraction rings, while aberrated images are not smoothed by atmospheric blurring.

The purpose of this paper is to describe the end-to-end modelling of a space-based astrometric telescope employing a Ronchi ruling as the metric. Our modelling codes function at the sub-microarcsecond level, and the algorithms are robust enough to allow for reasonable telescope perturbations. We begin this paper with the mathematical framework that describes the functionality of the ruling and noise characteristics of the detected signal. We then detail the implementation of a computer based end-to-end model as well as the plate constant model that is used to determine the astrometric precision of a field of stars. Finally, we present results of our modelling, including optical tolerancing of AIT and required integration times for several target stars.

2. Optical Model

The principal components of a Ronchi-ruling-based astrometric system are the telescope, ruling, focal plane apertures, and detectors. The telescope, described below, should have low distortion and a field-of-view sufficient for observing several reference stars surrounding the target star. The ruling is placed directly in the focal plane with no intervening optics. Behind the ruling are several movable apertures with fiber optics that carry starlight to the detectors. Each aperture collects the light from one star, and each photon-counting detector sees a periodic signal as the ruling is drawn across the field-of-view. Starlight is incident on the focal plane with a distribution given by $I(x, y)$. The Ronchi ruling is a periodic function given by $R(x - vt)$, where v is the velocity of the ruling and t is a relative time coordinate. The ruling is assumed to extend infinitely in the y direction, so that the power transmitted beyond the focal plane is given by the convolution of I with R:

$$P(t) = \int\limits_{-\infty}^{\infty} \int\limits_{-\infty}^{\infty} I(x, y) R(x - vt) \, dx \, dy \ . \tag{1}$$

The power reaching the detector, P_d, is the component of P that passes through the focal plane aperture and is transmitted by the fibers.

The detected signal is given by

$$s(t) = \sum_{i=1}^{N} \delta(t - t_i) \tag{2}$$

where t_i is the arrival time of the ith photon. The probability of detecting a photo-electron at a given time follows Poisson statistics, with a mean rate determined by the instantaneous power $P_d(t)$.

The ruling serves as a spatial filter of the incident starlight. In the Fourier domain, one can see that all of the intrinsic information of the detected light is localized to the first few Ronchi ruling harmonics. The Fourier transform of the detected signal is

$$\tilde{s}(f) = \sum_{i=1}^{N} \exp\{-2\pi j f t_i\}. \tag{3}$$

After averaging over positional and temporal statistics, the mean value is

$$\langle \tilde{s}(f) \rangle = \bar{N} \frac{\tilde{I}(f)\tilde{R}(f)}{\tilde{I}(0)\tilde{R}(0)} \tag{4}$$

where $\bar{N}$ is the total number of photons collected, I is now taken to be the line spread function $I(x) = \int I(x,y)\, dy$, and the tilde over a symbol is used to indicate the Fourier-transformed quantity. x is expressed as a temporal quantity via $t = x/v$.

Equation 4 demonstrates several important characteristics of detection with a Ronchi ruling. First, since the Fourier transform of a periodic function is a series of delta functions (the harmonics), all of the positional information must be derived from the harmonics. Since the true centroid is uniquely given by the slope of $\tilde{I}$ at $f = 0$, the Ronchi ruling cannot determine the true centroid unless it has an infinite period (thus, the harmonics sample $\tilde{I}$ infinitely close to the origin). Second, it is evident that a square-wave ruling is better than a sinusoidal ruling because the amplitude of $\tilde{R}$ at the first harmonic is higher by $4/\pi$ for a square wave, and the square wave provides additional harmonics. Third, one can see that beyond the maximum spatial frequency passed by $\tilde{I}$, there is no information. Finally, higher harmonics are attenuated by $\tilde{I}$. The last two points demonstrate the importance of diffraction limited imaging, for aberrations reduce the amplitude of $\tilde{I}$.

It can be shown (Goodman and Belsher, 1976) that the spectral density of the detected signal is given by

$$\left\langle |\tilde{s}(f)|^2 \right\rangle = \bar{N} + \left(\bar{N} \frac{\tilde{I}(f)\tilde{R}(f)}{\tilde{I}(0)\tilde{R}(0)} \right)^2 \tag{5}$$

so that one is left with the resulting white noise variance

$$\sigma_d^2 = \left\langle |\tilde{s}(f)|^2 \right\rangle - |\langle \tilde{s}(f) \rangle|^2$$
$$= \bar{N} \tag{6}$$

The characteristic white noise spectrum is used below in our centroid estimator.

Finally, we show that for a 50% duty cycle, the covariance of the harmonics is zero. It can be shown that the covariance of the noise is given by

$$\text{Cov}(\Delta f) \equiv \frac{1}{\sigma_d^2}(\langle \tilde{s}(f)\tilde{s}(f+\Delta f)\rangle - |\langle \tilde{s}(f)\rangle|^2)$$

$$= \frac{\tilde{I}(\Delta f)\tilde{R}(\Delta f)}{\tilde{I}(0)\tilde{R}(0)} \tag{7}$$

For a 50% duty cycle, only the odd harmonics exist. For a fundamental frequency of f_f, the Δf between harmonics is $2f_f$. But $\tilde{R}(2f_f) = 0$, so that there is no correlation. If, however, the duty cycle is changed, even harmonics appear. In that case, the correlation between adjacent harmonics is $\tilde{R}(f_f)$, which approaches $2/\pi$. Because of the large correlation between adjacent harmonics, there is little to be gained by using a duty cycle other than 50%.

3. End-to-End Model

Our model consists of several star fields selected from a list of nearby stars, a ray trace/diffraction program, a model of the Ronchi ruling, a photon noise generator, and an astrometric analysis program. Figure 1 shows a schematic of the end-to-end modelling process. All software is executed on a Sun Sparc 2 workstation.

3.1. IMAGING CODE

End-to-end modelling begins with a high-precision optical ray-trace and diffraction program called the Controlled Optics Modelling Package (COMP) (Redding and Breckinridge, 1991).

COMP's ray-trace has been verified against the commercially available optical program CODE V (Optical Research Associates, 1993). For AIT, we have shown that COMP's numerical precision is 20 nano-arcseconds for a bundle of 2600 rays traced through the system at field angles up to 10 arcminutes. The accuracy of the ray trace in terms of the centroid of the ray bundle is estimated by using larger and larger numbers of rays until the centroid no longer shifts. The interesting result of this study is that a small number of rays (e.g., 2000) gives accurate *relative* centroids at the sub-micro-arcsecond level, while the *absolute* centroids differ only by a multiplicative constant. In other words, by using a reasonably small number of rays, the system magnification is slightly off (by a fraction of a percent), but the quadratic and higher order distortion terms are accurately determined.

The diffraction calculation begins by tracing the rays backwards to the exit pupil of the system. A spherical phase term, with radius centered on the chief ray position in the image plane, is removed from the optical path of each ray. The remaining optical path defines the phase of the electric field at each grid point. The distribution is transformed by FFT to the image plane, where it is squared to yield the diffraction image. Typically, a 512×512 array is employed. Diffraction images are accurate even in the presence of large aberrations in off-axis images.

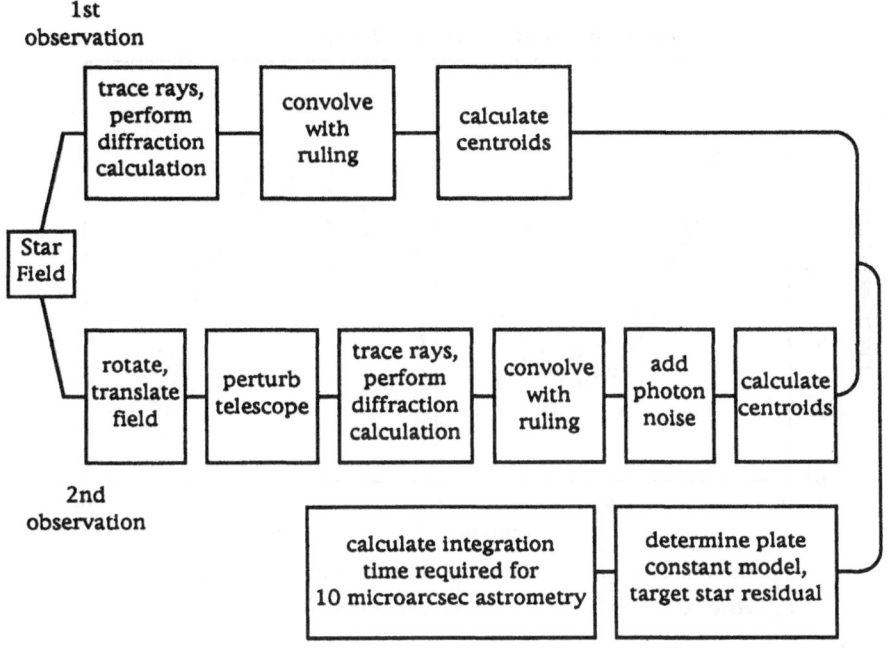

Fig. 1. Block diagram of end-to-end modelling process.

We find that diffraction centroids follow closely the ray centroids at the 30 μas level (after removal of a small linear magnification term). In theory, an image calculated by Fourier Transform of the exit pupil should have exactly the same centroid as the ray distribution (Lawrence *et al.*, 1991). We believe that computational approximations account for this difference, which in any case is accommodated in the end-to-end model; the quadratic plate constant model described below accounts for this effect.

In Figure 2, we show how the chief ray, ray centroid, and diffraction centroid compare across AIT's 16 arcminute field-of-view.

3.2. RONCHI RULING MODEL

The Ronchi Ruling (RR) is modelled as a one-dimensional one-zero function whose period and length are both powers of 2. In this way, the harmonics are integers in the discrete frequency domain. A 50% duty cycle is used, but other duty cycles could easily be simulated.

As noted above, the true image centroid is given by the derivative of the image FT evaluated at the spatial frequency origin. For a shifted, but otherwise symmetric image, phase is directly proportional to spatial frequency. Any spatial frequency (i.e., any ruling harmonic) can be used to estimate the centroid. However, for

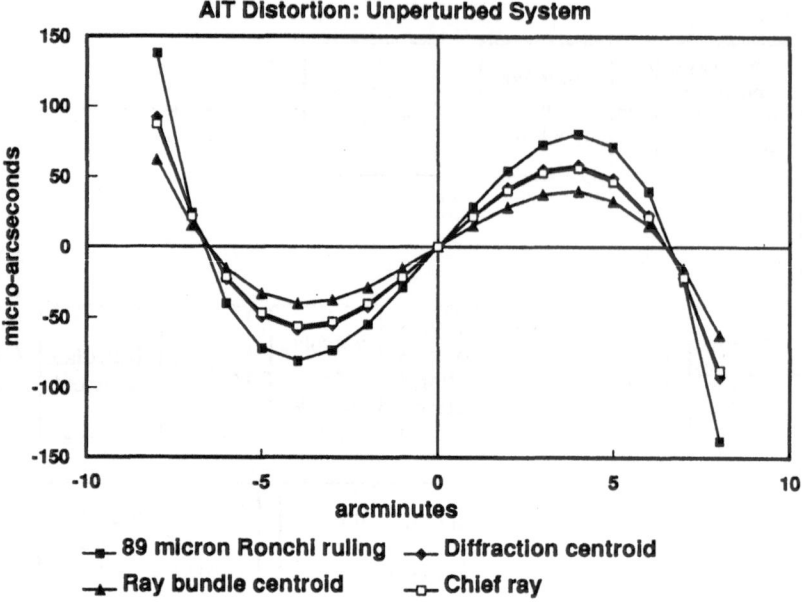

Fig. 2. Distortion in the Astrometric Imaging Telescope. A linear term (system magnification) has been removed.

asymmetric images, such as those containing coma, the centroid is still given by the phase slope at the origin, but the phase is no longer linear. (See, e.g., Figures 9 and 10 of Lawrence *et al.*, 1991). Spatial frequency samples made away from the origin contain both centroid and asymmetry information. Since the RR is periodic, it passes only discrete spatial frequencies. There is no direct way (using a linear estimator) to determine the phase slope at the origin. Non-linear estimators, such as matched filters, could be used, but they assume some *a priori* knowledge of the optical aberrations and are sensitive to misalignments or deformation of the optics.

We have chosen to implement a simple linear estimator in which the phase slope is estimated as the weighted least-squares line passing through the first several ruling harmonics. Weighting is determined by the amplitude of the modulation. In practice, this is implemented by integrating the image in one dimension, computing its FFT, then multiplying by the FFT of the ruling. A least-squares routine is used to determine a line passing through the origin and the first several harmonics.

The noise of the centroid estimate can be derived from Equations 6 and 7, where it is evident that the noise in the frequency domain is white and uncorrelated. At a given harmonic, the signal-to-noise ratio of the amplitude is given by

$$SNR_A = \sqrt{N} \frac{\tilde{I}(f)\tilde{R}(f)}{\tilde{I}(0)\tilde{R}(0)} \tag{8}$$

where N is the number of detected photoelectrons.

In the approximation $SNR \gg 1$, the phase noise (in radians) is then given by

$$\sigma_p = 1/SNR_A. \tag{9}$$

This approximation is always valid for integration times of several seconds or more on any of the 20 brightest stars in the field (as long as the ruling is neither too wide nor too narrow). The phase noise is used by the Ronchi ruling least-squares routine to obtain a formal error on the phase slope. The phase slope and error are then converted to centroid position and centroid error σ_c. When a single harmonic is used, the centroid error is

$$\sigma_c = \frac{\sigma_p}{2\pi}T \tag{10}$$

where T is the period of the ruling. For wide rulings, where higher harmonics are still well below the cutoff frequency of $\tilde{I}$, additional harmonics decrease σ_c by approximately $\sqrt{H}$, where H is the number of harmonics used. For narrow rulings, only the first harmonic makes a significant contribution.

As shown below, the optimum spacing in terms of the SNR is too sensitive to aberrations. Because of this sensitivity, we are forced to use a relatively wide ruling, increasing integration times above the theoretical limit.

3.3. ASTROMETRIC MODEL

AIT performs relative astrometry, measuring the motion of a target star relative to a background frame. Over its lifetime, it will measure each target star several times per year. With each subsequent observation, the pointing, roll, focus, focal plane position, and optical components will change at some small level. The reference stars allow one to make an affine transformation between frames. We have found that a 3-term linear model is too sensitive to aberrations to be useful. Instead we use a 6-term quadratic model given by

$$x' = a_0 + a_1 x + a_2 y + a_3 xy + a_4 x^2 + a_5 y^2 \tag{11}$$

where x and y are the coordinates of a star in the original frame, and x' is the coordinate in subsequent frames. y' is measured in separate observations. A least-squares routine is used to perform the affine transformation. Each star is weighted according to its brightness and image quality. At least 6 reference stars are required for this model. Error propagation using this model has been discussed by Eichorn and Williams (1963).

The 6-term model is more light-efficient than a third order model that could account for the standard distortion term. We have found that the designed third-order distortion (about ± 100 micro-arcseconds) is satisfactorily reduced by the quadratic model; target star errors are below 1 micro-arcsecond assuming that the telescope is repointed to within 10 arcseconds of the original frame.

4. Optical Design

The ideal astrometric telescope has zero distortion and forms perfectly symmetric images across the field-of-view. No two-mirror design can achieve this, but a special class of Ritchey–Chretien designs can eliminate third-order distortion, spherical aberration, and coma. The equations for determining this design have been given by Korsch (1990). The AIT design is driven by additional factors, such as the desire for a small secondary mirror to reduce sidelobes for another instrument, constrained overall length, and the need for a large collecting area (Pravdo *et al.*, 1993). The current design has a 1.5-m primary, 44-cm secondary, and a 22.6-m effective focal length.

5. Telescope Tolerances

As designed, AIT's ultimate astrometric limit is better than 10 micro-arcseconds. Motions of the secondary mirror, mirror contamination, mirror deformation, and background stars all affect the delicately balanced image symmetry and induce astrometric errors.

To test our sensitivity to these perturbations, we generate two observations: a first observation, with the target star centered in the field, and all optics operating perfectly; and a second observation where a part or parts of the system are perturbed. In the second observation, we assume a pointing error of 2 arcseconds and roll error of $1°$. The system is perturbed by modifying the COMP telescope prescription before tracing rays.

Figure 3 shows the field distortion when the secondary is decentered by 100 microns perpendicular to the telescope axis. The true centroids are well behaved, despite significant aberrations. The Ronchi ruling "centroid", however, now has 2 milli-arcseconds of error. This is due to the non-symmetric interaction of the field-independent coma with design astigmatism. A 100-micron decenter causes less than 0.01 wave rms of aberration, an indication of the sensitivity of the ruling to misalignments. To achieve 10-micro-arcsecond precision (assuming no in-flight calibration), decenter of the secondary must be maintained to 10 microns, while tilt must be maintained to 7 arcseconds.

Fortunately, astrometric precision is highly insensitive to "breathing" of the metering truss. When the truss expands, the primary-secondary spacing changes, but image symmetry is not affected. The plate-scale model accounts for the focal-plane scale change. Secondary motions of up to $\pm$ 0.5 mm are permitted.

We have also found that the end-to-end model is rather insensitive to the conic constants of both the primary and secondary mirrors. For this simulation, the conics are assumed to have an error in both the initial and final observations. We find that the conics can be in error by more than $\Delta k > 0.01$ on both mirrors.

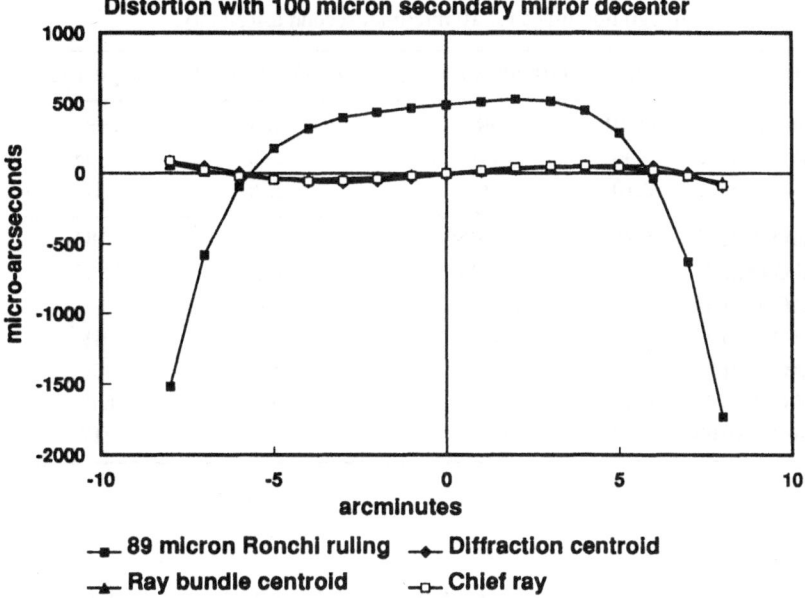

Fig. 3.   Distortion when the secondary mirror is displaced by 100 microns perpendicular to the optical axis. Chief ray, ray centroid, and diffraction distortion remain well behaved. The Ronchi ruling, however, is sensitive to the field-independent coma of the decentered system.

The Ronchi ruling used for these simulations has a period of 89 μm, corresponding to 0.8 arcseconds in the focal plane. This is 7 times larger than the optimal (for integration time) ruling, increasing integration time by a factor of 7.

To improve tolerancing requirements, in-flight calibration can yield information on focal-plane distortion. The general idea of the scheme is to observe a set of three or more bright stars separated by $\approx$ 1 arcminute at several points in the field. The apparent star separations are used to estimate the field dependence of the distortion. Tolerance requirements are relaxed by a factor of $\approx$ 5.

6. Integration Times for Several Target Stars

A set of 10 sample fields was chosen from a list of nearby stars. The set represents a wide range of magnitudes of both target and reference stars. The brightest 25 stars within an 8-arcminute radius of the target star are used to define the reference frame. We used the first 3 harmonics to estimate the centroid of each star. Table I gives the integration time required to reduce the target star positional error to 10 micro-arcseconds. Columns 3 and 4 indicate the average number of detected photons for the target star and reference frame, respectively. These numbers are

TABLE I

Integration times for 10 micro-arcsecond astrometry.

Star	m_v	Targ. Phot/s × 10^3	Ref. Phot/s × 10^3	Req'd T (s)
Ross 128	11.10	48	3534	7992
SAO 122963	9.54	210	3592	1303
SAO 065525	8.10	764	1326	812
SAO 082706	4.26	25309	786	5184
SAO 062377	7.49	1328	495	3387
SAO 157844	4.74	17509	1687	2143
SAO 080104	5.14	12114	1788	918
SAO 177866	4.93	14564	1515	756
SAO 062738	6.45	3336	1637	645
SAO 200163	4.62	19199	3338	317

based upon instrumental throughput of 2% which includes optical losses in the telescope, ruling and fiber optics, and the quantum efficiency of the detectors.

The target stars are divided into two groups. In the upper group, the target star is fainter than the reference frame. In the second group, the reference frame limits integration time. One can draw the general conclusion that brighter reference frames are better, but that is not always the case. For example, SAO 082706 requires a significantly longer integration than SAO 062377. The discrepancy is due to the light distribution in the reference frame.

7. Conclusion

The model described herein demonstrates that a Ronchi ruling space-based experiment such as AIT can achieve astrometric measurement accuracy of 10 micro-arcseconds. Most importantly, this is not an idealized model. Even in the presence of pointing errors and optical aberrations, the experimental design is sufficiently robust to perform at the design accuracy.

Acknowledgements

We would like to acknowledge many useful conversations with H. Kadogawa, E. Levy, R. Terrile, A. Buffington, A. Nonnenmacher, K. Shu, and B. Levine. This work was carried out at the Jet Propulsion Laboratory, California Institute of Technology, under contract with the National Aeronautics and Space Administration.

References

Buffington, A. and Geller, M. A.: 1990, *P.A.S.P.* **102**, 200.

Eichorn, E. and Williams, C.A.: 1963, *Ap. J.* **68**, 221.

Gatewood, G.D.: 1987, *A.J.* **94**, 213.

Goodman, J.W. and Belsher, J.F.: 1976, Technical Report RADC-TR-76-50.

Korsch, D.: 1990, in: *Astrometric Telescope Facility (ATF) FY'89 Final Report*, Jet Propulsion Laboratory, (JPL internal document), D-7113, 23.

Lawrence, G.N., Huang, C., Levy, E.H. and McMillan, R.S.: 1991, *Opt. Engineer.* **30**, 598.

Optical Research Associates, 1993: CODE V is a proprietary product of Optical Research Associates, 550 N. Rosemead Blvd., Pasadena CA 91107.

Pravdo, S. *et al.*, 1993, this volume.

Redding, D. and Breckinridge, W.: 1991, *J.G.C.D* **14**, 1021.

Shao, M. and Colavita, M.M.: 1992, *Astron. Astrophys.* **262**, 353.

AUTHOR INDEX

Astrophysics and Space Sciences **212**: 465–466, 1994.

466

SUBJECT INDEX

LIST OF PARTICIPANTS

Last Name / First Name	Affiliation	Address 1 / Address 2 / City State Zip	Country	Telephone / E Mail / Fax
Allen / Lew	Jet Propulsion Laboratory	1040 S. Arroyo Blvd. / / Pasadena, CA 91105	USA	(818) 441-4400
Allen / Mark	Jet Propulsion Laboratory	Planetary Sciences / M/S 170-25 / Pasadena, CA 91109	USA	(818) 354-3665 / mercu1::maa
Angel / R.	Steward Observatory	University of Arizona / / Tucson, AZ 85721		
Backman / Dana	Franklin & Marshall Coll./NASA-Ames	Physics & Astronomy Department / PO Box 3003 / Lancaster, PA 17604	USA	(717) 291-4132 / d-backman@acad.fand / (717) 291-4143
Basu / Dipak	Univ. of West Indies, Trinidad	Physics Dept. / St. Augustine / Trinidad, West Indies	West Indies	809-663-1369 / / 809-663-9686
Beckwith / Steve	MPI f. ASTRONOMIE	Konigstuhl 17 / D-6900 / Heidelberg 1	Germany	06221-528210 / svwb@astro0.mpihd.mp / 06221-528246
Belton / Michael	Kitt Peak Observatory	P.O. Box 26732 / 950 N. Cherry Ave. / Tucson, AZ 85726	USA	(602) 327-5511 / / (602) 325-9360
Benedict / G. F.	McDonald Observatory	University of Texas / / Austin, TX 78712	USA	(512) 471-3448 / fritz@astro.as.utexas.e / (512) 471-6010

Astrophysics and Space Science **212**: 471–486, 1994.
© 1994 *Kluwer Academic Publishers. Printed in Belgium.*

Last Name **First Name**	**Affiliation**	**Address 1** **Address 2** **City State Zip**	**Country**	**Telephone** **E Mail** **Fax**
Beust Herve'	Service d' Astrophysique	DAPNIA, Centre d' Etudes de Saclay, F91191 Gif-sus-Yvette Cedex France	France	+33-1 6908 5025 BEUST@FRIAP51 +33-1 6908-9266
Black David	Lunar and Planetary Institute	3600 Bay Area Blvd. Houston, TX 77058	USA	(713) 486-2180 nasamail::dblack (713) 486-2173
Borucki William	NASA/ARC	MS 245-3 Moffett Field, CA 94035-1000	USA	
Boss Alan	Carnegie Institution	5241 Broad Branch Road, NW Washington, DC 20018	USA	(202) 686-4402 boss@ciw.ciw.edu (202) 364-8726
Bouvier Jerome	Observatoire de Grenoble France	Laboratoire d'Astrophysique Universiti J. Fourier, B.P. 38041 38041 Grenoble Cedex	France	33-76-51-42-01 BOUVIER@GAG.OBSE 33-76-44-88-21
Boyarchuk A.A.	Institut of Astronomy of the Russian Academy of Sciences	48, Pyatnitskaya St. Moscow 109017	Russia	(095) 231-0924 all 576 asconsu (telex) (095) 230 2081
Brooke Tim	Jet Propulsion Laboratory	ms 169-237 4800 Oak Grove Drive Pasadena, CA 91109	USA	(818) 354-2112 tyb@scn6.jpl.nasa.gov (818) 393-4619
Brown Robert	Space Telescope Science Institute	3700 San Martin Dr. Baltimore, MD 21218	USA	(410) 338-4700 (410) 516-7450

Last Name First Name	Affiliation	Address 1 Address 2 City State Zip	Country	Telephone E Mail Fax
Burke Bernard F.	Massachusetts Institute of Technology	Rm. 26-335 Cambridge, MA 02139	USA	(617) 253-2572 bburke@athena.mit.edu (617) 258-7864
Burke Jim	Jet Propulsion Laboratory	ms 301-490 4800 Oak Grove Drive Pasadena, CA 91109	USA	(818) 354-4569 JDBURKE@JULIET.CA (818) 793-5528
Cameron Al	Harvard University	Harvard College Observatory 60 Garden St. Cambridge, MA 02138	USA	(617) 495-5374 cameron@cfa.harvard.
Capps Dr. Richard	Jet Propulsion Laboratory	4800 Oak Grove Dr. MS 180-603 Pasadena, CA 91109	USA	(818) 354-0720 rwcapps (818) 354-7354
Carlstrom J.E.	California Institute of Technology	M/S 105-24 Pasadena, CA 91106	USA	(818) 365-4970
Chahine Moustafa	Jet Propulsion Laboratory	M/S 180-904 4800 Oak Grove Dr. Pasadena, CA 91109	USA	(818) 354-6057
Chapman Simon	University of Wales	Department of Physics and Astronomy Cardiff, CF2 3YB, United Kingdom		
Cheng K.P.	NRC/NASA/GSFC	NASA/GSFC Code 684 Greenbelt, MD 20771	USA	(301) 286-3019 STARS::CHENG (SPAN)

Last Name	Affiliation	Address 1		Telephone
		Address 2		E Mail
First Name		City State Zip	Country	Fax
Ciupik	Adler Planetarium	1300 S. Lake Shore Drive	USA	(312) 322-0316
Larry				
		Chicago, IL 60605		
Clampin	Space Telescope Science Institute	3700 San Martin Dr.	USA	(410) 338-4700
M.				
		Baltimore, MD 21218		
				(410) 516-7450
Cochran	McDonald Observatory	The University of Texas, Austin	USA	
William				
		Austin, TX 78712		
Colavita	Jet Propulsion Laboratory	ms 169-214	USA	(818) 354-7835
M. Mark		4800 Oak Grove Drive		
		Pasadena, CA 91109		
Connes	Service d'Aeronomie		France	
Pierre				
		Verrieres 91371		
Cordes	NAIC/Cornell University	520 Space Sciences Bldg.	USA	(607) 255-0608
Jim				cordes@astrosun.tn.co
		Ithaca, NY 14853-6801		(607) 255-8803
Doyle	SETI Institute/NASA Ames	244-11	USA	(415) 604-1372
Laurance				doyle@gal.arc.nasa.go
		Moffett Field, CA 94035		(415) 968-5830
Drimmel	University of Florida	Department of Astronomy	USA	(904) 392-2052
Ronald		211 SSRB		l:drimmel@astro.ufl.edu
		Gainsville, Florida 32611		(904) 392-5089

Last Name	Affiliation	Address 1		Telephone
		Address 2		E Mail
First Name		City State Zip	Country	Fax
Duncan	Univ. of Chicago/Adler Planetarium	1300 S. Lake Shore Drive	USA	(312) 322-0316
Douglas				duncan@oddjob.uchica
		Chicago, IL 60605		
Elachi	Jet Propulsion Laboratory	m/s 180-704	USA	(818) 354-5673
Charles		4800 Oak Grove Drive		
		Pasadena, CA 91109		(818) 354-2946
Evans	University of Texas	Astronomy Dept.	USA	512-471-4396
Neal J.				nje@astro.as.atexas
		Austin, TX 78712-1083		512-471-6016
Fajardo	SUNY Stony Brook	Dept. of Earth & Space	USA	(516) 632-8229
Sergio				sfajardo@sbast1.ess.s
		Stony Brook, NY 11794-2100		(516) 632-8240
Fierro	UNAM, Mexico	Instituto de Astronomia, UNAM	MEXICO	+ 6-22-39-06
Julieta		Ap. Postal 70-264		julieta@alfa.astroscu.u
		C.P. 04510, D.F.		+ 5-48-37-12
Fisher	Ruffner Associates	P.O. Box 7070	USA	415-854-8564
Philip C.				
		Menlo Park, CA 94026		
Friedemann	Univ. Observatory Jena	Schillergaesochen 2	GERMANY	/82 22637
C.		0-6900 Jena		po1@rz.uni-jena.def
		Germany		03641/425039
Ftaclas	Hughes Danbury	100 Wooster Heights Road	USA	(203) 797-5248
Christ				
		Danbury, CT		(203) 797-6259

Last Name / First Name	Affiliation	Address 1 / Address 2 / City State Zip	Country	Telephone / E Mail / Fax
Giampapa Mark	NSO/NOAO	950 N. Cherry Ave. P.O. Box 26732 Tucson, AZ 85726-6732	USA	(602) 325-9236 giampapa@noao.span (602) 325-9278
Goldsmith Donald	Interstellar Media	2153 Russell St. Berkeley, CA 74705	USA	(510) 848-1989 (510) 848-4001
Gould Andrew	Institute for Advanced Study	Princeton, NJ 08540	USA	609-734-8057 GOULD@IASSNS.BITN 609-252-0738
Grady Carol	Applied Research Corporation	8201 Corporate Drive Suite 112D Landover, MD 20785	USA	(301) 459-8442 GRADY@FOSVAX.ARC (301) 731-0765
Grinin V.P.	Crimean Astrophysical Observatory	Crimea 334413 Nauchny	Ukraine	
Gulkis Dr. Samuel	Jet Propulsion Laboratory	4800 Oak Grove Dr. M/S 169-506 Pasadena, CA 91109	USA	(818) 354-5708 jplrag::gulkis (818) 354-2946
Gurtler J.	Jena University Observatory	Schillergasschen 2 0-6900 Jena	Germany	++ (3641) ++ (3641) 8222345
Hale Alan	Southwest Institute for Space Research	P.O. Box 4189 Alamogordo, NM 88311-4189	USA	(505) 434-0695 ahale@nmsu.edu (505) 434-0695

Last Name / First Name	Affiliation	Address 1 / Address 2 / City State Zip	Country	Telephone / E Mail / Fax
Harvey Paul	University of Texas	Astronomy Department Austin, TX 78712	USA	(512) 471-3452 pmh@astro.as.utexas.
Heacox William	University of Hawaii	Space Science Center Univ. of Hawaii at Hilo Hilo, HI 96720	USA	(808) 933-3382 heax@einstein.uhh.haw (808) 933-3693
Helmer Dr. Leif	Copenhagen University Observatory	Brorfeldirej 23 DK-4340 Tollosi Denmark	DENMARK	+45-53 488195 +45 53 488055
Henning Th.	MPG-WG "Star Formation	Max Planck Society, WG Schillergasschen 2-3 0-6900 Jena	Germany	++49-161-53164 pbs@PHYSIK.UNI-JEN ++49-3641-425039
Herrera Miguel Angel	PUIDE-UNAM	PUIDE. UNAM Mexico, DF 04510	Mexico	++622-8559 MIKE@SOLEDAD.AST ++548-2497
Huntress Wesley	NASA	Code SL 300 E. St. S.W. Washington, DC 20546	USA	(202) 358-0292 (202) 358-3097
Johnson Fred	Cal State University Fullerton	Fullerton, CA	USA	(714) 773-3366
Jones Dayton	Jet Propulsion Laboratory	M/S 238-600 4800 Oak Grove Dr. Pasadena, CA 91109	USA	(818) 354-7774 dj@bllac.jpl.nasa.gov (818) 393-4965

Last Name First Name	Affiliation	Address 1 Address 2 City State Zip	Country	Telephone E Mail Fax
Koerner David	California Institute of Technology	M/S 170-25 Pasadena, CA 91125	USA	(818) 356-6477 david@satur1.6ps.calte
Kolvoord R.A.	Univ. of Arizona	LPL Tucson, AZ	USA	
Korechoff Robert	Jet Propulsion Laboratory	ms 169-314 4800 Oak Grove Drive Pasadena, CA 91109	USA	(818) 354-0083
Koresko Chris	Cornell University	Space Sciences Building Ithaca, NY 14853	USA	(607) 255-5896 koresko@astrosun.tn.c (607) 255-5875
Kuiper Tom	California Institute of Technology/JPL	ms 169-506 4800 Oak Grove Drive Pasadena, CA 91109	USA	(818) 354-5623 kuiper@kuiper.jpl.nasa. (818) 354-8895
Langer William	Jet Propulsion Laboratory	M/S 169-506 4800 Oak Grove Dr. Pasadena, CA 91109	USA	(818) 354-5823
Latham David	Center for Astrophysics	60 Garden St. Cambridge, MA 02138	USA	(617) 495-7215 latham@cfa.harvard.ed (617) 495-7467
Leger A.	University of Paris	IAS - bat 121 University of Paris 11 91 405 Orsay	FRANCE	+ 33 - 16985 + 33 - 16985 8675

Last Name / First Name	Affiliation	Address 1 / Address 2 / City State Zip	Country	Telephone / E Mail / Fax
Lestrade Jean-Franc	Observatoire de Meudon/Derad	92195 Meudon Principal Cedex FRANCE	FRANCE	(33) 1 45 07 76 17670::lestrade(span); (33) 1 45 07 79 39
Levy Eugene H.	Univ. of Arizona	Lunar & Planetary Lab. Dept. of Planetary Sciences Tucson, AZ 85721	USA	(602) 621-6962 ehlevy@ccit.arizona.ed (602) 621-4933
Mahoney M.J.	Jet Propulsion Laboratory	m/s 168-327 4800 Oak Grove Drive Pasadena, CA 91109	USA	kam@betapic.jpl.nasa. 818-393-4683
Marsh Ken	Jet Propulsion Laboratory	m/s 168-327 4800 Oak Grove Drive Pasadena, CA	USA	kam@betapic.jpl.nasa. 818-393-4683
Mazeh Tsevi	Tel Aviv University	Wisc Observatory Ramat Aviv Tel Aviv 69978	Israel	++972-3- mazeh@wise7.tau.ac.1 ++972-3-6408179
McMillan Robert	LPL/Univ. of Arizona	Lunar & Planetary Lab University of Arizona Tucson, AZ 85721	USA	(602) 621-6968 rsmcmillan@nasamail.n (602) 621-4933
Mezger Peter G.	MPI fur Radioastronomie	Auf dem Hugel 69 5300 Bonn 1 GERMANY	GERMANY	+ 228 525 287 + 228 525 435
Mikhail Dr. Joseph	Nat'l Resrch Inst. of Astro. & Geoph.	National Research Institute of Astronomy and Geophysics Helwan, Cairo	EGYPT	788800 782683

Last Name First Name	Affiliation	Address 1 Address 2 City State Zip	Country	Telephone E Mail Fax
Monin Jean-Louis	Observatoire de Grenoble	Groupe d'Astrophysique BP 53X 38041 Grenoble	FRANCE	+ 33 76 51 42 14 monin@gag.observ.gr.ft + 33 76 44 88 21
Morris David	I.R.A.M.	300 Rue De La Discine, Domaine 38406 St. Martin D'Heres France	France	++33-76824930 morris@iram.grenet.fr ++33-76515938
Mumma Michael J.	NASA/GSFC	Lab. for Extraterrestrial Physics Greenbelt, MD 20771	USA	
Nakagawa Yoshi	University of Tokyo	Geophysical Institute Univ. of Tokyo Tokyo 113 Japan	Japan	++81-3-3812-21 yoshi@gpnws.geoph.s. ++81-3-3818-3247
Neuhauser Ralph	Max Planck Institut fur Extraterr. Physics	RPE, Giessenbachstrasse D-W-8046 Garching bei Munchen	GERMANY	+ 49 - 89 - 32993 rue@hpth03.mpe-garchi + 49 - 89 - 32993 569
Nickle Neil	Jet Propulsion Laboratory	M/S 180-703 4800 Oak Grove Dr. Pasadena, CA 91109	USA	(818) 354-8244 (818) 393-1492
Nonnenmacher Andreas	Hughes Danbury	100 Wooster Heights Road Danbury, CT	USA	(203) 797-5248 (203) 797-6259
Nuth Joe	NASA/GSFC	Code 691 Greenbelt, MD 20771	USA	(301) 286-9467 SPAN::u1jan (301) 286-3271

Last Name	Affiliation	Address 1		Telephone
		Address 2		E Mail
First Name		City State Zip	Country	Fax
Ohashi	Nobeyama Radio Observatory		JAPAN	+81-267-63-4391
Nagayoshi		Minamimaki, Minamisaku		ohashi@nro.hao.ac.jp
		Nagano 384-13		+81-267-98-2884
Owen	Insitute for Astronomy, UH	University of Hawaii	USA	(808) 956-8007
Toby		2680 Woodlawn Drive		owen@uhifa.ifa.edu
		Honolulu, HI 96822		(808) 988-2970
Papanastassiou	California Institute of Technology	Div. of Geological & Planetary Sciences	USA	(818) 356-6179
D. A.		ms 170-25		
		Pasadena, CA 91109		(818) 568-0935
Peale	Univ. of California, Santa Barbara	Dept. of Physics	USA	
S.J.		Santa Barbara, CA 93106		
Perez	CSC/IUE Observatory	Code 684.9	USA	(301) 286-3573
Mario		NASA/GSFC		iuesoc::perez (decnet)
		Greenbelt, MD 20771		(301) 286-7642
Phillips	California Institute of Technology	M/S 105-24	USA	
J.A.		Pasadena, CA 91106		
Pravdo	Jet Propulsion Laboratory	MS 168-222	USA	(818) 354-3131
Steven		4800 Oak Grove Dr.		shpravdo@osfvax.jpl.n
		Pasadena, CA 91109		(818) 354-8887
Rahe	NASA Headquarters	Code SL	USA	(202) 453-1597
Jurgen		300 E. St. S.W.		
		Washington, DC 20546		(202) 426-1023

Last Name **First Name**	**Affiliation**	**Address 1** **Address 2** **City State Zip**	**Country**	**Telephone** **E Mail** **Fax**
Rapp Donald	Jet Propulsion Laboratory	MS 4800 Oak Grove Dr. Pasadena, CA 91109	USA	(818) 354-4931 (818) 393-4868
Renzetti Nick	Jet Propulsion Laboratory	M/S 303-401 4800 Oak Grove Drive Pasadena, CA 91109	USA	818-354-4517 NASA MAIL - 818-393-6228
Robb David		285 University Drive Menlo Park, CA 94025	USA	(415) 322-5249 (415) 367-0675
Robinson R. Steve	University of Arizona	P.O. Box 43536 Tucson, AZ 85733	USA	rsr@ccit.arizona.edu
Roettger Dr.	USRA	NASA/SGFC Code 693 Greenbelt, MD 20771	USA	301-286-1528 yseer@lepvax.fsfc.nas 301-286-1629
Ruskol Eugenia	ITP/UCSB			
Ruzmaikina T.V.	Univ. of Arizona	Lunar & Planetary Lab. Dept. of Planetary Sciences Tucson, AZ 85721	USA	
Safronov V.S.	O. Yu. Schmidt Inst. of Physics of the Earth	Moscow 123810	Russia	

Last Name First Name	Affiliation	Address 1 Address 2 City State Zip	Country	Telephone E Mail Fax
Sagan Carl	Cornell University	Lab. for Planetary Sciences Space Sciences Bldg. Ithaca, NY 14853	USA	(607) 255-4971 york@astrosun.tn.corn (607) 255-9888
Sandhu Bhagwant	Charles University Prague	Institute of Astronomy Charles University Svedska-8, Praha-5 Czechoslovakia	Czechosl ovakkia	+42(2) 535764 +42(2) 8551848
Sargent Anneila	California Institute of Technology	Astronomy Dept Caltech 105-24 Pasadena, CA 91125	USA	818-356-6622 AFS@PHOBO 818-568-9352
Sasaki Sho	University of Tokyo	Geological Institute University of Tokyo Bunkyo-ku, Tokyo 113	JAPAN	+81-3-3812-211 sho@tansei.cc.u-tokyo +8l1-3-3815-9490
Schneider J.	CNES/Paris Observatory	Observatoire 92 195 Meudon	FRANCE	17670::schneider + 33-1-4507-7971
Shaklan Stuart	Jet Propulsion Laboratory	m/s 169-214 4800 Oak Grove Drive Pasadena, CA 91109	USA	(818) 354-0105
Shao Michael	Jet Propulsion Laboratory	M/S 169-214 4800 Oak Grove Dr. Pasadena, CA 91109	USA	(818) 354-7834
Snowden Michael	MOOREA	B.P. 1004 Papetoai MOOREA	FRENCH POLYNE	(689) 56-24-59 (689) 56-24-59

Last Name / First Name	Affiliation	Address 1 / Address 2 / City State Zip	Country	Telephone / E Mail / Fax
Stone Edward	Jet Propulsion Laboratory	m/s 180-904 4800 Oak Grove Drive Pasadena, CA 91109	USA	(818) 354-3405
Sullivan Woody	University of Washington	Dept. of Astronomy FM-20 Seattle, WA 98195	USA	(206) 543-7773 woody@uwaphast.bitne (206) 685-0403
Sylvester Roger	University College London	Dept. Physics & Astronomy Gower Street London, WCIE 68T UK	United Kingdom	++44-71-387-70 RJS@UK.AC.UCL.STAR ++44-71-380-7145
Tarter Jill	SETI Institute	 Mountain View, CA 94035	USA	(415) 604-5727 tarter@bkyast.berkeley
Terebey Susan	California Institute of Technology/JPL	IPAC M/S 100-22 4800 Oak Grove Dr.	Pasadena, CA	818-357-9506 st@ipac.caltech.edu. 818-397-9600
Terrile Richard	Jet Propulsion Laboratory	M/S 183-501 4800 Oak Grove Dr. Pasadena, CA 91109	USA	(818) 354-6158
The' Pik-Sin	Astr. Inst. Amsterdam	Kruislaan 403 1098 SJ Amsterdam Netherland CA	NETHER LAND	+31 20-525 7496 PSTHE@ASTRO.UVA.N +31 20-525 7484
Thronson Harley	Wyoming IR Workshop	Univ. of Whyoming Box 3905 Laramie, WY 82071	USA	(307) 766-6150 edison@corral.uwy.edu (307) 766-2652

Last Name / First Name	Affiliation	Address 1 / Address 2 / City State Zip	Country	Telephone / E Mail / Fax
Tsai John C.	NASA/Ames	NASA/Ames Research Center Moffett Field, CA 94035-1000	USA	(415) 604-3385 jcht@pan.arc.nasa.gov (415) 604-6779
Van Cleve Jeffrey	Cornell University	222 SpaceSciences Ithaca, NY 14853	USA	(607) 255-5901 vancleve@astrosun.tn. (607) 255-5875
Velusamy T.	Jet Propulsion Laboratory/NASA	m/s 169-506 4800 Oak Grove Drive Pasadena, CA 91109	USA	(818) 354-6112 velu@kuiper.jpl.nasa (818) 354-8895
Walter Dolores	Ukirt Hawaii	Joint Astronomy Centre 660 N. Aohoku Place Hilo Hawaii 96720	USA	808-961-3765 imc@jach.hawaii.Edu. 808-961-6516
Wasserburg G. J.	California Institute of Technology	Div. of Geological and Planetary Sciences ms 170-25 Pasadena, CA 91125	USA	(818) 356-6139 kathie@legs.gps.caltec (818) 568-0935
Weedman Daniel	NASA			
Werner Michael	Jet Propulsion Laboratory	4800 Oak Grove Dr. M/S 169-327 Pasadena, CA 91109	USA	(818) 354-0146 werner@jplrag.jpl.nasa.
Wetherill George	Carnegie Institution of Washington	DTM 5241 Broad Branch Rd NW Washington, DC 20015	USA	(202) 686-4325 (202) 364-8726

Last Name / First Name	Affiliation	Address 1 / Address 2 / City State Zip	Country	Telephone / E Mail / Fax
Wolszczan / Alex	Penn State Univ.	Dept. of Astronomy & Astrophysics / 515A Davey Lab / University Park, PA 16802	USA	(814) 865-2918 / alex@astro.psu.edu / (814) 863-3399
Yu / Jeffrey	Jet Propulsion Laboratory	m/s 169-214 / 4800 Oak Grove Drive / Pasadena, CA 91109		(818) 354-9179

LIST OF FORTHCOMING PAPERS

S. Ya. Braude, K. P. Sokolov, and S. M. Zakharenko: Decametric Survey of Discrete Sources in the Northern Sky. XI. *The Results of the UTR-2 Very Low-Frequency Sky Survey in the Declination Range 41° to 52°.* (Received 8 February, 1993.)

Scott R. Chubb: Sun-Induced Variations in Time in the Global Positioning System. (Received 20 July, 1993.)

Yu. A. Fadeyev: On Properties of the Fourier Harmonics in the Limit-Cycle Models of Classical Cepheids. (Received 19 July, 1993, in revised form 1 November, 1993.)

T. J. Kalvouridis: Particle Motions Around, Between or Outside the Two Dipoles of a Magnetic Binary System. (Received 10 November, 1993.)

S. C. F. Rossi, B. Barbuy, G. Pineau des Forêts, and P. Benevides-Soares: Cooling by Grains and H_2 in Collapsing Clouds of Different Metallicities. (Received 17 November, 1993.)

Somenath Chakrabarty: Strange Quark Matter in a Strong Magnetic Field. (Received 23 November, 1993.)

Letters to the Editor:

L. Herrera: Non-Topological Solitons as the Possible Origin of Intrinsic Extragalactic Redshifts. (Received 29 July, 1993.)

Lingxiang Cheng, Tipei Li, Xuejun Sun, Yuqian Ma, and Mei Wu: Evidence of Pulsed X-Ray Emission from Radio Pulsar PSR1951+32. (Received 3 September, 1993.)

J. Vranješ: Gravitational Instability Problem of Nonuniform Media. (Received 28 September, 1993.)

N. C. Wickramasinghe and F. Hoyle: Absorption Properties of Astronomical Iron Whiskers: An Accurate Cryogenic Model. (Received 2 November, 1993.)

J. I. Katz: Failure of Magnetic Fields to Explain Rotation Curves of Spiral Galaxies. (Received December, 1993.)

Book Reviews:

Earl W. McDaniel, J. B. A. Mitchell, and M. Eugene Rudd (eds.), *Atomic Collisions: Heavy Particle Projectiles* (T. J. MILLAR)

D. F. Malin, *A View of the Universe* (F. D. KAHN)